Interdisciplinary Mobile Media and Communications:

Social, Political, and Economic Implications

Xiaoge Xu
The University of Nottingham, Ningbo, China, China

A volume in the Advances in Wireless Technologies and Telecommunication (AWTT) Book Series

Managing Director: Lindsay Johnston
Production Editor: Jennifer Yoder
Development Editor: Erin O'Dea
Acquisitions Editor: Kayla Wolfe
Typesetter: James Knapp
Cover Design: Jason Mull

Published in the United States of America by
Information Science Reference (an imprint of IGI Global)
701 E. Chocolate Avenue
Hershey PA, USA 17033
Tel: 717-533-8845
Fax: 717-533-8661
E-mail: cust@igi-global.com
Web site: http://www.igi-global.com

Library of Congress Cataloging-in-Publication Data

Interdisciplinary Mobile Media and Communications : Social, Political, and Economic Implications / Xiaoge Xu, editor.
pages cm
Includes bibliographical references and index. Summary: "This book sheds light on emerging disciplines in multimedia technologies and discusses the changes, chances, and challenges in the mobile world, covering areas such as mobile governance, mobile healthcare, and mobile identity "-- Provided by publisher.
ISBN 978-1-4666-6166-0 (hardcover) -- ISBN 978-1-4666-6167-7 (ebook) -- ISBN 978-1-4666-6169-1 (print & perpetual access) 1. Information technology--Social aspects. 2. Mass media--Technological innovations. 3. Mobile communication systems. I. Xu, Xiaoge.
HM851.I5683 2014
303.48'33--dc23

2014013040

This book is published in the IGI Global book series Advances in Wireless Technologies and Telecommunication (AWTT) (ISSN: 2327-3305; eISSN: 2327-3313)

British Cataloguing in Publication Data
A Cataloguing in Publication record for this book is available from the British Library.

Advances in Wireless Technologies and Telecommunication (AWTT) Book Series

Xiaoge Xu
The University of Nottingham Ningbo China

ISSN: 2327-3305
EISSN: 2327-3313

Mission

The wireless computing industry is constantly evolving, redesigning the ways in which individuals share information. Wireless technology and telecommunication remain one of the most important technologies in business organizations. The utilization of these technologies has enhanced business efficiency by enabling dynamic resources in all aspects of society.

The **Advances in Wireless Technologies and Telecommunication Book Series** aims to provide researchers and academic communities with quality research on the concepts and developments in the wireless technology fields. Developers, engineers, students, research strategists, and IT managers will find this series useful to gain insight into next generation wireless technologies and telecommunication.

Coverage

- Mobile Technology
- Wireless Technologies
- Wireless Broadband
- Wireless Sensor Networks
- Broadcasting
- Virtual Network Operations
- Grid Communications
- Global Telecommunications
- Network Management
- Mobile Communications

IGI Global is currently accepting manuscripts for publication within this series. To submit a proposal for a volume in this series, please contact our Acquisition Editors at Acquisitions@igi-global.com or visit: http://www.igi-global.com/publish/.

The Advances in Wireless Technologies and Telecommunication (AWTT) Book Series (ISSN 2327-3305) is published by IGI Global, 701 E. Chocolate Avenue, Hershey, PA 17033-1240, USA, www.igi-global.com. This series is composed of titles available for purchase individually; each title is edited to be contextually exclusive from any other title within the series. For pricing and ordering information please visit http://www.igi-global.com/book-series/advances-wireless-technologies-telecommunication-awtt/73684. Postmaster: Send all address changes to above address.

Titles in this Series

For a list of additional titles in this series, please visit: www.igi-global.com

Cognitive Radio Sensor Networks Applications, Architectures, and Challenges
Mubashir Husain Rehmani (Department of Electrical Engineering, COMSATS Institute of Information Technology, Pakistan) and Yasir Faheem (Department of Computer Science, COMSATS Institute of Information Technology, Pakistan)
Information Science Reference • copyright 2014 • 313pp • H/C (ISBN: 9781466662124) • US $235.00 (our price)

Game Theory Applications in Network Design
Sungwook Kim (Sogang University, South Korea)
Information Science Reference • copyright 2014 • 500pp • H/C (ISBN: 9781466660502) • US $225.00 (our price)

Convergence of Broadband, Broadcast, and Cellular Network Technologies
Ramona Trestian (Middlesex University, UK) and Gabriel-Miro Muntean (Dublin City University, Ireland)
Information Science Reference • copyright 2014 • 333pp • H/C (ISBN: 9781466659780) • US $235.00 (our price)

Handbook of Research on Progressive Trends in Wireless Communications and Networking
M.A. Matin (Institut Teknologi Brunei, Brunei Darussalam)
Information Science Reference • copyright 2014 • 592pp • H/C (ISBN: 9781466651708) • US $380.00 (our price)

Broadband Wireless Access Networks for 4G Theory, Application, and Experimentation
Raul Aquino Santos (University of Colima, Mexico) Victor Rangel Licea (National Autonomous University of Mexico, Mexico) and Arthur Edwards-Block (University of Colima, Mexico)
Information Science Reference • copyright 2014 • 452pp • H/C (ISBN: 9781466648883) • US $235.00 (our price)

Multidisciplinary Perspectives on Telecommunications, Wireless Systems, and Mobile Computing
Wen-Chen Hu (University of North Dakota, USA)
Information Science Reference • copyright 2014 • 305pp • H/C (ISBN: 9781466647152) • US $175.00 (our price)

Mobile Networks and Cloud Computing Convergence for Progressive Services and Applications
Joel J.P.C. Rodrigues (Instituto de Telecomunicações, University of Beira Interior, Portugal) Kai Lin (Dalian University of Technology, China) and Jaime Lloret (Polytechnic University of Valencia, Spain)
Information Science Reference • copyright 2014 • 408pp • H/C (ISBN: 9781466647817) • US $180.00 (our price)

www.igi-global.com

701 E. Chocolate Ave., Hershey, PA 17033
Order online at www.igi-global.com or call 717-533-8845 x100
To place a standing order for titles released in this series, contact: cust@igi-global.com
Mon-Fri 8:00 am - 5:00 pm (est) or fax 24 hours a day 717-533-8661

Editorial Advisory Board

List of Reviewers

Table of Contents

Section 6
Mobile Uses and Usability

Detailed Table of Contents

Section 1
Mobile Literacy and Learning

Based on the traditions of both media literacy and information literacy, we may define mobile literacy as the ability to access, understand, and create contextual, social, and location-based mobile information and communication in different contexts ranging from mobile journalism to mobile health communication. To describe, explain, and predict mobile literacy, it is necessary to locate contextual factors and the structural impact of economic, social, cultural, and symbolic resources associated with mobile devices in order to construct a model of mobile literacy, as investigated in Chapter 1. As one of the mobile communications in the context of education, mobile learning is another area of mobile studies. Few studies, however, have examined the instructional needs of mobile learning apps. Chapter 2 proposes a set of questions to evaluate the instructional needs of mobile learning apps. Another important contribution of this book is to provide a better understanding of the mobile learning situation in the Arab world, as demonstrated in Chapter 3, which has identified challenges and implications faced by Arab students, educators, and researchers after reviewing mobile learning practices in the Arab world.

Drawing on ethnographic research conducted with adolescents at a rural Australian high school, this chapter constructs a theoretical model for "mobile literacy." Mobile technologies, and their increasing technological capabilities, present emerging challenges for definitions and understandings of what precisely constitutes "literate practice," challenges which have not been wholly resolved though more disparate discussions of "electronically mediated communication." Such an understanding is important in order to develop approaches that effectively integrate mobile technologies into formal educational contexts. The model constructed in this chapter draws on different theoretical traditions where literacy is concerned, combining these with a sociological model developed by Pierre Bourdieu, to draw out the importance of the social dimension in mobile technology use. The ethnographic methodology results in findings that reveal the structuring impact of economic, social, cultural, and symbolic resources associated with these devices. Far from revealing that mobiles free us from a consideration of "place," this research demonstrates that to be "mobile literate" is to be even more finely attuned to the contextual factors for any mobile technology use.

With the phenomenal growth of mobile applications or apps used for teaching and learning, we are all challenged with determining which ones are effective and efficient in meeting our specific instructional needs. The use of mobile apps directly impacts students, teachers, administrators, trainers, and employees worldwide. Apps are used across all discipline areas in a variety of settings including applied interdisciplinary approaches. With this in mind, it is critical to have a workable set of app analysis questions based on current best educational practices to assist in making informed decisions on app selections to provide quality teaching and learning experiences. This chapter provides a mixed method research study combining class observations with results from three pilots in an effort to create a set of quality questions for quickly evaluating mobile apps for instructional implementation. After creating a set of questions for evaluating the quality of the apps based on current best instructional practices, the following three pilot studies were conducted. The first pilot allowed students to select an app of their own choice followed by a survey to evaluate the app using both quantitative and qualitative open-ended responses. The second pilot had all students examine the same app followed by the same survey to analyze potential differences in results and to gain additional insights. The third pilot study used the same questions, but this time rather than using it to evaluate the app, the students evaluated the quality of the questions used. During the third pilot study, students were looking strictly at the quality of the questions for instructional use. All study participants were graduate-level students in Instructional Design and Technology and were aware of best instructional practices. It is anticipated, post study, instructors and trainers can begin using the evaluation instrument, selecting those questions meeting their unique instructional needs.

Most Arab countries started their own e-learning and mobile learning initiatives in order to cope with global integration of latest educational technologies. The high mobile phone penetration among Arabs as well as availability of good mobile infrastructure are all important factors that can enhance the shift to mobile learning. Moreover, several studies indicate positive attitudes and perceptions toward mobile learning at different Arab learning institutions. However, specific challenges may act as barriers to mobile learning in the Arab world. This chapter reviews some of the current mobile learning practices in the Arab world and provides an overview of challenges faced by Arab students, educators, and probably researchers. A description of future mobile learning in the Arab countries is then provided.

Much easier, faster, and mightier than both offline and online, mobile identity can be developed and mobilized, temporary or permanent, public or private, and individual or institutional. A mobile identity can be used in different contexts for different purposes. Potentially, however, it can have different risks and challenges, such as security, surveillance, privacy issues and their corresponding consequences, as examined in Chapter 4. Closely related to mobile identity are the issues of gender and rituals in a mobile environment. Investigating how gender rituals and use of mobile technologies are mutually shaped, Chapter 5 explores the implications of mobile technologies on gender from the perspectives of gender rituals. Related to gender in mobile

local slogan, "Wahala don sele" (problem has started), is articulated in intimate conflicts. The chapter concedes that mobile communication plays major roles in daily affairs of young female adults, but it has its social consequences. Therefore, it suggests that future research should look beyond the normative view of mobile development, which only concentrates on governance, education, business, and health.

Section 3
Mobile Engagement and Movement

To engage people is to facilitate and foster civic conversation and participation, which has been part of the tradition of media, traditional or new. Switching from offline to online, media engagement does not change its function. To go mobile is another advance in media engagement. Instead of opening up a new space, mobile engagement extends what we already occupy, such as health, education, elections, debates, campaigns, journalism, banking, data collection, shopping, business, trade, gaming, entertainment, and narration. Speaking of narration, use of mobile is redefining narrative media engagement, which is investigated in Chapter 7. Besides being widely used in engagement, mobile has also been widely and increasingly used for civic movements not only in most liberal societies but also in most authoritative countries, which is unprecedented in history. For instance, as documented in Chapter 8, mobile communication has been used for seeking and sharing information, galvanizing support, and mobilizing action for political and social issues in Singapore. Mobile engagement and mobile movement go hand in hand in most cases. As examined in Chapter 9, in Malaysia, youth engagement is said to take the center stage of democracy, and young adults use mobile to mobilize members and resources for online and/or offline movements to seek social, economic, and political changes.

The evolution of mobile technologies with the capacity to act as conduits to narrative media has led to the emergence of increasingly transmedia and pervasive storytelling practices. This chapter interrogates a mobile phone-based pervasive drama, The Memory Dealer (Lander, 2010), and a focus group run with participants, focusing on two "shifts" within the performance: from internally focused narrative (via headphones) to externally focused narrative (via public speakers) and from an individual experience to a communal experience. By considering how The Memory Dealer's audience responded to such shifts, it becomes possible to consider the inherent nature of being "an audience" of transmedia narrative. It is argued that the experimental pervasive forms facilitated by mobile technologies raises the potential to critically re-examine the nature of media engagement and the technological, social, and cultural dynamics that shape and are shaped by that engagement.

Youth engagement is often said to be at the heart of democracy, though the extent of such efforts in affecting policymaking remains highly debatable. Nonetheless, there have been heightened attempts by "ordinary citizens" to reform the country's state of politics and to improve society's living conditions. Malaysia's authoritarian democracy has been a crucial motivation for young adults to "have their say" in challenging the current regime. This chapter highlights the various ways in which young adults use mobile media to activate and participate in civic, community, and political engagement whilst taking into account the many restrictions that are set up by the ruling government to monitor and control such engagements. Discussed alongside youth definitions of nationalism, citizenship, and activism that are embedded within the interviews, the findings are juxtaposed with present post-election discourses taking place within the country. The relationship between mobile media and youth engagement further affirms the idea of a new generation of mobile users that are not just technologically savvy but are using their knowledge to affect significant societal changes.

At Hong Lim Park in February 2013, the "No to 6.9 Million Population" protest saw 5,000 Singaporeans expressing their unhappiness with the government's Population White Paper. Touted to be the largest demonstration since Singapore's independence, it bore witness to digital technologies' mobilization effects. Personal and organization websites, discussion forums, blogs, and social media provide viable spaces for individuals and marginalized groups to circumvent offline media regulations and participate in counter-hegemonic discourse. Surveys indicate that Singaporeans are increasingly leveraging mobile communication for utility purposes—seeking and sharing information—and for networking. This chapter identifies digital bottom-up movements that took place in recent years, the anatomy of these movements, and how digital technologies were used. What is evident is that groups championing different causes are using a wide range of digital platforms to galvanize support and mobilize action for political and social issues. However, the link between that and mobile communication remains unclear. This chapter concludes by presenting recommendations for future studies on mobile communication and bottom-up movements.

Section 4
Mobile Children and Parenting

Mobile children are those who are using mobile as an indispensable part of their daily lives. Besides enabling children to stay connected with peers and parents, mobile has also ushered in new forms of interaction and coordination among children for flexible and fluid social networking. Mobile has become a tool for children for peer identification, interaction, engagement, and communication. Earlier studies regarding use of mobile among children largely involved feature phones. With the advent of smartphones, more and more children are replacing their feature phones with smartphones. Instead of a mere change of device, that shift opens up a whole new world for children. Beyond phone calls and text messaging, they use smartphones for, among others, chatting, gaming, surfing the Internet, music, taking and sharing photos, and/or staying social. Equipped with advances in mobile technologies such as mobile location-based services, smartphones have brought children not only new opportunities but also new risks. Besides identifying both new opportunities and new risks, Chapter 10 examines major implications of the use of smartphones among children and sheds light on new areas for research on smartphones for children. Closely related to mobile children is naturally mobile parenting. Using mobile as a tool for parenting, parents extend their attention, care, and surveillance when they are away from their children. With an increasing number of migrant workers leaving their children in their home countries, how to parent them has become a bigger challenge for migrant parents. Chapter 11 investigates the practice of mobile parenting and its related demographic, social, economic, and technological challenges and implications through a case study of Filipino migrant mothers in Singapore.

The shift from mobile phones to smartphones, which integrate mobile communication, social media, and geo-locative services, expands the scope of mobile communication and opens up new opportunities for young people, as well as new risks. However, research on the adoption and use of smartphones among young people is still sparse. Therefore, the consequences of smartphone use on children's communicative cultures and relational practices remain yet to be explored. Drawing on the consistent literature on children and mobile communication and on the findings of preliminary research on younger users' domestication of smart mobile devices, the present chapter discusses the implications of mobile media use amongst young people in the light of the changes associated with smartphones and the mobile Internet.

This chapter focuses its attention on Filipino mothers in a diaspora and their mediated parenting as it tackles the centrality of mobile communication in transnational parenting. Apart from describing Filipino migrant mothers' mobile parenting practice in terms of the occurrence and content of their mediated parenting efforts, this chapter discusses mobile parenting by looking into the intersections of the socio-demographic, socio-economic, and socio-political landscapes of Filipino migrant mothers' mobile parenting with the social and technological dimensions of mobile parenting. In describing the landscapes, dimensions, and practice of Filipino migrant mothers' mobile parenting, evidence from interviews of 32 Singapore-based Filipino migrant mothers were supplemented by evidence from the literature on Filipino migrant parents and their transnational parenting efforts, their children left behind, and their children's caregivers.

Section 5
Mobile Marketing and Advertising

Both marketing and advertising have gone mobile, be it "mobile too," "mobile first," or "mobile only." While mass media are declining, mobile is increasing in popularity as a platform to catch attention, increase awareness, instill desires, and increase sales. Equipped with advanced technologies and connected with positioning technologies, mobile enables users to access Mobile Location-Based Services (MLBS). While enjoying MLBS benefits, mobile users are exposed to privacy issues, which are examined socially, economically, and politically in Chapter 12. Although mobile location-based advertising has all the benefits of location awareness, sociability, and spatiality, it does have constraints, as investigated in Chapter 13, which also discusses potentials of MLBA-enabled advertising in China. For marketing, mobile location-based services have also been employed in a game to create hybrid spaces, in which mobile users can be more engaged and empowered through brand experience, productive play, and value co-creation. Although it may be accompanied by some unintended consequences, mobile location-based games can prove to be effective in marketing, as demonstrated in Chapter 14.

for the brand. It also points to unintended consequences of the campaign, evidenced in consumers' deconstruction and reconstruction of the marketer-designed game process and their distortion of space. More generally, it reveals the importance of understanding the opportunities and challenges in addressing creative consumers in utilising mobile location-based game for marketing purposes.

Section 6
Mobile Uses and Usability

Mobile has been widely used in increasingly diversified areas, such as learning, identity-building, gender rituals, solution to intimate conflicts, audience engagement, civil movement, parenting, location-based marketing, and location-based advertising. More uses of mobile include gaming, healthcare, and journalism, three areas examined in this section. Behind different mobile uses lies the same need to enhance mobile usability. In this section, location-based mobile gaming, mobile healthcare, mobile news, and mobile usability are the selected topics for investigation. Mobile gaming is one of the most popular mobile activities. With the introduction of mobile location-based technologies, mobile gaming has substantially changed the way mobile gamers interact with other people and objects in a gaming environment. After critically reviewing earlier studies, the author of Chapter 15 investigates how mobile location-based games reconfigure people's relationships with other people and objects in their environment, and presents a model to understand this reconfiguration. Mobile has been actively used in news reporting and writing by both professional and amateur journalists. Equally, mobile has become a popular device for users to seek and follow news on a regular basis. When and under what circumstances, however, do mobile users use mobile as a device for news? The answer can be found in Chapter 16. As a fast-growing market, mobile health has become one of the big businesses. Looking into the use of mobile for healthcare in the context of global wirelessly connected health sector, the author of Chapter 17 offers timely and valuable observations of and insights on the exponentially expanded market as well as recommendations from a socio-economic and political standpoint for responsible and effective future industry growth. Mobile usability is a worldwide concentration of efforts to enhance mobile use for all purposes. To facilitate efforts to enhance mobile usability, the authors of Chapter 18 explain various usability testing methods, contextual complexity, audio interfaces, eye and hands-free interactions, augmented reality, and recommendation systems. The chapter also examines methodological challenges in mobile usability as well as the social and economic implications of mobile usability.

This chapter explores the opportunities of mobile games to critique and constitute the networks of which they are a part, attending particularly to location-based games. It discusses how these kinds of mobile games reconfigure people's relationships with other people and objects in their environment. In order to understand this reconfiguration, a model is put forward that clarifies the various ways in which people and objects are presented to the mobile game player. Using this model, examples are discussed of games that make interactions available that are disruptive of a social or political order, arguing that this disruption may be drafted into socio-political critique. Other examples demonstrate how mobile games bring everyday life within a capitalist logic, monetizing leisure and the mundane. This suggests that mobile gaming as a technology, practice, or product is neither fundamentally emancipatory nor fundamentally regressive but rather can be employed in various ways.

Based on a theoretical framework drawn from the diffusion of innovation theory, the expectancy-value model, and the technology-acceptance model, this chapter presents an empirical study of technological and informational factors as predictors of the use of second-generation mobile phones as a news device. The study differentiates the initial adoption of a mobile phone as a technology innovation from second-level adoption, which refers to the acceptance of a distinctive new function of a technology device serving a communication purpose different from that for which the device was originally designed. The study found that technology facility and innovativeness were significant predictors of mobile phone use as a news device; further, it partially confirmed the model of predictors of mobile phone use for news access. However, the two informational factors—perceived value of information and news affinity—were found to have no direct effect on mobile phone use as a news device. The study departs from the traditional approach of adoption research and offers a novel perspective on examining the adoption of new media with multiple evolving functions.

Wireless connected health is the most current, inclusive phrase to describe healthcare that incorporates wireless technologies and/or mobile devices. It represents one of the fastest growing sectors in the global mobile and wireless ecosystem, with extraordinary change occurring daily. According to the World Health Organization, 80 percent of people in greatest medical need live in low- to middle-income countries. Not enough has been written about how they will afford wireless connected health, or how it can bring positive benefits to patients everywhere with non-lethal chronic illnesses. It also remains to be seen whether people outside the healthcare industry, without any special interest in science, technology, medicine, or illness prevention, will adopt new and future behavior-changing connected health technologies. This chapter provides a current overview of the global health crises created by noncommunicable diseases, explains the evolution of the global wireless connected health sector, includes information about BRICS nations, and offers observations, insights, and recommendations from a socio-economic and political standpoint for responsible and effective future industry growth.

Context and the pervasive environment play a much greater role in mobile technology usage than stationary technology for which usability standards and methods were traditionally developed. The examination of mobile usability shows complex issues due to the ubiquitous and portable nature of mobile devices. This chapter presents the current state of mobile usability testing. More specially, topics covered are various usability testing methods, contextual complexity, audio interfaces, eye and hands-free interactions, augmented reality, and recommendation systems.

Foreword

Readers who know their Greek mythology will know that the Lernaean Hydra was the second of Hercules' 12 labors. The multi-headed hydra would sprout a new head for each one that was cut off. For the mobile communication researcher, the metaphor of an ever-growing number of functions and research directions resonates. Just when you think you understand the dimensions of interpersonal mobile communication, along come smart phones. Just when you think that you have a handle on mobile Facebook, along comes Snapchat. And the story goes.

In this book, Xiaoge Xu as editor does a Herculean job of giving the reader a perspective on this dynamic area of research. As with the field of mobile communication studies, this book is seemingly boundless in the geographic areas that are covered, in the dynamic nature of the device and the number of applications for which mobile communication is the locus.

The book covers a significant number of the areas associated with mobile communication. The book is divided into 6 sections including chapters on the mobile's effect on literacy and learning, identity and gender, political engagement, the status of children and parenting, location-based gaming, and finally on uses and usability.

This book covers a lot of territory both literally and in terms of the way it conceives of mobile communication. When thinking about geographic coverage, there are authors from many points of the globe. The authors touch on mobile communication in Arab countries, the Philippines, Nigeria, Malaysia, Singapore (where Filipino workers use the mobile to connect with their children), China, and the BRIC countries (Brazil, Russia, India, and China). Thus, while Taylor reminds us in his analysis of mobile literacy, there is a sense in which mobile frees us from particular locations. However, there is tension in this idea since there is very much a contextual side to mobile use. This is experienced both at the local level (facilitation of learning, parenting, identity formation, etc.) and at the international level. Thus, while there is place, there is also mobility. The authors in this volume help us to understand how mobile communication illuminates the tension between these two.

In addition to the geographical use of mobile communication, this technology is where we deal with a wide variety of social processes. These chapters reveal how the mobile device that we have in our pocket is becoming the locus for all manner of interactions with our world. There is seemingly no end to the functions and social dynamics that employ mobile communication devices. As we would expect, the authors help us to understand how the mobile phone and mobile communication facilitate interpersonal communication. Beyond that, the mobile phone has become the site of a seemingly endless array of social and commercial processes. In these pages, Holland describes how students and educators in a variety of settings use mobile apps. Al-Shehri describes the work of mobile learning initiatives in the Arab world. Brizel discusses the challenges and implications associated with the provision of mobile healthcare. Li describes our interest in the use of mobile news, and Evans provides us with an analysis of mobile narratives including film, television, theater, radio, and games. The issue of games is also described in the works of Zhao and Martin. Mobile devices that incorporate location-based sensors (see the article by Jiow) are being used for gaming, social networking, and—as descried by Wu and Yao—for various types of marketing. Moving in a different trajectory, Lim and also Soon and Cheong help us to understand how mobile devices are being used for political agency and "bottom up" movements.

Moving from more institutional interactions to that of individual identity and the familial sphere, mobile communication is a central point in our identity work and the coordination of our interactions. Fehír describes mobile identity while Soriano examines gender rituals as mediated by the mobile phone, and Afolayan studies how mobile technology allows liberation for young women in Nigeria. Thus, gender is a social phenomenon that is mapped onto the use and usefulness of the mobile phone. Looking at the family, Soriano has studied how the mobile phone is blurring the role of family life and leisure against access to a world of information. Manceroni and San Pascual examine how the mobile phone is the nexus of interaction between parents and their children. Finally, Gallant, Boone, and LaRoche help us to understand the role of mobile usability.

It is clear from this work that the mobile telephone is the locus of a wide variety of processes. It is also clear from these chapters that it is dynamic. The authors speak of emerging challenges and phenomenal growth. We see how mobile communication is ushering in a new era that has a dramatic as well as a dynamic trajectory. We are indeed fortunate that Xiaoge Xu has taken the job of bringing together this cornucopia of scholarship that helps us to get, at least temporarily, a handle on this Lernaean technology.

Rich Ling
IT University of Copenhagen, Denmark

Rich Ling *(PhD, University of Colorado, Sociology) is the Shaw Foundation Professor of Media Technology, Wee Kim Wee School of Communication and Information, Nanyang Technological University, Singapore. He works at Telenor Research and has an adjunct position at the University of Michigan. Ling has studied the social consequences of mobile communication for the past two decades. He has written The Mobile Connection (Morgan Kaufmann, 2004), New Tech, New Ties (MIT, 2008), and most recently, Taken for Grantedness (MIT, 2012). He is a founding co-editor of Mobile Media and Communication (Sage) and the Oxford University Press series Studies in Mobile Communication.*

Preface

Surpassing any other medium in penetration, usage, and impact, mobile has become the most popular and powerful medium for activities, ranging from mobile gaming to mobile governance. Equally superior to any other form of communication in penetration, usage, and impact, mobile has become the most popular and powerful communication, ranging from mobile journalism to mobile health communication.

Mobile has redefined space and presence (Fortunati, 2002), creating connected presence (Licopp, 2004), absent presence (Gergen, 2002), and modified presence and absence of individuals in social space (Fortunati, 2002). Enabling inter-personal and mobile-human interactions across time and space, mobile can be real, virtual, or mixed in its communication of all kinds. Dramatically and exponentially, mobile has been changing our world and our lives, redefining the way we live, the way we work, the way we communicate, and the way we interact with each other and with the world.

While mobile has surpassed desktops or laptops in many developed countries, it has been widely used as a replacement of desktops or laptops in developing or underdeveloped countries. Mobile has become a preferred medium for both individuals and institutions. Switching from "mobile, too" to "mobile first" or even "mobile only" strategies, governments, companies, institutions, and individuals around the world are all leveraging mobile to enhance their respective communications and performances.

Never-ending new forms, features, and functions of mobile keep arriving, shaping our present and future. For instance, as predicted by Stewart Wolpin (2014) on the eve of the 2014 Mobile World Congress, Bluetooth 4.0 will turn our room entry keyless and NFC will enable us to go ticketless and cashless. His other predictions include biosensors being able to monitor our health and fitness constantly, anytime, and anywhere, biometric security enabling us to get away with passwords or swiping patterns, wireless charging, and ubiquitous and automatic WiFi (Wolpin, 2014).

As the key driver of big data globally (Leonhard, 2014), mobile is playing a big role in bringing about big business worldwide. There is big business in mobile money and mobile banking. The big business also goes to mobile health, mobile travel, and mobile shopping. When it comes to mobile entertainment, mobile gaming, and mobile creative industries, the business is equally big. And big business also embraces mobile trade, mobile commerce, and mobile education.

As best described by Rich Ling in his foreword to this volume, mobile is going Lernaean. As a medium, mobile is constantly changing its forms, features, and functions. The moment the old dies, the new is born. Equally Lernaean as a communication, mobile replaces the old with the new the moment the old dies, be it a form, a feature, or a function of a communication. Naturally, the Lernaean nature of mobile has led to increasingly diversified studies of mobile media and communications in ever-changing areas across different disciplines.

STUDIES OF MOBILE

Mobile has been examined as a medium used in wide-ranging areas from different disciplinary perspectives. Key studies include, for instance, convergence, divergence, and the role of mobile media (Hjorth & Kim, 2005), mobile capacity to reconfigure users' relation to actual and virtual space and their experience (Richardson, 2007), the niches of mobile media in space and time for news in the interstices (Dimmick, Feaster, & Hoplamazian, 2011), the relationship between location-aware mobile media and urban sociability (Daniel & de Souza e Silva, 2011), gamification and personal mobility in location-based social network (Frith, 2013), mobile social space and news sociality (Lee, 2013), and mobile Internet adoption and diffusion (Damásio, Henriques, Teixeira-Botelho, & Dias, 2013). As shown by these representative studies, mobile has been examined as a medium at the micro level from different disciplinary perspectives, such as designing, gamifications, social space, diffusion of innovations, social networking, location-based services, and niche theory.

Studies of mobile as a communication have examined, for instance, diffusion and success factors of mobile marketing (Scharl, Dickinger, & Murphy, 2005), the influence of mobile phones on the public sphere (Gordon, 2007), the impact of mobile multimedia journalism on news (Martyn, 2009), mobile news (Westlund, 2013), transnational mothering (Chib, Malik, Aricat, & Kadir, 2014), and the effects of smartphone use on engagement in civic discourse in China (Wei, 2014). As shown by the these research articles, mobile has been examined as a communication in different but separate disciplines, such as journalism, advertising, marketing, health communication, governance, political communication, and family studies.

In studies of mobile published in the form of books, most scholars took various macro approaches in their examinations of mobile media and communications largely from different disciplinary approaches such as sociology, psychology, cultural studies, social networking, ethics, social policy, politics, and interaction between users and technologies. In their investigations of mobile as a medium, for instance, some scholars examined social and interactional aspects of the mobile age (Brown, Green, & Harper, 2001), the impact of mobile on society (Ling, 2004), global mobile media (Goggin, 2010), mobile media as cultural technologies and communication in the case of iPhone (Hjorth, Burgess, & Richardson, 2012), and mobile technology and place (Wilken & Goggin, 2013).

In the case of studies of mobile as a communication, scholars have investigated, for instance, private talk and public performance in mobile communication (Katz & Aakhus, 2002), the global and the local in mobile communication (Nyíri, 2005), renegotiation of the social sphere in mobile communications (Ling & Pedersen, 2005), mobile communication in everyday life (Höflich & Hartmann, 2006), mobile communication and the transformation of social life (Katz, 2006), how mobile communication is reshaping social cohesion (Ling, 2010), dimensions of social policy in mobile communication (Katz, 2011), and applications and innovations for the worldwide mobile ecosystem (Bruck & Rao, 2013).

MOBILE STUDIES AS AN EMERGING FIELD

After more than a decade of investigating mobile, a new field has emerged. It was first called Mobile Communication Studies when the Center for Mobile Communication Studies was established in June 2004 at the School of Communication and Information, Rutgers University, USA. Under the leadership of Prof. James E. Katz, the center helped to develop mobile-related courses for the Department of Communication and also provided advice for public and private sectors. After Prof. Katz moved to Boston University, he launched a center bearing the same name in the College of Communication, Boston University. Different in location, but same in principle, the center is designed to investigate "human uses, meanings, and co-construction that arise from" "the distribution by, or the exchange of, data from devices that can be moved physically with relative ease" from broad ranging disciplinary interests including those of psychology, sociology, philosophy, and political science (Center for Mobile Communication Studies, 2013).

Another milestone in the development of this emerging field is the launch of a new journal by SAGE, *Mobile Media & Communication*, co-edited by Rich Ling (IT University of Copenhagen, Denmark), Veronika Karnowski (Ludwig-Maximilians-Universität München, Germany), Thilo von Pape (University of Hohenheim, Germany), and Steve Jones (University of Illinois at Chicago, USA), which positions itself as investigating mobility in communication from interdisciplinary approaches. *Mobile Media & Communication* has become an icon of this emerging field being recognized worldwide.

As an emerging field, although it is often referred to as mobile communication studies (see Katz, 2008), mobile studies is a better term to accurately and fully describe the subjects, scope, and strategies of studies of mobile as a medium or a communication from interdisciplinary approaches. Mobile communication studies can be easily confused with studies of mobile communication from mobile computing science or design perspectives. For instance, in terms of academic journals from the science side, we have the *Journal of Mobile Communication*, the *International Journal of Mobile Communications*, *Personal and Ubiquitous Computing*, and the *International Journal of Mobile Computing and Multimedia Communications*. These journals largely publish studies of mobile from the designing and mobile computing perspectives.

First used by Kristóf Nyíri in 2007 in his book *Mobile Studies: Paradigms and Perspectives*, the term "mobile studies" was also used to name the global institute for mobile media and communication education, research, and services, which I founded in August 2012. The Institute's name was changed to Mobile Studies International by mid-2013 to match its international orientation, operation, and reputation. In Glossary on the Website of Mobile Studies International, I define mobile studies as follows:

"Being interdisciplinary and applied, mobile studies examines forms, features, functions, processes, experiences, and effects of mobile as a medium or a communication used in different areas for different purposes. Major areas of mobile studies include, not limited to, the following: mobile experience, mobile content, mobile services, mobile users, mobile governance, mobile marketing, mobile advertising, mobile education, and mobile journalism (Xu, 2012)."

In my definition, mobile consists of two dimensions. One is the medium side and the other is the communication side. My definition not only identifies fundamental components of mobile studies but also summarizes its subjects, scope, and strategies.

As an emerging field, whether it is called mobile studies or mobile communication studies, it should be "integrally connected to the study of media and communication more broadly" (Campbell, 2013, p. 11), "putting people at the center of future mobile communication research" (Wei, 2013, p. 54), and "look[ing] to the changes in the technical, social, political, regulatory, and other forms of infrastructures" (Horst, 2013, p. 147).

MOBILE STUDIES AS AN EMERGING DISCIPLINE

In response to the needs of both the emerging field of mobile studies and the rising dominance of the mobile industry, there have been efforts to launch mobile studies as an emerging discipline in institutions of higher learning.

After being examined in the social sciences and the humanities for more than a decade, mobile has come of age as a new discipline with an increasing amount of scholarship accumulated over the years. The results of mobile studies from interdisciplinary approaches were summarized in *Mobile Studies: Paradigms and Perspectives* (Nyíri, 2007), which constitutes another milestone of the development of the new discipline.

As "the world's first academic unit to focus solely on social aspects of mobile communication," the Center for Mobile Communication Studies "will educate Boston University students, especially those of the College of Communication, to become the next generation of experts in understanding the role and consequences of mobile communication for human society in all its dimensions" (Center for Mobile Communication Studies, 2013).

At Mobile Studies International, one of its global initiatives is Mobile Campus, which is "designed to offer customized courses from different perspectives. First of its kind, Mobile Campus is also created to help hosting universities to develop mobile curriculum, to train instructors, and to co-organize Mobile Workshops" (Mobile Studies International, 2014). Another initiative is to co-run Institute for Mobile Studies in universities around the world. Among its goals are (1) promoting mobile studies as a discipline in institutions of higher education around the world and (2) staying productive, relevant, and excellent in mobile media and communication education, research, and services. Its major activities include: (1) co-conducting research projects related to mobile media and communications, (2) co-organizing international conferences on mobile studies, (3) co-running workshops for mobile professionals, (4) co-supervising PhD students in mobile studies, and (5) co-running graduate programs in mobile studies (Mobile Studies International, 2014).

As part of the curriculum development at Mobile Studies International, core courses for the graduate programs include (MS01) Mobile Foundations, (MS02) Mobile Theories, (MS03) Mobile Methods, (MS04) Mobile Creativity, (MS05) Mobile Economy, (MS06) Mobile Content, (MS07) Mobile Services, (MS08) Special Topics in Mobile Studies (Mobile Studies International, 2014).

INVESTIGATING MOBILE: INTERDISCIPLINARY WAY

Against the backdrop of mobile studies' emergence as both a field and a discipline, seriously lacking are efforts to put mobile under interdisciplinary scrutiny, despite the increasing amount of scholarship on mobile media and communications. The interdisciplinary approaches are designed to get a holistic picture and a better understanding of ever-changing forms, features, functions, processes, experiences, and effects of mobile media and communications. As mobile has been used in increasingly different areas for different purposes, resulting in massive and complicated situations, no single discipline can do the job of fully and accurately describing, explaining, and/or predicting mobile uses, impacts, and implications.

The interdisciplinary approaches to mobile studies are being widely recognized but not widely practiced. For instance, according to the journal positioning statement of *Mobile Media & Communication*, the journal "is a peer-reviewed forum for international, interdisciplinary academic research on the dynamic field of mobile media and communication." "While the center of gravity lies in social sciences and humanities, the journal is open to research with technical, economic, and design aspects, provided they help to enlighten the social dimensions of mobile communication" (Ling, Karnowski, von Pape, & Jones, 2013).

Driven by interdisciplinary approaches to mobile studies, this volume investigates mobile locally, regionally, and globally. Its scrutiny covers more than 18 countries across 5 continents. Locally, it examines the impact of mobile on gender rituals in the Philippines, mobile location-based advertising and marketing in China, the role of mobile in bottom-up movements in Singapore, and the mobile engagement and empowerment among the youth in Malaysia. Its scrutiny also goes to mobile literacy in Australia, mobile engagement in UK, use of mobile as a news device in US, and the role of mobile in liberating young female Adults in Nigeria. Regionally, it investigates the use of smartphones among children in Denmark, Italy, Romania, UK, and Ireland. Under its microscope are the implications of mobile learning in the Arab world as well as mobile healthcare in BRICS countries (Brazil, Russia, India, China, and South Africa). Its global investigations focus on implications of mobile location-based services on privacy, challenges, and implications of mobile usability, and the reconfiguration of social-spatial relations in mobile gaming.

In its investigation of mobile in different geographical, social, cultural, political, and economic environments, this volume focuses its scrutiny of mobile media and communications by combining mobile studies with education, gender studies, identity studies, political communication, family studies, games studies, computing studies, marketing, advertising, healthcare, and journalism. To make interdisciplinary investigation possible, I have invited 22 researchers and scholars from 10 countries across 5 continents with diversified academic interests and approaches in exploring and examining the social, political, and economic implications of mobile media and communications.

Topics investigated in this volume are (1) mobile literacy, (2) evaluation of instructional apps, (3) mobile learning in the Arab world, (4) digital mobilization and identity, (5) implications of mobile technologies on gender, (6) social consequences of mobile use on young female adults in intimate conflicts, (7) dynamics of transmedia pervasive narratives, (8) mobile use in youth engagement, (9) mobile bottom-up civil movement, (10) implications of smartphone use among children, (11) mobile parenting amid global mobility, (12) social-spatial relations in mobile gaming, (13) implications on privacy of mobile location based-services, (14) mobile location-based advertising, (15) mobile location-based games in marketing campaigns, (16) implications of wireless technologies and mobile devices on healthcare, (17) mobile as a news device, and (18) mobile usability.

These topics were examined in different frameworks and concepts. For instance, they include the combination of different theoretical traditions of literacy with Pierre Bourdieu's sociological model, merging educational theory with technology tools and human interactions, mobile learning theories, concepts of digital mobilization and identities, ritual and gender studies, mobilization, collective actions, domestication, mobile location-based services, social-spatial relations, diffusion of innovation theory, expectancy-value model, technology acceptance model, and mobile usability.

The results of investigating the above-listed 18 topics are presented in 18 chapters in 6 sections: (1) "Mobile Literacy and Learning" (Chapters 1-3), (2) "Mobile Identity and Ritual" (Chapters 4-6), (3) "Mobile Engagement and Movement" (Chapters 7-9), (4) "Mobile Children and Parenting" (Chapters 10-11), (5) "Mobile Advertising and Marketing" (Chapters 12-14), and (6) "Mobile Uses and Usability" (Chapters 15-18). Each section starts with an editor's note to provide a preview of what is covered.

In the first chapter of the first section, "Mobile Literacy and Learning," drawing on different theoretical traditions of literacy studies, Pierre Bourdieu's sociological model, and the results of an ethnographic study conducted in Australia, Calvin Taylor constructed a model of mobile literacy, revealing both contextual factors and the structural impact of economic, social, cultural, and symbolic resources associated with mobile devices. In the second chapter, Janet Holland proposed a set of questions to evaluate instructional needs for mobile learning apps on the basis of the results of three pilot studies. In the third chapter, after reviewing mobile learning practices in the Arab world, Saleh Al-Shehri identified challenges faced by Arab students, educators, and researchers.

After examining definitions and dimensions of mobile identity and its relationship with digital mobilization, Katalin Fehér, author of Chapter 4, offered observations and insights on mobile identity. In Chapter 5, Cheryll Soriano explored the implications of mobile technologies on gender through the lens of gender rituals and highlighted the importance of culturally embedded rituals in shaping and understanding the mobile phone's place in society. In Chapter 6, Gbenga Afolayan discussed the role of mobile communication technology in liberating young female adults from the perspective of socio-economic transformations and its wider transformative social impacts. These three chapters are grouped under "Mobile Identity and Ritual" in the second section.

In Chapter 7, Elizabeth Evans examined mobile engagement in relation to the technological and social dynamics of transmedia persuasive narratives. In Chapter 8, Joanne Lim discussed the use of mobile media by youth to put forward their own agency, to challenge the existing social and political structures, to strive for social change, and to construct their own identities. In Chapter 9, Carol Soon and Cheong Kah Shin identified digital bottom-up movements in Singapore as Singaporeans were increasingly leveraging mobile communication for seeking and sharing information, networking, and galvanizing support and mobilizing action for political and social issues. Chapters 7-9 fall under "Mobile Engagement and Movement" in the third section.

The fourth section, "Mobile Children and Parenting," starts with the tenth chapter, in which Mascheroni Giovanna discussed implications of the use of smartphones amongst children in the context of changes in mobile technologies and the mobile Internet. It ends with the eleventh chapter, authored by Ma. Rosel San Pascual, who argued that mobile communication has become an imperative part of parenting for the survival and growth of transnational families in the age of increasing global mobility after investigating how parenting was done among the Filipino migrant mothers in Singapore.

After charting the trajectory of mobile location-based services applications in the mass market, Hee Jhee Jiow, author of Chapter 12, captured and discussed the social, political, and economic implications of mobile location-based services on privacy through a proposed framework of measuring user-position or device-position focus and alert-aware or active-aware applications. In Chapter 13, through examining a mobile location-based game launched by The North Face (TNF) in China, Elaine Zhao argued that it was important to understand both opportunities and challenges in utilising mobile location-based games for marketing purposes. In Chapter 14, Mei Wu and Yao Qi investigated the dilemmas faced by location-based advertising in China through a case study. The three chapters are placed under "Mobile Advertising and Marketing" in the fifth section.

In the final section, "Mobile Use and Usability," Paul Martin proposed in Chapter 15 a model to examine the reconfiguration and contestation of social-spatial relations in mobile gaming. In Chapter 16, Xigen Li located significant predictors of mobile phone use as a news device. In Chapter 17, Florie Brizel offered her observations of and recommendations for mobile healthcare. In Chapter 18, Linda M. Gallant, Gloria Boone, and Christopher S. LaRoche examined challenges and implications of mobile usability in their critical investigation of usability testing methods, contextual complexity, audio interfaces, eye and hands-free interactions, augmented reality, and recommendation systems.

This volume serves the needs of mobile practitioners, policymakers, professors, and students around the world for updates and insights on the social, political, and economic implications of mobile media and communications. Shedding light on the changes, challenges, and chances in the mobile world, this volume contributes to the mobile world by introducing a model for studying mobile literacy, a model for investigating socio-spatial reconfigurations in mobile gaming, an evaluation instrument for mobile leaning apps, and predication of mobile use as a news device. More contributions include insights on digital mobilization and identity, the interaction between gender rituals and mobile technologies, the role of mobile in liberating young adults in intimate conflicts, implications of mobile location-based services on privacy, audience engagement in a mobile environment, civil movement via mobile, domestication of smartphone among children, and mobile parenting. Further contributions lie in insightful observations and recommendations on mobile location-based marketing, advertising, and healthcare.

Serving as a steppingstone, this volume invites more global collaborative efforts in locating further social, political, and economic implications of mobile media and communications around the world.

Xiaoge Xu
The University of Nottingham Ningbo China, China

REFERENCES

Brown, B., Green, N., & Harper, R. (Eds.). (2001). *Wireless world: Social and interactional aspects of the mobile age*. New York, NY: Springer-Verlag.

Bruck, P. A., & Rao, M. (Eds.). (2013). *Global mobile: Applications and innovations for the worldwide mobile ecosystem*. Information Today, Inc.

Campbell, S. W. (2013). Mobile media and communication: A new field, or just a new journal? *Mobile Media & Communication*, *1*(1), 8–13. doi:10.1177/2050157912459495

Center for Mobile Communication Studies. (2013). *What is mobile communication studies?* Retrieved from http://www.bu.edu/com/academics/emerging-media/cmcs/

Chib, A., Malik, S., Aricat, R. G., & Kadir, S. Z. (2014). Migrant mothering and mobile phones: Negotiations of transnational identity. *Mobile Media & Communication*, *2*(1), 73–93. doi:10.1177/2050157913506007

Damásio, M. J., Henriques, S., Teixeira-Botelho, I., & Dias, P. (2013). Social activities and mobile Internet diffusion: A search for the Holy Grail? *Mobile Media & Communication*, *1*(3), 335–355. doi:10.1177/2050157913495690

Daniel, M. S., & de Souza e Silva, A. (2011). Location-aware mobile media and urban sociability. *New Media & Society*, *13*(5), 807–823. doi:10.1177/1461444810385202

Dimmick, J., Feaster, J. C., & Hoplamazian, G. J. (2011). News in the interstices: The niches of mobile media in space and time. *New Media & Society*, *13*(1), 23–39. doi:10.1177/1461444810363452

Fortunati, L. (2002). The mobile phone: Towards new categories and social relations. *Information Communication and Society*, *5*(4), 513–528. doi:10.1080/13691180208538803

Frith, J. (2013). Turning life into a game: Foursquare, gamification, and personal mobility. *Mobile Media & Communication*, *1*(2), 248–262. doi:10.1177/2050157912474811

Gergen, K. (2002). Cell phone technology and the realm of absent presence. In J. E. Katz, & M. Aakhus (Eds.), *Perpetual contact* (pp. 227–241). New York, NY: Cambridge University Press.

Goggin, G. (2010). *Global mobile media*. New York, NY: Routledge.

Gordon, J. (2007). The mobile phone and the public sphere: Mobile phone usage in three critical situations. *Convergence*, *13*(3), 307–319. doi:10.1177/1354856507079181

Hjorth, L., Burgess, J., & Richardson, I. (2012). *Studying mobile media: Cultural technologies, mobile communication, and the iPhone*. New York, NY: Routledge.

Hjorth, L., & Kim, H. (2005). Being there and being here: Gendered customising of mobile 3G practices through a case study on Seoul. *Convergence Journal*, *11*(2), 49–55.

Katz, J. E. (2006). *Magic in the air: Mobile communication and the transformation of social life*. New Brunswick, NJ: Transaction Publishers.

Katz, J. E. (Ed.). (2008). *Handbook of mobile communication studies*. Cambridge, MA: The MIT Press. doi:10.7551/mitpress/9780262113120.001.0001

Katz, J. E. (Ed.). (2011). *Mobile communication: Dimensions of social policy*. New Brunswick, NJ: Transaction Publishers.

Katz, J. E., & Aakhus, M. (Eds.). (2002). *Perpetual contact: Mobile communication, private talk, public performance*. Cambridge, UK: Cambridge University Press. doi:10.1017/CBO9780511489471

Lee, D. H. (2013). Smartphones, mobile social space, and new sociality in Korea. *Mobile Media & Communication*, *1*(3), 269–284. doi:10.1177/2050157913486790

Leonhard, G. (2014, February 26). *Big data, big business, big brother?* Retrieved from http://Ed.cnn.com/2014/02/26/business/big-data-big-business/index.html

Licoppe, C. (2004). Connected presence: The emergence of a new repertoire for managing social relationships in a changing communications technoscape. *Environment and Planning: Society and Space*, *22*, 135–156. doi:10.1068/d323t

Ling, R. (2004). *The mobile connection: The cell phone's impact on society*. San Francisco, CA: Morgan Kaufmann.

Ling, R. (2010). *New tech, new ties: How mobile communication is reshaping social cohesion*. Cambridge, MA: The MIT Press.

Ling, R., Karnowski, V., von Pape, T., & Jones, S. (2013). About the title. *Mobile Media & Communication*. Retrieved from http://www.sagepub.com/journals/Journal202140

Ling, R., & Pedersen, P. E. (Eds.). (2005). *Mobile communications: Re-negotiation of the social sphere*. New York, NY: Springer.

Linke, C. (2013). Mobile media and communication in everyday life: Milestones and challenges. *Mobile Media & Communication*, *1*(1), 32–37. doi:10.1177/2050157912459501

Mobile Studies International. (2014). *Mobile campus*. Retrieved from http://www.mobile2studies.com/mobile-campus.html

Nyíri, K. (2005). *A sense of place: The global and the local in mobile communication*. Vienna, Austria: Passagen Verlag.

Nyíri, K. (Ed.). (2007). *Mobile studies: Paradigms and perspectives*. Vienna, Austria: Passagen Verlag.

Richardson, I. (2007). Pocket technospaces: The bodily incorporation of mobile media. *Continuum: Journal of Media & Cultural Studies*, *21*(2), 205–215. doi:10.1080/10304310701269057

Scharl, A., Dickinger, A., & Murphy, J. (2005). Diffusion and success factors of mobile marketing. *Electronic Commerce Research and Applications*, *4*, 159–173. doi:10.1016/j.elerap.2004.10.006

Wei, R. (2013). Mobile media: Coming of age with a big splash. *Mobile Media & Communication*, *1*(1), 50–56. doi:10.1177/2050157912459494

Wei, R. (2014). Texting, tweeting, and talking: Effects of smartphone use on engagement in civic discourse in China. *Mobile Media & Communication*, *2*(1), 3–19. doi:10.1177/2050157913500668

Westlund, O. (2013). Mobile news: A review and model of journalism in an age of mobile media. *Digital Journalism*, *1*(1), 6–26. doi:10.1080/21670811.2012.740273

Wilken, R., & Goggin, G. (Eds.). (2013). *Mobile technology and place*. New York, NY: Routledge.

Wolpin, S. (2014, February 23). *7 new smartphone features that will help define your future*. Retrieved from http://Ed.cnn.com/2014/02/23/business/future-smartphone-mobile-world-congress/index.html?iref=allsearch

Xu, X. (2012). *Mobile studies*. Retrieved from http://www.mobile2studies.com/glossary.html

Section 1

Mobile Literacy and Learning

Based on the traditions of both media literacy and information literacy, we may define mobile literacy as the ability to access, understand, and create contextual, social, and location-based mobile information and communication in different contexts ranging from mobile journalism to mobile health communication. To describe, explain, and predict mobile literacy, it is necessary to locate contextual factors and the structural impact of economic, social, cultural, and symbolic resources associated with mobile devices in order to construct a model of mobile literacy, as investigated in Chapter 1. As one of the mobile communications in the context of education, mobile learning is another area of mobile studies. Few studies, however, have examined the instructional needs of mobile learning apps. Chapter 2 proposes a set of questions to evaluate the instructional needs of mobile learning apps. Another important contribution of this book is to provide a better understanding of the mobile learning situation in the Arab world, as demonstrated in Chapter 3, which has identified challenges and implications faced by Arab students, educators, and researchers after reviewing mobile learning practices in the Arab world.

Chapter 1
Mobiles, Movement, and Meaning-Making:
A Model of Mobile Literacy

Calvin Taylor
John Monash Science School, Australia

ABSTRACT

Drawing on ethnographic research conducted with adolescents at a rural Australian high school, this chapter constructs a theoretical model for "mobile literacy." Mobile technologies, and their increasing technological capabilities, present emerging challenges for definitions and understandings of what precisely constitutes "literate practice," challenges which have not been wholly resolved though more disparate discussions of "electronically mediated communication." Such an understanding is important in order to develop approaches that effectively integrate mobile technologies into formal educational contexts. The model constructed in this chapter draws on different theoretical traditions where literacy is concerned, combining these with a sociological model developed by Pierre Bourdieu, to draw out the importance of the social dimension in mobile technology use. The ethnographic methodology results in findings that reveal the structuring impact of economic, social, cultural, and symbolic resources associated with these devices. Far from revealing that mobiles free us from a consideration of "place," this research demonstrates that to be "mobile literate" is to be even more finely attuned to the contextual factors for any mobile technology use.

INTRODUCTION

The genesis for this research has been the emerging popularity, ubiquity and pervasiveness of mobile technologies across widely divergent cultures, especially amongst the young. In particular, for the field of education, mobile technologies have proven to be a particularly problematic technology, in part due to the disruption they cause to power relationships traditionally central to curriculum and pedagogy. Studies of contemporary and emerging literacy practices have typically grouped mobile technologies (such as smartphones and iPods) into a collective group with other emerging ICTs and electronically-mediated texts.

DOI: 10.4018/978-1-4666-6166-0.ch001

However, this research contended that the very 'mobility' of these devices, was what singled these devices out as a potential seismic shift in literate practice. There was also an emerging field of research examining the literacy potential and value of SMS-language: an approach which was somewhat conservative and restrictive in terms of the potentials of mobile technologies for future communication practices. Within a short amount of time, society had gone from mobile phone handsets which could make phone calls and send SMS-messages, to phones with cameras, calendars, games, music, MMS, Bluetooth, Wi-Fi, internet, email, and other downloadable applications (consider the 'iPhone Apps Store' for instance). Additionally, because these devices were used at the everyday level, across a wide range of social situations, the role of context in the meaning of any mobile-related text, needed to be considered. What are the contexts for both sender and recipient when the text "? U" is sent between two people? Because without an understanding of their variable situations, as well as their relationship, the text is relatively meaningless; such considerations of audience and context are central concerns of literacy scholarship. However, there was at the time of this research, no direct, detailed and methodological study of literacy as it pertained to mobile technologies.

This paper is based upon research conducted for PhD studies at Monash University, Australia (Taylor 2011). Being based upon data collected across 2007 and 2008 – does create a degree of time-restriction related to its historical context. Since this data collection there have been a number of advancements that had the potential to change the outlook and possibly results. However, by focusing on patterns of *human* behaviour rather than technical devices in and of themselves, it is hoped that this research will develop a theoretical model which can speak to wider experiences beyond the limitations of particular societies and time periods.

This chapter offers an overview of an attempt to speak to a gap in research and theory. In offering a model of mobile literacy, it seeks to help build pathways into educational reform: mobiles are coming to our classroom and our pedagogies – in some senses, they're already there – and educators need to find uses for them that are beneficial for learning.

THE CHALLENGE OF MOBILE TECHNOLOGIES

They're everywhere: the increasing ubiquity and pervasiveness of mobile technologies throughout increasing societies (regardless of geographic or socioeconomic) has resulted in changes to the structure of social, cultural and interpersonal communication practices.

Across the vast swathe of research and literature addressing the impact of mobile technologies, a central assertion underpins the drive to understand: the unprecedented rapidity with which mobile technologies have disseminated widely throughout the world (see Katz & Aakhus, 2002; Ling & Pedersen, 2005). The significance for research and scholarship is the way that this globally-popular behaviour is manifested at the local level. As Castells, Fernández-Ardèvol, Qui and Sey point out in their global survey of wireless communication: 'Wireless communication has diffused faster than any other communication technology in history. But it has done so differentially' (2007, p. 7). Whilst it is clear that globally, mobile communication technologies have had a pervasive impact – across societies as diverse as Europe, Africa and Oceania – variation at the local level paints a slightly different approach to adoption and domestication in each case.

It is beyond the scope of this restricted section to fully outline the complexity of the effect mobile technologies have had on many different aspects of human experience: this is in fact the purpose of this very collection of papers and others like it. The focus of this discussion is specifically on offering a framework to describe the relationship between literacy and mobile technology use, in order to suggest ways forward for research and educational development.

Research around the impact of mobile technologies on various aspects of the human condition is an area of scholarship that is continually enriched and enriching with possibilities. As a field of research and academia, investigations into mobile technologies and their effects straddle a wide range of disciplines and fields of interest: from medicine, to sociology, from politics, to economics, from demographics and education, to engineering and cultural studies. Similarly, research into the impact of mobile devices transcends geographic and cultural boundaries, with research being conducted in widely divergent countries, from Australia, China, Singapore, Finland, Switzerland, Uganda, South Africa, Jamaica, India and global trends.

The true impact of mobile technologies is not so much on the meta-structures of societies in general, though this is an important point to consider, but in the impact on the everyday lives of individuals. The impact of these technologies *emerge* in a bottom-up fashion, developing practices from the local up to the global. There is no more significant example than the emergence of SMS – or txtng (Crystal 2008) – and the rules of this negotiated text form.

As a specific form of mobile technology, mobile phones have been implicated in significant changes to human communication practices, freeing individuals from a dependence upon place, and locating a locus for individual contact at the level of the individual body, even at vast distances (see Höflich, 2005; Katz, 2003; Katz, 2005; Rule, 2002). The capacity to be constantly contactable (at least ideally) has resulted in increased negotiation and feelings of security: a softening of time and space (Castell, et al, 2007): 'the reality is more prosaic and more complex' than a simple belief that both 'socially and psychology, the "digital generation" is seen to operate in quite different ways from the generations that preceded it' (2007, p. 75). This is not only influenced by individual factors, but by technological advancements in themselves, which as we have seen, is a dialogic process, rather than one of uncomplicated adoption. The process of convergence for instance, Buckingham suggests, will contribute to the breakdown of established social hierarchies, even those around age differences (Buckingham, 2007, p. 81). Indeed, Kennedy, Judd, Churchward, Gray and Krause (2008b) conducted research into first year university student use of web 2.0 technologies, finding that they did not typically fit the concept of 'digital native': the reality was more complex and varied, just as Buckingham (2007) suggested above (see also Kennedy, Dalgarno, Gray, Judd, Waycott, Bennett, Maton, Krause, Bishop, Chang, & Churchward, 2007; Kennedy, Dalgarno, Bennett, Judd, Gray & Chang, 2008a).

The very ability of individuals who would otherwise be separated by distance, combined with mobile phones being identified with personal locus, rather than a location of place (Castells, et al, 2007), enables and encourages the intensification of social connections though communication as 'gift-giving' (Johnsen, 2003), and helps to build 'social capital' amongst the whole group (Ling, 2008; see also Shirky, 2008).

Mobile phones – as the most pervasive of mobile technologies – are further implicated in substantial changes in the organisation of social interactions and meetings, encouraging an ad hoc, on-the-go, just-in-time, negotiated approach (Castells, et al, 2007; Colbert, 2005; Katz & Aakhus, 2002a; Ling & Haddon, 2008; Ling & Yttri, 2002; Rheingold, 2002). One of the most useful and persuasive perspectives developed to explain this new strategy of social coordination is Ling and Yttri's concepts of 'micro-coordination' and 'hyper-coordination' (2002). Whilst the first concerns mundane coordination of social interactions, the second concept adds a dimension of expression and emotional attachment, connecting it more firmly with identity and group dynamics.

Crang, Crosbie and Graham (2006) also paint a more complex picture of usage patterns of ICTs related to the 'digital divide' in urban environments. Whilst for both the affluent and marginalised neighbourhoods examined in this case study, the '*absence* of ICTs is both feared and craved' (Crang, Crosbie & Graham, 2006, p. 2566), variability of access and use of ICTs generally was related to socioeconomic status, with 'affluent neighbourhoods' being associated with 'pervasive ICT use' of 'multiple ICT systems' (p. 2565), yet this was not the case for the marginalised, working-class neighbourhood. Therefore, their finding that 'pervasive ICT users do seem to have much more individualised usage practices than episodic users' (p. 2566) indicates that being a 'digital native' is an identity that is written by a diversity of factors, some beyond a child's control (their socioeconomic status), and as such, doesn't necessarily apply to all individuals, but rather, erases their individual patterns of experience.

RE-THINKING LITERACY FOR MOBILITY

Mobile technologies are changing the topographies of our social lives. As such, they offer new and emerging opportunities for how we construct meaning (both as articulation and interpretation), and the modal resources used for such purposes. Whilst recent scholarship examining electronically-mediated texts and the reconfiguration of literacy models with respect to such changes provide some insight into the impact of mobile devices, this paper argues that the very 'mobility' of these devices provides a paradigmatic shift that is significant enough to require a reconceptualization of our understanding of literacy to account for the particular technologies of mobile devices.

The purpose of this section is to broadly sketch the literacy domain as it is relevant to mobile technology use. Scholarship around emerging literacies is continuously expanding; as such this review is not exhaustive, rather, it traces relevant and promising trajectories of research for a focus on mobile technologies. It examines approaches to understanding literacy under three paradigms: traditional, everyday and concept-based. That this draws mainly from the NLS perspective is due primarily to the fact that this is where the use of mobile technologies currently occurs, the place in our lives that they are positioned. As such, this research is located within that field of literacy studies that sees it as a multimodal, increasingly electronically mediated component of social practice, concerned with meaning-making using a range of semiotic resources, modes and mediums. The nature and structure of these semiotic resources and their relationships to each other constitute an inadequacy in research which this study will address.

Debates over literacy and what this actually constitutes and involves are often vigorous and fiercely fought. In Australia for instance, this took the form of a debate between critical-literacy and back-to-basics approaches played out in the Australian media, and explored by Ilana Snyder in her examination *The Literacy Wars*. Inherent in the debate over 'new literacies' and their place within formal educational contexts, are a number of tensions: between discrete skills and literacy as social practice (Collins & Bolt, 2003; New London Group, 2000); between traditional and progressive concepts of literacy (Carrington & Marsh, 2005); between a 'return to basics' and a 'technologised' perspective on literacy (Snyder, 2008) In recent years, arguments about the 'damage' txtng language is doing to the literacy standards of our adolescents have been somewhat debunked by more considered examination of this language form, and consideration of the functional and symbolic role that it plays within communication frameworks (Carrington 2005; Crystal 2008).

The fact that different people – and therefore different students – have differential access to a range of technologies (and this will vary across life trajectories), means we need a broad understanding of literacy. Literacy then, is a complex endeavour, both in theory and practice. In seeking to provide an overview of the field of research and scholarship, this researched developed a triple-paradigm structure, into which the main areas of literacy scholarship were organised.

Traditional-Literacy Paradigm

Perspectives that see literacy as being the ability to code and decode written texts, comprise the 'traditional literacy paradigm'. This perspective designates the formal and 'legitimate' understanding of literacy that is often taught and assessed in schools (cf. Carrington & Luke, 1997). This literacy model makes very little room for practices that occur beyond the classroom.

The 'traditional literacy paradigm' is clearly elaborated by Brian Street in his concept of 'autonomous literacy' (1984). This autonomous model of literacy emerges from the distinction between written and oral language, and the historical privileging of the former over the latter. Literacy here concerns adherence to certain conventions as set down by those with the power to make this definition:" 'What is being tested is often the social conventions of a dominant class, rather than universal logic. The convention most often mistaken for logic is explicitness, which, he shows, is not the same thing at all'. (Street, 1984, p. 27) Thus, whilst 'autonomous' literacy practices appear to be about getting things right, what they are really concerned with is the inculcation and mastery of a particular set of conventions and beliefs about what 'proper literacy' is.

It seems an almost ironic contradiction that despite a lack of formal rules associated with the evaluation of 'reading & writing' associated with mobile devices, formal assessment tasks have still made use of SMS, or other 'mobile-based' text forms. For example, this was seen on the 2005 Victorian Certificate of Education final English exam, where students were required to compare a letter from John Keats to his love Fanny, with an SMS-text message professing love. One wonders in such a case about the ideological underpinning of the question, and how many teachers explicitly addressed the text form of an SMS-message during the year of instruction with their students. Even if students use texting for communicating themselves, and perhaps have some kind of intuitive understanding of the form, its rules and guiding principles, this does not necessarily translate into a rigorous and academic understanding which would enable them to effectively address the text in a comprehensive and insightful manner.

In practice, proponents of traditionalist perspectives on literacy instruction, commonly adopt approaches advocating phonetics or linguistics, teaching the rules of grammar and appropriate forms of writing (Snyder, 2008). Direct connections can be drawn from these linguistic perspectives, to traditionalist notions of literacy, such as phonetics, and offer potential pathways into productive and critical classroom uses of this genre (Crystal 2008).

Everyday-Literacy Paradigm

The 'everyday literacy paradigm' concerns Street's 'ideological model of literacy', (1984), more fully realized in the New Literacy Studies (NLS) perspective (Beavis, 1999; Burke, 2006; Cope & Kalantzis, 2000; Duncum, 2004; Gee, 2003; Gee, 2004; Gee, 2008; Kist, 2005; Knobel, 1999; Knobel & Lankshear, 2007a; Kress, 2000a; Kress, 2000b; Kress & Van Leeuwen, 2001; Pahl & Rowsell, 2005; Snyder, 2002). From this perspective, literacy is not something merely restricted to the classroom, or to written language in and of itself. Rather, literacy practices concern the meaning-making practices individuals engage in throughout all aspects of their daily life, engaging with multiple modes and semiotic resources for meaning-making. Literacy then, is more than simply the ability to code and decode written text, but rather, includes 'negotiating a multiplicity of discourses' (New London Group, 2000, p. 9). In this respect, language and literacy only carry a part of the meaning in any text or communicative exchange, and one must examine its relationship to context and other modes used (cf. Kress, 2003).

The problem with this everyday-literacy paradigm is that it does not translate neatly to school and educational institutions. As the meaning-making practices of our daily lives, they are tied heavily to identities, context and communities of practice, which do not always intersect neatly with education in its institutional form. In their seminal work developing a 'pedagogy of multiliteracies' the New London Group (2000) broke down the concept of 'multiliteracies' to six distinct design parameters which intersect and interrelate within the multiliteracies framework. Specifically, these are: linguistic, visual, aural, gestural, spatial and multimodal designs.

Defining just what literacy involves, what modes and mediums, is a contested space, even within the 'everyday literacy paradigm'. Whilst some researchers seek to retain the connection between literacy and written language (see Kress, 2003), others seek to expand the concept to include other modes: 'The practices of students with new technologies – even as prosaic as they are in school – require a broader re-conceptualisation of literacy as multimodal design.' (Jewitt, 2006, p. 8) This research adopts this perspective, conceptualising literacy as concerning how multiple modes for meaning-making – different semiotic resources – are used across a range of everyday and formal contexts.

Concept-Literacy Paradigm

In popular discourses it has become somewhat fashionable to describe competency in any particular area/discipline of human endeavour as some type of 'literacy'. Whilst this has angered some researchers and academics within the fields as contributing to a degeneration in the rigor of this theoretical term, this approach to discussing literacy does have some merit, particularly through its connection with the 'everyday-literacy' paradigm. If we understand each of these models as a focused version of multiliteracies more generally, then they find a place within established theories of literacy and literacy education.

The 'concept-literacy paradigm' is essentially a measure of competency with regard to a specific area of human endeavour or interaction. Discipline-based literacy theories are a combination of traditional and everyday literacy perspectives. They concern specific knowledge and actions related to a particular field of study, area of life, or aspect of social interaction: examples include, 'technological literacy' (Luke, 1997), 'visual literacy' (see Anstey & Bull, 2000; Kress, 2003; Wilhelm, 2005), 'media literacy' (see Hobbs, 1998; Penman & Turnbull, 2007), 'bit literacy' (see Hurst, 2007), 'economic literacy' (see Forsyth, 2006; Stigler, 1970), 'digital literacy' (see Braga & Busnardo, 2004; Eshet-Alkalai, 2004; Faigley, 1999; Gibson, 2008; Gilster, 1997), 'postmodern literacy' (Willinsky, 1991). The usefulness of these concepts lies within the extent to which they can be used to focus attention, discussion and research on a particular field of human activity. This discussion does not aim to be exhaustive in explaining different theories of this type – for they can be defined by scholars and commentator for specific purposes – but rather, to sketch examples which explain what this paradigm involves.

This model of 'mobile literacy' does not advocate for one of the particular literacy paradigms over another: rather, they should be understood as ideologically-bounded perspectives on understanding meaning-making practices. Whilst models of literacy help us to understand mobile technology uses at some levels, the very 'mobility' of these devices throughout the different spaces of our lives suggests the need for a framework to describe the social practice aspect of mobile device use at an intricate level. It is here that the theories of Pierre Bourdieu become useful in understanding the dynamics of social interactions, over which we can lay theoretical understandings of literacy.

ENGAGING ETHNOGRAPHICALLY

In order to emphasise the importance of social interaction in literacy surrounding mobile technology usage, the sociological theories of Pierre Bourdieu were used, particularly his writings concerned with subjective everyday practices within the context of objective social structures (Bourdieu, 1977; 1980; 1986; 1991).

Bourdieu's sociological theories are used to draw out the 'social practice' dimension of mobile technology use and articulate this in an explicit fashion. In this way, emergent meanings associated with mobile technology usage patterns can be explored with regard to how they relate to the practices

Bourdieu's concept of 'habitus' has been reworked for the realm of mobility by Kress and Pachler (2007). Here they define a habitus for the new world of m-learning as: "The habitus has made and then left the individual constantly mobile – which does not refer, necessarily, to a physical mobility at all but to a constant expectancy, a state of *contingency*, of *incompletion*, of moving towards completion, of waiting to be met and 'made full'. (Kress & Pachler, 2007, 27). In searching for the mobile habitus in the experiences of the Riverton adolescents, this research will therefore be concerned with exploring patterns which are emergent rather than fixed and stable.

Where Bourdieu's theories have been applied to literacy, there has been a common trend. This is particularly associated with examining how the position of 'legitimate literacy' (associated with the 'traditional literacy paradigm') is maintained within educational contexts, through the exclusion of out-of-school literacies through *symbolic violence* (Albright & Luke, 2008; Carrington & Luke, 1997; Heller, 2008; Hill, 2008). As mobile technologies are caught up in this juncture between 'legitimate' and everyday literacy practices, it is important that this approach be flexible, which has been achieved through working with Bourdieu's central concepts of 'capital', 'field' and 'habitus'.

This research also sought to reconfigure Bourdieu's concept of field, in order to more adequately describe the ways in which mobile devices are augmenting of and augmented by the fields (or contexts) which they intersect. Here a 'mobile field' represents the possibilities offered by mobile technologies to augment any intersecting field through which they travel: the 'mobile field' is one potential, always waiting to be realised, such as when one receives a phone call during an important meeting after forgetting to put it on silent.

Bourdieu's concepts of capital are outlined in Table 1 with a description and examples drawn from mobile devices. These concepts form the central feature of his theories which this research works with. Bourdieu's concepts of capital describe the various types of resources that individuals have access to within the realm of social practice (1986). Each type of capital has a structural impact upon the literacy practices that one uses to create meaning, at both the point of articulation (speaking and writing) and interpretation (reading, viewing, listening) (Kress, 2003). Each of the main forms of capital – economic, social, cultural and symbolic – will be explored in relation to their role within the mobile usage patterns of Riverton adolescents, understood as literacy.

OUTLINE OF RESEARCH METHODOLOGY

In seeking to understand the culture associated with youth mobile technology use, this project made use of an ethnographic methodology. It drew from a critical ethnographic perspective (Madison 2005) which emphasises an examination of lived experience in tension with different ideological

Table 1. Bourdieu's forms of 'capital' (Bourideu, 1986) as they relate to mobile technology usage

	Description	Specific Examples
Economic Capital	Financial Resources, Money	Own Money, Parent's Money, Friend's Money, Mobile Phone Credit
Social Capital	Social Networks and Connections	Mobile phonebook
Institutional Cultural Capital*	Network Membership	Network Contract
Objective Cultural Capital	Mobile devices	Smartphone, iPods, PSPs, mobile phones
Embodied Cultural Capital	Usage Patterns	Digital texts, performance texts
Symbolic Capital	Prestige Associated with Possession or Use	Situationally-dependent readings

* Institutional Cultural Capital in Bourdieu's traditional sense seems to have limited relevance, but may be understood as membership of certain organisations that permit technology use, and at higher levels qualifications that allow differential access to mobile devices or services.

frameworks. The ethnographic model also drew from Marcus' mobile ethnography (1998); multi-sited ethnography urging the researcher to "Follow the people … follow the thing … follow the plot, story or allegory … follow the life or biography, [and/or] follow the conflict." (Marcus, 1998, pp. 90-95). Another key component of the ethnographic perspective of this research is based on understanding 'culture as a verb' (Heath & Street, 2008). If we think of culture as a verb, then it is something that is constantly moving, changing and evolving: this perspective sees culture as 'unbounded, kaleidoscopic and dynamic' (p. 7). It also evokes the importance of reading 'performance' in the ethnographic research framework (see Denzin, 1997; Madison 2005). The ethnographic structure was also informed by the work of Denzin's "Interpretative Ethnography" (1997) and Willis' "Ethnographic Imagination" (2000). The ethnography is directed at exploring the contextualised richness of a local experience of a global phenomenon.

This study was undertaken in a rural township over 250 kilometres from Melbourne: Riverton (a pseudonym). The township consists of around 10,000 people, with a district population of over 21,000. Riverton High (a pseudonym), a school of just under 1,000 students, was the focus site for the study. This is appropriate as the research is concerned with the intersection between in- and out-of-school practices around mobile technologies. The school ground also provides a nexus point for social interaction, where adolescents from dispersed geographical locations meet: in this case students travelled anywhere from 75 kilometres to a couple of hundred metres to reach school each day.

The data collection instruments used were conventional for an ethnographic methodology and included: artefact collection, interviews, participant observation and a focus group. The research initially started with a unit of study which contributed to the student's learning in their English class, on the topic of 'Mobile Technology and Me.' Students completed a number of activities as part of this learning and this was used as the first point of data collection. The work of these students was used to generate questions and focus for the subsequent observations, interviews and focus group. There were 13 participants whose work and interviews form the basis of this ethnographic examination.

A MODEL OF MOBILE LITERACY

This model of mobile literacy draws together understandings of literacy and a theoretical perspective on social practice suggested by the work of Bourdieu. Whilst models of literacy enable deep description of the meaning-making practices that are occurring, as both articulation and interpretation, concepts such as habitus, field and different types of capital (Bourdieu, 1986) contextualise these literacy practices within a dynamic framework of social interactions.

This section of the discussion will concern drawing connections between literate behaviour and the different forms of capital (economic, social, cultural and symbolic) that both structures and is emergent from mobile device usage. In order to explore and unpack the experiences of Riverton's students, the discussion section will make use of selected students and their experiences as exemplars of different narratives of experience.

Economic Capital: Pre-Paid Literacy

The dimension of economic capital describes the influence that financial cost has on the literacy practices one engages in: modes and mediums used, and the structure of different texts both at the point of articulation and interpretation.

Mobile devices cost money, not just on a one-off basis, but continuously. Whether it is paying for phone usage, buying digital content, repairs or upgrading a device model, as products of a system of consumerism, young people are increasingly positioned and targeted as consumers (Kenway & Bullen, 2001). Whilst there is nothing new about adolescents being targeted as an audience for consumption, the pervasiveness of these devices, combined with the - at times - high cost of devices and features, creates concerns about financial management among people who have limited disposable income.

Nina Weerakkody (2007) demonstrated through research that young people - particularly adolescents - are increasingly experiencing financial stress as a result of mobile phone contracts. Whilst this may apply to some sections of the community, all except two of the participants in this project were on a pre-paid '1-cent' text plan with Telstra. The majority of students then deferred to sending SMS messages as their default form of communication. Curiously, this did not result in 'squeezetext' (Carrington 2005) or other language innovations associated with SMS as a language form (Crystal 2008). The very cheapness of the individual texts was indicated by participants as a reason why they sent large volumes of text messages: it was not uncommon for students to indicate that they went through 10 dollars of credit (equivalent to 1000 SMS messages) in a week, although beyond anecdotally, this could not be verified.

As minors, adolescents and students, they had restricted access to their own financial resources. Although some of them did have part-time or casual jobs, their capacity to earn economic capital was limited by their age, skill level and capacity for employment commitment. As such all participants sought ways to minimise their costs, including those associated with mobile technology use.

All participants in the project made use of parental income to pay for their mobile technologies and their uses. Participants received gifts of mobile devices for birthdays and other occasions. Participants received money for credit for their phones. Few students had phones that their parents owned the contracts for and paid for individually: the only participant in this project was Bailey.

Significantly for literacy, all participants made use of 'pranking' to contact people for pre-arranged or pre-understood purposes, as a way of cost-shifting. This intersection between economic and social capital offers an interesting demonstration of a significant literacy text linked to mobile devices. Links bodily hexis to digital hexis, links narratives across time, communicates emotional content, moves through space – security – location dependent software.

Contract knowledge: participants have a choice of two different models of phone contract: pre-paid and post-paid contract. The majority of participants were on the Telstra pre-paid 1-cent txt plan, and indicated that amongst their peers this was the preferred payment structure. An economy in pre-paid credit has actually emerged among peers, where individuals could send each other credit for their mobile phones, though they did indicate that this was based on reciprocity, with Tom clearly indicating that a friend who asked for credit without ever giving was excluded from his circle of financial exchange. Here is another point where economic and social capital interrelate through the mutual exchange of phone credit.

Social Capital: The Literacies of Mobile Socialization

CT: Those people you contact via SMS…your friends you have via SMS and the friends you have at school … do you think … because you can contact them in all sorts of different times, has made the friendships stronger?

O: Yeah. Oh I think … yeah, I think it might because you seem to distance yourself away from people you don't talk to for a long time. So if you don't see 'em at school … oh well, it doesn't apply so much to people you see at school, because you see them every day and you just talk to them there … but people you don't see at school … if you don't message them you might lose the relationship. So like, for example, if I met someone at *St Mark's*, on Saturday night, I message them Sunday, I could keep talking to them and have a relationship/friendship with them.

CT: But if there's a big gap between…?

O: But if I stop talking Saturday, and didn't talk to them for a couple of weeks and didn't see them, then it's gone. (Owen)

Mobile devices – particularly those associated with communication – are tied to the processes of socialisation and social networks. In this way modern communication technologies are involved in the establishment and maintenance of social capital as something which is held collectively by groups:

"Social capital is not the possession of an individual but of a collective. It is a characteristic of the social situation in which the individual finds him or herself. As individuals, we can contribute to the development of social capital by helping to maintain the group." (Ling, 2008, 26)

Mobile phone phonebooks can be understood as an embodiment of social capital, in that the names and contact details contained therein delineate the social network connections of any individual, though importantly, not necessarily whether they utilise these connections. As is indicated by the work of Bourdieu and Ling, social capital is only realised and developed through reciprocal exchange between participants in the network: basically, if an individual does not contribute to the development of social capital by 'gift giving', then they lose access to that social capital. This same phenomena as Owen – above – talking about the need for frequent contact to develop a friendship, or Tom's acknowledgement that he did not lend phone credit to an acquaintance, because that person never reciprocated in kind. So whilst phone books are useful to *suggest* the social connections of an individual, they do not describe their social capital as a functional aspect of their life; nor do they describe the nature of particular social interactions and the effect they have on literacy practices.

The development of 'grouping' capacities for contacts through smart phones does provide an interesting advancement which suggests possible avenues of inquiry for exploring how different social networks structure literacy practices in different ways. The students of this research already indicated how they would change their communication method not just to suit the purpose of the communication, but to suit the audience.

Mobile phones have the capacity to mediate the maintenance of social relationships through increased capacity for communication on an ad hoc basis. As Owen indicates above, frequency of connection was related to the depth of friendship. This, combined with cultural capital – outlined below – contributed to a 'monopoly-membership dynamic' whereby those adolescents who were not connected to the Telstra network (and therefore cheap to contact, and provided with access to the most extensive infrastructure), were excluded from social interactions. A number of participants indicated that membership with any other telecommunications provider would result in a degree of social exclusion from mobile communication.

Cultural Capital: Cultures and Mobile Meaning Making

The adoption of mobile technologies across varied societies and cultures has resulted in particular configurations of meaning-making practices becoming connected to cultural discourses and narratives. As 'cultural artefacts' (Goggin, 2006), devices are imbued with meanings connected to the culture in which they are deployed, such as The Cult of iPod (Kahney, 2005), SMS-language variations across cultural and language groups (Crystal 2008).

Objective cultural capital was understood as the physical devices themselves: whether this was a mobile phone (however complicated or 'smart'), iPod, PDA or other specialised or generalised device, they all had a physical form. In this form, we can examine mobile devices themselves for the meanings that are contained within their physical parameters and features. The iPod (and subsequently the iPhone, iPad, even the 'i' prefix) has been a significant cultural phenomenon, not just at the economic level, but with social and cultural impacts as well (Jones, 2005; Kahney, 2005; Levy, 2006).

Most students were pragmatic in their attitude to devices, and little materialistic concern was demonstrated by them. Most were happy with a basic model which they purchased themselves, or asked for (and received) them as gifts from relatives or friends. There was also an exchange of second-hand devices amongst peers when devices were no longer needed. As can be seen in the small selection of participant devices in Figure 1, there was wide diversity in the types of devices individuals carried. A couple of students personalised their devices with covers, decals or even a *Simpson*'s sock on Peter's phone. However, the main way in which a mobile phone was personalised – where this feature was technically available – was through the setting of specific images or photos as a background. Indeed, digital texts associated with mobile devices can also function as a form of objective cultural capital.

At the level of the peer-group, phones and digital texts as objective cultural capital can be articulated in an effort to gain symbolic capital, as a form of social display (embodied cultural capital) resulting in an elevation of social status. In this way, the use of objective cultural capital (devices) at the level of bodily hexis was a 'presentation of self' (Goffman, 1959) which communicated alliance with certain trends and fashions. "It's also becoming a fashion item." (Bailey) The various technologies of participants in this research can be seen at Figure 1 indicating not just a variety of devices, but degrees to which individuals personalised their devices: this could either be done to the physical object, such as different covers, or digitally. Many of the participants spoke of using pictures taken on their phone of friends as background images, thereby displaying and writing a particular identity and social connections.

Objective cultural capital gains its meaning at the point of embodiment: when they are *used* as part of social practice, when it is articulated as a message (either conscious or unconscious) or interpreted for meaning. This has both a physical and digital manifestation: phones as objects were typically prohibited in most classrooms of Riverton High, and so students typically kept them in their pockets or otherwise hidden when in classrooms. Though there were exceptions. Observations revealed that despite the school rule, some students still used or kept their devices out in the open in some classrooms. This was variously policed by different teachers, but the extent

Figure 1. Devices of Riverton students. Top: Jennifer's phone; Erin's phone; Erin's Glucometre; Middle: Bailey's phone; Sarah's iPod; Brad's new phone; Bottom: Josh's personalised phone; Peter's phone with cover; Brad's old analogue phone

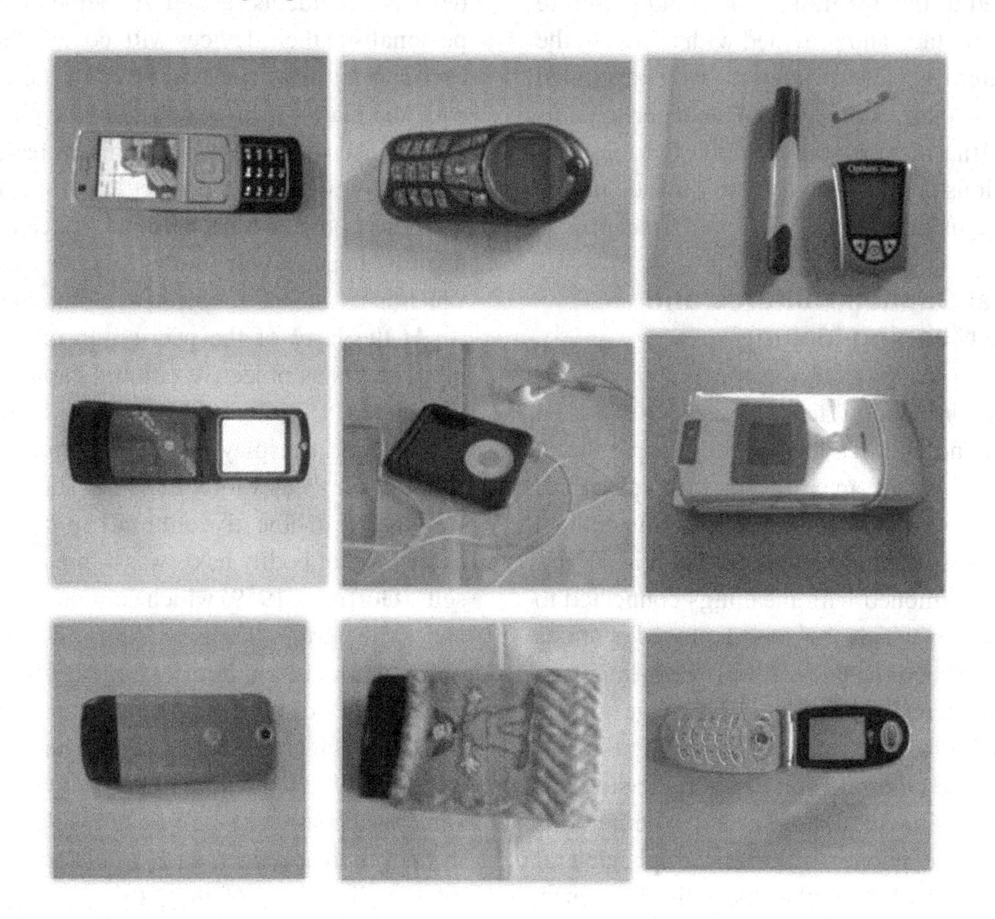

and nature of this surveillance by staff was not a focus of this research. Still, the physical form of these devices is written as a performative element within classrooms, with students making efforts to continue digital connections during class time.

It was a notable (though not unexpected) feature that all the participants, and almost all other students of this age group encountered in the course of the research, had their own mobile device, typically a phone, sometimes and iPod, and at times both. There was also anecdotal evidence of some students (an unmeasured amount) who had more than one mobile phone; this experience was represented by Bailey, who had between two and three phones throughout her participation in this research.

The most significant type of cultural capital in terms of an understanding of literacy is 'embodied cultural capital', because it is here that identity and meaning is inscribed at the bodily hexis (or its digital equivalent). It is at this level – the level of empirical experience – that the influence of local factors structure how global concerns and trends are experienced.

At one level, communication practices were structured by the rural location (Taylor 2009a). Gaps in coverage, and the market monopoly of one telecommunications provider due to their superior infrastructure (Telstra), bore a structural relationship to the texts that students used to express meaning. In this case, it was firstly important to have the correct institutional cultural capital in the form of connection with the Telstra network. As reported by the Australian Communications and Media Authority (2008), mobile phone reception and gaps in coverage remain a continuing concern and problem for rural consumers. All participants spoke about coverage drop-outs beyond the limits of the township, making synchronous verbal phone calls unreliable: this resulted in a preferences for SMS as a form of asynchronous communication. A message could be created and sent, but would only go through when the phone came back within range of a communications tower.

The 'articulation' of meaning through a range of modes and mediums (Kress, 2003) was used by participants to signify particular identities and social connections. Whilst communication primarily took the form of simple SMS, denser, more complex modes were used to express other aspects of identity. This included the creation, consumption and swapping of digital texts such as photos, videos and music. However, it also concerned patterns of how mobile devices were used at the level of bodily hexis, understood as social performance. How technologies are used in daily life, how they are articulated as part of practice is the point at which the objective cultural capital an individual student has access to, becomes embodied at the level of bodily or digital hexis. In the process of 'articulating' a meaning (Kress, 2003) – in the preferences for modes, the awareness of context – individuals imbue the embodied cultural capital with traces of their identity, or at the very least the accent of their habitus.

One of the study participants – Owen – uncovered a curious feature of mobile device usage patterns connected to their objective physical forms, which had a direct influence on whether a student used 'squeezetext' (Carrington, 2005) or other SMS-language innovations. In completing his classwork assignment on mobile phone use rules Owen surveyed his classmates and discovered that they were divided exactly 50/50 around whether the student used predictive-texting features on their phone. The significance of this for the language features of their messages was only revealed later in the interviews when two of the participants indicated that using the predictive-text features meant that sending 'squeezetext' was actually more difficult and cumbersome. This demonstrates that the physical features of a phone, combined with an individual's understanding of and capacity to use them, can have a structural influence of the texts that emerge.

Exploring the function of mobile device use as cultural capital is best achieved by exploring the peer-group level, primarily because educational institutions don't have a culture that values mobile devices. The meaning of mobile technology uses at the peer-group level was something that was not mandated, but rather negotiated in an on-going fashion. As opposed to the clearly stated rules for classrooms, appropriate use of devices with their friends and family was something that needed to be determined from individual situations: the appropriateness of embodied cultural meanings was something that one did, supporting the idea of 'culture as a verb' (Street & Heath, 2008). The rules of use however were vague and indistinct, tacitly negotiated, neither written in stone, nor completely irrelevant.

Although there were some variations, the study participants were all aware of school rules governing mobile technologies. There was even some agreement over family rules, with 'no phones at the dinner table' being a common one. However, all the participants had difficulty explaining rules or guidelines that governed their behaviour with peers, especially when and where to use a phone when with others.

This multitasking required an understanding of the 'affordances' of particular modes (Kress, 2003), with respect to the impact on those physically co-present:

Well, MP3 players, not really. But with ... you know, we can multitask, listen to that and talk at the same time, but we all get annoyed if we're trying to talk to someone and they're just ignoring us and they're on the phone. Like if we're just sitting there and no one's really saying much, or no one's trying to talk directly to you though, like, we don't care. But if someone's trying to like talk directly to you and you're just ignoring them flat-out, then... (Jennifer)

Whether to communicate using a mobile technology, and which mediums to use, are caught up in the dynamics of social interactions. Amongst peers, competing literacy practices have to be prioritised and balanced in a way that affords symbolic value. Whether the symbolic value realised is valuable at the school level or the peer level, is a matter of personal priorities connected to individual habitus: the disposition to act in a certain way and believe certain things. Many students spoke about giving appropriate attention to peers during interactions during lesson breaks (recess and lunchtime), such as not using a phone when having a conversation with a friend, but at the same time spoke about balancing relative communication priorities: deciding about the relative values of different communications on a moment by moment basis. Such a balancing act suggests something about personal priorities when students are prepared to engage in surreptitious and hidden mobile phone usage during class times; it tells us something about their personal priorities towards that particular educational communication paradigm (the classroom and lesson).

The importance and emotional attachment of students to mobile phones in particular was demonstrated at the end of all classes. My observations are riddled with daily observations of multiple students 'walking and txtng' often after leaving class. A common behaviour observed in students as they left classrooms was to very quickly pull out mobile phones and engage with them (what they were actually doing could not be discerned). I also observed many students around the school grounds sharing iPods or music players, by wearing one ear-bud each. These performance patterns were enough to write a narrative of constant connection, where classrooms exist as obscuring forces to this valued process.

The use of photos as part of social practice was a display of social connections or personal tastes: it amounted to an 'articulation' (Kress, 2003) of embodied cultural capital, a writing of identity. Students set photos as backgrounds for their phones (Jennifer, Rebecca, Josh), recorded images or videos to share with friends (Tom and Peter), took photos of things of personal interest (Bailey) or of funny events (Josh, Peter); this can be seen in the screen shots of Tom's bike crash (Figure 2). He was enormously proud of this video, and actually had two different versions of two similar crashes whilst going over a bike jump. He was in the process of learning how to upload this to YouTube, having already shown it to all his friends at school. This was a moment where the prestige associated with his cultural capital in the form of the video, was not one which carried value at the educational level. The experience of this video mirrors the 'Jackass' phenomena of people recording dangerous stunts in an effort to attract online fame, though for Tom, this was a secondary consideration after impressing his real life friends. In this way identity as related to a particular cultural text type, was recorded and displayed using visual and audio-visual texts on mobile phones.

Symbolic Capital: The Prestige of Mobile Technology Use

I don't reckon that teachers will ever allow it [mobile technology] to be used in class. (Bailey)

I use mine a bit, but I'm conscious of being rude and not doing it when, you know, someone's trying to talk to you or whatever... (Jennifer)

Figure 2. Tom's bike-jump crash: stills taken from a video, numbered sequentially

The Symbolic Capital that was of most interest for this research was that associated with the educational field (specifically 'education-as-institution'), though there were many other spaces traversed by these participants that imposed particular symbolic structures on them. The symbolic value of mobile technologies and their use across various spaces and contexts is a measure their prestige. This measure of is the essential transformation underpinning just what counts as 'good mobile literacy'. Significantly, as the symbolic value of mobile devices usage patterns changes across different fields and situations, what counts as 'good' or 'effective' mobile literacy therefore changes according to variable prestige associated with the appropriateness of mobile technology use for particular fields of human endeavour.

The key conflict at the heart of mobile literacy in schools is the different symbolic value assigned to the literacies from students' everyday lives, and those educational institution values (cf. Carrington & Luke, 1997). The participants' experiences with mobile technologies present a particular challenge, not just because they transcend the spaces of home and school, but because their particular ubiquity for many adolescents embodies the distance between real life and school as an institution (see Buckingham, 2007). Essentially, these conflicts reveal the constructed and arbitrary nature of schooling. This was demonstrated through the fact that all participants saw their mobile technologies as separate from school learning. Significantly, when asked directly about the educational potential of these technologies, the participants gave negative assessments. Part of this is the process of 'symbolic violence' (Bourdieu, 1991).

The ways in which individuals use mobile technologies as part of their social practice, can be read as an inscribed performance of identity. This links Goffman's understanding of performance in everyday life (1959) with studies of identity around ICT use (specifically mobile technologies) and the gestural and spatial design modes laid out in the New London Group's (2000) pedagogy of multiliteracies, but also with the notion of 'conspicuous consumption' (Levinson, 2004, p. 74), in that the use of these devices is itself is an indication of individual connections with consumer culture, and one specifically valued by their peers. Because mobile technologies intrude across multiple fields, and are both structured and structuring of such fields – as a 'mobile field' – their physical use at the level of bodily and digital hexis has a sociocultural meaning beyond traditional understandings of text.

Drawn from the work of Bourdieu (1991), 'symbolic violence' is the process by which institutions maintain the system of distinction around particular behaviours: it is the exercise of domination of one ideology over another. Through the encouragement and/or punishment of different practice, institutions structure the symbolic systems by which prestige can be gained. Individuals subject to symbolic violence participate in its perpetuation through their 'misrecognition' of the arbitrary nature of the system, and their acceptance of the prescribed perspective as objective. The 'everyday literacy paradigm' is involved in this process, at the level of lived experience, as Bourdieu (1991) indicates:

Thus the modalities of practices, the ways of looking, sitting, standing, keeping silent, or even of speaking... are full of injunctions that are powerful and hard to resist precisely because they are silent and insidious, insistent and insinuating. (p. 51)

Riverton High, as an educational institution, positioned as valuable particular social practices, literacy skills and technologies. This extended from the school administration (setting school policies and rules) to the individual classroom, where teachers negotiated arrangements around technologies (either explicit or tacit). It was not so much the devices as objective cultural capital that were intrinsically forbidden (although at times they were, where safety was an issue) but rather their 'articulation' (Kress, 2003) as embodied cultural capital, when students used them in particular ways. Although students spoke of mobile technologies as separate from their school lives, as being objects banned in classrooms, they likewise continued to use what they saw as necessary technologies within educational fields, either with permission or without. Individual attitudes and usage patterns about technologies are of significant importance when considering how to integrate mobile technologies into learning environments. Being designed for single-person use, mobile devices suggest different approaches to learning from traditional collective learning environments, like classrooms.

Well phones and that are supposed to be off during class and you are allowed to use them in the school grounds. (Peter)

The most significant issue for mobile technology use in school environments is the way that its use is 'interpreted' (cf. Bourdieu, 1991; Kress, 2003). In literacy terms: it concerns the way that mobile literacy practices are 'read' within the school context. The specific ideological structures of individual schools, indeed, individual classrooms, can position identical usage practices in very different ways.

Significantly for education, specifically pedagogy, it is the modes of performance that emerge as significant for mobile literacy, rather than any concern over linguistic skills. Aural, gestural and spatial modes all play a significant role in the successful negotiation of symbolic values associated with proper school behaviour, and the ideal student, whilst maintaining practices and behaviours that align with a habitus formed in the more technologically-saturated world beyond the school gates. This all involves an understanding of the potential offered by various devices and their affordances, something demonstrated in Josh's informative writing assignment: an awareness of the multimodal nature of mobile phone use, and how this can impact on social standing. He suggests that inappropriate use can be trouble across social situations, not just in school.

Some of the study participants demonstrated a desire to be seen as good students; polite, attentive and studious (see Goldstein, 2008). This is linked to individual student dispositions regarding schooling and the behaviour expectations that field engenders. As part of this role, mobile technologies were typically managed in accordance with school rules and teacher expectations, either consciously and conscientiously, or through a congruency between school expectations and individual student dispositions related to the process of inculcation. A large grouping of the students, including Jennifer, Bailey, Erin, Sarah, Josh, Brad and Owen, expressed this conforming identity. Jennifer's particularly useful for considering the role of mobile literacy for compliant students, as throughout the study she demonstrated a particular concern for being seen as polite in her social interactions.

In certain circumstances, the use of particular mobile technologies was permitted. Erin's use of her diabetic device – or 'glucometer' as she called it – was obviously permitted in the school grounds for medical reasons. However, predominately within the school, use of mobile phones within classrooms was not encouraged and teachers were asked to support this policy. All contact between students and others was to occur through the official school channels, that is, through the school's General Office. Mobile phones in particular were positioned as detrimental to the integrity of the classroom field, in that they offered the potential to disrupt the strict rules of time and space which governed lesson times and directed instruction.

Of the participants, Tom was the only one who seemingly regularly had problems with his phone being confiscated. He indicated that his phone had been confiscated on occasion, though still maintained a strong self-belief in his ownership of the device and its contents. This is a situation where the student valued personal and peer-based texts over the prestige assigned to 'lack of mobile phone' within school classrooms: in essence, he recognised the 'symbolic violence' (Bourdieu 1991) being exercised against him by the school, and refused to comply. The result was that his uses of mobile devices within the school context were typically seen as disruptive and counter to learning goals.

In order to circumvent punishment for mobile technology use within classrooms, some participants engaged in a performance of an 'attentive student', using multimodal resources to fit within the symbolic capital framework of classrooms. Although there was the general appearance of 'attentive silence' (Goldstein, 2008) and studious work from many students, this did not always reflect their actual reality. The small size of these devices, held at the level of bodily hexis, meant that at times they were used, whilst a performance was concurrently articulated, which gave the impression of a 'good student'. A key component of this is the use of technical and performative elements to give the impression to any external 'reader' that the student is not doing what they actually are doing. At the most basic level, this meant obscuring as many modes related to the mobile technology and its use as possible.

In order to play the role of a good student, individuals had to make use of a range of multimodal resources in order to maintain an appearance fitting that role. This bears a similarity to Fortunati's (2005b) examination of mobile phone users on Italian trains, where she worked with Goffman (1959) in developing the notion of the 'stage' upon which individual performances are intentionally and unintentionally presented. The shift here involves a consideration of modes not typically associated with literacy – namely aural, gestural and spatial designs – and the multimodal interaction between all elements (New London

Group, 2000). The particular design of the school field influenced the design of these modes associated with the bodily hexis in a fashion designed to obscure mobile usage: this included setting the phone to silent (sound), allowing minimal body language to display usage (gesture), and making use of furniture and lines of site to obscure usage (spatial). Fortunati (2005b) would talk of this as a student maintaining a performance on their 'stage' that successfully hides the backstage: they perform attention to their work, whilst also directing it towards their private communications.

I can txt without looking, so it's pretty easy. I'm still trying to learn to txt in my pocket. (Tom)

Although there were some classes where students would simply not risk texting, the practice of breaching rules was not an aberration, but common amongst study participants. For instance, the researcher observed Erin secretly txtng in the feared Ms Higgins' class, and asked her about this risky behaviour. Her response was simply: 'Yeah, I haven't been caught yet, so that's quite good'. The researcher also observed students obviously txtng, out in the open, in the class of another teacher, who appeared to not regulate the behaviour. Breaching school rules then takes the form of an inaccurate match between a particular usage pattern and the structural demands of a particular field: but this also extends beyond the classroom, to breaching social norms and expectations, but also practicalities:

I did notice the other ... when was it? I was ... riding my bike home and I noticed I pulled my phone out just to check and it's like: "What am I doing? What if a car comes or something!" So, yeah, I just put it back and it was just like: concentrate! [laughs] (Brad)

Within the classroom as an educational and institutional field, there are certain expectations about behaviour, or more specifically, the interpersonal literacy games that individuals plan. The purpose of the classroom structure - including the power differential between teacher and student - is to regulate and maintain this position of symbolic domination. This form of "symbolic violence' (Bourdieu 1991) whilst often not recognised by students, seems to becoming more visible in the age of mobile devices.

Bailey for instance made use of performative elements - gesture, visual and spatial design - to 'perform' the role of attentive and dutiful student. During one interview she was anxious to demonstrate how she could write and send a text message whilst not looking at her phone. In this sense, whilst she maintains contact with her peers via mobile device, she also gives the 'appearance' of attentive student, even when she is not being so. This is a point where the digital and bodily identities of students are written separately and differently for different audiences, in a synchronous fashion.

The educational field that is most important for considering the place of mobile literacy within educational institutions is the classroom. Clearly, there are some meaning making practices associated with mobile devices that would be inappropriate for this field. Put another way, the 'mobile field' has the potential to augment the classroom field in either beneficial or negative ways, depending on just how meanings are embodied, using particular modal and technological resources. Taking a phone call from a friend in the midst of a class is typically positioned as inappropriate, and therefore lacking symbolic value, whilst 'secretly' texting, whilst one's phone is set to 'silent' may be technically in breach of school rules, but as it's quiet, asynchronous and at times unnoticed, has to potential to realise significant symbolic value (at least for the person on the other end of the texting-conversation). Bailey indicated how this could be done when she demonstrated that she could type and send a text whilst holding her phone beside her thigh (under the desk), and not looking at the phone or screen. Whilst she may have been exhibiting the performance of an attentive and dutiful student (therefore maintaining her prestige at the classroom level through the use of visual, gestural and spatial design), she was simultaneously able to maintain her digital communication (by using particular arrangements of linguistic, visual and aural elements, thereby maintaining prestige at the level of digital communication as well, feeding into the maintenance of social capital mediated by mobile devices.).

It is those students who intentionally or carelessly disregard the symbolic structures of schools when using their mobile devices (e.g. Tom), that provide the impetus of arguments about inappropriate or ineffective use of mobile technologies. Likewise, student use of design modes suited to a mobile field within another field, bring about concerns about literacy standards. A notable example of this is students making use of 'txtng' language – such as abbreviations and squeezetext (Carrington 2005) – within formal assessment tasks. This in itself is not an example of bad literacy; to the contrary, as Crystal argues, students must actually have sophisticated literacy skills in order to use txt language effectively. The problem in this case actually relates to the student mistakenly using a linguistic features suited for one form of text (SMS) in a text form for which it is not suited (e.g. a formal essay).

DISCUSSION

The data collection for this research occurred in 2007 and 2008, meaning to some extent the technologies participants used are by now somewhat outdated. This also pre-dated the significant release of the iPhone in Australia, the iPad and the rise in smartphones within Australian society since 2010: technologies which through the use of a touch-screen interface has importance for any model of 'literacy' developed with respect to these technologies.

This research demonstrates a connection between text modes and mediums, and the various form of advantage to be gained through the successful use of such texts as resources for social interaction. Understanding text forms structured in relation to a system of resources for exchange (Bourdieu's concepts of capital) provides a framework for understanding the heterogeneous local experiences of mobile technologies, and offers a shared concept structure for such research and commentary. Table 2 illustrates the relationships between different forms of capital, and associated mobile technology issues, literacy practice and text forms heavily structured by that form of capital.

An important test of the relevance of this research can be found through applying its structure to new technologies and usage patterns. Drawing on the triple-paradigm of literacy, as well as Bourdieu's concepts of capital, this understanding of what constitutes effective contemporary 'literate practice' is capable of describing such practices as: App usage, gestural interfaces, reality augmentation, sexting, cyberbullying and advertising.

Monopoly-membership dynamic – where there is a historic or cultural narrative that features a particular provider having a monopoly on the marketplace, such structures can produce effects in terms of behaviours. The importance of being with this particular network provider was emphasised by its description in terms of all forms of capital: it cost less (economic) to communicate with friends and family (social) using the same telecommunications network (cultural capital), therefore membership of that particular provider was preferred for both social prestige and pragmatic reasons (symbolic capital).

The most significant finding to emerge from this research related to the importance of situational context for the meaning of any articulated or interpreted literacy practice. The 'freedom' that mobile devices offer from a dependence on particular places is a double-edged sword: it is an illusion that we are therefore freed from a consideration of the spaces we inhabit. In fact, the opposite is true: users of mobile devices must be all the more carefully attuned to their surroundings

Table 2. Summary of Riverton mobile literacy practices

Capital	Mobile Techs	Literacy	Texts
Economic	*Cost factors structure access to mobile technologies.*	*Cost-minimisation influences use of cheapest content.*	*SMS, pranks*
Social	*Mobile technologies are used for and among social groups.*	*Practices structured by the social relationships concerned.*	*SMS, photos*
Cultural	*Discourses of convenience, entertainment and constant contact permeate their world.*	*Variety of written, visual, hybrid, intertextual and rich texts created and consumed.*	*SMS, photos, videos, music, devices, performance, games*
Symbolic – peer level	*Use of mobile devices is valued and encouraged as part of interaction and 'being' an adolescent.*	*Emergent patterns of practice, based on tastes and dispositions, linked to individuals and different groups.*	*SMS, pranks, photos, videos, music*
Symbolic – school level	*Limited or negotiated usage patterns, based on centralised control.*	*Performances designed to obscure individual usage patterns.*	*Performance*

and the context of their usage patterns. The news media and internet are replete with examples of faux paus associated with mobile device use at the expense of awareness of surroundings; people walking into things whilst txtng, sending a private message to the incorrect recipient, accidents caught on mobile phone cameras. There are also more serious situations: using a mobile phone whilst driving is illegal in Australia, having been implicated in a number of fatal road accidents; GPS and hacking opens up issues around stalking and cybersecurity; sexting is emerging as a problem of moral and legal concern across diverse societies. All of these experiences point to a need for more nuanced readings of experience as related to mobile use; the model of 'mobile literacy' laid out above sketches out a potential framework for understanding and discussing the relationship between mobiles and meaning-making, not just within educational frameworks, but well beyond.

CONCLUSION

Whilst it may appear on the surface that mobile technologies present an extension of 'electronic communication' practices which can be understood in the light of contemporary scholarship associated with such practices; this research demonstrates that the very 'mobility' of these devices present a paradigm shift in how we understand what counts as 'good literacy', now and into the future.

There is a dynamic and mutable relationship between the economic, social, objective cultural, embodied cultural and symbolic resources that individuals can draw on or realise though their use of mobile devices. If we take just the instance of SMS – and in particular the social panic about 'SMS-language' – then the participants use of the more controversial contractions and innovations of this language form was influenced not just by their individual dispositions, but by the cost of sending a text message (if they were cheap there was no need for significant condensation of language)

On its surface, the disconnection of literate practices from specific spaces enabled by mobile technologies seemed to promise a dwindling relevance of spatial considerations for communicative exchanges (either in articulation or interpretation): what this research demonstrates is that this is far from the case. In fact, the reverse seems to be the case.

When we consider what constitutes good 'mobile literacy' a constant sensitivity to and awareness of the structures of different spaces, means that a mobile user must be constantly aware of space, constantly evaluating and reevaluating their communicative practices to ensure that they are the most 'apt' for that particular space. Because mobile device traverse and transcend different spaces, the user must likewise negotiate variations in symbolic capital associated with these different fields of human endeavour; they must make use of modes and mediums for meaning making that allow them, from moment to moment, to achieve a positive level of prestige.

As we move into a future where computing is truly mobile, the potentials offered by wearable computing (Google Glasses) and augmented reality (GPS-based services) continues to demonstrate the importance of a model of literacy that explicitly thinks in terms of the mobility of these devices.

ACKNOWLEDGMENT

This research was completed as a PhD study at Monash University, Australia, under the supervision of Professor Jane Kenway and Dr. Scott Bulfin (Taylor, 2011).

REFERENCES

Albright, J., & Luke, A. (Eds.). (2008). *Pierre Bourdieu and literacy education*. New York: Routledge.

Anstey, M., & Bull, G. (2000). *Reading the visual: Written and illustrated children's literature*. Sydney: Harcourt.

Australian Communications and Media Authority. (2008). *Telecommunications today: Report 3: Farming sector attitudes to take-up and use*. Canberra, Australia: Australian Communications and Media Authority (Commonwealth of Australia). Retrieved on 26 December 2010, from www.qrwn.org.au/pdfs/technologytoday/2007_telecomms_today_report.pdf

Beavis, C. (1999). Magic or mayhem? New texts and new literacies in technological times. In *Proceedings of Annual Conference of the Australian Association for Research in Education*. University of Queensland.

Bourdieu, P. (1977). *Outline of a theory of practice*. Cambridge, UK: Cambridge University Press. doi:10.1017/CBO9780511812507

Bourdieu, P. (1980). *The logic of practice*. Stanford, CA: Stanford University Press.

Bourdieu, P. (1986). The Forms of Capital. In J. G. Richardson (Ed.), *Handbook of theory and research for the sociology of education* (pp. 241–258). New York: Greenwood Press.

Bourdieu, P. (1991). *Language and Symbolic Power*. Cambridge, MA: Polity Press.

Braga, D., & Busnardo, J. (2004). Digital Literacy for Autonomous Learning. In I. Snyder, & C. Beavis (Eds.), *Doing Literacy Online: Teaching, Learning and Playing in an Electronic World* (pp. 45–68). Cresskill, NJ: Hampton Press.

Buckingham, D. (2007). *Beyond Technology: Children's Learning in the Age of Digital Culture*. Cambridge, UK: Polity Press.

Burke, A. (2006). Literacy as Design. *Orbit (Amsterdam, Netherlands)*, *36*(1), 15–17.

Carrington, V. (2005). Txting: The end of civilization (again)? *Cambridge Journal of Education*, *35*(2), 161–175. doi:10.1080/03057640500146799

Carrington, V., & Luke, A. (1997). Literacy and Bourdieu's Sociological Theory: A Reframing. *Language and Education*, *11*(2), 96–112. doi:10.1080/09500789708666721

Carrington, V., & Marsh, J. (2005). Editorial Overview: Digital Childhood and Youth: New texts, new literacies. *Discourse: Studies in the Cultural Politics of Education*, *26*(3), 279–285.

Castells, M., Fernández-Ardèvol, M., Qiu, J. L., & Sey, A. (2007). *Mobile communication and society: A global perspective*. Cambridge, MA: MIT Press.

Colbert, M. (2005). Usage and user experience of communication before and during rendezvous. *Behaviour & Information Technology*, *24*(6), 449–469. doi:10.1080/01449290500043991

Collins, J., & Bold, R. K. (2003). *Literacy and Literacies: Texts, Power, and Identity*. Cambridge, UK: Cambridge University Press. doi:10.1017/CBO9780511486661

Cope, B., & Kalantzis, M. (2000). *Multiliteracies: Literacy Learning and the Design of Social Futures*. London: Routledge.

Crang, M., Crosbie, T., & Graham, S. (2006). Variable Geometrics of Connection: Urban Digital Divides and the Uses of Information Technology. *Urban Studies (Edinburgh, Scotland)*, *42*(13), 2551–2570. doi:10.1080/00420980600970664

Crystal, D. (2008). *Txtng: The Gr8 Db8*. Melbourne, Australia: Oxford University Press.

Denzin, N. K. (1997). *Interpretive Ethnography: Ethnographic Practices for the 21st Century*. Thousand Oaks, CA: Sage Publications. doi:10.4135/9781452243672

Duncum, P. (2004). Visual Culture Isn't Just Visual: Multiliteracy, Multimodality and Meaning. *Studies in Art Education*, *45*(3), 252–264.

Eshet-Alkalai, Y. (2004). Digital literacy: A conceptual framework for survival skills in the digital era. *Journal of Educational Multimedia and Hypermedia*, *13*(1), 93–106.

Faigley, L. (1999). Beyond Imagination: The Internet and Global Digital Literacy. In G. Hawisher, & C. Selfe (Eds.), *Passions and Pedagogies and 21st Century Technologies* (pp. 129–139). Logan, UT: Utah State University Press.

Forsyth, A. (2006). Economic Literacy – An Essential Dimension in the Social Education Curriculum for the Twenty-First Century. *Science Educator*, *24*(2), 29–33.

Gee, J. P. (2003). *What video games have to teach us about literacy and learning*. New York: Palgrave Macmillan.

Gee, J. P. (2004). *Situated Language and Learning: A critique of traditional schooling*. New York: Routledge.

Gee, J. P. (2008). *Social Linguistics and Literacies: Ideology in Discourse* (3rd ed.). New York: Routledge.

Geser, H. (2004). Towards a Sociological Theory of the Mobile Phone. *Sociology in Switzerland: Sociology of the Mobile Phone*. Retrieved 22 December 2010, from http://socio.ch/mobile/t_geser1.pdf

Gibson, M. (2008). Beyond literacy panics: Digital literacy and educational optimism. *Media International Australia: Culture and policy (Digital Literacies)*, *128*, 73-79.

Gilster, P. (1997). *Digital literacy*. New York: John Wiley & Sons.

Goggin, G. (2006). *Cell Phone Culture: Mobile technology in everyday life*. London: Routledge.

Goldstein, T. (2008). The capital of attentive silence and its impact on English language and literacy education. In J. Albright, & A. Luke (Eds.), *Pierre Bourdieu and Literacy Education* (pp. 209–232). New York: Routledge.

Greenfield, A. (2006). *Everyware: The dawning age of ubiquitous computing*. Berkley, CA: New Riders.

Heath, S. B., & Street, B. V. (2008). *On Ethnography: Approaches to Language and Literacy Research*. New York: Teachers College Press.

Heller, M. (2008). Bourdieu and literacy education. In J. Albright, & A. Luke (Eds.), *Pierre Bourdieu and Literacy Education* (pp. 50–67). New York: Routledge.

Hill, M. L. (2008). Towards a pedagogy of the popular: Bourideu, hip-hop, and out-of-school literacies. In J. Albright, & A. Luke (Eds.), *Pierre Bourdieu and Literacy Education* (pp. 136–161). New York: Routledge.

Hobbs, R. (1998). The Seven Great Debates in the Media Literacy Movement. *The Journal of Communication*, *48*(1), 16–32. doi:10.1111/j.1460-2466.1998.tb02734.x

Höflich, J. R. (2005). The mobile phone and the dynamic between private and public communication: Results of an international exploratory study. In P. Glotz, S. Bertschi, & C. Locke (Eds.), *Thumb Culture: The Meaning of Mobile Phones for Society* (pp. 123–135). London: Transaction Publishers.

Hurst, M. (2007). *Bit Literacy*. New York: Good Experience Press.

Jewitt, C. (2006). *Technology, Literacy and Learning: A multimodal approach*. New York: Routledge.

Johnsen, T. E. (2003). The Social Context of the Mobile Phone Use of Norwegian Teens. In J. E. Katz (Ed.), *Machines That Become Us: The Social Context of Personal Communication Technology* (pp. 161–169). New Brunswick: Transaction Publishers.

Jones, D. (2005). iPod, Therefore I Am: Thinking Inside the White Box. New York: Bloomsbury.

Kahney, L. (2005). *The Cult of iPod*. San Francisco: No Starch Press.

Katz, J. E. (Ed.). (2003). *Machines That Become Us: The Social Context of Personal Communication Technology*. New Brunswick: Transaction Publishers.

Katz, J. E. (2005). Mobile communication and the transformation of daily life: The next phase of research on mobiles. In P. Glotz, S. Bertschi, & C. Locke (Eds.), *Thumb Culture: The Meaning of Mobile Phones for Society* (pp. 171–182). London: Transaction Publishers.

Katz, J. E., & Aakhus, M. (Eds.). (2002). *Perpetual Contact: Mobile Communication, Private Talk, Public Performance*. Cambridge, UK: Cambridge University Press. doi:10.1017/CBO9780511489471

Kennedy, G., Dalgarno, B., Bennett, S., Judd, T., Gray, K., & Chang, R. (2008). Immigrants and natives: Investigating differences between staff and students' use of technology. In *Hello! Where are you in the landscape of educational technology? Proceedings ascilite Melbourne 2008*. Retrieved on 7 January, 2011, from www.ascilite.org.au/conferences/melbourne08/procs/kennedy.pdf

Kennedy, G., Dalgarno, B., Gray, K., Judd, T., Waycott, J., Bennett, S., et al. (2007). The net generation are not big users of Web 2.0 technologies: Preliminary findings. In *ICT: Providing choices for learners and learning: Proceedings Ascilite Singapore 2007*. Ascilite. Retrieved on 7 January, 2011, from: www.ascilite.org.au/conferences/singapore07/procs/kennedy.pdf

Kennedy, G. E., Judd, T. S., Churchward, A., Gray, K., & Krause, K. L. (2008). First year students' experiences with technology: Are they really digital natives? *Australasian Journal of Educational Technology*, *24*(1), 108–122.

Kenway, J., & Bullen, E. (2001). *Consuming Children: Education-entertainment-advertising*. Buckingham, UK: Open University Press.

Kenway, J., Kraack, A., & Hickey-Moody, A. (2006). *Masculinity Beyond the Metropolis*. Houndsmills, UK: Palgrave Macmillan. doi:10.1057/9780230625785

Kist, W. (2005). *New Literacies in Action: Teaching and Learning in Multiple Media*. New York: Teacher's College Press.

Kress, G. (2000a). A Curriculum for the Future. *Cambridge Journal of Education*, *30*(1), 133–145. doi:10.1080/03057640050005825

Kress, G. (2000b). Multimodality. In B. Cope, & M. Kalantzis (Eds.), *Multiliteracies: Literacy Learning and the Design of Social Futures* (pp. 182–202). London: Routledge.

Kress, G. (2003). *Literacy in the New Media Age*. London: Routledge. doi:10.4324/9780203164754

Kress, G., & Pachler, N. (2007). Thinking about the 'm' in m-learning. In N. Pachler (Ed.), *Mobile learning: towards a research agenda* (pp. 7–32). London: The WLE Centre, Institute of Education, University of London.

Kress, G., & Van Leeuwen, J. (2001). *Multimodal Discourse: The modes and media of contemporary communication*. London: Arnold.

Levy, S. (2006). *The Perfect Thing: How the iPod Shuffles Commerce, Culture, and Coolness*. New York: Simon & Schuster.

Ling, R. (2008). *New Tech, New Ties: How Mobile Communication is Reshaping Social Cohesion*. Cambridge, MA: The MIT Press.

Ling, R., & Haddon, L. (2008). Children, Youth and the Mobile Phone. In K. Drotner, & S. Livingstone (Eds.), *The International Handbook of Children, Media and Culture* (pp. 137–151). London: Sage. doi:10.4135/9781848608436.n9

Ling, R., & Pedersen, P. E. (Eds.). (2005). *Mobile Communications: Re-negotiation of the Social Sphere*. London: Springer-Verlag.

Ling, R., & Yttri, B. (2002). Hyper-coordination via mobile phones in Norway. In J. E. Katz, & M. Aakhus (Eds.), *Perpetual Contact: Mobile Communication, Private Talk, Public Performance* (pp. 139–169). Cambridge, UK: Cambridge University Press.

Luke, C. (1997). Technological Literacy. Melbourne, Australia: Adult Research Literacy Network (Language Australia Limited).

Madison, D. S. (2005). *Critical Ethnography: Methods, Ethics, and Performance*. Thousand Oaks, CA: Sage.

Marcus, G. E. (1998). *Ethnography Through Thick and Thin*. Princeton, NJ: Princeton University Press.

New London Group. (2000). A Pedagogy of Multiliteracies: Designing social futures. In B. Cope, & M. Kalantzis (Eds.), *Multiliteracies: Literacy Learning and the Design of Social Futures* (pp. 9–37). London: Routledge.

Pahl, K., & Rowsell, J. (2005). *Literacy and Education: Understanding the New Literacy Studies in the Classroom*. Los Angeles, CA: Sage.

Penman, R., & Turnbull, S. (2007). Media Literacy—Concepts, Research and Regulatory Issues. Canberra, Australia: Australian Communications and Media Authority (Australian Government).

Prensky, M. (2001a). Digital Natives, Digital Immigrants. *Horizon*, *9*(5). doi:10.1108/10748120110424816

Prensky, M. (2001b). Digital Natives, Digital Immigrants, Part II: Do They Really Think Differently? *Horizon*, *9*(6). doi:10.1108/10748120110424843

Rheingold, H. (2002). *Smart Mobs: The Next Social Revolution*. Cambridge, MA: Perseus Books.

Rule, J. B. (2002). From mass society to perpetual contact: Models of communication technologies in social context. In J. E. Katz, & M. Aakhus (Eds.), *Perpetual Contact: Mobile Communication, Private Talk, Public Performance* (pp. 242–254). Cambridge, UK: Cambridge University Press.

Shirky, C. (2008). *Here Comes Everybody: The Power of Organizing Without Organizations*. New York: The Penguin Press.

Snyder, I. (2008). *The Literacy Wars: Why teaching children to read and write is a battleground in Australia. Crow's Nest*. Australia: Allen & Unwin.

Stigler, G. J. (1970). The Case, If Any, for Economic Literacy. *The Journal of Economic Education, 1*(2), 77–84. doi:10.1080/00220485.1970.10845301

Street, B. V. (1984). *Literacy in theory and practice*. Cambridge, UK: Cambridge University Press.

Taylor, C. (2009a). Choice, Coverage and Cost in the Countryside: A topology of adolescent rural mobile technology use. *Education in Rural Australia, 19*(1), 53–64.

Taylor, C. (2009b). Pre-paid literacy: Negotiating the cost of adolescent mobile technology use. *Engineers Australia, 44*(2), 26–34.

Taylor, C. A. (2011). *The Mobile Literacy Practices of Adolescents: An Ethnographic Study*. (Doctoral dissertation). Retrieved from Monash University Research Repository. (Identifier: ethesis-20110810-160841)

Weerakkody, N. N. (2007). *Framing the Discourses of Harm and Loss: A Case Study of Power Relations, Mobile Phones, and Children in Australia*. Paper presented at the Australia New Zealand Communication Association (ANZCA) 2007 Conference: Communications, Civics, Industry. Retrieved on 2 March, 2009, from http://www.latrobe.edu.au/ANZCA2007/proceedings/Weerakkody.pdf

Wilhelm, L. (2005). Increasing Visual Literacy Skills With Digital Imagery. *T.H.E. Journal, 32*(7), 24–27.

Willinsky, J. (1991). Postmodern Literacy: A Primer. *Interchange, 22*(4), 56–76. doi:10.1007/BF01806966

Willis, P. (2000). *The Ethnographic Imagination*. Cambridge, MA: Polity Press.

KEY TERMS AND DEFINITIONS

Capital: Drawn from the work of Pierre Bourdieu (1986) the resources (physical and ideological) that form the core of social interactions configured as a system of exchange.

Field: Drawn from the work of Pierre Bourdieu (1977, 1980), primarily delineates the context (physical, social, cultural, ideological) for any social practice, including the way in which this context structures the nature of a social practice.

Habitus: Drawn from the work of Pierre Bourdieu (1977, 1980), this term offers a way of understanding personality and dispositions to act, behave or think in certain way; the 'structuring structures' of individual behaviours.

Medium: The communicative structure used to convey information, such as books, radio, website, television (audio-visual), etc.

Mode: The form of a text, specifically: linguistic, visual, aural, gestural, spatial or multimodal.

Monopoly-Membership Dynamic: The structuring of social interactions around mobile devices as a result of market dominance by a single commercial organisation or interest.

Multiliteracies: A theoretical perspective on literacy associated with the New Literacy Studies approach, which understands literacy as something that is integral to social practice, made of multiple modes, expressed in different mediums and existing at the everyday level.

Chapter 2
Mobile Learning Apps:
Evaluating Instructional Needs

Janet Holland
Emporia State University, USA

ABSTRACT

With the phenomenal growth of mobile applications or apps used for teaching and learning, we are all challenged with determining which ones are effective and efficient in meeting our specific instructional needs. The use of mobile apps directly impacts students, teachers, administrators, trainers, and employees worldwide. Apps are used across all discipline areas in a variety of settings including applied interdisciplinary approaches. With this in mind, it is critical to have a workable set of app analysis questions based on current best educational practices to assist in making informed decisions on app selections to provide quality teaching and learning experiences. This chapter provides a mixed method research study combining class observations with results from three pilots in an effort to create a set of quality questions for quickly evaluating mobile apps for instructional implementation. After creating a set of questions for evaluating the quality of the apps based on current best instructional practices, the following three pilot studies were conducted. The first pilot allowed students to select an app of their own choice followed by a survey to evaluate the app using both quantitative and qualitative open-ended responses. The second pilot had all students examine the same app followed by the same survey to analyze potential differences in results and to gain additional insights. The third pilot study used the same questions, but this time rather than using it to evaluate the app, the students evaluated the quality of the questions used. During the third pilot study, students were looking strictly at the quality of the questions for instructional use. All study participants were graduate-level students in Instructional Design and Technology and were aware of best instructional practices. It is anticipated, post study, instructors and trainers can begin using the evaluation instrument, selecting those questions meeting their unique instructional needs.

DOI: 10.4018/978-1-4666-6166-0.ch002

INTRODUCTION

Fast emerging technologies such as mobile apps pose unique challenges to traditional research methodology, when there is not a sufficient amount of time to effectively research thousands of mobile apps being released each year. With new apps, it is difficult to find the appropriate literature to determine their effectiveness for instructional purposes. Where immediate specific classroom data is needed about the effective use of educational apps, a needs analysis with pilot studies is the best option. Compiling the results from many small pilot studies then offers the additional benefit of fleshing out key issues to be examined later in greater detail using a full range of research tools to study the selected educational apps with the promise of extending theory or scientific practices.

When searching for existing needs analysis resources there are several books on how to develop surveys in general, but none found which have been tested, specifically targeting evaluating mobile apps for teaching and learning with the best instructional practices as the underlying bases. Instructors and trainers need a set of workable questions to make quality quick assessments to meet their instructional needs, goals, and objectives for just in time teaching and learning with emerging technologies. Since the instructors' priorities may shift with each lesson, module, or unit, the assessment instrument needs to be flexible with choices for the questions, selected so they directly correlate to the intended instructional purposes. By structuring questions based on solid instructional practices it can support the selection and implementation of quality resources. Therefore, the bases for the questions generated for the apps needs analysis are derived from the literature based current best teaching practices grounded in instructional design and technology theory and practice. By effectively identifying the learning needs, instructors can better facilitate memorable knowledge acquisition.

LITERATURE BASED CURRENT BEST EDUCATIONAL PRACTICES

Mobile applications or apps are a great way to invigorate online learning or blended learning environments. We are finding apps can be used to embrace new digital learning tools by having learners actively research, collaborate, innovate, and share their ideas. Many of the apps have collaborative tools built in, used to increase knowledge acquisition quickly and efficiently while making global connections for broader perspectives. Providing meaningful integration of new technologies through the careful selection of quality tools aligning to best instructional practices can alter how learners and instructors engage with concepts and each other to achieve powerful learning. The following sections provide some background knowledge on the current best instructional practices used as the bases for the formulation of the needs analysis questions developed to evaluate mobile apps selected for teaching and learning.

Mobile Learning

One of the biggest trends in education is the ability to be mobile. Thereby, driving the huge surge in app development and use in education. According to Apple's released reports, more than 40 billion apps were downloaded for the iPhone, iPad and the iPod Touch, in 2012. It is hard to deny the success realized with approximately 83 million iPads sold by the third fiscal quarter of 2012, Nations (2013). To put the impact into perspective, iPads have now surpassed Mac OS sales with the new mobile iOS, Caulfield (2011). The sales reflect the strong consumer demand for this new media. There are many iPad contenders such as Amazon, Archos, Disgo, Acer, Asus, HTC, Google, Android, Motorola, Toshiba, BlackBerry, Sony, Samsung, Microsoft, Dell, Vizio, HP, and the e-book readers including the Kindle, Kobo, and Nook. Users are drawn to the sleek design, small portable size, long battery life, in store support, inexpensive, intuitive natural interface, with a vast number of quality content applications or apps to run on it. In addition, many of the apps can be used on computers and mobile phones as well. Learning can then be extended beyond the classroom to working from home, on the go, and in the field. "We really have reached the point where we do have magic, and thus we have the opportunity to ask what we should do with it" (Quinn, 2012). In the corporate environment educational applications range from training, performance support, increased access, collaboration, and learning. In the educational setting learners are gaining new content, communicating, capturing information, analyzing data, presenting, sharing, and even using location based activities. "To have mobile learning work well, power has to shift from instructors and manages to the learners themselves" (Woodill, 2011). It is a self-directed or do-it-yourself (DIY) approach to learning.

Authentic Problem-Based Learning

Problem-Based Learning is an instructional method in which learners, usually working in teams, are given complex authentic problems or challenges then asked to solve them. This approach is often used to increase learner interactions by working together collaboratively. Teams determine the needs, then, work through the steps to solve the problem. Barrows (1986) describes problem-based learning as a way to motivate students' solutions through self-directed explorations while gaining additional practice. Problem-solving models of instruction are based on contributions from Dewey (1916, 1938). Dewey defined a problem as anything giving doubt or uncertainty. His active learning experiences included providing an appropriate learning topic, which was important, and relevant.

Inquiry Learning

The researchers Bigge and Shermis (2004), Holcomb (2004), Joyce and Calhoun (1998), Van Zee (2001), and others define inquiry learning as capitalizing on students' interest in discovering something new or finding alternatives to unsolved questions or problems. Learners often work together to conduct research, experiment, synthesize, classify, infer, communicate, analyze, draw conclusions, evaluate, revise, and justify findings. In inquiry learning, students are responsible for problem solving, discovery, and critical thinking to construct new knowledge through active experiences. "Inquiry teaching requires a high degree of interaction among the learner, the teacher, the materials, the content, and the environment. Perhaps the most crucial aspect of the inquiry method is that it allows both students and teachers to become persistent askers, seekers, interrogators, questioners, and ponderers" (Orlich, Harder, Callahan, Trevisan, and Brown, 2007).

Motivating Learning

Keller's (1983) ARC (attention, relevance, confidence, satisfaction) model of motivation provides insight into providing motivating instructional learning environments. In general, gaining attention involves capturing learner interest, stimulating inquiry, and maintaining it. Relevance includes identifying learner needs, aligning them to appropriate choices and responsibilities, and building on prior experiences. Confidence includes building positive expectations, support, competence, and success. Satisfaction includes providing meaningful opportunities to apply new knowledge and skills, reinforcement, and positive accomplishments. In (Gagne, 1985) "Conditions of Learning", he indicated it is necessary to gain students' attention before they will be able to learn. Studies in the field of educational motivation continued to grow with work by Wlodkowsky (1999), Brophy (1983, 1998), and others. Some additional traits of motivated learners include the desire to learn, work, meet a need, personal value, reach a goal, complete task, engaging, curiosity, successful effort or ability, achievement, and personal responsibility. In a constructivist framework, motivation includes both individual and group generated knowledge and concepts.

Communication, Collaboration, and Social Interactions for Learning

Building professional relationships through collaborating, coaching, and mentoring are all social interactions directed towards learning to share ideas, give and receive feedback, and offer support (Carr, Herman, and Harris, 2005). The concepts of social learning can be traced to Bruner (1961) and Vygotsky (1978) and others. Quality instructional design directed towards technology-enhanced learning requires a great deal of student interaction. Promoting learner-to-learner interactions can increase engagement through negotiations, reflections, and shared understandings. The interactions allow students to expand viewpoints and build social connections to each other. Dialogue directed towards learning can provide students a way to expand ideas, extend concepts, and apply theory in authentic ways to solve challenges. "The focus of this work is ongoing engagement in a process of purposeful inquiry designed to improve student learning" (Carr, et al., 2005). "Collaboration forms the foundation of a learning community online-it brings students together to support the learning of each member of the group while promoting creativity and critical thinking" (Palloff, & Pratt, 2005). Some of the constructivist contributing to social learning included Piaget (1969), Jonassen (1995), and Brookfield (1995). Social presence creates the "feeling of community and connection among learners, has contributed positively to learning outcomes and learner satisfaction with online courses" (Palloff & Pratt, 2005). Researchers finding a strong connection between social presence and improved learning, interaction, and satisfaction include Picciano (2002), Gunawardena and Zittle (1997), Kazmer (2000), Murphy, Drabier, and Epps (1998). With the wide range of collaborative apps available for communications and collaboration, we have the tools to form the perfect foundation for social interactions and collaboration directed towards learning.

Multimedia and Multimodal Learning Environments

Multimedia refers to the use of text, graphics, sound, video, animation, simulation, or a combination of media. By appropriately aligning rich media to the content message, it can provide additional clarity and increase student focus rather than detract from it. Using a variety of media can increase interest and motivation while allowing unique opportunities to reach diverse learners. Mayer conducted many studies comparing lessons presenting content with words, to lessons presenting content with words and relevant visuals, Mayer (2001), Clark & Mayer, (2003). The results have consistently demonstrated the positive impact of appropriate instructional visual selections. "Rich media can improve learning if they are used in ways that promote effective cognitive processes in learners" (Reiser & Dempsey, 2007). Whether an educator prescribes to the 30's learning principles of Skinner - changing behavior, 70's cognitive psychology focus on memory and motivation, 80's constructivist or real world application, or a mixture of approaches, multimedia used effectively can help students to learn. Some media considerations include, gaining and keeping attention, being memorable, using an appropriate speed, level of difficulty, comprehension, placement, easy access, media matching the purpose, image content value, discovery, and level of interaction to improve effectiveness. "Ultimately good learning environments begin with the principles of learning and instruction, but require evaluation, revisions, and fine tuning to balance these competing values and ensure that the benefits are accrued for all intended learners" (Alessi & Trollip, 2001). Multimodal learning can include a wide range of multimedia and interactive tools used to engage learners thereby providing multiple modes of interfacing within the system. One of the strengths of the apps is working in a multimodal media rich learning environment.

Reaching Diverse Learners

Students learn in different ways with unique abilities and preferences on how they best acquire new information. The exceptionalities in intellectual ability, communications, sensory, behavioral, physical, and combinations, sometimes require special learning accommodations. One benefit apps can provide is the unique interface differing from traditional computing with gesture controlled navigation, the offering of computer-assisted programs, ability to increasing the size and contrast for text, images, audio, audio readers, audio text recording, audio commands, video media, interactive and collaborative tools to target specific learning needs. In addition, there is an increase in multi-language support. This can include assistance for both special needs, low and high, as well as the ever-increasing diversity of learners from all over the world joining our classes and workplaces.

Globalized Sharing

With the tremendous increase in travel, immigration, and communication technologies, the world is becoming more diverse, connected, and interdependent. Globalization has accelerated the exchange of ideas and perspectives thereby increasing the overall knowledge base. Current apps provide increased opportunities for extending content and perspectives to transform knowledge into innovative ideas shared. Using integrated curriculums, team teaching, media rich instructional technologies, forming partnerships, and fostering innovation, we can create knowledge and skills to prepare learners to work in future markets. Success in global markets we are now experiencing demand successful interactions with a diverse range of individuals and cultures. It begins with intercultural knowledge, skills, and respect for our combined contributions and strengths. As educators we need to become international stewards sharing insights and preparing learners for the future. The dramatic increase in mobility and digital communications now "connects people and facilitates transnational understanding" in ways not previously possible (Bryan & Vavrus, 2005). As a result, the International information infrastructure allows learners to interact and share multimedia resources easily with anyone across the globe. Current apps easily allow for original creations and global sharing.

Hands-On Active Learning

Hands-on refers to the learning activity involving practice on actual equipment, or in this example the use of apps. The learning activity is designed with the goal of promoting the transfer of knowledge through application. In an active learning environment students are active, working in teams, and socializing directed toward learning productively. "Students must be actively involved in the learning process if their classroom experience is to lead to deeper understandings and the building of new knowledge. Students (and adults, as I have discovered) need to hear it, touch it, see it, talk it over, grapple with it, confront it, question it, laugh about it, experience it, and reflect on it in a structured format, if learning is to have any meaning and permanence" (Nash, 2009). The dialogue provides time for learners to digest new information, exchange ideas, and engage with others in authentic, active hands-on way for expanded perspectives, and memorable learning experiences.

Creative Thinking and Original Projects

Open-ended apps allowing for original solutions to problems or challenges provide the perfect environment for creative thinking. Students can demonstrate understanding through a wide variety of app resources to present and share their unique solutions. It is critical to develop learners who can think beyond the box and lead us to new innovations. Simply reading and testing over material will not develop the creative original thinking needed to move our society forward. Instructors often use Bloom's Taxonomy (1956) to ensure inclusion of high-level knowledge and skills found in original creative work.

Increased Content Knowledge with Practice

The main consideration when selecting content resource apps is the relevance to the desired topic, and how clear the main ideas are communicated to learners. Providing learners with a graphic organizer is a nice way to show what will be studied by providing a brief overview of the content. By isolating facts, concepts, and generalizations it makes it easier to understand new content. The higher levels of knowledge integration teaches learners how items are related, similar, different, and compare so they can understand more complex relationships. Interactions with the content and others can provide additional practice to better retain new information. Some instructional activities are designed to provide learners with opportunities for review of previously learned information through repetition. Some of the apps provide the needed practice activities using repetition to ensure retention into long-term memory. It is important to identify the objectives and align them with the learning activity.

Assessment, Feedback, and Support

By providing learners with timely information about their actions, they will know how they compare to the desired level of criteria. "We should ensure that they receive feedback about their success and failure, are appropriately resourced with support to ultimately succeed, and ideally can share tasks and learning with one another" (Quinn, 2012). Learner feedback can take many different forms such as traditional instructor exams with rating scales or comments for students. Another alternative is to use student self-evaluations using checklists or rubrics for individual or group work to learn to monitor their own success. Sometimes instructors will also use checklists or rubrics for evaluation and providing student feedback. Instructors can use a pretest to assess learners' current level of knowledge, diagnostic test to assess areas of strengths and weakness, formative assessments to measure ongoing progress, and summative letter grade assessments to make judgments on the quality and completion of projects. The data gathered by the instructor can be used monitor learning and make adjustments as needed as the course progresses or for changes to be made before teaching the lesson, unit, or module again. It is important to identify the desired learning of "behaviors, activities, and knowledge you will be evaluating" (Orlich, et al., 2007). Instruction can include the teaching of knowledge, performance skills, and attitudes such as found in collaborative group work. Another consideration is whether the learning goal aligns to standards and provides feedback in this regard to students, parents, and administrators as needed.

Goals, Objectives, and Standard Driven

Mager's (1975) model for objectives, indicates quality objective includes the following three elements: 1) statement of the conditions or context of performance, 2) statement of the task; and 3) measurable way to evaluate the performance. Meaningful objectives are the backbone for instructors to create learning activities designed for knowledge to be retained, transferred, and applied to similar situations. It is accomplished by providing a specific statement of what learners will be able to do when they complete the lesson. A measureable performance objective statement describes the behavior student will demonstrate at the end of the lesson, the conditions under which they will be demonstrated, and the criteria for acceptable performance. Identifying the objectives became the guiding force for the selection of appropriate apps to get to the desired learner outcomes.

Flipped Classroom

Flipped classrooms are a more recent trend used to transform the way instructors are providing information by inverting traditional classroom lectures into online video and screencast presentations, so learners can view them prior to attending class. At home, learners can watch step-by-step explanations of concepts with visual examples to better understand complex concepts. The digital presentations allow each student to learn at their own pace with the ability to pause and replay as much as needed, on their own personal schedule when they are the most receptive to learning, to acquire the needed foundation knowledge. Class time is then flipped, so students complete homework and practice activities applying the new concepts in class. When attending class, students are engaged in student-to-student interactions, collaborations, and critical thinking with the instructor serving as a facilitator to support learners, as needed. The classroom is transformed into an active, authentic, learning environment where students can deal with complex issues related to the content topic. The Flipped classroom can be an alternative to traditional lecture-based models or can be used as a blended learning environment to engage student learning. Screencast technology is often used to leverage learning outside of class, so a teacher can spend more time facilitating project-based learning during class. This is most commonly being done using teacher-created videos students view outside of class time. Then, spend class time on problem solving, thereby increasing interactions between students and instructors. With the tremendous growth of mobile devices it has increased learners ability to view the videos on their own time. Then, class time can be spent on problem-based collaborative learning apps.

Applying Educational Theory to App Evaluation Questions

The merger of educational theory, technology tools, and human interactions can be a positive force in curriculum innovation in a variety of learning environments. By providing instructional guidance in selecting appropriate tools aligned to teaching and learning goals students can excel in new digital learning environments. We are seeing such a phenomenal growth in new tools and instructors can definitely use a helping hand in determining which quality resources work best for the intended educational purpose.

After examining the current best educational practices a set of app evaluation questions were developed based on it, then evaluated for instructional utility. The goal is to have a set of quality questions from which instructors can select the ones meeting their instructional needs and goals. As a result three pilot studies were conducted, 1) test questions with students choosing their own app, 2) test questions with the instructor selecting one app for all students, and 3) testing the quality of the questions.

EVALUATION OF APP ASSESSMENT QUESTIONS

One of the driving objectives of the three pilot studies is to design, develop, and test the questions

Table 1. Summary of mixed method pilot studies

Mixed Methods	Data Collection
Literature Review	Apps Needs Analysis
Literature Review	Instructional Design and Technology
Class Observations	Apps For Teaching and Learning
Quantitative Survey	Questions on Apps for Teaching and Learning
Qualitative Feedback	Open-Ended Feedback on Apps for Teaching and Learning

used to evaluate mobile apps used for instruction to make wise educational decisions directed towards the effective and efficient integration of app technologies. The following questions were the basis for the immediate action research pilots designed to optimize student learning when working with apps.

The guiding pilot study research questions included:

1. What app assessments can be found through an extensive search meeting current best instructional practices?
2. Defining current best instructional practices to determine how quality apps can be selected for implementation into instruction while aligning to the targeted curriculum goals and objectives?
3. What data would students' quantitative responses provide?
4. What data would students' qualitative open-ended questions provide and how would it compare to the quantitative data in regards to selecting quality apps?

The questions listed have become the basis for the three pilot studies "design most appropriate for the question[s] being studied" (Shavelson, 1981) as seen in Table 1. The pilot studies were conducted by instructional practitioners with the goal of improving just in time classroom teaching and learning, specifically when using emerging technologies. Applied research, as found in these pilot studies, can be defined as "research designed to develop and test predictions and interventions that can be used directly to improve practice" (Gall, Gall & Borg, 1999). "Action research enables teachers, administrators, school counselors, and other education practitioners to investigate and improve their performance in systematic, personally meaningful ways" with "the purpose of improving local practice rather than producing theory or scientific generalizations" (Gall, et al., 1999). The quality of the research depends on how well the project serves immediate, local needs. Table 1 provides a quick overview of the mixed method approach used in the pilot studies.

One-problem instructors, students, and trainers face when selecting mobile apps is, finding quality sites to assist in making appropriate selections to align with desired goals and objectives. Learners need assistance learning how to navigate app resources, especially with so many options to pick from. As Jenson stated, "we are witnessing a radical shift in how we establish authority, significance, and even scholarly validity" (Jensen, 2007). The paradigm shift from independent learners to the opposite end of the spectrum, crowd sourcing with the co-construction of knowledge requires a very different approach. With mobile apps, it is about a "change in focus to participation, user control, sharing, openness, and networking" (Eisenberg, 2008). Social learning networks are more open, participatory, conversational, and democratic, Cohen (2007). With learners and workers using new media content resources with questionable quality, it is even more important now to seek out multiple means of comparison for cross checking the sources, content, and analysis of each resource.

It is easy and comfortable to stay with what is familiar, but the payoff is much greater and worth the time invested to expand tool options for fostering creative thinking and expanded perspectives. The good news is many of these apps are free to use so the cost becomes less of an issue. The bad news, many instructors have not grown up using the new technologies, thus pointing to the need for professional training and support with some hands on experience to help embrace the path of continued life long learning. Let's face it, instructors and trainers are very busy and need all the assistance they can get, to create quality technology integrated lessons or workplace challenges. With the myriad numbers of mobile app choices available, it can be overwhelming just trying to decide where to start. Having a good needs analysis will help make the needed digital tool selections to align to the desired goals and objectives.

When selecting apps for student learning activities such as research, collaborations, presentations, it is important to select and implement appropriate apps. By supporting instructors in selecting appropriate mobile apps, they can be taught to develop the knowledge and skills needed to select and verify their app selections.

We begin with research of current best instructional practices found in the literature to align app quality selection to lessons based on targeted learning goals and objectives. In the following section, classroom observations are used to further refine and better align appropriate mobile apps with the desired learning tasks. The sections below are devoted to three classroom pilot studies used to determine the best app selections for immediate instructional implementation.

INSTRUCTIONAL PRACTICES

A good starting place for instructional planning is to begin with an analysis of learner needs, goals and objectives, to align the content and apps selected. Quality instructional practices often include an intellectual challenge to assist learners in thinking critically about the content. Intellectual rigor and engagement can be accomplished by allowing learners to build their own understandings, promoting meaningful collaborations, encouraging critical and creative thinking, and showcasing through authentic audiences, Dockter, Haug, and Lewis (2010). Many of the current apps align perfectly for creating compelling learning and workplace opportunities. Teaching learners to "effectively and creatively find, evaluate, analyze, use, and communicate information" are important lifetime skills, (Valenza & Johnson, 2009). One of the important hallmarks of the use of apps is the use of collaboration to build a richer expanded perspective within a social environment. By being socially and globally connected, learners can develop academic relationships with individuals and groups, thereby increasing the pace and knowledge base at the same time.

By targeting learning goals and objectives, instruction can be directed towards a wide variety of mobile apps to "increase student achievement, help meet standards, and capitalize on existing investments yields a excellent return on technology investments, whether measured by use or products or learner outcomes" (Baumbach, 2009).

It may feel pretty overwhelming, with the number of new technologies being released. However, by trying just one new app at a time, one can build up a nice repertoire over time. Start by looking at the needs of the learners, and matching the best tool to the learning goals. In the end, it is about what learners need to know and be able to do. "It is important for educators to find the appropriate tool to unlock learning possibilities" (Fredrick, 2010). To keep pace with the rapidly shifting landscape of digital tools, the best tip is "recognizing that change is no longer an option" and "we need to be willing to change, to be open to new ideas" (Brooks-Young, 2008). "Find one app you think has use in your classroom or work environment and use it. Start small. Start with a tool that compliments what you currently do either as a teacher or with your students. Start with something you are comfortable with" (O'Brien & Scharber, 2010). Integrating mobile apps effectively will have far reaching effects for preparing learners and workers for the realities of a globally connected world. Foster an inquiry-based approach to building a technology rich vision for teaching, learning, and working. Tap into students' desires to work together, expand ideas, and work efficiently.

APP QUESTION SELECTIONS

Mobile app options can vary widely, it will be up to the instructor to determine the primary objectives used to narrow the questions selected for evaluating the quality of the resources. Invigorating teaching, learning, and workplace challenges with new apps requires a mixture of vision and excitement to challenge learners. Newer apps are easier to use, more engaging and offer the potential for increased collaboration and communication. Part of the challenge is finding innovative ways to bring together instructors, students, mentors, and experts for social scholarship to create the "cyberinfrastructure needed for increased innovation, globalization, and knowledge networking" (Greenhow, 2009). Apps providing opportunities for collaboration are game changers for the "way learners can retrieve, share and evaluate information, and create knowledge" (Benson & Brack, 2009). By designing learning challenges within the social environment, one can promote cognitive engagement through the co-construction of knowledge and exchange of ideas. By making learning interesting, fun, and challenging it becomes a great way to capture learners' imaginations and passions for the topic while continuing the learning process.

RESULTS

Needs Analysis Using Apps for Teaching and Learning

An extensive search was completed looking for existing instruments to evaluate mobile apps to be used in the classroom for teaching and learning, for training in the workplace, and personal learning; none could be found. Since no tested questions could be located for evaluating the educational value of the apps, a clear need was identified and addressed in three pilot study results below. Lacking that set of tools, a subsequent review of current best educational practices led to the development of a set of questions used to determine the efficacy of each mobile app with respect to their educational application. One challenge identified, early on, the rapid evolution of mobile apps left very little time to validate their actual educational potential. Therefore, these questions are the best available and may now be considered collectively as a rapid, just in time, assessment instrument. Selective use of these questions in the assessment of apps will give one a standardized approach to determining what apps to use in which situations. Specifically, the questions are good as a quick evaluation of a specific app or, generally, for the comparison of apps. Once the learning goals have been identified, relevant question items can be selected for the evaluation process. To provide a level of validation to the questions, an on-going study was initiated.

The study, so far, has included three pilot surveys with results obtained from graduate level courses on Instructional Design and Technology. The first pilot study was conducted using the students' own choice on the application or app selection used for teaching and learning. It included 15 master degree students in one course from a small Midwestern University. The graduate students were provided a link to participate in an online survey. The ages ranged from 18 to 55 years old. The participants included 5 male and 10 female students. The pilot surveys consisted of 19 question items evaluated using a five-point Likert scale, with items rated as *Very Low* (1), *Low* (2), *Average* (3), *High* (4), *Very High* (5). All surveys concluded with open-ended questions to gather additional information on students' feedback about using apps for teaching and learning.

The second pilot study used one app called iBooks Author ®™, rather than being an open-ended choice like the first pilot. The study included 14 master degree students from a small Midwestern University. The graduate students were provided a link to participate in an online survey. The participants included 14 graduate students in one course. The ages ranged from 18 to 55 years old. The participants included 5 male and 9 female students. Based on student feedback during the first pilot three additional questions on usability were added. Since it was later determined the questions were directed towards usability rather than based on best instructional practices they did not apply to the scope of this study when considering the educational value of the app. Even though usability issues are important, it was determined to be a separate issue from evaluating the educational value of the apps selected. Since each instructor will have different usability needs, this could be considered on an individual level. Or, a future study examining usability issues should be conducted. The usability feedback could also be very valuable information for app developers in making them user-friendly for instructional purposes.

The third pilot study had the graduate students evaluate the questions used to select apps for teaching and learning. The study included 14 master degree students from a small Midwestern University. The graduate students were provided a link to participate in an online survey. The participants included 14 graduate students in one course. The ages ranged from 18 to 55 years old. The participants included 5 male and 9 female students. Since the students were not evaluating an app but rather evaluating the questions used, the number of open-ended question items was reduced.

The results of the data are presented in visual charts, below, so one can view the three different pilot studies at a quick glance. In the first pilot, participants selected an app of their own choice. In the second pilot, one high quality app called iBooks Author ®™ was selected for comparison purposes to gauge if there were differences in responses. The subsequent data indicated the questions were working as intended with variances in scores present. The third pilot analyzed the questions themselves to confirm whether any questions should be eliminated, or new ones added. The data did not point to a need for any questions to be eliminated. The participant suggestions for additions pertained to usability, determined to be a separate issue from evaluating the educational value of the app for teaching and learning. The overall data results demonstrate how the questions are showing variances as it should and makes a good set of baseline questions for evaluating the educational value of the apps selected. As one can see in the visual data displays, below, some apps scored higher in different areas when targeted for specific purposes. The data also revealed some apps were highly rated across all three pilot studies indicating the potential for application across a wider range of applications. This is thought to be due to current educational trends or apps with multiple features reaching a broader audience. Figure 1-19 provide a quick visual overview of the results of all three pilot studies.

Figures 1 through 19: Summary of Student Quantitative Results (Pilot 1, N=15 and Pilot 2 & 3, N=14)

1. What level of student interaction/engagement does the app provide?

The mean scores for pilot 1 (M=4.33), pilot 2 (M=4.07), pilot 3 (M=3.93). For question one, across all three pilots, student interaction and engagement was rated high overall. Scores rating the questions so closely across all three pilot studies indicate the app may have a wider scope of use across multiple apps (Figure 1).

Figure 1. Pilot 1=grey (diamond), pilot 2=black (square), pilot 3=white (triangle)

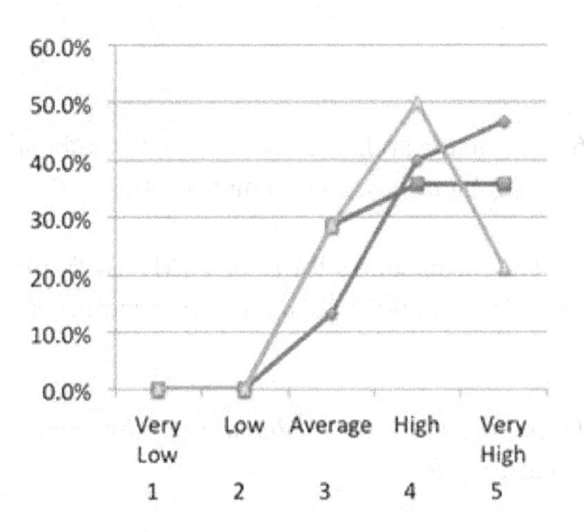

2. What level of communication does the app provide between students?

The mean scores for pilot 1 (M=3.73), pilot 2 (M=3.14), pilot 3 (M=3.36). For question two, the responses were more spread out reflecting the differences in the use of the application as sometimes communications between students are needed and sometimes they may not be required (Figure 2).

Figure 2. Pilot 1=grey (diamond), pilot 2=black (square), pilot 3=white (triangle)

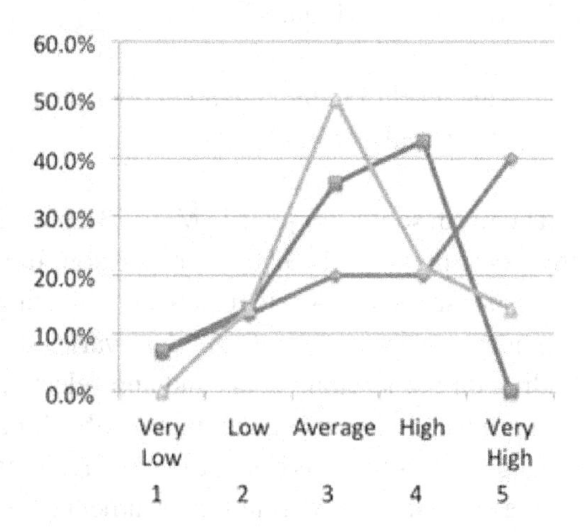

3. What level of communication does the app provide between students and the instructor(s)?

The mean scores for pilot 1 (M=3.53), pilot 2 (M=3.29), pilot 3 (M=3.36). For question three, the responses put the open-ended app peaking in the average range, while the digital publishing app selected by the instructor peaked in the high to very high range. Based on a simple review of

Figure 3. Pilot 1=grey (diamond), pilot 2=black (square), pilot 3=white (triangle)

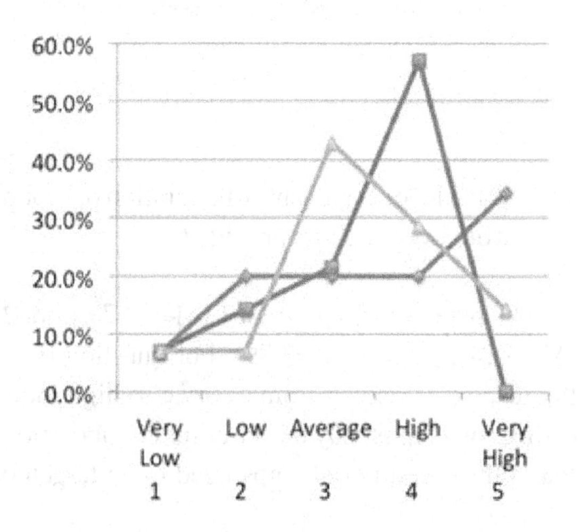

the question and responses, the instructor selected app clearly supported students sharing their work with the instructor (Figure 3).

4. What level of communication does the app provide outside of class?

The mean scores for pilot 1 (M=3.73), pilot 2 (M=3.64), pilot 3 (M=3.50). For question four, the responses show a wider spread indicating communications outside of the class varies depending on the learning goals and app selected. The high peak when using the digital publishing app appears to show it is easier to share creations outside of the specific learning environment when using this app (Figure 4).

Figure 4. Pilot 1=grey (diamond), pilot 2=black (square), pilot 3=white (triangle)

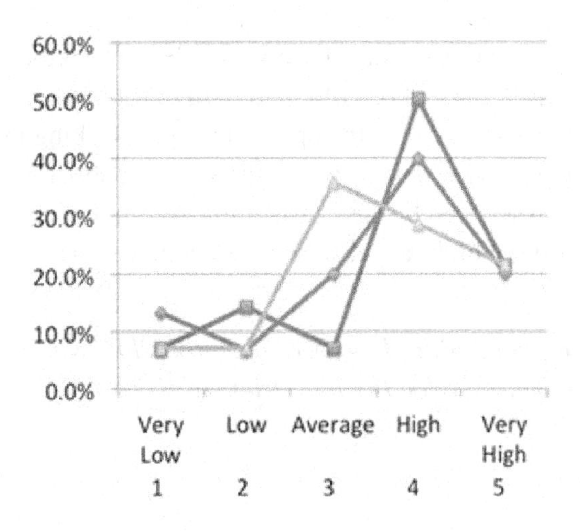

5. What level of student collaboration or group work does the app provide?

The mean scores for pilot 1 (M=3.87), pilot 2 (M=3.14), pilot 3 (M=3.79). For question five, the responses ranged from average to high indicating some apps may be better for collaboration than others and not all apps need to be targeted for group work as sometimes individual work is desired (Figure 5).

Figure 5. Pilot 1=grey (diamond), pilot 2=black (square), pilot 3=white (triangle)

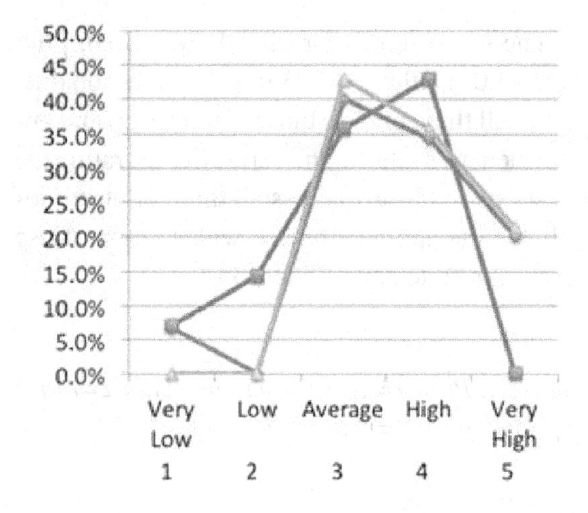

6. Rate the ability to use the app for authentic problem-based learning activities?

The mean scores for pilot 1 (M=3.47), pilot 2 (M=3.43), pilot 3 (M=3.43). For question six,

Figure 6. Pilot 1=grey (diamond), pilot 2=black (square), pilot 3=white (triangle)

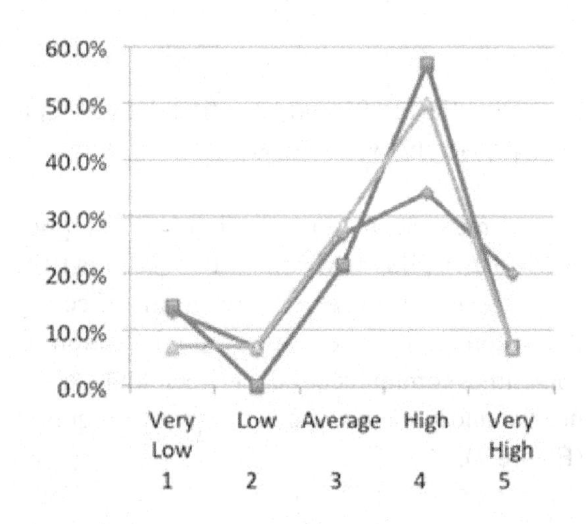

the responses display average to high scores pretty closely aligned indicating authentic problem-based learning is a desired trait across many different applications (Figure 6).

7. Rate the ability to use the app to accommodate diverse learning needs?

The mean scores for pilot 1 (M=3.73), pilot 2 (M=4.00), pilot 3 (M=4.07). For question seven, the responses indicate apps designed to accommodate diverse learning needs rate high across multiple applications (Figure 7).

Figure 7. Pilot 1=grey (diamond), pilot 2=black (square), pilot 3=white (triangle)

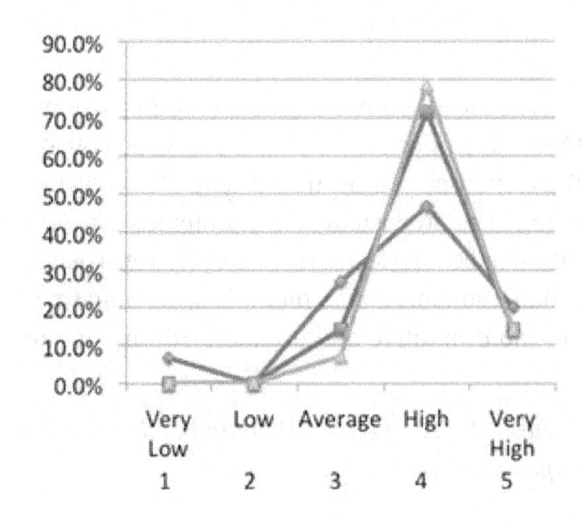

8. Rate how effectively you can connect the learning goals or objectives with the use of the app?

The mean scores for pilot 1 (M=3.80), pilot 2 (M=3.86), pilot 3 (M=3.93). For question eight, the high scores closely match up demonstrating the generally agreed importance of aligning the learning goals and objectives with the app selected across multiple settings (Figure 8).

Figure 8. Pilot 1=grey (diamond), pilot 2=black (square), pilot 3=white (triangle)

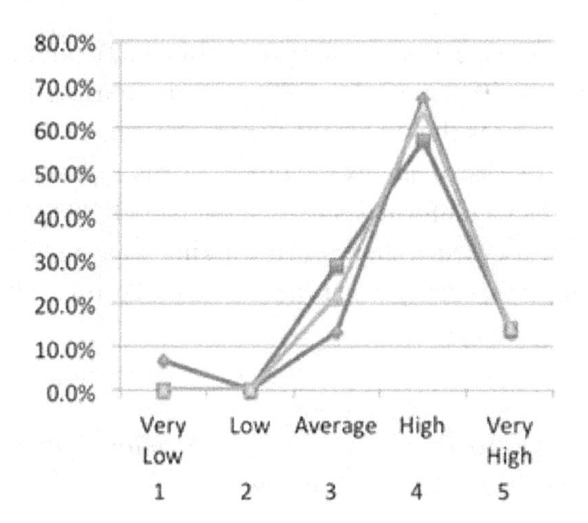

9. Rate the level of feedback or summary data the app provides?

The mean scores for pilot 1 (M=3.60), pilot 2 (M=3.07), pilot 3 (M=3.21). For question nine, the responses show an average peak and spread out scores indicating feedback or summary data is not always needed (Figure 9).

Figure 9. Pilot 1=grey (diamond), pilot 2=black (square), pilot 3=white (triangle)

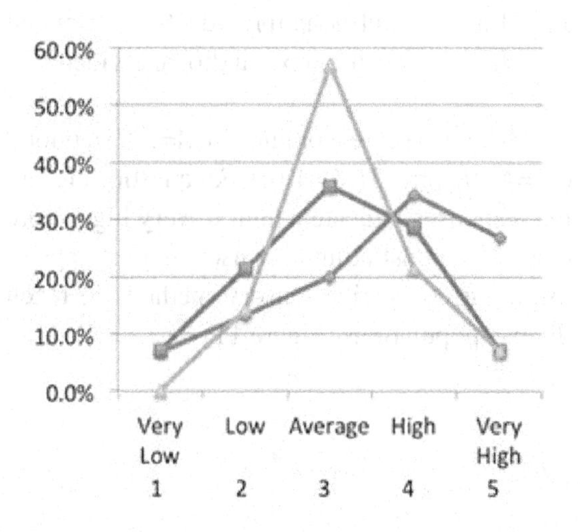

10. Rate the quality of the multimedia used in the app?

The mean scores for pilot 1 (M=3.67), pilot 2 (M=4.50), pilot 3 (M=3.79). For question ten, the responses show a high to very high rating overall for the use of quality multimedia. This appears to reflect a uniform desire for a media rich interactive user experience when working with mobile apps; especially when one selects the app oneself (Figure 10).

Figure 10. Pilot 1=grey (diamond), pilot 2=black (square), pilot 3=white (triangle)

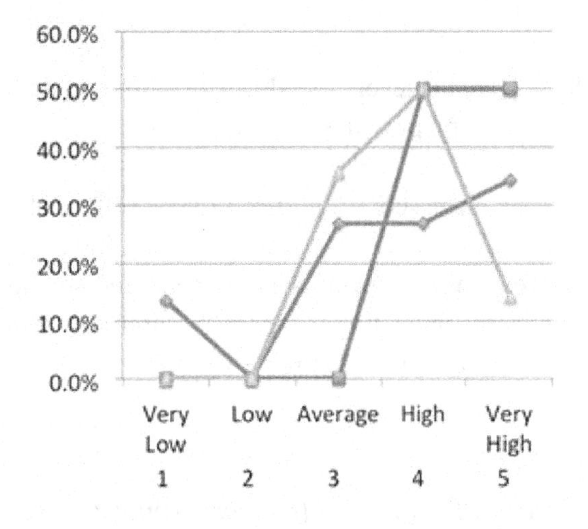

11. Rate the level the app provides for multimodal learning such as text, audio, and video?

The mean scores for pilot 1 (M=3.53), pilot 2 (M=4.50), pilot 3 (M=4.14). For question eleven, the responses indicate a high to very high score overall demonstrating how users expect apps to use a variety of rich sensory methods to reach diverse populations (Figure 11).

Figure 11. Pilot 1=grey (diamond), pilot 2=black (square), pilot 3=white (triangle)

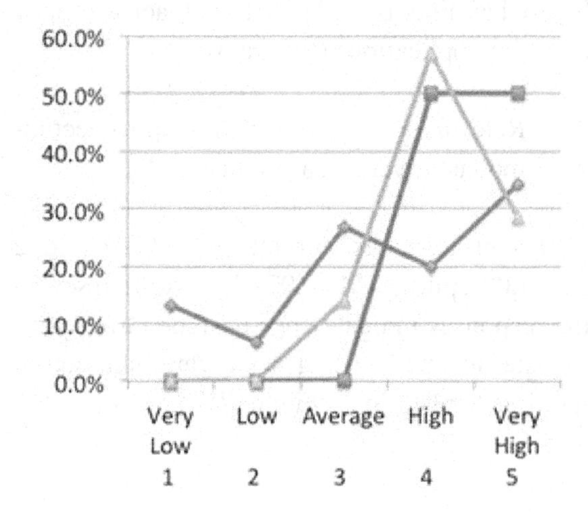

12. Rate the level of appropriate support for use of the product?

The mean scores for pilot 1 (M=3.60), pilot 2 (M=3.50), pilot 3 (M=3.64). For question twelve, the responses demonstrate a range from average to high and are visually closely aligned across all pilots, indicating learners' desires for user-friendly product support across multiple apps. Some apps are more intuitive and do not need further sup-

Figure 12. Pilot 1=grey (diamond), pilot 2=black (square), pilot 3=white (triangle)

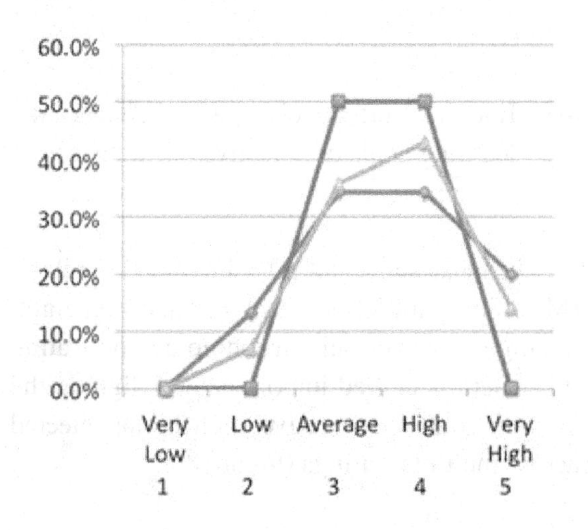

port while others require additional information to support learners efforts (Figure 12).

13. Rate the ability to use the app in a flipped classroom?

The mean scores for pilot 1 (M=3.53), pilot 2 (M=4.29), pilot 3 (M=3.86). For question thirteen, the responses range from average to very high indicating most apps could be used for flipped teaching and learning, but it is not always needed depending on the learning goals (Figure 13).

Figure 13. Pilot 1=grey (diamond), pilot 2=black (square), pilot 3=white (triangle)

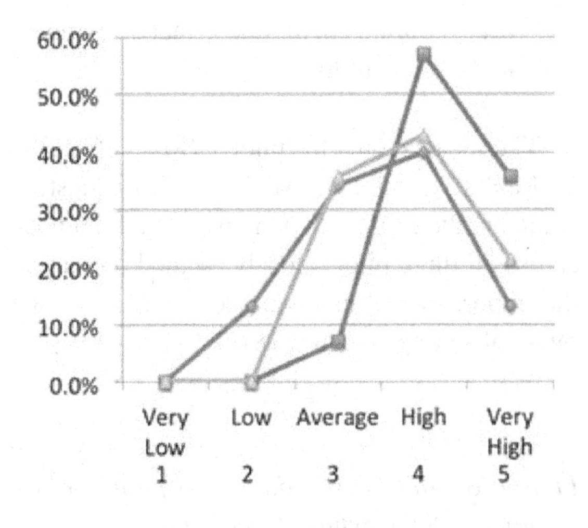

14. Rate the ability of the learners to create their own content using this app?

The mean scores for pilot 1 (M=3.70), pilot 2 (M=4.64), pilot 3 (M=4.21). For question fourteen, the responses range from average to very high indicating the creation of content is not always needed as sometimes other learning activities are desired. The very high peak score demonstrates the difference when the app selected required creating content was used (Figure 14).

Figure 14. Pilot 1=grey (diamond), pilot 2=black (square), pilot 3=white (triangle)

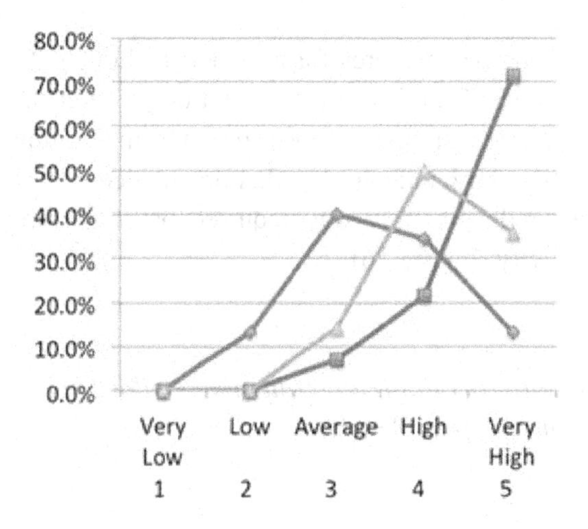

15. Rate the level the app allows for learner inquiry research or experiments?

The mean scores for pilot 1 (M=3.27), pilot 2 (M=3.21), pilot 3 (M=3.50). For question fifteen, the responses are more spread out and range from very low to high. Sometimes inquiry research or experiments are needed and sometimes other approaches are needed (Figure 15).

Figure 15. Pilot 1=grey (diamond), pilot 2=black (square), pilot 3=white (triangle)

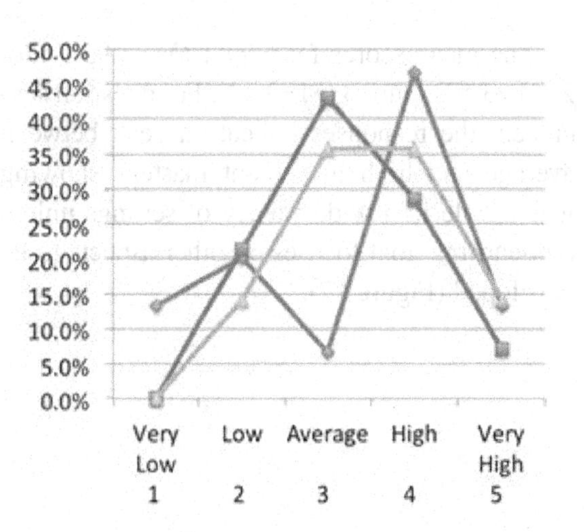

16. Rate the level the app facilitates authentic practice?

The mean scores for pilot 1 (M=4.06), pilot 2 (M=3.57), pilot 3 (M=3.36). For question sixteen, the responses show a spread from very low to very high indicating authentic practice may be desired, but is not always required for the learning activity (Figure 16).

Figure 16. Pilot 1=grey (diamond), pilot 2=black (square), pilot 3=white (triangle)

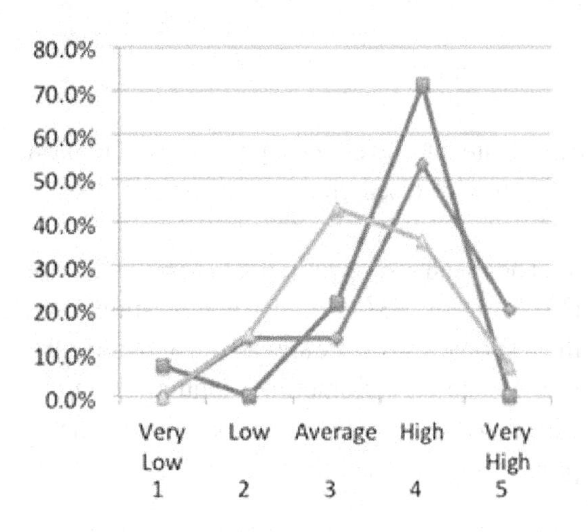

17. Rate the ability of the app to demonstrate content mastery?

The mean scores for pilot 1 (M=3.93), pilot 2 (M=3.57), pilot 3 (M=3.43). For question seventeen, the responses indicate a peak between average and high in content mastery showing it is usable in a wide variety of settings unless the learning goal focuses on other aspects to be developed (Figure 17).

Figure 17. Pilot 1=grey (diamond), pilot 2=black (square), pilot 3=white (triangle)

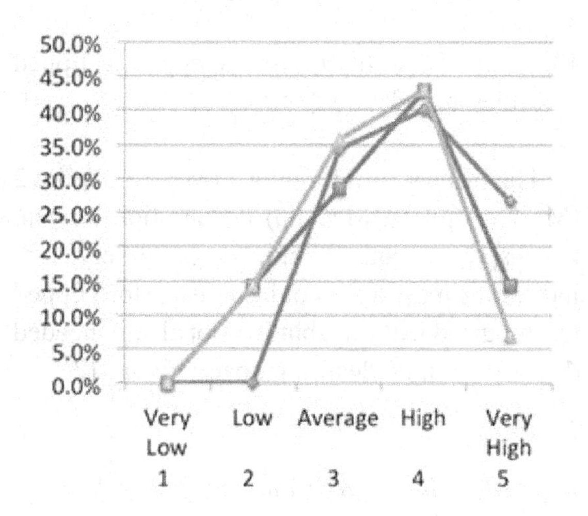

18. Rate the ability of the app to increase student hands-on activities?

The mean scores for pilot 1 (M=3.86), pilot 2 (M=4.07), pilot 3 (M=3.79). For question eighteen, the responses demonstrate a high peak across all three pilots closely aligned indicating the importance of hands-on activities in a wide range of settings (Figure 18).

Figure 18. Pilot 1=grey (diamond), pilot 2=black (square), pilot 3=white (triangle)

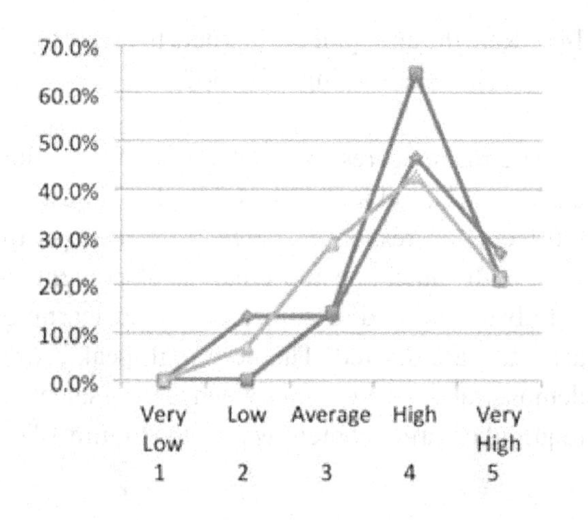

19. Rate the ability of the app to provide open-ended creative solutions?

The mean scores for pilot 1 (M=3.06), pilot 2 (M=4.21), pilot 3 (M=3.86). For question nineteen, the responses are more spread out indicating open-ended creative solutions are not always needed and may depend on the specific learning activity. The high peak was when a creative app was selected (Figure 19).

Figure 19. Pilot 1=grey (diamond), pilot 2=black (square), pilot 3=white (triangle)

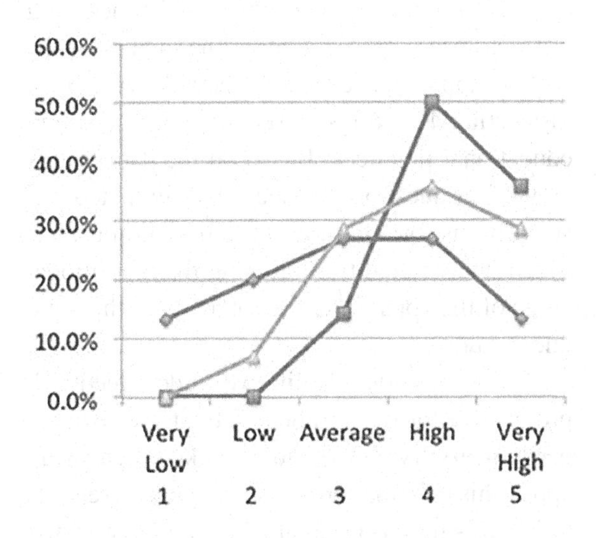

CONCLUSION

After an extensive search of the literature no tested appraisal instruments for evaluating the educational value of mobile applications or apps could be found. Instructors are left to their own examinations without any guidance from current best instructional practices. Instructors lacking appropriate training on current best instructional practices are frequently left on their own to try to make these decisions. Some may go through a process of trial and error, look at non-validated ratings, non-evaluated questions, or suggestions from others to make app selections. This lack of tools to help find worthy apps was the reason the three pilot studies were conducted, to begin the process of being able to make better app selections using a tested set of questions based on current best instructional practices. The results of the studies indicated the questions developed form a good set of baseline questions. As educational practices change, survey questions can also evolve and change, as needed, with additional studies. It is my hope additional studies will be conducted to further refine the assessment instruments, to increase the validity with increased numbers of participants, to increase the diversity of geographic regions, and subject area differences. With the growth of mobile applications used for instruction it is critical to be able to make informed decisions, quickly due to the rapid rate of technology innovations.

As noted at the beginning of this chapter- with the phenomenal growth of mobile applications or apps used for teaching and learning, we are all challenged with determining which ones are appropriate and meet our specific instructional needs. This chapter presents a number of questions that can be asked about every mobile application one might consider using in an instructional setting. The questions have been applied in three different settings: 1) Students reviewing their own individual applications for use, 2) All students evaluating a single common application, 3) Having students look directly at the prepared questions to evaluate them, then making additional suggested changes.

Generally, for the whole list of questions, there is sufficient evidence from the pilot studies to consider these questions as a good start with respect to using them when evaluating apps in an educational setting. The questions were generally accepted by all users, as noted from a review of all the participant results. The positive trends indicate the questions can be used in future appraisals of applications for use in educational settings. The instructor, trainer, or students could use the entire set of questions developed or select those questions they deem appropriate for their unique educational needs.

In the future, the actual makeup of the groups using these questions could be broken down by professional setting (business, education, other), age group, or other statistical variables to further refine and perfect the questions for the appraisal of the rapidly changing mobile applications hitting the market and educational settings.

Specifically, from the first pilot study the three most important student comments were suggestions for additions to the needs analysis questions. As a result, in the subsequent studies the following three questions were included. 1) Is the cost or value benefit appropriate for the app? 2) Is the interface quick and easy for users to access? 3) Can this app be used on a single device with the entire classroom effectively? It was determined the participant suggestions, from the open-ended feedback, were based on usability rather than evaluating the educational value of the app; as a result, they were eliminated. Although usability issues are important, they were considered separate topic for a future study. This type of information would be especially valuable to app developers to create user-friendly apps.

The second pilot study was conducted with a new group of students in one class using one app selected by the instructor rather than open-ended like it was during the first pilot, to better compare the student responses and to increase the overall number of participants and feedback included in the study. From the list of software identified by the students in the first pilot study, the iBooks Author ®™ app was selected for use in the second pilot as it could be applied across a wide range of different subject areas and fields. Students were directed to try out the iBooks Author ®™ app then completes the online survey to evaluate it. In comparing the open-ended app selection in the first study to the iBooks Author ®™ app selected by the instructor during the second study, the results indicated it appeared to be working by showing variances in the data results. From personal experience I knew iBooks Author ®™ was a high quality app, so it assisted in making comparisons between the quantitative number data results showing the needed variances in results.

The third pilot study was conducted using the same group of students as in the second pilot study to then rate the quality of the questions used for evaluating apps and to see if there were any suggestions for questions to be added or eliminated. The goal was to develop a set of questions which an instructor or trainer could easily select from, as needed based on their own specific learning goals, to use when evaluating the quality of the mobile apps selected across a wide variety of instructional settings based on current sound educational practices. Based on the data results none of the questions needed to be eliminated. All suggestions for additions were based on usability and not related to evaluating the educational value of the apps, so it did not apply to the topic under study.

The open-ended qualitative student feedback provided a lot of comments about the features participants liked and did not like when using apps. This information is also valuable in regards to features important to app developers working towards making apps engaging and user-friendly. Some of the desired features mentioned included; quality content, import/export ability in a wide variety of file formats, cross platform integration, multi-device access, free or affordable apps, easy to use, ability to share, collaborate, universal access, creative tools, create projects, no registrations, no junk e-mail, intuitive layout and navigation, globalized, searchable, references, vocabulary, ability to edit, file storage, rich multimedia, timeline, evaluation, feedback, practice, ability to make comparisons, review, quiz, individual and group work, use on a single device with the entire class, tutorials, solve problems, discovery learning, interactive, and increase student achievement. From this open-ended student list of suggestions one can see overlapping issues aligning with the questions presented in the study based on current best educational practices with some ideas for additional questions extending beyond the app evaluation itself and entering into the overall usability. It seems like consumer and educator expectations continue to raise the technology bar. It is important to continue collecting data to seek feedback from educators and learners themselves, in order to continue improving our selections of mobile apps used for teaching and learning.

Since all three pilot tests were limited in numbers and restricted to one discipline area, it would be good to conduct a full research study with a larger sample size to see if the trends hold true across additional discipline areas. However, if there isn't time available to do a full research study, as is the problem with constantly evolving apps, one may continue to perform additional pilot studies. The additional studies are needed to increase the number of participants sampled and compare their results to this set of data, which can now be classified as a baseline data set. The comparisons are needed to determine if the review questions maintain their usefulness for rating applications, whether further questions need to be added, or whether the existing questions themselves need to be revised or deleted as the technologies and educational practices evolve over time.

The three pilot study surveys consisted of both qualitative and quantitative data about selecting apps for teaching and learning. These types of pilot studies were needed to provide baseline data since no past data analysis existed. When class sizes are too small for a full research study, multiple studies can yield the needed information. In some instances, the time needed to conduct a full study can cause the data to be out of date due to the pace of changes in emerging technologies. In addition, full research studies could not provide the needed immediate classroom feedback for just in time learning. Therefore, it is recommended, in those instances where immediate specific classroom data is needed, a needs analysis with pilot studies is the best approach.

Technology is transforming our lives with 24/7 access so it will be important to consider the "ease, efficiency, and effectiveness" of how we will use it (Nicol, 2013). Whether needed for business to become more productive or to gain access to information within educational and social contexts for learning, we need an "understanding of the strategic value" of staying "relevant in the midst of this new technology shift" (Nicol, 2013). Identifying the learning goals will help to define the educational objectives so we can "establish the appropriate measurements to ensure that the desired goal has been achieved" through "measurable performance indicators" (Nicol, 2013). With so many tech savvy users today, consumer expectations are very high for easy to use engaging mobile technologies. The next generation of apps is anticipated to become even more transformative in design to "predict the next best action" reducing the interface cognitive load, to improve the efficiency of knowledge based tasks at hand (Nicol, 2013). Mobile learning is seeing the "convergence of Social, Cloud, Mobile, and Big Data analytics" with mobile applications offering greater opportunities to leverage intelligence within an increasingly globalized setting (Nicol, 2013). A crucial last step, "testing across a variety of devices and environments will be important to ensuring the quality end users expect" (Nicol, 2013). This means, development of effective learning experiences requires identifying the needs of the users, the learning goals, having access to current best practices, and measurement instruments to access and facilitate the desired learning outcomes.

It is apparent the growth of mobile devices and applications are continuing to expand as our lifestyles change to a more globalized society. The future looks very exciting with emerging technologies in wearable devices, augmented reality, touch-less or gesture-based computing interacting with the physical senses in a more authentic way, and many more, all of which require apps to run them. It will be important to continue to examine current best instructional practices all along the way in light of the changing mobile landscape.

REFERENCES

Alessi, S., & Trollip, S. (2001). Multimedia for learning: Methods and development (3rd Ed.). Allyn & Bacon, A Pearson Education Company.

Baumbach, D. (2009). Web 2.0 and you. *Knowledge Quest*, *37*(4), 12–19.

Benson, R., & Brack, C. (2009). Developing the scholarship of teaching: What is the role of e-teaching and learning? *Teaching in Higher Education*, *14*(1), 71–80. doi:10.1080/13562510802602590

Brooks-Young, S. (2008, May). *Web tools: The second generation.* District Administration, Professional Media Group.

Bryan, A., & Vavrus, F. (2005). The promise and peril of education: The teaching of in/tolerance in an era of globalization. *Globalisation, Societies and Education*, *3*(2), 183–202. doi:10.1080/14767720500167033

Carr, J., Herman, N., & Harris, D. (2005). *Creating dynamic schools through mentoring, coaching, and collaboration.* Association for Supervision and Curriculum Development.

Caulfield, B. (2011). *Apple now selling more iPads than Macs, iOS eclipses Dell and HP's PC Businesses*. Retrieved March 17, 2013 from http://www.forbes.com/sites/briancaulfield/2011/07/19/apple-didnt-just-sell-more-ipads-than-macs-ios-has-now-eclipsed-dell-and-hps-pc-business-too/

Cohen, L. (2007, October). Information literacy in the age of social scholarship. *Library 2.0: An Academic's Perspective.*

Cooley, B. (2013). *CBS This Morning, Gadgets and gizmos galore*. Retrieved Jan. 8 from http://www.cbsnews.com/video/watch/?id=50138517n

Dockter, J., Haug, D., & Lewis, C. (2010, February). Redefining Rigor: Critical engagement, digital media, and the new English/Language Arts. *Journal of Adolescent & Adult Literacy*. doi:10.1598/JAAL.53.5.7

Eisenberg, M. (2008). The Parallel Information Universe. *Library Journal*, *133*(8).

Farkas, M. (2007). *Social software in libraries: Building collaboration, communication, and community online*. Information Today, Inc.

Fredrick, K. (2010). In the driver's seat: Learning and library 2.0 tools. *School Library Monthly*, *26*(6).

Gall, J., Gall, M., & Borg, W. (1999). *Applying educational research: A practical guide* (4th ed.). Addison Wesley Longman, Inc.

Greenhow, C. (2009). Social scholarship: Applying social networking technologies to research practices. *Knowledge Quest*, *37*(4), 42–47.

Jensen, M. (2007). The new metrics of scholarly authority. *The Chronicle*, *53*(41).

Nash, R. (2009). *The active classroom: Practical strategies for involving students in the learning process.* Corwin Press.

Nations, D. (2013). *How many iPads have been sold?* Retrieved March17, 2013 from http://ipad.about.com/od/iPad-FAQ/a/How-Many-iPads-Have-Been-Sold.htm

Nicol, D. (2013). *Mobile strategy: How your company can win by embracing mobile technologies*. IBM Press & Pearson Education, Inc.

O'Brien, D., & Scharber, C. (2010). Teaching old dogs new tricks: The luxury of digital abundance. *Journal of Adolescent & Adult Literacy*, *53*(7).

Orlich, D., Harder, R., Callahan, R., Trevisan, M., & Brown, A. (2007). *Teaching strategies: A guide to effective Instruction* (8th ed.). Houghton Mifflin Company.

Palloff, R., & Pratt, K. (2005). *Collaborating online: Learning together in community*. Jossey-Bass.

Quinn, C. (2012). *The mobile academy mlearning for higher education*. Jossey-Bass.

Reiser, R., & Dempsey, J. (2007). *Trends and issues in instructional design and technology* (2nd ed.). Pearson, Merrill Prentice Hall.

Shavelson, R. (1981). *Statistical reasoning for the behavioral sciences* (3rd ed.). Allyn and Bacon.

Valenza, J. K., & Johnson, D. (2009, October). Things that keep us up at night. *School Library Journal*.

Woodill, G. (2011). *The mobile learning edge: Tools and technologies for developing your teams*. The McGraw-Hill Companies.

KEY TERMS AND DEFINITIONS

Action Research: Allows testing for the purpose of improving local practices.

Applied Research: Designed to test interventions used to directly improve instruction.

Apps: A short version of the term application often associated with mobile apps used on smartphones and tablets.

Flipped Classroom: Videos lectures are used for homework and class time is used for active problem-based activities.

Instructional Design: Teaching effectively through the use of best practices.

Instructional Technology: Imparting knowledge or skills through the effective use of technology.

Mixed Methods: Quantitative and qualitative data is collected to produce better research results.

Mobile Apps: Represent applications used on mobile devices.

Multimodal: Provides users with multiple modes of interfacing with a system.

Needs Analysis: Used to define the learners needs based on the tasks, goals, and objectives.

Chapter 3
Mobile Learning in the Arab World:
Contemporary and Future Implications

Saleh Al-Shehri
King Khalid University, Saudi Arabia

ABSTRACT

Most Arab countries started their own e-learning and mobile learning initiatives in order to cope with global integration of latest educational technologies. The high mobile phone penetration among Arabs as well as availability of good mobile infrastructure are all important factors that can enhance the shift to mobile learning. Moreover, several studies indicate positive attitudes and perceptions toward mobile learning at different Arab learning institutions. However, specific challenges may act as barriers to mobile learning in the Arab world. This chapter reviews some of the current mobile learning practices in the Arab world and provides an overview of challenges faced by Arab students, educators, and probably researchers. A description of future mobile learning in the Arab countries is then provided.

INTRODUCTION

Education systems in the Arab world are adopting different ideologies that differ from, say, Arabian Gulf countries to Morocco and Algeria in the west. While some systems tend to be more liberal and democratic, others have more focus on the study of religion and traditional societal values. However, most Arab learning institutions have relatively similar agendas that aim at incorporating latest ICT and e-learning trends. Thus, several e-learning, and later mobile learning, initiatives have been established as an attempt to cope with educational systems at the developed world. Additionally, most Arab countries are experiencing economic difficulties and are attempting to provide better but cheaper education. Thus, mobile and distance learning were meant as solutions to income shortage and/or geographical conditions at these countries (see Almarwani 2011; Al-Shehri, 2012). In other parts of the Arab world, the investment of technology into learning came as a result to the very high penetration of mobile devices particularly among young population. Moreover, the availability of modern mobile communication services as well as affordable mobile

DOI: 10.4018/978-1-4666-6166-0.ch003

internet made the integration of mobile learning an essential educational step. In this chapter, an overview of current mobile learning practices in the Arab world is provided. Challenges faced, and are being faced, by students, educators, policy makers, and possibly researchers are reviewed and synthesized. An outlook of future mobile learning in the Arab world is discussed later in the chapter.

CURRENT PRACTICES

The integration of mobile technology into educational systems in the Arab world came as a result to the global shift that implemented more technological solutions in education. Ministries of education as well as universities in the Arab world are enthusiastically working on improving their academic processes and learning outcomes. Several e-learning initiatives and practices were found as a result to cope with contemporary education systems. For example, schools and universities have already provided technological infrastructure and invited all educators to take part in those promising initiatives. Some universities have also trained their staff to use technology effectively, and provided online courses for students in different mediums.

According to Sawsaa, Lu, and Meng (2012), mobile learning has not been widely adopted in Arabic countries. However, "several attempts have been made to identify and discover the importance of m-learning and its use for improving the educational services and developing the existing systems" (p. 172). One major factor that made mobile learning a suitable and effective choice in the Arab world is the widespread penetration of mobile devices, mobile phones in particular, among Arab young students (Al-Shehri, 2012). Saudi Arabia Consumer Electronic Report Q3 (2013), for instance, reports that mobile handset sales reached US $ 1.2bn in 2013 and expected to grow to US $ 1.6bn by 2017 due to the strong demand and popularity of internet-enabled mobile devices. Almutawwa (2012) also reports that mobile phone penetration in Saudi Arabia exceeded 200 percent in 2012. SHUAA Capital issued a report on the Gulf Countries telecom sector and states that United Arab Emirates has one of the highest mobile penetration rates in the world exceeding 230 percent (Mubasher, 2012). The mobile market in Kuwait, another Gulf country, experienced strong growth in mobile penetration to 175.9 percent in 2012 (Kuwait Telecommunications Report Q4, 2012). Mobile penetration at other lower-income Arab countries such as Egypt has also passed 100 percent and reached 100.79 percent in 2011 (Telecompaper, 2012) and 113.2 percent in 2013 (Eid, 2013). However, Eid reports a decline of mobile phone penetration in Egypt with 3.74 percent in 2013. This might be attributed to the current political events taking place in Egypt. Mobile penetration at other "unstable" Arab countries such as Iraq is still only 80 percent in 2013 (Al-Khalidi, 2013). Nevertheless, high mobile phone penetration in the Arab world can be attributed to the high improvement in smart phone sales in the region, particularly Apple and Samsung products (see Figure 1).

Mobile short messaging, mobile social media, and video sharing seem to be the most apparent mobile behaviors among Arab users. In Saudi Arabia, for instance, WhatsApp Messenger tops the list of mobile applications (see Table 1). As for devices, 55 percent of Arab population are Apple fans. However, an analysis of mobile social media usage, conducted by Alwagait and Shahzad, indicates that Apple is losing market share particularly at non-English speaking countries including the Arab world (SamMobile Report, 2013).

There are several attempts in Arab educational systems to cope with the mobile learning shift. The motivation for people to use mobile learning at the Arab world differ from one country to the other. For example, young Arabs at the Arab Gulf countries are so immersed in mobile social media and mobile video sharing than others at low-income countries. Moreover, mobile learning

Figure 1. Mobile devices penetration (producers) in the Arab world, based on a study by Ipsos (from, Bin Yahya, 2013)

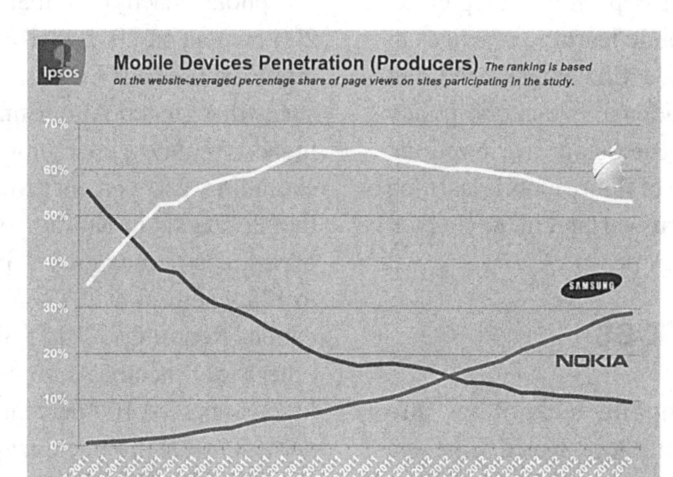

at other Arab States was embraced as a cheaper and efficient learning solution, whereas it is merely an option at other States that was adopted to cope with the global shift to mobile learning. Hence, both the application and implementation of mobile learning differ from one Arabic country to the other due to several factors discussed in the next section.

Saudi Arabian universities were among the first Arab universities to invest in educational technol-

Table 1. Most popular mobile application in Saudi Arabia (Ethos Interactive, 2012)

RANK	PAID	FREE	GROSSING
1	▲1WhatsApp Messenger	YouTube	Clash of Clans
2	▼1App3ad | آب-عاد	Podcasts	Badoo – Meet New People, Chat, Socialize
3	▲1Tweetbot for Twitter (iPhone & iPod touch)	Find My Friends	▲2 ممالك الأبطال
4	▲1 FIFA 13 by EA SPORTS	iTunes U	▼1 SayHi! – Find People Nearby!
5	▼2تفاصيل	Find My iPhone	▼1 Middle East – iGO primo app

ogy and adopt blended learning and e-learning solutions (Al-Shehri, 2012). Saudi universities established deanships of e-learning that cater for online and mobile courses as well as students and staff training. King Khalid University, for instance, started in 2011 to use mobile devices such as iPads and iPods at selected courses for learning and assessment purposes. Limitless Knowledge is also an innovative mobile learning project launched in 2013 by Ibn Khaldoon Schools and Riyadh Valley Company. The project was launched after a three-year cross-sectional study that investigated students' usage of electronic curriculum already available on iPads and Android tablets. The study found that over a million Saudi students accessed mobile learning curriculum although it was not a requirement by their institutions. Manager of the schools states that mobile curriculum will not be a requirement at this stage, and will not replace regular textbooks. However, mobile curriculum, he continues, has been found an effective learning medium and a powerful source of information and instructions for both students and parents (Sabq, 2013). Different e-learning and mobile learning practices have also been found at many Arab institutions and universities.

Communication technologies and technology infrastructure at other Arab countries are acceptable. Overall, technology affordances particularly for learning in the developing countries cannot be compared to the learning situations in the developed countries. For example, Alsanaa (2012) states that although an oil-rich country like Kuwait has a positive technology and communications infrastructure, public higher education in Kuwait did not adequately adopt effective e-learning or mobile learning initiatives.

There are several case studies on mobile learning conducted at different Arab contexts. For example, Al-Fahad (2009) investigated attitudes and perceptions of female arts and medicine students' toward the effectiveness of mobile learning at King Saud University, Saudi Arabia. The study indicates that students' perceptions and attitudes towards mobile learning were mostly positive, and that the future of mobile learning in Saudi Arabia is brilliant. Al-Fahad also asserts that university students are making the most of their mobile phones and mobile social media. Another study by Seliaman and Al-Turki (2012) also explored the potential of mobile phones and tablets for learning using an extended Technology Acceptance Model. Their study also attempted to explore how university students use their mobile phones to access course materials, search for information related to their discipline, share knowledge, and conduct assignments. Seliaman and Al-Turki concluded that "perceived innovativeness does not show high positive correlation with perceived usefulness of m-learning" (p. 393). This indicates that although students can be professional users of mobile technology, there seems to be lack of awareness among students of the potential of mobile technology for learning. Chanchary and Islam (2011) conducted another study at Najran University to explore students' perceptions as well as possible challenges towards mobile learning. Their study concluded that the available mobile network conditions in Saudi Arabia support the implementation of mobile learning. In addition, students' attitudes towards mobile learning were encouraging. Another study by Nassuora (2013) adopted a Unified Theory of Acceptance and Use of Technology (UTAUT) model to determine what could be the factors that can affect the use of mobile learning among students of Al-Faisal Private University. Nassuora indicates that students had good perceptions and acceptance of mobile learning although they were not familiar with the concept of mobile learning. In other words, students' familiarity with mobile technology can effectively enhance successful mobile learning practices. Al-Shehri (2012) carried out a PhD project at King Khalid University to explore how authentic and contextual learning can be maintained through the use of mobile phones and social media, on the one hand, and how they can provide English language learners with more

Figure 2. Three layers of M-learning system of higher education provider based network architecture (Sarrab & Elgamel, 2013, p. 1414)

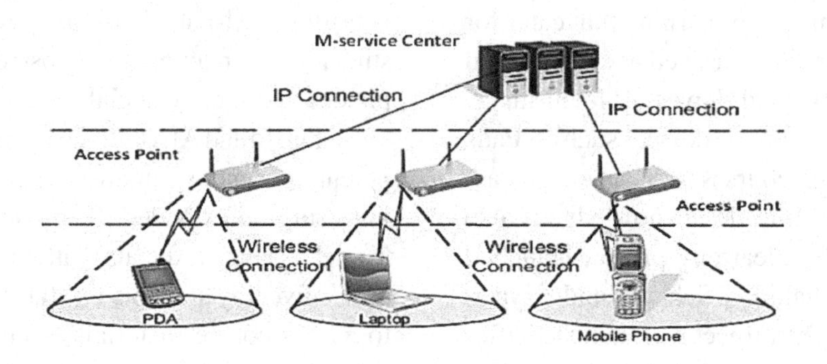

opportunities to practice the language outside the classroom. His study found that mobile social media helped in contextualizing the learning task, and helped students to stay engaged after-class. However, students listed some technological limitations of mobile learning including the lack of good network coverage especially in remote areas. As a result, the researcher reports that there was a tendency among students to use their computers in preference to their mobile phones. The interference between learning and non-learning materials on social media, i.e. mobile Facebook, was also a limitation for some students.

At Sultan Qaboos University, Oman, Al Aamri (2011) also investigated students' practices, attitudes, and challenges with mobile learning and concluded that the use of mobile phones in the classroom is still limited. Al Aamri suggests that Omani students need to reconsider the importance of mobile learning provided that both teachers and students need to "gain confidence in using it and allowing it in the classroom respectively" (p. 151). Another study from Oman by Sarrab and Elgamel (2013) proposed three layers of mobile learning system based on network architecture and provides a framework of contextual mobile learning system for the Omani context. Layers in the proposed framework include user's mobile device, Wi-Fi access point and m-service center (Figure 2). Researchers argue that an intelligent agent or personal assistant can be installed on the student's mobile device, whereas Wi-Fi access points should be deployed around the campus to facilitate students' access to the mobile learning service. Finally, m-service center can act as a web service provider.

Blackboard® Mobile Learn was launched at King Khalid University in 2011 so that students started to access learning materials via their mobile devices as well as iPads provided by the university. Similarly, University of Abu Dhabi, at the United Arab Emirates, signed an understanding contract in 2010 with Blackboard Inc. to provide innovative learning medium for students through their mobile phones. Officials of the university considered this as a mobile learning initiative at the UAE (Economics, 2010). An initial iPad pedagogy has also been developed in 2012 at the UAE by the federal higher education system as a national adoption of new educational computing platform. According to Cavanaugh, Hargis, Munns, and Kamali (2012), the goal of such an adoption was to improve learning and to support national development goals of the National Higher Education. The goals were to:

Figure 3. Student's perspective of the future of mobile learning (Mohammad & Anil Job, 2013, p. 17)

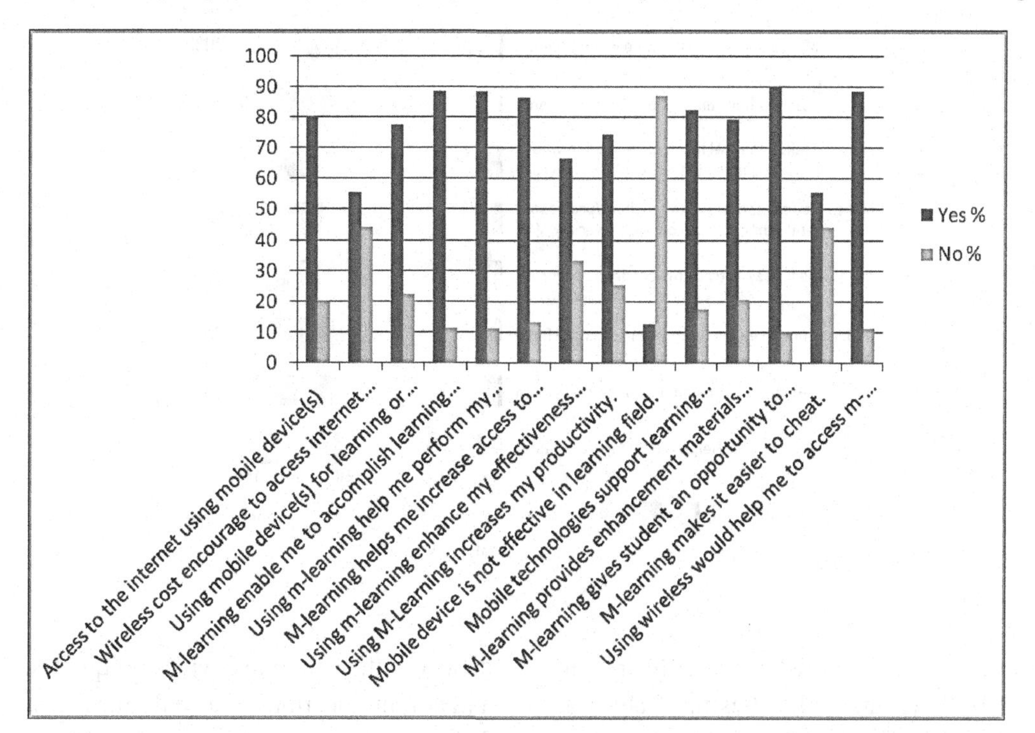

- "Achieve individualized student learning consistent with "Post PC Era" trends;
- Introduce challenge-based learning or other progressive classroom pedagogy;
- Increase student participation and motivation;
- Enhance opportunities for cross-institutional collaboration between faculty members;
- Increase faculty collaboration through cross-institutional repositories of learning objects; and
- Facilitate the migration to e-books." (p. 1).

The UAE iPad pedagogy was inspired by the Apple stages of technology adoption: Entry, Adoption, Adaptation, Infusion, and Transformation. After the adoption process, the majority of faculty responses were positive, and that the preparedness to engage with the iPad initiative was high, despite the fact that the timeline between the decision to adopt and testing the results was short (Cavanaugh et al., 2012).

A study was conducted at Arab Open University, Bahrain branch, to measure the acceptance of mobile learning as a blended learning tool. Mohammad and Anil Job (2013) surveyed students to explore their perceptions of future mobile learning as well as their routine usage of mobile phones. The study indicates that the vast majority of students used their mobile phones for messaging, conversations, and to play games. About half of the surveyed students indicated that they used their mobile phones to serve the web. About 37 percent of students stated that they did not use their mobile phones for educational purposes. However, majority of students anticipated that the future of mobile learning in the region is promising (see Figure 3). The study concludes that students agreed on the role that mobile learning played in improving access to learning materials and resources, and helping them to learn in smarter and easier ways.

Figure 4. Tasks made easier with mobile devices (Dahlstrom & Warraich, 2012, p. 5)

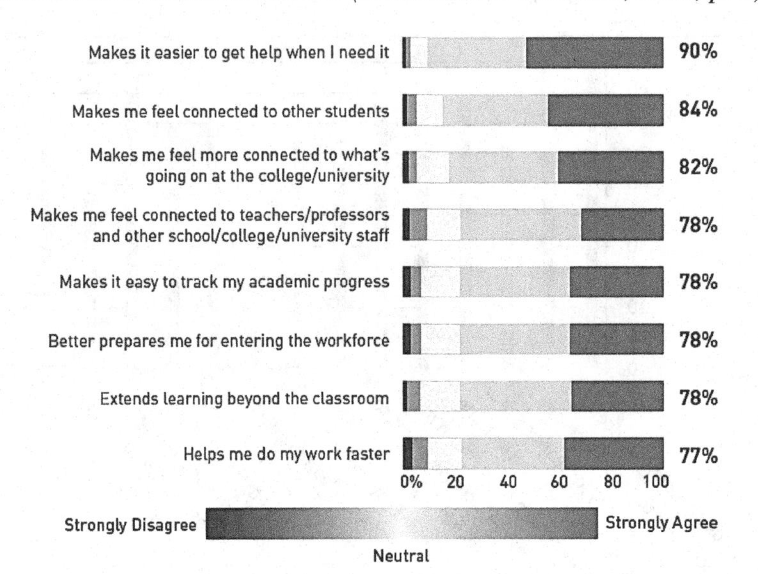

Qatar has also established a well-known ICT infrastructure and authorities that cater for e-learning and mobile learning initiatives in the country. ICT Qatar was established in 2004 in order to facilitate the integration of technology in all sectors, particularly in education. Mobile learning was also a focus by ICT Qatar and educational institutions. For example, Sidra Medical and Research Center, as cited by Weber (2010), is investigating the potential of mobile phones to deliver e-health information in an attempt to support clinical trials, and to implement e-health through mobile phones as almost everybody in Qatar owns one. According to Agonia (2012), a study was conducted by Texas A&M Qatar and US-based non-profit association EDUCAUSE with 369 students from seven Qatari universities. The study found that there is a need for more investment in mobile learning since "mobile computing has created new teaching methods that allow educators to convey content differently, create more student engagement and introduce more efficient campus operations" (p. 23). However, the study revealed that the majority of students still consider laptops as significant devices for learning particularly for sophisticated learning process. In contrast, students use their mobile devices for quick searches and communications, the study further revealed. This is in line with Al-Shehri (2012) who also found that students sometimes prefer PC or laptop access particularly for fulfilling more complicated processes that mobile devices cannot do. This includes downloading/uploading large files from/to the internet, photo and video editing, and so on. In Figure 4, Dahlstrom and Warraich (2012) summarize attitudes and conceptions of Qatari students toward mobile learning, and how mobile technology can facilitate learning. This summary is also based on the study of Texas A&M Qatar and EDUCAUSE.

Funded by the Qatar National Research Fund (QNRF), Ally, Samaka, Impagliazzo, and Abu-Dayya (2012) conducted a research project investigating the use of innovative mobile technology, and how it can facilitate Qataris training to be prepared for the 21st century workforce. The project specifically aims at improving trainees' communicative skills in English in order to maintain effective communication in the workforce. The study revealed that trainees' performance has improved by 16 percent after the mobile training session. It also concluded that all trainees indicated that they

Figure 5. E-learning main trends in Egypt within 2-10 years (El-Gamal, 2012, p. 3)

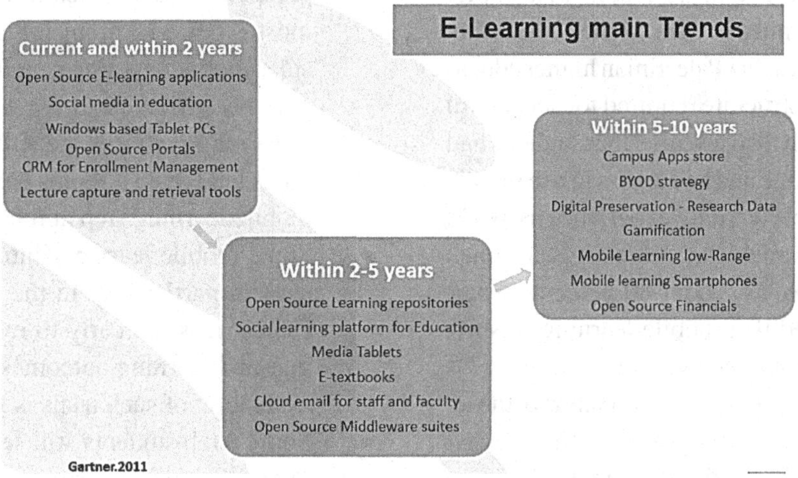

benefited from the session, and that they liked the innovative nature of the mobile training course.

Kuwaiti students' acceptance of incorporating innovative technology including mobile technology in higher education was examined by Alsanaa (2012). In her study, Alsanaa explored the attributes and influences of emerging technologies at the social, economic, political, and above all, educational levels. The study drew upon Technology Acceptance Model as a conceptual framework. It was revealed that the majority of students owned mobile and/or mobile devices, whereas about 40 percent of students did not. The researcher attributed this to monetary issues. The study also revealed that most students were frequent users of the social media. In addition, about half of the students indicated agreement for the use of educational mobile applications and virtual learning environments such as Moodle and Blackboard. Overall, the study indicates positive willingness among Kuwaiti students to use e-learning and mobile learning.

Al-Zoubi, Jeschke, and Pfeiffer (2010) examined the development of mobile learning into engineering education in Jordan. Researchers developed a five-phase mobile learning design comprising of analysis, design, development, implementation, and evaluation. First, analysis of needs, target audience, and technology issues was carried out. Then, the task was designed so that complete storyboard screens were developed and then implemented. Screens had both mobile and web versions. The learning task was then conducted and evaluated. Researchers conclude that "mobile devices should only be used for short-period courses, and one should start with simple knowledge courses that everyone needs to know like English language and computing basics", and that "a campaign is needed to promote mobile learning" (p. 6). In other words, the study suggests that mobile learning is still facing some challenges in the Arab world, and that both students and educators must be aware of the potential of mobile learning. Again, the study revealed that cost and speed of mobile internet is still a concern in Jordan.

In the Palestinian context, Alzaza (2012) investigated the awareness of mobile learning among

Palestinian university students. The study also attempted to explore students' readiness to implement mobile learning at their higher education. The study indicates that Palestinian higher education had the infrastructure required for successful mobile learning. It also revealed that students had adequate knowledge and awareness to use mobile technology in their learning experience as Wi-Fi networks were available across most Palestinian universities. However, about 80 percent of participants indicated that mobile learning was not provided by their universities yet.

In Egypt, El-Gamal (2012) analyzed the e-learning and mobile learning trends within the next ten years. El-Gamal anticipated that e-learning practices are the current trends that could last for two-five years. Practices include open source learning applications, social media, e-textbooks, and so on. Mobile learning is not expected to be fully adopted in Egypt before five years (see Figure 5).

El-Gamal (2012) asserts that the educational system in Egypt needs a faster upgrade and reform. Curricula, teaching methods, qualifications frameworks, training, and quality assurance are all important factors that should be modernized in order for better adoption of mobile learning, he further emphasizes.

CURRENT CHALLENGES

Previous case studies and literature review on the current implementation of mobile learning in the Arab world indicate that:

- The majority of studies investigated students and/or educators perceptions and readiness to implement mobile learning in their learning experience and profession. Few experimental studies looked at the actual practice of mobile learning, e.g., Al-Shehri (2012) and Al-Zoubi et al., (2010),
- There is no adequate evidence of the actual practice of mobile learning provided that most Arab e-learning initiatives still consider mobile learning as a future step. This may be attributed to the fact that current e-learning practices are still in their beginnings, and thus, it seems hard to shift to the mobile learning step at this stage,
- Some mobile learning initiatives have just started, particularly in the Gulf countries. Hence, it is too early to expect their findings and learning outcomes, and an overall evaluation of such trials is yet to be done,
- Some Arab students still see mobile learning as a luxurious or unnecessary learning option although; (a) they can be considered as professional users of mobile technology, and (b) the penetration of mobile phones among Arab students is high in most countries. Yet, the awareness of the potential of mobile learning in the Arab world needs to be improved,
- Arab educators and policy makers are still facing some barriers. Barriers or difficulties that affected and still affecting the integration of mobile learning in the Arab world education systems can be classified into technological, social-political, personal, and economical barriers. Below is a discussion of these barriers.

According to Al-Shehri (2012), students still find some technological limitations that may lead to less-successful incorporation of mobile learning. For example, Al-Shehri indicates that some students are still in favor of PC-browsing rather than the use of mobile phones. Students also raised concerns about slow upload and download times, and small screen size. Alsanaa (2012) also found that students seem to find VLEs relatively hard to use. Thus, it can be argued that mobile-based VLEs might be much more harder in such contexts. From another technological perspective, Almarwani (2011) states that large national ICT projects in the Arab world can be challenges particularly at large countries such as Saudi Arabia. Geographical distance, she asserts, makes it hard for local education systems to adopt innovative mobile learning projects. Additionally, mobile learning in such contexts might be a short-term solution due the current shortage of qualified instructors and good infrastructure, she further notes.

The social factor is another important challenge to mobile learning in the Arab world. Smala and Al-Shehri (2013) assert that Arab societies, particularly in the Gulf region, can highly be considered as conservative; specific societal values and yet fully established freedom of speech rights might be a concern for Arab young students "when education leaves the classroom and is accessed on mobile devices (p. 317). Additionally, governments and authorities are still being considered by Arab people; restrictions are being applied on mobile and social media communications. New technologies are inspected by religious and cultural norms in such societies (Almarwani, 2011). Thus, Arab young students still need to adopt specific mobile behavior that does not contradict with local legislations. Criticism of governments or religious issues is not permitted at most Arab countries. In Saudi Arabia, for instance, local authorities banned some mobile-based applications like Viber, with recent rumors that users of the social media will need to connect their accounts with their national IDs, so that the government would have more control on the freedom of speech. Such legislations would have a vague effect on future mobile learning. Moreover, people attempting to approach other people from the opposite sex need to be cautious, as most Arab traditions would not allow for a male-female interaction outside marriage, or possibly work. Such social-political factors may not allow for more democratic mobile learning practices, and therefore, ultra-formal student-student and teacher-student mobile interaction is anticipated. However, it is important to note that restricted freedom of speech and conservative legislations do not exist at all Arab countries. Lebanon could be the only Arab country that has a reasonable history and good current practice of democracy, particularly in education.

Resistance to use mobile learning among Arab students and teachers can be another challenge. Although most of the previously reviewed studies indicate positive perceptions towards mobile learning (see for example, Al-Fahad, 2009; Alsanaa, 2012; Cavanaugh et al., 2012; Mohammad & Anil Job, 2013), both students and educators still need to be aware of the potential of mobile learning. On personal level, not all students are keen to use their private space i.e. mobile phones, for educational purposes (Stockwell, 2008). Thus, the distinction between the personal space versus the learning space in mobile technology should be made clear. Teachers may still need some training and awareness to embrace the mobile learning shift. Furthermore, few Arab institutions have integrated mobile learning initiatives, whereas it seems teachers' self-initiatives at other institutions. Self-initiative mobile learning practice is just a matter of "leaving the policy making for classroom use in the hands of individual faculty members who may be unaware of the potential benefits and have many of the same fears as K-12 teachers and administrators" (Pollara, 2011, p. 5). Hence, an acute awareness of the potential of mobile learning to improve current learning practices in the Arab world needs to be established. Furthermore, current e-learning initiatives should dedicate more attention to mobile learning and conduct more training for both students and teachers. This is important since approximately 40 percent of teachers in a country like Qatar did not receive any ICT training (Weber, 2010). The situation for other Arab countries could be worse. Mobile learning opportunities should also be provided so that Arab institutions should not wait until students are somehow ready for this shift.

The studies of Al-Shehri (2012) and Al Aamri (2011) indicate that the cost factor is still a concern for Arab students in terms of mobile phones internet. Yet, fees of mobile phones internet are dropping these days due to the high competition between Arab mobile phone companies. Therefore, cost-efficient mobile phone internet will be at the reach of most Arab students very soon. On larger scale however, it seems that some Arab universities still facing financial hardship to adopt innovative e-learning and mobile learning. Modern mobile devices, for instance, cannot be afforded by all institutions particularly those at lower-income countries. Thus, a Bring Your Own Device (BYOD) approach would be the best option for such institutions in the near future (El-Gamal, 2012). Students and teachers will need to use their own mobile devices to cope with the global mobile learning trend.

Other significant challenges, as Weber (2010) emphasizes, "lie ahead in the region such as course accreditation, the creation of Arabic language learning objects, the equitable distribution of technology goods and services, and the tempering of idealistic expectations of what e-learning can actually accomplish" (n. p.). Again, as long as expectations about e-learning are not obvious enough, more efforts need to be exerted with respect to mobile learning in the Arab world. Overall, global revolution in mobile technology and social media should gain attention by Arab educators and learning institutions to stay in the loop, and to cope with contemporary mentality of Arab youths.

FUTURE OF MOBILE LEARNING

One of the major challenges of mobile learning in the Arab world is that most Arab educators, policy makers, and probably students still consider it as the *future* of e-learning. Generally, Arab proponents of mobile learning seem to ignore that mobile learning is not the future as it was supposed to be, but rather, it became a reality and a realistic learning practice in most parts of the world. However, it is anticipated that mobile learning would witness huge progress in the Arab region due to several factors:

- Arab countries, particularly in the Gulf region, started ambitious learning initiatives to reform their educational systems. One aspect of this reform is by sending thousands of their citizens to selected international institutions to pursue their higher education and professional training in all fields including mobile learning,
- Almost all studies reviewed so far imply positive perceptions and expectations of future mobile learning in the Arab words as well as good behavioral intention to use mobile learning once it is available as a learning option,
- Factors like extrinsic influence and university commitment, as Mohammad and Anil Job (2013) affirm, play a crucial role in making students realize that mobile learning as *current* rather than *future* learning practice. Thus, Arab institutions will need to adopt this strategy in an effective way,
- Social media has played and is still playing an undeniable role in the recent Arab Spring revolutions. Arab young population, the motor for demonstrations and revolutions, found in the social media and mobile technology effective channels to organize their activities and to report them to the world. They also found them successful channels that could challenge governments' restrictions on freedom of speech. Hence, successful mobile learning initiatives would highly consider the impact of social media, and how it became an integral part of almost all Arab students,
- Due to the high demand on mobile learning in the near future, Arab education systems would establish research centers and subsidize intensive research activities dedicated to mobile learning. It is worth noting that countries like Saudi Arabia and Qatar

have already addressed this issue and developed centers for e-learning and distance learning (Al-Shehri, 2012; Weber, 2010). Particularly, Saudi Electronic University and National Center for E-Learning and Distance Learning are recent institutions in Saudi Arabia that were meant to cater for e-learning and mobile learning initiatives in the country.

- Access of information is a significant factor that Arab mobile learning educators and policy makers need to consider. According to Rosenberg (2000), "knowledge management supports the creation, archiving, and sharing of valued information, expertise, and insight within and across communities of people and organizations with similar interests and needs" (p. 66). This highlights the facilitation of knowledge and learning as well as enhancement of both students and teachers performance using mobile technology for academic purposes.

Overall, the future of mobile learning in the Arab world seems brilliant given that there is good infrastructure at most Arab countries for mobile learning. In addition, both educators and policy makers already had some awareness of the potential of mobile learning. Last but not least, Arab students are making the most of their mobile devices and mobile social media. Consequently, a promising near future of mobile learning is coming ahead.

CONCLUSION AND FURTHER RESEARCH

Mobile leaners around the world are taking the same train leading to knowledge. The difference is that some are enjoying the first class cabin with all functions and affordances available. Others are taking the business class cabin having some, but not all, functions offered for first class passengers. Arab mobile learners are relatively close in their situation to the economy class passengers; they are taking the same train, yet, with limited awareness of available facilities. Analysis of the current situation of mobile learning in the Arab world indicates that Arabs are still planning for the future. Studies reviewed so far (see for example, Al-Fahad, 2009; Alsanaa, 2012; Chanchary & Islam, 2011) explored perceptions and readiness among students and/or educators to implement mobile learning. It is important to note that in order to get the most of mobile learning, actual practices, rather than cautions investigations, are needed. In other words, mobile learning in the Arab world needs to embrace practical rather than theoretical implications. From practical point of view, researchers could determine which theoretical framework would work and which ones would not. This is true since we are in a rapidly changing era that does not wait for students and/or teachers to be somewhat ready for mobile learning.

Besides conducting practical experiments of mobile learning in the Arab world, intensive investigations of Arab students' behavior are needed. We need to know what are the different usage patterns of mobile devices, mobile phones in particular, and mobile social media among Arab young users and how they affect their daily life. The correlation between mobile phones as personal devices and as devices that had a tremendous effect on recent Arab revolutions needs further research. In addition, the world needs to know how mobile technology as well as mobile social media already made the change, and how they can revolutionize learning practices in the Arab world.

REFERENCES

Agonia, A. (2012). *Qatar students view mobile tech as an aid to education* (p. 17). Qatar Tribune.

Al Aamri, K. S. (2011). The use of mobile phones in learning English language by Sultan Qaboos University students: Practices, attitudes and challenges. *Canadian Journal on Scientific & Industrial Research*, *2*(3), 143–152.

Al-Fahad, F. N. (2009). Students' attitudes and perceptions towards the effectiveness of mobile learning in King Saud University, Saudi Arabia. *The Turkish Online Journal of Educational Technology*, *8*(2), 111–119.

Al-Khalidi, S. (2013). *Interview: Zain Iraq sees double-digit growth this year*. Retrieved from http://uk.reuters.com/article/2013/06/06/iraq-zain-telecoms-idUKL5N0EH11Q20130606

Al-Shehri, S. (2012). *Contextual language learning: The educational potential of mobile technologies and social media*. (Unpublished doctoral dissertation). The University of Queensland, Brisbane, Australia.

Al-Zoubi, A. Y., Jeschke, S., & Pfeiffer, O. (2010). Mobile learning in engineering education: The Jordan example. In *Proceedings of the International Conference on E-Learning in the Workplace 2010* (pp. 8-15). New York: Academic Press.

Ally, M., Samaka, M., Impagliazzo, J., & Abu-Dayya, A. (2012). Use of emerging mobile computer technology to train the Qatar workforce. *Qatar Foundation Annual Research Forum Proceedings: 2012, CSP6*. Retrieved from http://www.qscience.com/doi/abs/10.5339/qfarf.2012.CSP6

Almarwani, A. (2011). ML for EFL: Rationale for mobile learning. In *Proceeding of the International Conference ICT Language learning* (4th ed.). Florence, Italy: Academic Press.

Almutawwa, K. (2012). *An international expert: Mobile phone penetration in Saudi Arabia reaches 200 percent*. Retrieved from http://www.alsharq.net.sa/2012/06/05/324934

Alsanaa, B. (2012). Students' acceptance of incorporating emerging communication technologies in higher education in Kuwait. *World Academy of Science. Engineering and Technology*, *64*, 1412–1419.

Alzaza, N. S. (2012). Opportunities for utilizing mobile learning services in the Palestinian higher education. *International Arab Journal of e-Technology*, *2*(4), 216-222.

Bin Yahya, E. (2013). *Ipsos's study about the internet penetration and usage in the Arab world*. Retrieved from http://www.tech-wd.com/wd/2013/03/15/ipsos-report-2012

Cavanaugh, C., Hargis, J., Munns, S., & Kamali, T. (2012). iCelebrate teaching and learning: Sharing the iPad experience. *Journal of Teaching and Learning with Technology*, *1*(2), 1–12.

Chanchary, F. H., & Islam, S. (2011). *Mobile learning in Saudi Arabia: Prospects and challenges*. Retrieved from http://www.nauss.edu.sa/acit/PDFs/f2535.pdf

Dahlstrom, E., & Warraich, K. (2012). Student mobile computing practices, 2012: Lessons learned from Qatar. *A report by EDUCAUSE*. Retrieved from http://www.educause.edu/library/resources/student-mobile-computing-practices-2012-lessons-learned-qatar

Economics. (2010). *Itisalat supports learning via mobile phones*. Retrieved from http://www.albayan.ae/economy/1277246021486-2010-10-22-1.295970

Eid, N. (2013). *Egypt's ICT ministry: 3.7% decline in mobile penetration march-end*. Retrieved from http://www.amwalalghad.com/en/investment-news/technology-news/19204-egypts-ict-ministry-37-decline-in-mobile-penetration-march-end.html

El-Gamal, H. R. (2012). The power of e-learning for Egypt: A spot light on e-learning. *A report by GNSE Group*. Retrieved from http://www.gnsegroup.com/news/The_Power_of_elearning_for_Egypt_ver01.pdf

Ethos Interactive. (2012). *Most popular mobile application in Saudi Arabia*. Retrieved from http://blog.ethosinteract.com/2012/09/27/most-popular-mobile-applications-in-saudi-arabia/

Kuwait Telecommunications Report Q4. (2012). Retrieved from http://www.companiesandmarkets.com/Market/Telecommunications/Market-Research/Kuwait-Telecommunications-Report-Q4-2012/RPT1108596

Mohammad, S., & Anil Job, M. (2013). Adaption of M-Learning as a Tool in Blended Learning - A Case Study in AOU Bahrain. *International Journal of Science and Technology*, *3*(1), 14–20.

Mubasher. (2012). *UAE ranks among highest mobile penetration countries worldwide – SHUAA*. Retrieved from http://english.mubasher.info/DFM/news/2189413/UAE-ranks-among-highest-mobile-penetration-countries-worldwide-SHUAA

Nassuora, A. B. (2013). Students acceptance of mobile learning for higher education in Saudi Arabia. *International Journal of Learning Management Systems*, *1*(1), 1–9. doi:10.12785/ijlms/010101

Pollara, P. (2011). *Mobile learning in higher education: A glimpse and a comparison of student and faculty readiness, attitudes and perceptions*. (Unpublished doctoral dissertation). Louisiana State University, Baton Rouge, LA.

Rosenberg, M. J. (2000). *E-Learning: Strategies for delivering knowledge in the digital age*. McGraw-Hill.

Sabq. (2013). *Launching the iPad project (Limitless Knowledge) at Saudi Schools*. Retrieved from http://www.sabq.org/AcCfde

SamMobile. (2013). *Apple losing market share among non-English speakers (Apple vs. Samsung)*. Retrieved November 15, 2013, from http://www.sammobile.com/2013/02/07/apple-losing-market-share-among-non-english-speakers-apple-vs-samsung/

Sarrab, M., & Elgamel, L. (2013). Contextual m-learning system for higher education providers in Oman. *World Applied Sciences Journal*, *22*(10), 1412–1419.

Saudi Arabia Consumer Electronic Report Q3. (2013). Retrieved from http://www.market-research.com/Business-Monitor-International-v304/Saudi-Arabia-Consumer-Electronics-Q3-7691595/

Sawsaa, A., Lu, J., & Meng, Z. (2012). Using an application of mobile and wireless technology in Arabic learning system. In Z. J. Lu (Ed.), *Learning with Mobile Technologies, Handheld Devices, and Smart Phones* (pp. 171–186). Hershey, PA: IGI Global. doi:10.4018/978-1-4666-0936-5.ch011

Seliaman, M. E., & Al-Turki, M. S. (2012). Mobile learning adoption in Saudi Arabia. *World Academy of Science, Engineering, and Technology*, *69*, 391–393.

Smala, S., & Al-Shehri, S. (2013). Privacy and identity management in social media: Driving factors for identity hiding. In J. Keengwe (Ed.), *Research Perspectives and Best Practices in Educational Technology Integration* (pp. 304–320). Hershey, PA: IGI Global.

Stockwell, G. (2008). Investigating learner preparedness for and usage patterns of mobile learning. *ReCALL*, *20*(3), 253–270. doi:10.1017/S0958344008000232

Telecompaper. (2012). *Egypt passes 100% mobile penetration*. Retrieved from http://www.telecompaper.com/news/egypt-passes-100-mobile-penetration--853147

Weber, A. S. (2010). *Web-based learning in Qatar and the GCC states*. Georgetown University. Retrieved from http://cirs.georgetown.edu/publications/papers/120276.html

KEY TERMS AND DEFINITIONS

Arab Gulf: Or Gulf Cooperation Council (GCC), is a term refers to Arab Gulf states consisting of Bahrain, Kuwait, Oman, Qatar, Saudi Arabia, and United Arab Emirates.

Arab Spring: A term refers to the revolutions and demonstrations that began in 2010 at some Arab countries as a dissatisfaction with the rule of local governments.

Arab World: A combination of the Arabic-speaking 22 countries stretching from the Atlantic Ocean to the Arabian Sea, and from the Mediterranean Sea to the Horn of Africa and the Indian Ocean.

Conservatism: A trend that promotes retaining traditional social values. Tribal and religious values are primarily followed by conservatives Arabs.

Education Reform: The demand to improve educational and learning practices and outcomes.

Knowledge Management: The process of capturing, developing, sharing, and using knowledge inspired by online technology and social media.

Mobile Phone Penetration: The number of people who buy a mobile phone in a given period, divided by the size of the market population.

Mobile Social Media: Human-human interaction and exchange of information fulfilled via mobile technology and mobile devices.

Technology Acceptance Model: An IS (information systems) theoretical framework that describes how users come to accept/reject and use technology.

Section 2
Mobile Identity and Ritual

Much easier, faster, and mightier than both offline and online, mobile identity can be developed and mobilized, temporary or permanent, public or private, and individual or institutional. A mobile identity can be used in different contexts for different purposes. Potentially, however, it can have different risks and challenges, such as security, surveillance, privacy issues and their corresponding consequences, as examined in Chapter 4. Closely related to mobile identity are the issues of gender and rituals in a mobile environment. Investigating how gender rituals and use of mobile technologies are mutually shaped, Chapter 5 explores the implications of mobile technologies on gender from the perspectives of gender rituals. Related to gender in mobile communication is how to deal with intimate conflicts between males and females via mobile. As a case study of how female adults handle their intimate conflicts, Chapter 6 offers insightful observations on the wider social consequences of mobile technologies on female adults since mobile communication plays a major role in their daily affairs.

Chapter 4
Digital Mobilisation and Identity after Smart Turn

Katalin Fehér
Taylor's University, Malaysia

ABSTRACT

Users have digital and digitally mobilised footprints. The online data sets are defined and identified within their different networks. These result in data sets and interactive markers via personal media such as digital/mobile/smart devices. The compass-devices support to manage everyday life and to define digitally mobilised identities in online networking society, culture, and business. This chapter studies the digitally mobilised identitiy in a multitasking context, digital trends that shape personal/organizational mobile identity, and emerging technologies, services, and cases that support these changes and challenges.

DIGITAL MOBILISATION AND IDENTITY

Users have offline and online identities. They have gender, nationality, generation, and they are living in a culture. They have a qualification, job/profession, and hobbies and they have family members, friends and enemies. They are under social control because of all these parameters and due to other identity factors. They live under a wider social control via platforms of online publicity. They are also consumers of digital devices and digital contents. Consumer-driven online identity produces user-generated content and interactive communication design. Users are living through online networks: networks of devices and platforms, links, data and contents, networks with other digital identities. Offline life is represented online via digital screens. Users' self-expressions are not identical with that of the offline way. However, they are akin to it. Feelings, stories, projects could be the same but the online environment and linked digital networks – that is, the online framework – provide alternative routes. The choices and tools are determined by the media. Interoperability and a strict separation among closed systems, semi-public scenes and publicity deploy mass media and the rites of interpersonal communication. However, mobile devices provide an opportunity for communication

DOI: 10.4018/978-1-4666-6166-0.ch004

real time and on the move in these networks and layers of publicity.

Users have digital and digitally mobilised footprints and these online data sets are defined via personal or social /community or corporate/ organizational identities within their different networks. The result consists of so-called data sets which generate digital identification in online systems and networks, and can respond to one another using digital tools and platforms to create or traffic contents. Users define their identities and their social networks in/on/via/for new media and surveillance technology networking between the two poles of "me"/consciousness/selfbrand on the one hand, and, on the other hand, that of other users, namely, other personae, social communities, instant communities (Castells et al., 2007, pp. 244-249), governments or companies. Actually the media are "us", "you", "me" and "them" together trapped within fragmented and temporal networks. Our online and interactive markers and their lack depict our appearance, our self-representation operated and processed by communities and by official or other sites.

Users need devices for each of these activities. The mobile devices, mainly smart phones, are personal media (Aaltonen, Huuskonen & Lehikoinen, 2005), however, provide a compass for users. A *compass* we need so as to manage everyday life: for example, to navigate in space, to find people or to select and order a service. Smart devices are compasses to reach different programs, applications and software for every sphere of life. Users furnish them with applications and services, and label them with covers and other material markers. The mobile device belongs to Me 2.0 (Schawbel, 2009), who is living online to come out on the net and represent "offline Me 1.0" for any bet, or, otherwise, Me 1.0 gets lost for online data traffic.

Personal digital identities define digital corporate/organisation communication nodes and vice versa. Online communication trends and tools are repositioning personal identities related to corporate identity via algorithm-based networks and platforms. These identities are not centralised but connected: the focus is the network (Håkansson, 2002). These networks exist without centres. Their substance is constituted by points and lines where users and data/contents/platforms/programs/devices/virtual harbours are the points and the strong or weak connectivity and data traffic are the lines. Digital identities reach some mainly constant networks like personal relationships offline and online and they have a possibility to join temporary and/or local networks real or virtual. They can share all these via some networks online. How can users organise the points and lines after mobilisation? The device and the platforms and other tools provide access to the networks. Perception becomes mobile-virtual where "we are the message" (McConnell & Huba, 2006) in our networks and in our wireless connections. The "wireless communication networks are diffusing around the world faster than any other communication technology to date" (Castells et al., 2007, p. 1) and the intensive dynamic they put users through compels users to change their communication tools and habits.

What digital trends shape this personal/ organisational mobile identity? What emerging technologies, services, cases support these changes and challenges? Our brief overview is looking for answers via depicting global trends and via providing instructive insights.

THE WORLD OF BIG DATA AND DIGITAL TECHNOLOGY

Mobile and smart devices primarily imply a sense and a possession: users can identify with it or they may also refuse to. The furnished mobile has a design, a mobile phone allows to customize the visual and audio signals and smart phones extend these with applications/services, photo, video and music repertory and with more digital contents. This mobile/smart device is only the base, a starting point, a port for users. It is a tower for broadcasting or selfcasting, to share, to project or to code/decode. Users are moored to a mobile device via real time public location or via the lack of it. They mediate their similar or different messages and are waiting for other people's messages in every six minutes: users check their mobiles at an average of 150 times a day, where given a waking day with 16 hours it means about 6 minutes' checking slots. Voice calls, short messages, e-mail and internet use, GPS navigations, alarm, playing games, changing songs, taking pictures and other functions belong to this activity (Nokia global study 2013, http://discussions.nokia.com). Boundaries of dependence are shifting, always to be on/line could be a requirement.

Users are leaving digital footprints, they are observable via this conduct. Everybody sees/watches same-but-different screens and sets of contents and they are looking for their identification in these service systems. Digital tableaux reflect and constitute digital identity/ies. Mainstream tools and services provide common platforms and patterns. Online information design and content networks' interiorization have been increasing. „This form of customisation not only operated as a constant reminder of the intimate, but also, being an object outside the often publicly visible phone, it signalled a type of public intimacy trenchantly addressed" (Berlant, 2000).

"Every day, we create 2.5 quintillion bytes of data — so much that 90% of the data in the world today has been created during the last two years alone. This data comes from everywhere: sensors used to gather climate information, posts to social media sites, digital pictures and videos, purchase transaction records, and cell phone GPS signals to name a few. This data is big data." ("What is Big Data", http://www-01.ibm.com). Global mobile data traffic grew 70 percent and mobile network connection speeds doubled in 2012. Mobile data traffic generates a new perspective in mobilisation: mobile-connected devices are exceeding globally in 2013 (tablets will reach the 10% in 2015) and the world's mobile data traffic will be 75 percent in/on video by 2017. The Middle East and Africa will be showing the strongest mobile data traffic growth in the next couple of years ("Cisco Visual Networking Index: Global Mobile Data Traffic Forecast Update, 2012–2017", http://www.cisco.com). Business Insider's prognosis is "future is mobile": fixed digital devices are growing slowly but mobile ones increase extremely till 2015: before 2010 the fixed and mobile devices ratio was 50%-50%, for 2015 the forecast marks three times as many mobile devices than fixed ones in G-20 ("The Future of Digital", Business Insider 2012). The rate of our mobilisation affects our offline and online identities and vice versa. If mobile and smart phones and tablets become part of users' lifestyle, the permanent change of place, environmental and social impacts give a chance for continuous presence and sharing, for conscious or unconscious configuration and usage, as well as for hiding, too. In meantime the digital identity tries to use the big data environment for his/her goals. The boundary is constituted by privacy settings. "Protecting privacy of individuals behind the data is obviously the key reason for access and usage limitations of big data" (Laurila et al., 2012, p. 8).

The first step for identification was made in the early mobile ages and is still valid in current non-smart mobile areas: this constitutes the physical appearance with covers, stickers or with other material things. The mass product became unique, recognizable, and identifiable. Sound design and optional wallpaper on colour screen confirm these markers. The mobile owner after this ritual can feel a connection with his/her device.

Smart phones and tablets mean a new dimension after "dull mobile" ages: if somebody chooses a device, he or she chooses an operational system and platform to mask the outside – and for the outside. This is genuinely mobile and is customized by applications, services and with contents. Manufactures and distributors are engaged in a fierce competition concerning devices, operational systems, standards and services. Their market distributions show different results on smartphone and on tablet markets. The different functionalities and services are suitable both for social and professional lifestyles with divergence. However synchronized devices and common platforms generate convergence in usage. Finally mainstream smart devices and smart platforms traffic together with emergent and disappearing ones (see updating sources by www.businessinsider.com and www.gartner.com).

However, a mass of start-up companies try to find the way to develop applications ready to spread like viruses in response to generated or recognized needs in order to satisfy unique demands. Along with this we have to set up cemeteries of good-for-nothing hardware and software. Costumers recognise huge digital noise and try to choose problem solutions. A perfect example is the application, "Umbrella". This simple program answers our typical question with "yes" or "no": 'do I need an umbrella today or not?' While a weather application contains more information, a user test can reveal that this information is sufficient in everyday routine. This is, however, only a function. A design or an unusually rapidly spreading application can gear emotions and engagement. "Personalized" or "uniqe," or "prestigious" feelings provide the owner with the sense of being special/unique. It has become a necessity in prototype design, testing and continuous feedback via companies and in users' emotional attachment: they all cater for the choice for personalisation and identification tools, contents and genres converge towards mobile multiplatforms and they have homogenized these devices, and their visual and ergonomical patterns towards best practices, which ultimately resulted in an illusion for the extension of identification and personalization. The pure examples are screens with icons and tiles of smart phone screens as platform architectures match. Difference is only a single factor for identification: the "lovemark" effect (Roberts, 2005). The built company brand and the corporate philosophy of Apple Inc. are workable in marketing, design and in product prestige price. This is one of the most important issues for users: phenomenological user experience (Manovich, 2001, p. 66) to be in interaction with information design and with further digital points. Usability, plastic elements, comfort and familiarity claim active users.

Finally users line up in front of the stores to have access to the most recent product. This is their identification with devices in digital ages. Functionality and business design provide another solution for users and engagement, like Windows Phone. Finally, identification connects to a logo and to usability.

Some new developments mark a new potential paradigm for the future in mobile device design, in personalisation, and they also mark an even closer relationship and immersion. The common design logic implies that we do not have to furnish our mobile device. It is the other way round: they dress us up and/or they become even more prosthetic for perception and of self-extension, or at least become more transparent and flexible. Design and functionality are personal and suitable/adapted for different lifestyles (Figure 1).

Google OK Glass and similar developments are particularly relevant for us in this context. First we watch videos later we can wear our video and timeline (see above Cisco data of video traffic in the future). The approach speaks for itself on Google Glass official site with three elements: "How it feels + What it does + How to get one" ("Google Glass", http://www.google.com/glass/start/). They tell us the story in videos and through widescreen pictures "naturally". However, at the time of writing this paper there is a Google Glass application available that has been designed to read emotions (http://www.googleglass.gs/new-google-glass-app-can-help-you-with-discerning-emotions/). The analysis of facial expressions and vocal patterns recognizes happiness, sadness, frustration and further emotions. This application can help to recognize for example autistic communication but each usage may imply some ethical issue in the future. Mass media and expert bloggers have expressed some concern with reference to this:

Figure 1. Design for mobile devices: innovation and fashion design together
Sources: ©Ciccaresedesign, ©Thomas Lænner ©Google Inc. Used with permission of designers and followed the Google permissions policy (www.google.com/permissions).

recognizable emotions assume – for users – a greater impact on and a stronger control over different fields from commercials to other forms of manipulation.

The effect of augmented reality image capture and transmission, the record of mobile lifestyles, the potential privacy repercussions raise heated debates pro and contra. If identity and identity's views are "on air" his or her life and networks become more transparent and vulnerable by digital system sets. Appropriate analysis software can summarize the incoming data and draw a more accurate profile of identity than ever. Raytheon's company has presented how the Rapid Information Overlay Technology (Riot) software uses photographs or other digital information to analyse data on social networks with latitude and longitude details (http://www.raytheon.com/). And this software is one among the many. Future digital devices and analytic software solutions will be in close connection and they will be unavoidable because of public services or other social functions. Digital identities will be "on air", "on/line" between points of networks and mobilized via different devices. They are involving and immersing into augmented and virtual scenes where services are the guides to organise their lives and narratives. This will be the standard. Challenges are multitasking lifestyle, divided attention, real time decision in identification and in networks, protocols for staying standby.

BLENDED, PROSUMER AND RISKY EXISTENCE

Using mobile devices implies an always online and off blended existence. The concept of blended identity refers to online self-presentations that include both online and offline aspects of individuals. To understand the process of the creation of blended identity is to know how people (a) derive identities online related to their offline experiences and the online community they have joined, and then (b) migrate from online to offline bringing with them the online identities that they then introduce to others whom they have met first online (Baker, 2009, p. 15).

To move with devices increase this blended factor. Parallel footprints are generated by locality and by activity/passivity/interaction/sharing. Localisation, geo tagging and providing information on location (Arminen, 2006, p. 320) depict users' online environment, temporally anchoring and connecting with other digital identities: this is one of the default functions in mobile devices. However, companies, institutes, services and brands can create digitalised spaces of their own and a localisation that is permanent, regular and occasional for users via flashmobs or other events to share. "Here I am" or "Here we are" are identifications with various goals like prestige, (self)brand design, altruism and mutual support, in order to get a discount price/gifts/game/other surprises, or just for fun. Foursquare, Tripadvisor, Nike Running application, FlightRadar24 and other localisation services support availability, collective local knowledge sharing, and motivation for activities, for alternative routes and for our access to megasystems. In this the main drivers are: commuting, (mega)city lifestyles, created scenes like a festival. They find relevant information concerning geographical proximity, current position in offline and online networks, tourism or other navigation or planning journeys. It is fundamental to determine whether a user is somewhere or if s/he is not there. With mobile devices s/he has two layers in localisation: offline with a body and offline and/or online in his/her perception and mind. Where the user is in his/her mind and activity, this is the real space for him/her. These footprints are realistic: they are data sets of identity.

Activity/passivity/interaction/sharing can be synchrony and asynchrony in time. Synchronicity assumes connection with digital systems and networks. Technology, data transmission services, platforms and applications imply frameworks for users and a stress for interactivities and action chains. Timeline contents reinforce and increase each other. Asynchrony is a standby between two synchronicities. Real time information flow, event creations, planning of meetings in common schedule platforms, to query the shortest and fastest route in navigation set up time data collection of identity. The independent spaces for identity's content are generated by cloud computing: we are also identified via cloud services and our content, just like in case of the collection of music of our own or that of other contents on tablets and smart phones. Mobile devices can get access to clouds more and more efficiently.

When synchronicity and locality are in direct connection offline and online an augmented reality is constituted. In both online and offline existence special features navigate via layers between online and real time existences. Augmentation connects reality with the online, that is, with the virtual, interactively in 3D (Azuma, 1997). Every day activities are related to symbols, logos, icons via superimposed/projected 2D or 3D signal systems. If somebody or a brand or some other entity has a tagged/shared ID, users can check its trafficking identification here and now. Users' markers when checked in mediatised on augmented scenes become visible (Figure 2).

Digital identities and their extensions and alternatives (e-citizenship, virtual identity in online role games etc.) take decisions in technological and mediatised environments. Localised-mobilised, synchronic-asynchronic entities are vulnerable due to data sets and to surveillance technology, but they become a part of the plenitude of networks – and it goes with some reputation: the *prosumer* (producer + consumer) status they have gained fills digital frameworks with content, that, the other way round, also define them. Mediatised and augmented Me 2.0 is visible, recognisable and identifiable via their personal media. Advanced mobile identity management (Satchell et al., 2011) and validation belong also to mobile identity. Machine readable documents and biometrics already play a crucial role in public services, for police and for law enforcement agencies, while, at the same time, safeguarding the principles of data privacy and its possible infringements have a priority. Identity verification is a public service and can be used in official enquiries. The digital version is Identity 2.0: s/he is the one who uses the internet and emerging user-centric technologies and, consequently, has an Open ID with official face photo capture, original signature, fingerprints, demographic details (e.g. www.avalonbiometrics.com, Figure 3). Her/his identity "virtually" is a platform and/or an interface.

How can we or how do we have to live with digital footprints via mobile and other devices? This is a *global learning process*. The mobile device phenomenon and internet penetration rate, the global and regional diffusion of mobile telephony (Castells et al., 2007, pp. 7-11), the

Figure 2. Screen among elements and identities: digital augmentation
Source: ©Layar, www.layar.com. Used with permission of Layar.

social media networking and activity, privacy and security settings determine conscious / controlled and accidental / incidental cases.

Reference points, observation points, checkpoints have already been mobile devices for digital identities with authority. User centric and targeted messages are reachable, targeting groups' data collections are analysed and they provide feedback via personal media.

Digital technology highlights a topic with reference to the issue discussed above, namely: *risk*. Digital systems are vulnerable: they entail a technical and transactional risk. Devices and big data trafficking generate a gap for hacking, surveillance and identity theft. Users, device companies and service providers need to define risks and find the tools to protect.

First of all technical risk refers to the physical level with reference to devices and to other gadgets. For example if somebody steals a device with sensitive data that can generate big risk for hacking data collection. Samsung and LG give a 'red bottom' answer for it that kills 'switch on' programs. This kill switch application, when activated, renders the smartphone unusable. It could a way to reduce technical risks (source: http://www.koreatimes.co.kr/www/index.asp).

On the other hand, transactional risk is more dangerous. Sometimes the users have no idea about real time data hacking or stealing. A very sensitive question could be for example data on personal health stored on our devices. Hacking data concerning health might even cause somebody's death. This is just one of the possible risks. If we

Figure 3. Identity 2.0
Source: www.gemalto.com. Used with permission from ©Gemalto Group.

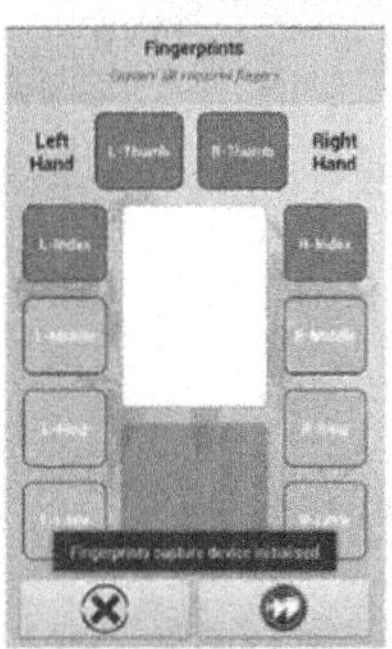

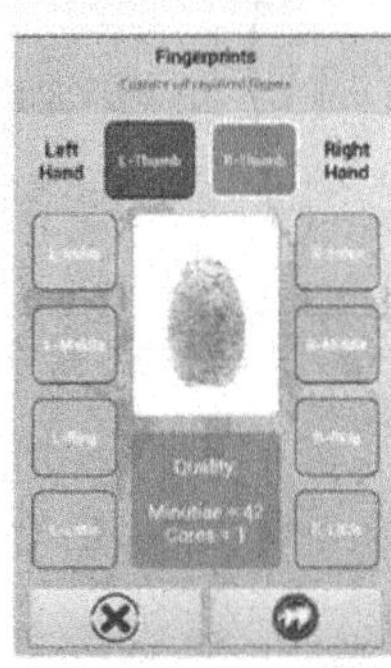

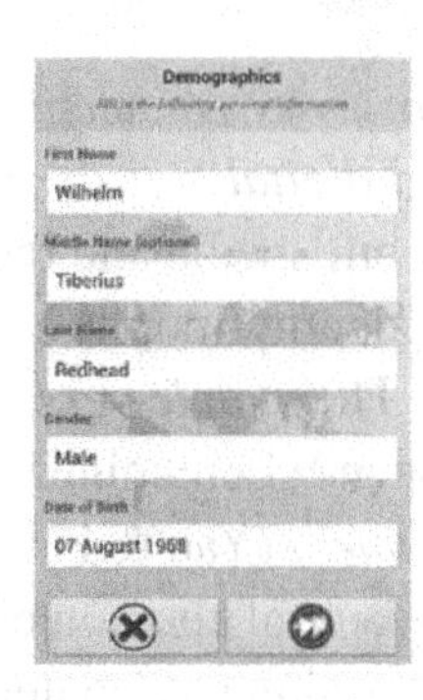

look around there can be quite a number of risks: "cybercrime or cyber-terrorism surges, the digital forensics (DF) of mobile communication devices still enormously lags behind computer forensics" (Chu et al., 2011, abstract).

The third question concerns the risk of identity theft. "It occurs when a malicious person uses an honest user's personal information such as the user's name, Social Security number (SSN), credit card number (CCN) or other identifying information, without his/her permission" (Bhargav-Spantzel et al., 2006, p. 277). Smart phones include a couple of personal/corporate sensitive data and wifi connections give generate risks to reach these.

To sum up: digital and mobile risks have an impact on our lives in a personal and in a global context. Mobile phones need "more control, security and privacy over digital identity than any other technology and thus users' mobile phones are an ideal site to maintain the bits of data that constitute personal digital identity" (Satchel et al., pp. 56-57).

MOBILE CONTEXT AND THE CONTENTS

Mobile or smart phone context is a mediatised and digitalised context in one. Digital screens and other digital interfaces determine our usage and thinking in communicative interactions and we are the message – and the vehicle of the message. Users exist in minimum two contexts at the same time: in an actual presence offline and in another with transmission and interactions via devices online. The users always can vary, mix and rotate these. They are on parallel contexts where "context information is any information, which characterizes the situation of an entity that is relevant to the usage situation'' (Dey, 2001, p. 4).

Aaltonen and Lehikoinen (2005) illustrated the context-sensitive view of users with invisible parts of contexts in locality. If we use a mobile device, it detects our status. "We are »always on«, and »always with«" (Aaltonen & Lehikoinen, 2005, p. 382). Their model helps to understand the extension of user's tracking when the person is moving (see Figure 4).

The location information, the dynamic changing contexts, layers of visual structure and actual views move together and in connection with mobile services like call logs or social media activities in this complexity. Aaltonen and Lehikoinen give us a simple model to understand the perception in mobile context where we are always on and we

Figure 4. Context sensitive view in locality, r= visibility range: observed area/context size, α= visible portion of, YAH= you are here, current location
Source: Aaltonen & Lehikoinen 2005, p. 384.

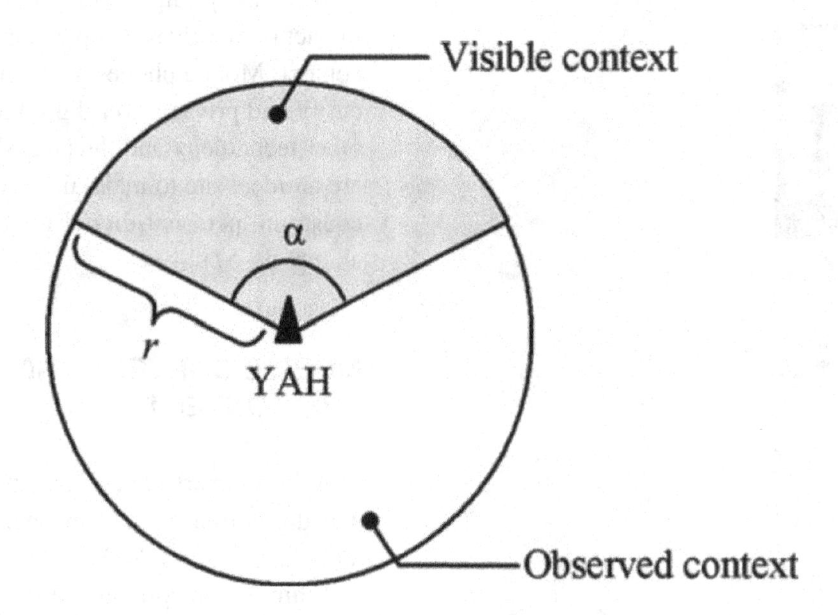

leave digital footprints with only one activity: switching on. "Space of the telecommunicative connection can also be labelled as phonespace" (Hulme & Truch, 2005, p. 466). That Meyrowitz (1985, 2005) interpret this with an opposite perspective: if the user turns into non-place this is socially meaningless. Telecommunicative space as such has no identity (Villi, 2010). These points depend on the perspective.

Physical and symbolic dimensions together assume social interactions. With Michael Bull's metaphor (2005, p. 178) this is an immersion into "mobile media sound bubbles of communication" and without places this can be of lose significance. The social and architectural affordances belong to communication and data transactions (Papacharissi, 2010, pp. 305-325) where affordance means quality of an object or an environment to allow an action for an individual, such as the direct perception tools. Chan et al. (2006) propose six important affordances to mobile technology, including *portability, social interactivity, individuality, context sensitivity, connectivity, and bridging the physical and digital worlds*. These are the projections of mobile media and the most important principles in identity-mobile connection.

The *mobile personal media* (Aaltonen, Huuskonen & Lehikoinen, 2005) and *smart space environment* (Yau & Fariaz, 2004) assumes contents in each of the contexts. Useful, interesting and entertainment contents are common in online content consumption. The rapid and instant contents are rather represented for fast messages or for neutral gears via mobile. "Smart phone performs only the minimum amount of work for filtering and utilizes the results of the information calculated on the desktop." (Lee & Choi, 2012, abstract).

The main traffic means texts/chats, pictures and videos to represent digitally mobilised identities and their connections with other entities and data/ messages. Sharing and sending photos is one of the most often researched fields because photo messaging reasserts mobile phone as a medium (Scifo, 2005, pp. 367–368).

Sharing photos symbolically keeps in touch, enforces social bonds mainly with family members and friends in a physical co-presence (Kato et al., 2006, 306; Oksman 2005) From the point of view of digitalised and mobilised identity it is also easy to understand the importance of the use of photos via mobile communication: a photo points at an identity while a name implies one. However cameraphone photography recalls self-presentation and expression that support social encounters and ongoing 'face-work' (House & Davis, 2005, p. 33). (Motion) pictures provide a display identity: „each image can accumulate meanings which develop semiotically and can be read in many ways. I regard these semiotic changes as textual transformations and have shown how their impact on people's perceptions and experience of everyday matters." (Davies 2007, p. 562) Finally the view is local and situational, contextual and is determined by device view – mainly as mobile camera view. Finally some views are changing because of mobile phone usage view like when in our leisure time we send the message, "here I am" and "I see this", we share "selfies" or further close captures, with feedback to the Google Glass view in mobile technology.

However, the photo is the first step in digital identification (as results by the author's work-in-progress research on "digital identity" for generation Y segment and CEOs/decision makers segment via 30 semi-structured interviews indicate): similar names or similar contents can identify less. Users' pictures give the highest level of identification. Closely related with this result we are working on a research to study 670 Facebook profiles of Generation Y and Z: users represent their own face on profile photo in 77 percent and they prefer professional photos to be displayed in 20 percent (this is a work-in-progress project as well, hosted by Eötvös Loránd University in Hungary, research leader is Ákos Csókay in collaboration with the author).

The photo becomes more frequented in identification on digital and mobile screens than earlier. This can be a first step to recognize each other before users meet for the first time and the change to be followed real time, later. The name becomes the second identification element according to the author's digital identity work-in-progress research (see above). Face recognition and visual analysis software developments provide new consequences for digital and mobile environment and perhaps for users' behaviour. As we have already mentioned a Google Glass application can read emotions as well. We can forecast a complex sensor and filter to detect other users by smart devices for the future. Mobile identification to create or to perceive data sets implies real time and temporally localised entities adapting different online and offline contexts. The author has referred to this as *streamed identity* that would be permanent within its fluent contexts.

The textual parts of any mobile content would be very short just like in microblogs and chatrooms. Blogging identity is also a context-driven representation of a person (Dennen, 2009). The text as a medium points out the identification of an entity and the cultural background as well. As Li's chat research (2007) demonstrated identities are more or less open to show personal and identification data. His research focused on English-Chinese distances on mobile chatting: "English chatters tended to be more "free" with information release online. This should not be too surprising since the doctrine of free speech is highly valued in individualist societies. Accordingly, it was found that English chatters provided more personal and individual information online than their Chinese counterparts, such as marital status, E-mail, personal latest news and homepage. On the other hand, since China belongs to strong uncertainty avoidance cultures, Chinese chatters offered more apparently fake information online, such as fake real name, fake age, and fake personal photo, than English chatters for the sake of avoiding uncertainty and risk on the Internet." (Li 2007, p. 132) Digital devices convert openness and cultural embeddedness to share personal information. And this may be more instinctive in context dependent, short-immediate and reflexive-interactive mobile communication.

Different contexts and content forms shape mobile identity in co-production. The mediatised and digitalised Me represents itself partly via offline personal-cultural communication and the view is partly determined by devices. With this the social dimension is changing as well.

Social Dimensions

The internet is a mediator of social activity (Nabeth, 2006, p. 3). The mobile personal device gives a chance to form rituals in this context for personal and semi-personal networks. The main point is the interaction that can be in open and closed communities or themed groups with common topic and social networks with focus for relationships. (Buss and Strauss 2009, pp. 12-18). Mobile phones extended boundaries 1) to interactions for social encounters and augmented flesh and blood meetings (Satchell & Graham, 2010, 2) to more interactions in micro-relationships (Gergen 2002, p. 237). "People on the move remain embedded in their personal social networks, despite of the varying distance. Japanese studies show that most mobile phone communication is done with a small circle of close friends and family, generally 2–5 others but no more than 10" (Ito 2004). Boberg's paper (2008, p. 135) found similar results in Finnish and French culture: mobile phone users focused mainly on friends and family matters. These closer relationships mean similar lifestyle patterns (Hjorth, 2009, p. 16) that support regular connectivity. The other networks are less represented just in full volume with weaker network connections and higher standard deviation in time factor.

Mobile networks and communities (Kozinets, 2010, p. 3) move together or they can operate both offline and online. The planned meetings are organised also by applications. Unexpected meetings can generate new networks with online confirmation. Networks are flaring or drop off. Online confirmations, for example an "add as friend" function and a positive response imply an engagement with someone and/or his/her networks. In turn a post about a relationship is a proclamation. Users have to decide about these in social networks as privacy settings and have to face the consequences.

Digitally mobilised identities need to redefine themselves and their routines again and again in real time, in online-mobile represented localisation and on different platforms as personal and professional networks, and, also, in terms of their goals. However, personal and professional networks overlap each other partly and people need to decide about transparency: who, when and where can reach updated information about them. Self-actualization (Ito et al., 2005, p. 176) is always on with visible or non-visible venues and actions. Identities try to avoid some communication interactions for example cyberbullying. Mediation of emotions (Vincent, 2011, p. 205) is one of the most important motors of social networks. The goal is subjective well-being and people stand for building a positive self-presentation with positive associations to skip negative responses (Kim & Lee, 2011, p. 363). People need to control and to realise this in every minute online. Mobile connections bring unexpected meetings, sometimes online and offline simultaneously and entity is identified in plurality. For example somebody checks in to Foursqure and suddenly he or she is in a flash meeting (Ito & Okabe, p. 2005) with some people from different personal and professional networks and maybe in ad hoc relationships (Cristofaro, 2011/2013, p. 50). Smart phone connects people with other people or to a network/group at the same time; and, also, to places. These smartphones may also warn users to refrain from other places and from meeting undesirable people/network/group.

A mobile is a *compass* for digital identity in social interactions. It is a compass to find or avoid people/group/event. Compass to develop a social cohesion (Ling, 2007, p. 7). We steer with it; the visible and observed views are changing, like changing partners in a dance. The permanent partner is the device – other connections are special. A famous video 'Forgot my Phone' has staged this: a girl is together with other people in different situations like on training, at a birthday party, at the theatre and so on. But these people hold mobiles in their hands such as voodoo. They are trafficking messages in these situations with people who are actually not there and finally the girl cannot contact with mobile users anymore (see Figure 5, picture cut from the video). The video became a viral video for millions – ironically, mainly via social media. Its conclusion is that social rituals are changing via emerging mobile solutions (Fortunati, 2005), and the video warns us about losing our reality. It seems to be a bit nostalgic that the digitalised mobile area could not return any more. Mobilisation and the mobilised users' lifestyle forward and determine the logic of the future. Our compass becomes an interface to further users' interfaces when contacting one another.

VIRTUAL IDENTITY, IDENTITY 2.0, AND CORPORATE IDENTITY

Finally some short reflections. Our main focus is the user: users' mobile identity. We need to mention some more aspects from a wider view to understand more impacts in mobilisation. We could not analyse all of aspects relying on these, so we shall share a few insights.

Firstly we did not focus on virtual worlds. These are important, but it has another aspect of social dimensions and mainly with an indirect impact. The role games, the strategic games and others with avatar connections mix less with offline identities. These are definitely not digital identities but virtual ones. They have had an impact on digital identity but players usually suspend their offline identities for the time virtual-being. Sometimes we find connections in offline-digital-mobile identity links – like when a game company prescribes special expertise for a job. But we are not studying these now because of digital identity focus.

The other thing is that Identity 2.0 could be a relevant topic for us here. As we mentioned earlier, Open ID, identity verification and e-citizenship are important fields in digitalisation. Governments need to identify citizens in online and mobile systems and they have to develop user-centric mobile applications for example to vote or to reach local services. Biometrical identification, localisation and other digital tools are fixing citizen information. The question is the accessibility of the citizen identity information for public sectors that depend on further factors with reference to the law to ethics. The answer could be „Security and accuracy of the identity information the government holds on the citizen, especially the risks caused by human errors or interventions" (Lips, 2010, p. 287). But it depends on culture, society, technological developments and it marks different outputs.

A mobile identity is refers its device and data to security. If identity participates in excessive scenarios, e.g. in a war or in anarchy the next issue is the freedom to protect oneself and to have an impact on what is happening. A mobile device owner would go against the government or against political power like it happened during Arab Spring. The device could be a gun or shield as well due to working networks. But let us return to a normal and peaceful situation.

A further question in social context is the "transparent society" (Tufekci, 2008, p. 34): Let us see an example from the U.S.A. If somebody likes to buy a flat s/he can check the environment and owner's criminal records online. A mobile device gives updated information about it during our apartment search procedure directly. Later we shall also feature on this map. These networks belong to social logic with a direct impact on users' life, real time and locality setting via mobile identity. Identity 2.0 and e-citizenship are special fields; this paper has provided only some impressions concerning it in social/network dimension. At this point we cannot get into details.

Furthermore, we must clear some aspects with reference to business context. Global and local services rate on mobile business arena from the big sharks to the millions of start-ups. The fastest-growing mobile startups are, among others, Imgur, Twitch, Rdio, Spotify, Snapchat, Kik, Social Point, Maaii and so on. Sharing photos/images, music subscription services, video broadcasting and chat communities for gamers, personal clouds, personalised distance education and much more providers try to reach customers. The B2B business mixes brick and click solutions or focuses just for online services. It operates multiple identities and special corporate IDs.

Digital identity has two main aspects in business: 1) personal entity that find business via offers on mobile directly 2) corporate/organisational identity with internal and external communication/networks. The first brings the marketing-sales-PR--personal data-security view, the second is more complex because its body is full of data. We saw which aspects are relevant for digital identities – from devices to social connections.

Companies can give more answers for digitalisation and mobilisation. They can learn new possibilities in effective work like ad hoc networking, co-working, collaboration, teamwork and leisure teams. These are flexible and working on tablets and smart phones effectively. The most important issue is vulnerability because of security and reputation management. Control and authentication are in focus on the field of corporate digital identity (Shadbolt, 2008).

The author of this paper has been working on a research concerning digital identity. The main focus is on strategies of digital identity. One of the research segments concerns CEO's and decision makers. The first results of this study show that companies keep under control employee's data transfers via mobile devices and/or they issues guidelines with reference to them. An increasing number of multinational companies have e-learning programs concerning online and mobile communication security and control, large and SME companies use guidelines and they rely on employee's responsibility (work in progress, planned to finish in 2014, more details will be available on www.katalinfeher.com and further sites).

OBSERVATIONS AND INSIGHTS

Notions like 'personal,' 'public' and 'common' have blurred their meanings. Indicators of digital identities are Identity 2.0 for verification, Me 2.0 for representation, e-citizen for government and public services, pseudonymity for mediatised roles online, virtual identity for simulated worlds. Users are remixed (O'Neil, 2008) and have blended identities within the given communication space (Weight, 2007, p. 164) via online and mobile-impact scenes. Users have multiple definitions themselves in digitalisation and mobilisation. Real and perceived control above mobile lifestyle where mobile is part of the body as a prosthesis and extension and part of the mind with platform patterns and organizing principles via algorithms. The >mobile< is an interface to contents and to our social-cultural networks in order to collaborate and to filter supplies collectively (Anderson, 2006). The mobile is a personal medium physically that is complemented/supplemented with services and with contents and features. Finally the mobile media provide the "freedom-enhancing dimension of the Internet" (Solove, 2007, p. 6.).

There is a case for personal anchor to mobile content and services design. An approach insight for mobile content extension with loyalty consists of two cases of the globally famous mobile and tablet game "Angry birds". Angry birds has by now modified its original Angry Bird media platform features (see: www.designbykatrin.com) and targets its planned bird prototype video portfolio (see: beepbird.com) at novel functions

Figure 5. Viral video about the impact of mobile devices
Source: https://www.youtube.com/watch?feature=player_embedded&v=OINa46HeWg8
Used with permission from © Miles Crawford.

that channel mass digital content consumption via mobile devices. Personal media content would also be personal real time and it would be part of an identity or that of an identity's profile offline.

And there is a personal anchor to mobile services with instant community activity and localisation at the same time. Foursquare Epic Swarm experiment was organised in Hungary in 2012. The self-organized event had a big goal in Foursquare gamification context where the common presence means extra points and badges. "More people at the same place" is honoured with badges from 50 users but this performance was general and easy project at the end of 2012. A greater number of simultaneous logins were the challenge. A university student community with young independent helpers turned toward the 1000 simultaneous logins, the Epic Swarm Badge. The main problem was the technology because of the overload of transmission towers – and wifi technology. The Epic Swarm experiment was successfully completed by careful preparation and enthusiasm. More than a thousand people have used the application Foursquare check-in at the same time. They represented himself/herself with a temporal network. This is a symbolic act for engagement via personal mobile media and a token of loyalty towards Foursqare brand and towards digitalised mobile technology.

The observed freedom-enhancing dimension of the Internet, the importance of identifying engagement and design, the activity-based online brand loyalty and experimentation are possible for a further extension of mobile identity. And another extension is the technological effect on how users learn: learn in school, in corporate context, according to interest and to self-organization or they learn of technology that may help them to acquire further knowledge. The essential question is how they can learn digital technology tools in order to learn of other things. We can interpret this as a metaphor of a box with a double bottom: to learn again and again the changing digital-mobile technology and to learn something via digital technology. Information technology generates a technical framework (a metalanguage of devices) in order to generate the object language of online/ digital scenarios. It assumes a lifelong learning concept connected to content networks and to learning communities. The notion of *connectivism* highlights the network focus to this phenomenon: "knowledge is distributed across a network of connections, and therefore that learning consists of the ability to construct and traverse those networks." (Downes, 2007). Network complexity generates self-organization and a community-organisation of knowledge. The bottom up and user generated content movements meet with conventional learning methods and, most of all, with institutional frameworks in a blended learning paradigm.

The *blended learning* paradigm is an educational potential via mixed formal and non-formal, online and offline methods. The reallocated resources and the customised solutions provide more flexibility and can react faster to technological and learning-environmental changes. Mobile devices used to give an answer to this with podcast content services (c.f. author's comments on this). At the time writing this paper mobile devices provide an extension for sharing and reaching elements of global knowledge (see the feature of big data above).

Mobile identity is connected to search engines, to social media, to institutional or auto-directed learning in lifelong learning contexts. Mobile language learning, the massive open online courses (MOOCs), databases of common knowledge are reorganized and are shared via digital platforms, and further digital-mobile learning services provide novel paradigms for learning and for e-learning. Mobile teaching sites and applications have heavy competition and they offer free, semi-free and payable educational services and this in turn also reorganizes further online networks and learning functions. "We have to recognise that attempts at identifying and defining mobile learning grow out of difference, out of attempts by emergent communities to separate themselves from some older and more established communities and move on from perceived inadequate practices" with a greater impact on society (Traxler, 2009, p. 2). Digitally mobilized identities can take part in different blended learning processes in different roles or they can mix and organize their auto-learning process via digital platforms parallel in real-time or delayed.

"Always on" (see above) is a lifelong learning process with transparency: to share knowledge, learning process, milestones and results. Courser.org as an MOOCs learning services via selected universities and courses is a simple example for this. The courses provide collaboration, peer-reviewing, joining a professional network and more. The certificates are visible for potential employers and recruiters – for example, automatic uploading to LinkedIn professional network. Users can buy applications for further services or they can pay for different form of participation. Leaving plenty of digital footprints is evidence in this case via learning activity, learning communities and learning results on different platforms and applications.

The big challenge of this e-learning and m-learning market is to develop learning environments that answer users' needs, such as real time searching questions and continuously changing contents and networks. The mobile identity is hungry for new and relevant information and learning process to feature in supported networks. However, identity management is raising questions about influence for planned learning processes (Smala & Al-Shehri, 2013, p. 304). Social media usage enables more enthusiastic and autonomous learning processes via authentic identification (i.e. Baghdasaryan post on http://karinebaghdasaryan.edublogs.org in 2010) and stimulates more activities. Finally, mobile identity can use a blended strategy for co-learning so as to employ different networks and platforms. This is a user's responsibility to find the relevant frameworks, applications and networks for their new needs. If they do not find a suitable solution they can develop it for example in an open source community and they can share a common solution. This will represent mobile identity uniquely.

Finally, we can realise how digital identities are looking for their choices for their choices in digital and mobilised environments. They are out of space and time and parallel in more, they are blended and remixed, networked and in instant online situations. The antagonism of this technology "that reflects and supports the broader ambivalence of presence and absence in the conditions of postmodernity: two (very) different but co-existing practical possibilities inherent in the management of social space in contemporary society. We want others to be present but claim the right to be absent; we seek for presence but reserve the possibility for absence, and social proximity in absence, in order to be able to access different spaces (and different intimacies) simultaneously. (Villi & Stocchetti, 2011, p. 109). Mobile identities are inside and outside simultaneously, sometimes it lost themselves, sometimes find one's way back and look for new alternative ways as internet network logic.

They live offline and online finally in both life. The Ancient Greek use for 'both life' the 'amphi' (αμφι) = both and bios (βιος) = life. These means together *amphibian*: an animal that lives both in the terrestrial and in the aquatic ecosystems. This metaphor flashes the dual nature of mobile identity: we need to define ourselves for offline and online roles, we come from offline just like amphibian come from the water and we adapt our offline experiences for online presence just like amphibian adapt for a partly different standard. This is the offline adaption for online presence. Our identities live on two layers: in real life and on screens and via interfaces. Mobile identity is a commuter among these and the memories of our lives prior to digital and mobile technology. The two layers are merging. Digital technology and mobile technology were looking for possibilities for new developments parallel earlier. The "smart" movement or the "smart" turn provided the merger of these ways via the organization of data sets. Digital mobilization is a new dimension for previously separated or semi-separated digitalisation and mobilisation. Merged technologies and merged services left earlier (semi-)isolated telecommunication and computing aspects. Digital mobilisation is structured via a smart turn: the complex devices and platforms with applications set up a personalised and lifestyle media for blended and display identity, for prosumers and for pseudonimity with rapid and instant contents in changing contexts. This marks the age of digitally mobilised identity, where offline and online identity meet – see the metaphor, "amphibian" - and navigate in real and online extended worlds with mobile devices – see the metaphor of the "compass."

RECOMMENDATIONS FOR MOBILE AND DIGITAL IDENTITY

Basically standards and platforms are the highlights with developing digital-mobile-network competences: networks of links, contents and users' identification networks are determined by operational systems, platforms and applications, users, communities, companies and organisations need to optimize and have joint decision based solutions and selections for effectiveness, in turn users need to top up their digital-mobile-network competences in continuously changing environment to apply standards and platforms.

The other point is that we need to focus on connections with offline: multitasking and mobile lifestyle reorganise the focus and offline flows again and again; anchoring gives a chance to immerse real reality, optimally temporally away online/mobile network systems, identification has non-digital and non-mobile dimension, the "dependence" category is becoming narrower but exists to reflect and understanding of the original roles and identity roots.

Then again there is the feature of data and content provide the main characteristic via personal mobile device: patterns of contents effect users' thread and when these are transmitted via algorithms or other users. Public, semi-public and closed information are questions of decision: control and possible positive reputation, professional or other relevant content representation "on air".

Furthermore it makes strategic decisions with reference to contexts of security and surveillance: big data and cloud and localisation and vulnerabilities of always being connected to systems and to data collection sets and the most sensitive point is a mobile device with its personality, intimacy, digitally mobilised identity design, protection and control.

Last but not least mobile identity means managing and controlling of social life, online selfbrand, collaboration, connectivity and mobile lifestyle: usage of a mobile device provides a compass function in everyday life where we need to find a balance between online represented digital footprints and real activity in face to face or other conservative forms of communication. Online channels can use different feedback tools or, sometimes, these may not have any tools for real feedback. We need to realise this. The amphibian needs to live healthy and normal life within both ecosystems – just like users do on and offline. Mobile lifestyle has made it possible leave behind our earlier media assets and get on with an increasing healthier lifestyle which is less harmful to the environment. This mobile environment supports economy and society to develop a new logic for life with users who are responsible for their own digital-mobile footprints and considering, also, other users

Finally we need to stay competitive: digitalisation and mobilisation means a continuously investment for competitiveness. Volume and frequency of investments correlate with goals and with a defined profile. In-progress orientation/observation/research/education is inevitable. Mobile identities connect to their extensions and compasses directly: they are in charge.

ACKNOWLEDGMENT

I am grateful to Budapest Business School where I was a senior research fellow when I started to work on this chapter. The academic framework of Budapest Business School Research Centre enabled me to do research within the international network of the authors and publishers who have contributed to this book.

REFERENCES

Aaltonen, A., Huuskonen, P., & Lehikoinen, J. (2005). Context Awareness Perspectives for Mobile Personal Media. *Information Systems Management*, *22*(4), 43–55. doi:10.1201/1078.10580530/45520.22.4.20050901/90029.5

Aaltonen, A., & Lehikoinen, J. (2005). Refining Visualization Reference Model for Context Information. *Personal and Ubiquitous Computing*, *9*(1), 381–394. doi:10.1007/s00779-005-0349-4

Anderson, C. (2006). *The Long Tail: Why the Future of Business is Selling More of Less*. New York: Hyperion.

Arminen, I. (2006). Social Functions of Location in Mobile Telephony. *Personal and Ubiquitous Computing*, *10*(5), 319–323. doi:10.1007/s00779-005-0052-5

Azuma, R. (1997). *A Survey of Augmented Reality Presence: Teleoperators and Virtual Environments*. Retrieved from http://www.cs.unc.edu/~azuma/ARpresence.pdf

Baker, A. J. (2009). Mick or Keith: Blended Identity of Online Rock Fans. *Identity in the Information Society*, *2*(1), 7–21. doi:10.1007/s12394-009-0015-5

Berlant, L. (Ed.). (2000). *Intimacy*. Chicago: University of Chicago Press.

Bhargav-Spantzel, A., Squicciarini, A. C., & Bertino, E. (2006). Establishing and Protecting Digital Identity in Federation Systems. *Journal of Computer Security*, *14*(3), 269–300.

Boberg, M. (2008). *Mobile Phone and Identity: A Comparative Study of the Representations of Mobile Phone among French and Finnish Adolescents* (Dissertation). Joensuu: University of Joensuu. Retrieved from http://epublications.uef.fi/pub/urn_isbn_978-952-219-103-8/urn_isbn_978-952-219-103-8.pdf

Bull, M. (2005). The Intimate Sounds of Urban Experience: An Auditory Epistemology of Everyday Mobility. In K. Nyíri (Ed.), *Sense of Place: The Global and the Local in Mobile Communication* (pp. 169–178). Vienna, Austria: Passagen Verlag.

Buss, A., & Strauss, N. (2009). *Online Communities Handbook*. Berkeley, CA: New Riders.

Castells, M., Fernández-Ardèvol, M., Linchuan Qiu, J., & Sey, A. (2007). *Mobile communication and society: A global perspective*. Cambridge, MA: MIT Press.

Chan, T. W., Roschelle, J., & Hsi, S., Kinshuk, Sharples, M., & Brown, T., et al. (2006). One-to-one Technology Enhanced Learning: An Opportunity for Global Research Collaboration. *Research and Practice in Technology-Enhanced Learning*, *1*(1), 3–29. doi:10.1142/S1793206806000032

Chu, H. C., Deng, D. J., & Chao, H. C. (2011). Potential Cyberterrorism via a Multimedia Smart Phone Based on a Web 2.0 Application via Ubiquitous Wi-Fi Access Points and the Corresponding Digital Forensics. *Multimedia Systems*, *17*(4), 341–349. doi:10.1007/s00530-010-0216-7

Davies, J. (2007). Display, Identity and the Everyday: Self-presentation through Digital Image Sharing. *Discourse. Studies in the Cultural Politics of Education*, *28*(4), 549–564. doi:10.1080/01596300701625305

de Cristofaro, E., Manulis, M., & Poettering, B. (2011/2013). Private Discovery of Common Social Contacts. *International Journal of Information Security*, *12*(1), 49–65. doi:10.1007/s10207-012-0183-4

Dennen, V. P. (2009). Constructing Academic Alter-Egos: Identity Issues in a Blog-Based Community. *Identity in the Information Society*, *2*(1), 23–38. doi:10.1007/s12394-009-0020-8

Dey, A. K. (2001). Understanding and Using Context. *Personal and Ubiquitous Computing Journal*, *5*(1), 4–7. doi:10.1007/s007790170019

Downes, S. (2007, February 6). *Msg. 30, Re: What Connectivism Is. Connectivism Conference*. University of Manitoba. Retrieved from http://ltc.umanitoba.ca/moodle/mod/forum/discuss.php?d=12

Fortunati, L. (2005). Mobile Telephone and the Presentation of the Self. In R. Ling, & P. E. Pederson (Eds.), *Mobile communications: Re-negotiation of the social sphere*. London: Springer. doi:10.1007/1-84628-248-9_13

Gergen, K. J. (2002). The Challenge of Absent Presence. In J. E. Katz, & M. Aakhus (Eds.), *Perpetual Contact: Mobile Communication, Private Talk, Public Performance* (pp. 227–243). Cambridge, UK: Cambridge University Press.

Håkansson, H., & Ford, D. (2002). How Should Companies Interact in Business Networks? *Journal of Business Research*, *55*(2), 133–139. doi:10.1016/S0148-2963(00)00148-X

Hjorth, L. (2009). *Mobile media in the Asia-Pacific*. London: Routledge.

Hulme, M., & Truch, A. (2005). Social identity: The new sociology of the mobile phone. In K. Nyíri (Ed.), *Sense of place: The global and the local in mobile communication* (pp. 459–466). Vienna: Passagen Verlag.

Ito, M. (2004). *Personal Portable Pedestrian: Lessons from Japanese Mobile Phone Use*. Paper presented at Mobile Communication and Social Change, the 2004 International Conference on Mobile Communication. Seoul, Korea.

Ito, M., & Okabe, D. (2005). Mobile Phones, Japanese Youth and the Replacement of Social Contact. In R. Ling, & P. Pedersen (Eds.), *Mobile communications: Renegotiation of the social sphere*. London: Springer. doi:10.1007/1-84628-248-9_9

Ito, M., Okabe, D., & Matsuda, M. (Eds.). (2005). *Personal, portable, pedestrian: Mobile phones in Japanese life*. Cambridge, MA: MIT Press.

Kato, F., Okabe, D., Ito, M., & Uemoto, R. (2005). Uses and possibilities of the Keitai camera. In M. Ito, D. Okabe, & M. Matsuda (Eds.), *Personal, portable, pedestrian: Mobile phones in Japanese life*. Cambridge, MA: The MIT Press.

Kim, J., & Lee, J. E. R. (2011). The Facebook Paths to Happiness: Effects of the Number of Facebook Friends and Self-Presentation on Subjective Well-Being. *Cyberpsychology, Behavior, and Social Networking*, *14*(6), 359–364. doi:10.1089/cyber.2010.0374 PMID:21117983

Kozinets, R. (2010). *Netnography*. London: Sage.

Laurila, J. K., Gatica-Perez, D., Aad, I., Blom, J., Bornet, O., Do, T., et al. (2012). The Mobile Data Challenge: Big Data for Mobile Computing Research. In *Proc. Mobile Data Challenge Workshop (MDC) in conjunction with Int. Conf. on Pervasive Computing*. Retrieved from https://research.nokia.com/files/public/MDC2012_Overview_LaurilaGaticaPerezEtAl.pdf

Lee, K. J., & Choi, D. J. (2012). Mobile junk message filter reflecting user preference. *Transactions on Internet and Information Systems (Seoul)*, *6*(11), 2849–2865.

Li, C. (2007). Online chatters' Self-Marketing in Cyberspace. *Cyberpsychology & Behavior*, *10*(1), 131–132. doi:10.1089/cpb.2006.9982 PMID:17305459

Ling, R. (2007). Mobile Communication and Mediate Ritual. In *Communications in the 21st century*. Budapest: MTA – T-Mobile. Retrieved from http://www.richardling.com/papers/2007_Mobile_communication_and_mediated_ritual.pdf

Lips, M. (2010). Rethinking Citizen – Government Relationships in the Age of Digital Identity. *Information Polity*, *15*(4), 273–289. DOI: 10.3233/IP-2010-0216

Manovich, L. (2001). *The language of new media*. Cambridge, MA: MIT Press.

McConnell, B., & Huba, J. (2006). *Citizen marketers: When people are the message*. Chicago: Kaplan Business.

Meyrowitz, J. (1985). *No sense of place: The impact of electronic media on social behavior*. New York: Oxford University Press.

Meyrowitz, J. (2005). The rise of glocality: New senses of place and identity in the global village. In K. Nyíri (Ed.), *A sense of place: The global and the local in mobile communication* (pp. 21–30). Vienna: Passagen Verlag.

Nabeth, T. (2006, May). Understanding the Identity Concept in the Context of Digital Social Environment. In T. Nabeth (Ed.), Del 2.2 Set of Use Cases and Scenarios, FIDIS Deliverable, (pp. 74–91). FIDIS.

O'Neil, J. (2008). *Remix Identity: Cultural Mash-Ups and Aesthetic Violence in Digital Media*. Retrieved from mcluhanremix.com

Oksman, V. (2005). MMS and Its 'Early Adopters' in Finland. In K. Nyíri (Ed.), *A Sense of Place: The Global and the Local in Mobile Communication* (pp. 349–362). Vienna: Passagen Verlag.

Papacharissi, Z. (Ed.). (2010). *A Networked Self: Identity, Community, and Culture on Social Network Sites*. New York: Routledge.

Roberts, K. (2005). *Lovemarks*. New York: PowerHouse Books.

Satchell, C., & Graham, C. (2010). Conveying Identity with Mobile Content. *Personal and Ubiquitous Computing, 14*(3), 251–259. doi:10.1007/s00779-009-0254-3

Satchell, C., Shanks, G., Howard, S., & Murphy, J. (2011). Identity Crisis: User Perspectives on Multiplicity and Control in Federated Identity Management. *Behaviour & Information Technology, 30*(1), 51–62. doi:10.1080/01449290801987292

Schawbel, D. (2009). *Me 2.0*. New York: Kaplan.

Scifo, B. (2005). The domestication of cameraphone and MMS communication: The early experiences of young Italians. In K. Nyíri (Ed.), *A Sense of Place: The Global and the Local in Mobile Communication* (pp. 363–374). Vienna, Austria: Passagen Verlag.

Shadbolt, N. (2008). A Crisis of Identity. *Engineering & Technology, 7*(6), 20. doi: doi:10.1049/et:20081000

Smala, S., & Al-Shehri, S. (2013). Privacy and Identity Management in Social Media: Driving Factors for Identity Hiding. In J. Keengwe (Ed.), *Research Perspectives and Best Practices in Educational Technology Integration* (pp. 304–320). Hershey, PA: IGI Global.

Solove, D. J. (2007). *The Future of Reputation: Gossip, Rumor, and Privacy on the Internet*. New Haven, CT: Yale University Press.

Study on Mobile Identity Management. (2005). In G. Müller & S. Wohlgemuth (Eds.), *Future of Identity in the Information Society by WP3, Albert-Ludwigs-Universität Freiburg*. Retrieved from http://www.fidis.net/fileadmin/fidis/deliverables/fidis-wp3-del3.3.study_on_mobile_identity_management.pdf

Traxler, J. (2009). Learning in a Mobile Age. *International Journal of Mobile and Blended Learning, 1*(1), 1–12. doi:10.4018/jmbl.2009010101

Tufekci, Z. (2008). Can You See Me Now? Audience and Disclosure Regulation in Online Social Network Sites. *Bulletin of Science, Technology & Society, 28*(1), 20–36. doi:10.1177/0270467607311484

Value of Our Identity. (2012). *Liberty Global, Inc. with permission of The Boston Consulting Group, Inc.* Retrieved from http://www.libertyglobal.com/PDF/public-policy/The-Value-of-Our-Digital-Identity.pdf

van House, N., & Davis, M. (2005). The Social Life of Cameraphone Images. In *Proceedings of Workshop on Pervasive Image, Capturing and Sharing: New Social Practices and Implications for Technology Workshop (PICS 2005) at the Seventh International Conference on Ubiquitous Computing (UbiComp 2005)*. Tokyo, Japan: UbiComp. Retrieved from http://people.ischool.berkeley.edu/~vanhouse/Van%20House,%20Davis%20-%20The%20Social%20Life%20of%20Cameraphone%20Images.pdf

Villi, M. (2010). *Visual Mobile Communication: Camera Phone Photo Messages as Ritual Communication and Mediated Presence*. Aalto University School of Art and Design.

Villi, M., & Stocchetti, M. (2011). Visual Mobile Communication, Mediated Presence and the Politics of Space. *Visual Studies, 26*(2), 102–112. doi:10.1080/1472586X.2011.571885

Vincent, J. (2011). *Emotion in the Social Practices of Mobile Phone Users*. PhD Thesis at University of Surrey. Retrieved from http://epubs.surrey.ac.uk/770244/1/Vincent_2011.pdf

Weight, J. (2007). Living in the Moment: Transience, Identity and the Mobile Device. In G. Goggin & L. Hjorth (Eds.), *Mobile Media 2007: Proceedings of an International Conference on Social and Cultural Aspects of Mobile Phones, Convergent Media and Wireless Technologies*. Sydney: University of Sydney.

Yau, S. S., & Fariaz, K. (2004). A Context-Sensitive Middleware for Dynamic Integration of Mobile Devices with Network Infrastructures. *Journal of Parallel and Distributed Computing*, *64*(2), 301–317. doi:10.1016/j.jpdc.2003.10.007

KEY TERMS AND DEFINITIONS

"Amphibian": A metaphor to reflect the dual nature of mobile identity with offline and online roles. C.f. ancient Greek for 'both lives' the 'amphi' (αμφι) = both and bios (βιος) = life. These have merged into the term, *amphibian*: an animal that lives both in the terrestrial and in the aquatic ecosystems. We associate this term with the feature of digital communication according to which we come from the offline domain just like the amphibian comes from the water and we adapt our offline habits to online operation just like the amphibian adapts to a partly different ecosystems. This is how the offline is being adapted to online operation. Our identities live on two layers: in real life and on screens and via interfaces.

Big Data: Excessive volume of digital data. This volume of data comes from sensors, social media posts, digital pictures and videos, purchase transaction records, and cell phone GPS signals to name but a few.

Blended Identity: Refers to both online and offline features of individuals. Identification has many different aspects for self-definition and for synchronous lifestyles in analogue-digital context.

"Compass": A metaphor for mobile devices to find or avoid people/groups/communities/ events / activities / brands or more.

Digital Identity: Is constituted by a set of online data, which defines personal and social identity, or some other type (e.g. corporate or organizational identity) of identity, to be identified and distinguished.

Digitally Mobilized Footprints: The data trail left by the interactions in a digital environment via mobile devices (smart phones, tablets and further digital devices/sensors). Digitally mobilized footprints consists of tracing location and time, hits, key words, contents, digital activities, social links and further data via mobilized digital identity.

Mobile Lifestyle: Encompasses all the activities that take place on the go via mobile devices: navigation, checking in an event, sharing information and several further digital activities.

Online/Digital Reputation Management: A strategic tool to manage and control the positive, relevant, identifiable reputation for different entities (users, companies, institutes, brands) via online data analysis, communication strategy and risk management.

Online/Digital Vulnerability: A term that marks the sensitivity of online data with reference to issues of privacy, policy, security and that of surveillance.

Prosumer: A "producer + consumer" status to fill digital frameworks with user generated content. Mediatised and augmented Me 2.0 is visible, recognisable and identifiable via personal media.

Smart Turn: The displacement of non-smart phones – with exclusive telephonic functions – by smart devices. Smart phones, tablets and further smart devices with computing capabilities provide users with new media platforms and network connectivity.

Chapter 5
Mobile Technologies and Gender Rituals

Cheryll Ruth R. Soriano
De La Salle University, Philippines

ABSTRACT

This chapter explores the implications of mobile technologies on gender through the lens of gender rituals. While maintaining social order and social roles, rituals also legitimate key category differences, ideologies, and inequalities. The increasing convergence of media and content in mobile devices, and the blurring of the spaces for work, family, and leisure amidst the landscape of globalization and mobility have important implications for the enactment of rituals, and in the performance of gender. The chapter discusses this mutual shaping of gender rituals and mobile technologies through a case study of the Philippines, with some broad implications for other contexts. The study finds that the personalization, mobility, and multitude of applications afforded by mobile devices offer many opportunities for the exploration of new possibilities for subjectivity that challenge particular gender stereotypes and restrictions while simultaneously affirming particular gender rituals. While exploring the implications of the mobile device on gender in a developing society, the chapter in turn highlights the importance of culturally embedded rituals in shaping and understanding the mobile device's place in society.

INTRODUCTION

This chapter explores the implications of mobile communication on gender and culture through the lens of rituals. Although a global technology, the mobile device has been shaped by local socio-cultural adaptations and its pervasive adoption has been accompanied with risk, opportunity, and adaptation. Moreover, with the increasing convergence of media and content facilitated by smartphones, the blurring of the spaces for work, family, and leisure has important implications on gender roles. The analysis of the implication of mobile media on the "maintenance of gender restrictions" or in the "liberation" of people from gender norms, can be understood with reference to specific gender expectations embodied in "*rituals*". While maintaining social order, rituals also legitimate certain key category differences and inequalities (Couldry, 2003; Ling, 2007). The social implications of mobile technology on gender can be observed in multiple sites and across varied

DOI: 10.4018/978-1-4666-6166-0.ch005

social spaces, and can be interpreted as affirming, challenging, or reconstructing *gender rituals*.

The present study builds on existing works that have investigated the relationship between mobile phones and rituals (Ling, 2007, 2008; Hoflich & Linke 2011; Doron, 2012; Soriano, Lim, & Rivera, forthcoming). The study focuses on the Philippine context as a case study, although the social, political and economic implications will find resonances in other contexts. While exploring the implications of the mobile phone on gendering in a developing society, the paper in turn highlights the importance of culturally-embedded rituals and relationships in shaping the mobile phone's place in society.

Mobile technology has experienced an explosive growth globally, illustrated by its emergence as a primary form of telecommunication. The number of mobile subscriptions globally has increased steadily over the years, from 2.7 billion in 2006, to approximately 6.8 billion in 2013 (ITU, 2013). This rise is sharpest in developing countries, where mobile cellular subscription increased from 1.6 billion in 2006 to 5.2 billion in 2013 (ITU, 2013). In the Philippines, where mobile subscription has reached almost 90% of the population, the local adoption of mobile technology has been profound, as mobile phones evolved from a lifestyle gadget for the rich into a technology that has important social, developmental, political and spiritual roles woven into the daily lives of the citizens. This broad reach is driven by the availability of low-cost phones, as well as prepaid, "sachet" (purchase of mobile credit at very small denominations) and "unlimited" promotions from telecommunication companies and Internet service providers. The labeling of the Philippines as the "sms capital of the world" (Mendes, Alampay, Soriano et al., 2007; Nagasaka, 2007) amidst a state of relative poverty and economic development makes it a compelling case for the analysis of mobile technology and societal interactions. Moreover, although gender divide (as with the digital divide), has been found to be especially large in low income countries, a Lirneasia study found that this was hardly evident in the Philippines, with the ratio of mobile access between men and women at 1:1 (Zainudeen, Iqbal, & Samarajiva, 2010). However, the differences in uses of the mobile phone between males and females was statistically significant, prompting the question of how gender rituals influence the differential use of mobile devices, and in turn, how the continued prominence of mobile media in this locale has implications on gender rituals. The availability of cheap, locally assembled android phones such as *Cherry Mobile* and *MyPhone* offered from US $40 per unit, as well as other low-cost tablets from China and Malaysia allowed the reach of smart devices with Internet and social networking capability to a broader segment of the population, reaching various economic classes and genders. While exploring the implications of the mobile device on gendering in a developing society, the paper in turn highlights the importance of culturally-embedded rituals and relationships in shaping the mobile phone's place in society.

GENDER RITUALS

Ritual

The term ritual refers to a type of symbolic expressive activity constructed of multiple behaviors that occur in a fixed, episodic sequence and that tend to be repeated over time (Rook, 1985, p. 252; Salamone, 2010). Previous analyses of rituals restricted ritual experience to religious and mystical contexts, thereby focusing on special ritual events that marked a person's significant life passages, such as baptism, circumcision, coming out parties, graduation, marriage, or death. And yet, in everyday life, people also participate in a variety of ritualized activities at home, work, and play, both as individuals and as members of some larger community. Bringing Durkheim's notion of rituals to analyze everyday interactions, Goffman (1967) explored rituals as small scale, individually authored interactions and everyday actions that shape the everyday sense of social order. Moreover, instead of seeing rituals as dictated merely by structures and institutions, Goffman's lens saw people as responsible for the actual staging and participation in a ritual interaction (Ling, 2007). Following Goffman's articulation of Durkheim's concept of a ritual in the everyday, Ling (2007) argued that mobile communication facilitates the re-enactment of such everyday rituals beyond co-present situations.

But what are the functions of rituals? Rituals define the "right way of doing things" and make symbolic statements about the social order (Rook, 1985). A number of public rituals are viewed as contributing to social cohesion (Ling, 2007), for binding a nation through symbolic nationalistic practices, and also as a means of regulating social conflict (Levy & Zaltman, 1975). Advancing a critical position, Couldry (2003) sees rituals as rights of institutions, which, while maintaining social order, also institute as natural and legitimate certain key category differences, ideologies, and inequalities. This essentially shifts the emphasis of ritual analysis from mere questions of meaning towards questions of power (Couldry, 2003, p. 12) that are embedded, maintained, or communicated through rituals. Rituals work because they reproduce categories and patterns of thought that bypass explicit belief through their repetitive form. Thus, it is their ritualized form that enables them to be successfully reproduced without being exposed to questions about their content (Couldry, 2003).

Gender Rituals in the Philippines

Gender "is a social construction and representation" (Fortunati, 2009, p. 23) and needs to be defined by its "economic and social functionality in respect to the whole system" (p. 24). Following Butler's (1993) theorization, gender is not a primary or 'real' category, but an attribute that is performed and shaped by social rules and the continued transformations that the social construction of femininity or masculinity undergoes (Soriano, et al., forthcoming). Scholars have argued that gender is heavily shaped by cultural rituals, and rituals in turn, are defined by gender roles (Salamone, 2010; Siapera, 2012; Parrenas, 2001, 2005; Aquiling-Dalisay, Nepomuceno-Van Heugten, & Sto. Domingo, 1995; De Castro, 1995; Gonzales-Rosero, 2000). Considering Couldry's (2003) conceptualization of rituals – that they affirm social order, naturalize arbitrary boundaries, and institute and reproduce categories and patterns of behavior, it becomes necessary to look at how gender is constructed, challenged or transformed through repetitive social behaviors.

Gender formulations and the concepts of femininity and masculinity have had a dramatic evolution throughout Philippine history. Scholars agree that pre-Spanish societies across the islands were culturally egalitarian, but that demarcations across gender lines may have existed anyway when it came to housework or community work (Gonzales-Rosero, 2000, p. 42; Quindoza-Santiago, 2010, p.109; Angeles, 2001, p.17). Where the divisions were drawn, however, remain ambiguous. Dramatic changes were brought during colonial times, shifting the role of women from public to private spheres (Gonzales-Rosero, 2000, pp. 44), through new differentiations of roles and duties according to new economic, political, religious and social systems.

Examining the "Spanish rulebooks" implemented during the Spanish colonization, Camacho (2010) found that they serve as windows into the cultural practices and underlying values influencing gender roles. The rulebooks were written and directed to the mothers of the house as scripts for how daughters and sons ought to be raised in accordance with particular values and role expectations. For example, modesty and chastity topped the list of expected feminine behavior, which included "interior quiescence, fostered piety, and the cultural practice and value of domesticity." (Camacho, 2010, p. 298) This classification of roles and the predominance of the related virtues intended to prepare women to develop an integral focus on the home and family. Domestic housework has also become so gendered that the hired help are usually always women (Gonzales-Rosero, 2000, p. 51). The reliving of such rituals in the present has been argued:

The normative ideal of reduced engagement (of females) with a wider world, particularly in regard to males (depending on women's socioeconomic status), continuous subjection to male authority, and limited options to develop talents other than those related to home, education, and personal service, were colonial realities... Today the feminine ideal studied here does not remain a thing of the past but still resonates in the present... (Camacho, 2010, pp. 312-313)

The differentiation of roles and responsibilities assigned to men and women historically gave rise to the re-articulation of such well- entrenched gender roles and expectations in contemporary contexts. As children, females are expected to play *bahay-bahayan* (playhouse) and are taught how to take on household and nurturing responsibilities at a young age. On the other hand, it would be more acceptable for boys to play outdoors. In the earlier years, only males are encouraged to go to school, in preparation for their being the breadwinners of the house. Specifically for families with limited financial capabilities, girls are trained to conduct domestic responsibilities and are discouraged from going to school "because they will get married anyway and become domestic managers". While young girls are at home, it is normal for young men to go out and belong to a *barkada*, or a group of friends who live in the same community or go to the same school. They hang out and participate in social activities like drinking, smoking, and encouraging each other to find girls to court or have sex with. De Castro (1995) describes it as a rite of passage for young men to realize aspects of their masculinity. The same cannot be said for women who partake in the same vices; women are expected to abide by the same gendered virtues that were indicated in the Spanish rulebooks. The same double standard exists when it comes to maintaining relationships: having two spouses or romantic partners implies a strong sense of masculinity in men; the same multiple partners would imply promiscuity in women (Angeles, 2001). Such ritualized gender distinctions are invoked in everyday life, household dynamics, and even in gift-giving (i.e. matchboxes and toy guns for boys, baby dolls or toy-kitchenette for girls).

Other studies explain the role of religion in the formation of gender rituals. In the Philippines, folk Christianity sees the mother or the woman as the central node that brings together not only the family, but the entire society. With *Jesus' mother, Mary,* as the model of Christian faith, the Filipina is expected to possess certain qualities such as kindness, piety, obedience, modesty, chastity, care and virtue (Libed, 2010; Soriano et al., forthcoming). As a result of gendered socialization through religious institutions, most Filipino women have adopted this Marian image (i.e. charity, purity) into their roles as wives or daughters (Aguilar, 1998; Camacho, 2010, pp. 304-305). Similar gender rituals define the role of women in other countries such as in India (Doron, 2012) and Japan (McVeigh, 2004), as well as in Latin America (Hondagneu-Sotelo & Avila, 1997; Carling, Menjivar, & Schmalzbauer, 2012) where women are identified as "good wives and mothers", characteristic of the maternal materialism well- entrenched within national consciousness.

In the dawn of a globalized and digitized world, more recent studies have explored how such gender norms and expectations are carried over or challenged in the present day (Parreñas, 2005; Cabanes & Acedera, 2012; Madianou & Miller, 2011; Tadeo-Pingol, 1999). Over time, women have adopted bolder roles. Philippine President Corazon Aquino became a feminist role model for Philippine women, inspiring the participation of women in other important realms of society such as the workforce and politics (Komisar, 1987;

Libed, 2010; Dela Cruz, 1988). Other challenges to traditional gender norms have been globalization and labor migration. The increasing number of Filipinas migrating to foreign lands as migrant workers led to the shift in the role of women from homemakers to economically active, transnational women (Soriano et al., forthcoming). Data shows the increasing feminization of migrant labor in the Philippines: in 1992 there were 129,000 Filipina female workers worked overseas versus 130,725 males, compared to 208,278 females versus 77,850 males in 2002 (POEA, 2012). This feminization of labor is also observed in other developing countries (Hjorth, 2009; Wajcman, Bittman, & Brown, 2008; Parreñas, 2005, 2001).

Connected to this is the reversal of traditional expectations of the male as the breadwinner and the woman as in charge of domestic matters. In the Philippines, a masculine society with well-entrenched gender roles, the impact of overseas labor migration of mothers is greater in terms of its impact towards the reversal of the "father's traditional role of breadwinning and the mother's traditional role of nurturing" (Cabanes & Acedera, 2012; Parreñas, 2005; Soriano et al.,forthcoming). Other studies found that despite the evolving picture of the Filipino family, various factors still end up limiting fundamental changes in power structures and gender relations as mothers are found to take on the role of both breadwinning and nurturing (Parreñas, 2005, 2001; Madianou & Miller, 2011). Migrant working mothers also carry over particular gender expectations towards their children. For example, they tend to trust their daughters with the responsibility of financing daily expenses in their absence (Parreñas, 2005). Daughters are assumed to be more responsible and capable of showing restraint because they are not expected to actively have *barkadas* (group of friends) or vices. As women, they are expected to echo qualities of their mothers – caring, self-sacrificing, oriented towards domestic work (Angeles, 2001, p. 23). Other studies explored the societal changes brought about by the emergence of business process outsourcing (BPO) in the Philippines (Cabrera-Balleza, 2005). BPOs in the city and those that operate at night have also challenged women to move from suburbs and rural areas to the city, breaking domestic bonds and duties in the process (Cabrera-Balleza, 2005, p.145).

Queers and Cultural Rituals

Philippine society might seem as one of the "LGBT-friendly" societies in Asia. The Philippines allowed the formation and candidature of an LGBT political party, purportedly the only in Asia. The first Gay Pride Parade in the region was also organized in the Philippines in 1994, which inspired similar Pride Parades in other parts of the region (Garcia L., 2008). However, despite such developments, LGBTs continue to be the objects of ridicule (Austria, 2007) and the branding of LGBTs as "immoral" or "threat to the youth" reflects the prejudice that many sectors hold towards the LGBT community (Soriano, 2014). Such prejudice also translates to physical violence and hate crimes towards the community (Philippine LGBT Crime Watch, 2012).

Garcia JNC (2000, pp. 267-268), argued that it is important to be reminded that sexual and gender subjectivity is locally and socially constructed. In the Philippines, these fall between religious and secular registers, a Filipino "psycho-spirituality" (Garcia JNC, 2008; Dumdum, 2010). Understanding of the self is strongly embedded in the teachings of folk Catholicism and it is for this reason that the lesbian, gay, bisexual, or transgender is conflicted on the exercise of sexuality, because the church, seen as the moral authority, only accepts union between a man and a real woman under the norms of procreation (Soriano, 2014). Being gay, lesbian, transsexual, or transgender therefore transgress traditional gender categories of male and female. Moreover, as queers grow up in societies where homosexuality is loathed and mocked, a homosexual growing up in this society begins to loathe himself or herself and adopts a negative view of being queer, thereby hiding activities such as partner seeking or associations with other queers (Austria, 2007, Garcia JNC, 2008; Soriano, 2014). Such characteristics also create stereotypes on the roles and actions that LGBTs can take in society.

MOBILE DEVICES AND GENDER RITUALS

Previous sociological works on mobile communication and mediated rituals identified instances wherein mobile devices mediate ritual interactions that dictate a sense of social order and solidarity within a locale (Ling, 2008; Dobashi, 2005; Silverstone & Hirsch, 1992; Katz & Aakhus, 2002; Hjorth, 2009; Ito, Okabe, & Matsuda, 2005).First, the ways we communicate over the phone, greet each other, and chat with our friends and families via mobile devices represent ritual interaction (Ling, 2008, p. 167). For example, Ito et al. (2005) described how the Japanese use sms – mediated ritualized interaction in a dating context; and Pertierra (2006) argued that ritualized greetings also underlie the introduction of strangers in mobile chat lounges. These imply the importance of exploring how such ritualized interactions over the mobile device are gendered. Moreover, according to Ling (2008, pp.170-171),

After meeting and establishing contact and exchanging mobile phone numbers the nascent couple engages in a more or less SMS-based courtship. In this period the form of the interaction is highly scripted in that the messages are carefully written and edited. This indirect form of interaction is calculated to allow the individuals to carefully work through their utterances and to cover over some of the pitfalls that might weigh heavily in the nascent period of the romance.

Secondly, seeing ritual interaction as "the glue holding society together", mobiles devices facilitate a "connected presence" (Licoppe, 2004; Ling, 2007) where they are used through small, constant messages and talks that create an illusion of presence. In this way, the moderate use of mobile devices may facilitate an active social life that can contribute to the maintenance of social ties (Ling & Haddon, 2003). Connected here is the ritual function of mobile phones in the amplification of collectivist cultures and in facilitating new modes of cooperation and social cohesion (Campbell & Kwak, 2010; Burrell, 2010; Sreekumar, 2011). Linked to this notion of rituals as facilitating social cohesion and family communication, mobile devices have been argued to facilitate a "full-time intimate sphere" (Matsuda, 2005). The mobile phone, for example, is used by women predominantly for "performing intimacy" with significant others so as to hold together the family and the community by building and maintaining relationships (Wajcman et.al, 2008, pp. 646-647). In the context where mothers may be physically distant, as in the Philippines where many women leave the country to seek work overseas, the concept of "remote mothering" over the mobile phone serves as the women's tool in continually maintaining family cohesiveness regardless of physical absence. On the other hand, other studies have pointed out how the excessive use of mobile devices connected to the Internet can be detrimental to social interactions, thereby undermining connectedness (Nie, 2001; Turkle, 2011). Other works have found that this connected presence represents a spillover between work and family, which has been found to have a negative effect on family satisfaction, especially for women (Wajcman et.al, 2008).

Third, the mobile phone enacts ritualized roles. Studies have found that women's uses of the mobile phone fit the spheres of activity and interests traditionally designated to them, such as taking responsibility for the emotional and material needs of husbands and children, the elderly, the handicapped, or the sick (Zainudeen et al., 2010), while men used phones for work or business

(Huyer, Hafkin, Ertl & Dryburgh, 2006). The mobile device also facilitates women's performance of well-entrenched domestic responsibilities, such as the micro-coordination of schedules and household tasks (Brown, Harper, & Green, 2002; Wajcman et al., 2008; Ling & Haddon, 2003; Rakow, 1992; Dobashi, 2005). Such arrangement of small, practical, and daily logistics underlie a mother and her family's daily ritual, for example, arranging the children's meals and other daily requirements, doing household chores, helping the family deal with health and safety, or paying family bills. A study in India (Doron, 2012, p. 422) found evidence on the uses of the mobile phone to promote gender rituals, whereas "for men, the mobile phone was a tool for work, communication with friends and relatives, and for entertainment, while for the women the mobile was viewed as a tool for 'basic conversations'... the woman's phone should be used only to communicate with her natal kin and husband," which again connects to women's ritualized roles as nurturers of the home (Doron, 2012, p. 422). In the same study, an Indian woman was prevented from visiting her ill husband in the hospital because she "is able to speak with the husband on the household mobile". This restriction in the young wife's physical movement (as mediated by the mobile phone) is in consonance with cultural classifications of gender roles where women, especially young wives, are expected to focus inside the household (420-422).

While the above studies assert that mobile phones function within the traditional gendered division of labor, others present it as a potential site of power struggle, or as a source of liberation from gender norms and restrictions (Lemish & Cohen, 2005, p. 512), where ritualized roles and expectations are transgressed (Pertierra, 2006; Pertierra, Ugarte, & Pingol, 2012). Some studies have found that while mobile phones contribute to the maintenance of patriarchal gender roles, in other contexts, they were found essential for broadening the women's social space, leisure, and other forms of personal gratification, thereby allowing greater possibilities in the performance of gender subjectivity (Wearing, 1999; Rivera, Walton, & Sreekumar, 2012; Hjorth, 2009). Studies have further argued that the emergence of the mobile phone has been accompanied by the increased subversive appropriation of the technology by the active female user (Pertierra, 2006).

Previous works have explored the relationship between queer subjectivities and the Internet (Pullen, 2010; Brickell, 2012; O'Riordan, 2007; Nip, 2004; Austria, 2007; Berry, Martin, & Yue, 2003). However, the implication of mobile media to the performance of queer subjectivity is relatively understudied. Nonetheless, some of the insights advanced in studies probing online communication and queer performativity can be useful, especially with the integration of Internet applications into mobile devices. For example, it has been argued that the Internet offers a key site in which queer subjectivities and collectives can be constructed (Pullen, 2010). Internet spaces, and the social practices embedded in them, coupled with mobility, constitute social life (Crampton, 2003, p. 3) as we become who we are through our interactions in these kinds of spaces (Brickell, 2012, p. 30). The use of the personal gadget such as the mobile device may facilitate the exploration of queer subjectivity beyond the ritualized gender binary of male/female and its accompanying expectations (Hjorth, 2003). For example, aside from facilitating an intensified self-exploration of one's interests and seeking partners, consuming online gay and lesbian erotica, which can be facilitated by mobile Internet, is an important resource for young people to learn about their own sexuality and connect with others who share the same interest. These sites can work to produce "specific sexualities, desires and modes of pleasure" for the LGBT community, which may not have been previously acceptable due to gender norms (Brickell, 2012, p. 40). Moreover, as males, females, or queers publicly reconstitute their sexualities using mobile Internet, they also resist regulative aspects of power and pre-existing social and gender inequalities.

Mobile devices can also facilitate the formation and maintenance of queer communities and social ties that can allow members to find belonging and establish connections to advance particular social and political causes (Gross, 2003; Pullen, 2010; O'Riordan, 2007; Soriano, 2014). The interconnectivity facilitated by mobile Internet enables the swift exchange of queer ideologies and networks across distant spaces, allowing queer individuals "to experience something of a queer community", and obtain advice and information about a variety of queer issues (Fraser, 2010, p. 31; Nip, 2004; Castells, 2010). These communicative spaces also serve as venues for expression and belonging even for those still "in the closet" (Gross, 2003; Pullen, 2010) or those who have "come out" but uncomfortable with public expression of their sexuality (Soriano, 2014). Nonetheless, like the offline world, the use of mobile devices for accessing Internet sites provides vehicles for heterosexism and other forms of prejudice.

CHANGES, CHALLENGES, CHANCES

This section discusses findings from a qualitative pilot study on mobile phone use and meanings for Filipino men, women, and queer. The discussion focuses upon ethnographic interviews with adult mobile users aged 25 and above and a focus group with mobile phone users aged 15-24 conducted between June to October 2013. Adult research participants were selected for in-depth interviews because they are in a good position to reflect on the changes in the enactment of gender rituals across time. The interviews were loosely structured, allowing the stories to spontaneously unfold (Mischler, 1986). They began with "grand tour" questions (Spradley, 1979), for example, "What do you use your mobile phone for?" because preliminary interviews showed that asking direct questions on gender rituals did not generate rich responses primarily because respondents did not realize the enactment of such rituals in the first place. The use of mobile devices for the performance of gender rituals were embedded in the participants' narratives of everyday use of their mobile devices. I coded the data and created descriptive analysis matrices (Miles & Huberman, 1994) to explore how their uses of mobile devices reflect the reinforcement or challenging of gender rituals identified in the literature. New categories were developed for insights divergent from previous literature. The data is discussed in the succeeding section according to emergent themes. The quotes have been translated from a mix of Filipino and English. The names of the respondents have been changed.

As shown in the paper's historical analysis of gender rituals, certain gender norms in the Philippines have become prescriptive. A mismatch between expectation and performance can merit social consequences, for example, a girl aggressively introducing herself to a potential male partner can be branded as *pakawala* (promiscuous) or *pokpok* (prostitute) because she is expected to maintain purity, chastity, and meekness, reserving herself to the man who will win her heart by wooing her. Similarly, the concept of gay men or women openly exploring partners publicly would be frowned upon because it is an inappropriate gender performance based on predominantly Catholic standards. The pilot narratives explored in this chapter locates the role of mobile phones in undermining or changing these discourses. Here, I explore the idea that technology may help people overcome both bodily and societal limitations that allow for a broader range of gender performance and subjectivity.

Mobile Devices and Gender Rituals of Courtship and Romance

The emergence of mobile phones has brought about interesting changes in the ritual of courting and relationship between couples. A male government employee, Andrew, narrated that nowadays, girls seem to have more courage to initiate the expression of their feelings towards men. He also emphasized that girls seem emboldened by the mobile phone to pursue the men whom they like. Drew explains that it has been a norm in the Philippines for women to wait to be courted by men. "Men have to woo the woman, sometimes the man even has to woo the entire family, right? I think it was very rare for women to do the first move. But now that has changed, sometimes you will get a text message from someone that she likes you." Drew thinks that the mediated form of communication gives women the courage to initiate the act because "it's less embarrassing" in comparison to doing it face to face. Drew also narrated how girls in romantic relationships seem to have become more persistent:

When a boy texts the girl, okay, that's it, the boy does not really expect an immediate response. But the girl expects an immediate response from the boy. It even becomes the cause of a lover's quarrel. It's very common. The girlfriend would say, 'I texted you but you did not reply'. It's as if you no longer care about her when you fail to reply immediately. Because it is much faster to communicate, and the mobile phone is a hand held device. So a girl would expect that the guy has no reason not to respond because the mobile phone is handheld.

Andrew further explained that texting for romantic couples follows a particular routine. "First there would be the morning text, 'good morning', or 'how was your sleep?'. Then during lunch, something like, 'have you had your lunch?', and then in the afternoon or *merienda* (snack), and then of course, at night. It does not matter what you text, sometimes it is just an emoticon." He added that either the boy or girl would initiate, but the girl would often initiate the texting. Such ritualized interactions amongst couples support the notion of connected presence, where it is not the content but the maintenance of constant connection that is important (Pertierra, 2007).

The narrative points out the phone's role in increasing the women's capability to take a more active role in courtship and romantic relationships. From the narrative, we see the changing landscape of female romantic expression, which the mobile device is deemed instrumental in facilitating. This finding also runs parallel to Pertierra's earlier findings on how "numerous women chatters ranging from self-professed virgins to the isolated girlfriends and wives of OCWs (Overseas Contract Workers) welcome these modes of communication as rare opportunities to reject patriarchal constraints and develop their sexual personas" (2006, p. 86). Other studies also found that the mobile phone had increased women's role in looking for partners, a vital aspect of their ritualized life, as they use their mobile phones to ring a wide circle of connections in the search for suitable grooms (Tenheunen, 2008, p. 524; Nagasaka, 2007, p. 118-120). Another study by Doron (2012, p. 428) found that the mobile phone allowed Indian women to take part in "illicit" nightly conversations with their partners, which is a way for them to "temporarily escape the restrictions imposed on couples prior to marriage." This shows that women's engagement in such activities over their mobile phones allow them to express themselves as active, desiring individuals.

However, mobile technologies of romance and pleasure appear to simultaneously empower and threaten their users, especially women. In some instances, chat lounges and "sex-texting" are apparently ridden with scammers or some new participants run the risk of unknowingly identifying their identities. In the age of intensified commercialism, mobile phones are also used to contact and promote girls and women involved in sex and tourism, thereby reinstating their role as mere providers of pleasure. Mobile devices are also used to harass women and to distribute videos taken for private purposes. Instances of publicity of 'sex videos' taken conspicuously via the mobile devices has also had more detrimental effect to the woman than the man involved, again because of traditional expectations for women to avoid engaging in "promiscuous sexual acts" (Gurumurthy, 2010).

Mobile Devices and Expression of Gender Identity

Aesthetics of the Mobile Device

Studies of mobile use by the youth (Green & Singleton, 2007; Ling & Donner, 2010) found that mobile devices represent the display or expression of the self, which also reflect the adoption of traditional gender ideologies in contemporary life. For example, the aesthetic decoration of the mobile phone is shaped by pre-identified gender expectations (e.g. colorful or fancy phone casing for girls, sleek black or silver for boys). This is supported by interviews with younger male and female mobile phone users. The female users showed their phones with generally pink, purple, or white casings embellished with various decorations. Representing ritualized images of social interaction, their phones also contained many photographs of their 'selfies' with their friends, family, or pets. The males, on the other hand, use casings with colors culturally attributed to males, such as blue, black, or green. The same is not articulated, however, by older users, where the males or females did not seem to care much about the color or aesthetic aspect of the mobile device. One adult female respondent mentioned that her iPhone is not confined to 'girly colors' and she can play around with colors of her phone depending on her mood. However, although not concerned with the visual appeal of their phones, the adult male respondents noted their aspiration for the devices with sophisticated features and capabilities.

An interesting finding amongst some married users of mobile devices is that the phone is shared between couples, particularly for those who maintain home-based businesses. This then influences the "neutral look" of some of the devices, contrary to previous studies of how female devices were heavily decorated while male devices took on "masculine" designs or colors. For example, a respondent, Shiela, narrated that she and her husband share the phone. Shiela appears to be financially capable of buying her own phone, but because they work in the same business and live in the same house, "there really is no need" for her to own a separate mobile phone. "It's really James who uses the cellphone. Uhm, in fact I don't use the phone often. Even for our friends, James contacts them for both of us, because we basically have the same set of friends. So we just have one number and our friends know that. If my friends have a message for me, sometimes I just ask him to respond for me. So we just share the phone". This shared nature of the mobile device for the couple implies that the phone does not represent any particular gender identity, unlike the younger users.

Gendered Rituals of Leisure and New Avenues for Self-Expression

Leisure, as experienced through mobile devices also appear to be altering gender norms, not only in terms of the availability of "gender-neutral" games, but the personalized and mobile nature of these devices allows the playing of games which traditionally would be considered as oriented towards another gender. Toys represent depictions of lived reality as well as gender rituals—as baby dolls and pink toy houses represent women's nurturing and child-rearing roles; toy guns, swords, and matchboxes signify men's role as the pillar of strength, speed, or security. In the Philippines where gender norms place a demarcation on the games that males and females play, gender-neutral games, and the freedom to play games that transgress gender expectations present an initial step in challenging pre-imposed categories of gendered life.

For example, Arvin, an IT executive, believes that gaming and social networking using a personal gadget such as the smartphone or tablet allows males, females, and queers to express themselves in ways that go beyond acceptable norms. He narrated that contradictory to norms where specific leisure activities are prescribed for each gender, the nature of mobile engagement facilitates the exposure to and eventual involvement in 'gender-neutral' games. Many of his childhood games and toys are highly gender differentiated. For example, he recalled that playing with dolls or plastic kitchen toys are culturally prescribed for girls. Boys found playing them would instantly be branded as *bakla* (gay) or are forewarned, "*Baka maging bakla ka!*" (you might turn into a homosexual). Similarly, girls are not encouraged to play with toy guns (will make them violent) or bikes (they might lose their virginity). On the other hand, *Candy Crush* or *Plants vs Zombies* are "gender neutral" and are played by both men and women. Mobile gaming also allows males to explore games that may be socially unacceptable for them to play otherwise. For example, Arvin explained that he would be able to play the mobile games "*Hair Salon*" or "*Cooking Mama*" without being ostracized; when he was a child, it would have been unacceptable for him to play with kitchen or hair salon-related toys. "In the first place, these plastic toys were often produced in pink, yellow, or red colors and therefore would be off-limits to boys!" Such conditioning of "acceptable gender behavior" even in the context of play facilitated the self-disciplining of social actors of their own preferences and actions that are carried over into adulthood. According to Arvin, he observes that mobile gaming also gives women, men and queer the capability to explore gaming themes that previously would have been "assigned" for a particular gender,

Temple Run is like an Indiana Jones game, an adventure game, while DOTA is a violent game. I think girls would traditionally not play that kind of game, but now, people of any gender would play it. And there are male and female avatars one can choose from. It's the mobility, privacy, and also the ease of access to these games. For example, would you go to a physical store and buy a hair-curler-toy? Of course not, there is too much effort, but because I can easily download or play it anytime without divulging my identity or gender, then I will download it. So it does not care whether I am a girl or a boy or a gay. I think that changed it. Now you are not limited by the norms on what you can play.

Other respondents also argued that in the recent times it is important to break gender lines whether in terms of games or domestic responsibilities. Moreover, whereas traditional leisure opportunities for women are limited, my interviews confirmed earlier findings that mobile gaming allows women, specifically working mothers, to enjoy leisure at "no-where places" and "no-when times" (Caronia, 2005, p.97), such as watching over the baby's sleeping or waiting for the food to cook. However, in these instances, such window for the challenging of gender norms co-exists with deeply entrenched gender roles.

Arvin's narrative points out that the complex nature of the mobile phone as being simultaneously "personal", "shared", and "mobile" facilitate this transgression of gender divisions in the context of play. Although those games can be played in computers, the personalized nature of mobile phones allows greater ease in self-exploration, while the mobility gives one the flexibility on when to exercise leisure. For example, I would play any game when I'm bored or when I'm waiting. Sometimes that allows me to explore any game in the gadget". Yet, although the mobile phone has the capability to be "personalized" in the Philippines, there are instances when the mobile phone is "shared". This sharing of the phone, Arvin explains, creates the possibility for exposure to broad interests and options. For example, because he shares his mobile phone to his 4-year old niece, it becomes more acceptable to download games such as *Cooking Mama* and *Hair Salon*. Yet, this also implies that although he considers the phone "personal", there remains a subconscious guilt of downloading "games for girls", which is explained by the need to justify having to download such games. Arvin also narrated that some of his office colleagues give smartphones to their male and female children for 'sharing', and this helps break pre-set boundaries of what games each child can play, regardless if their parents would dictate the playing of "gender-specific games".

Contrary to the notion of "shared mobile devices" is the "highly personal" device. An interesting insight shared by young female respondents is the active use of their mobile devices as their "diaries on the go". These respondents explained that they have been accustomed to documenting life events, day-to-day emotions or thoughts in their devices, which they are unable to express towards other people. Carla narrated, "I write anything, like when I feel good or when I'm really upset about something or someone, it's just easier to write them in my iPhone". Another participant explained that writing in her "personal" mobile device feels safer and more spontaneous, "at the time that I'm feeling happy or angry, I can write it in my personal notes or tweet about it, so that I don't forget". She narrated that she also used to have a paper diary but writing in her device feels more personal and safe than writing on a physical paper diary which she does not carry at all times and which can be seen by anyone just lying around the house. When I probed whether the mobile phone cannot be found lying around the use, she quipped, "No! I always hold on to my phone, no?!" Another common use by young females of their device is watching or creating YouTube videos and expressing their interests through creative works such as writing blogs or fan fiction. One respondent shared that she has written hundreds of fan fiction (her primary leisure activity), many of them about her K-pop idols, through her iPod touch. She writes fiction everyday which, as she narrates, has many followers. When asked why she opts to write using her mobile device, she explained that the personal computer is shared in her home and her personal mobile device allows her to write freely without fear of being judged by her mom or siblings. These narratives show how the mobile device has created new opportunities for young women to express themselves and explore their identities through this "highly personal" device.

Maintenance and Development of Social Networks

The mobile phone has been found to be an integral player facilitating a sense of belonging. The role of mobile devices in the maintenance and development of social ties that facilitate the construction of gender identity has dominated past literature on mobile communication. Findings from this study found support for this, particularly in the context of members from the LGBT community who are able to find new opportunities to meet new partners and friends through mobile applications:

In this conservative Catholic country, how could a gay person broadcast the search for a partner in a restaurant or a public place? It is not always easy to identify another gay in a bar or public place. My gay friend told me about the app Grindr, which allows the easy identification of a gay in a public space... Finally, it has become much easier for gays to find friends or partners.

The above quote points out that the smartphone's mobility and multitude of applications creates various possibilities for LGBTs to express their gender identity and build social networks in the context of a predominantly conservative culture. The term, 'finally' is used to emphasize a history of repression and challenge amongst the LGBT community to openly find opportunities for association and belonging.

In the same way, mobile devices are actively used across genders to socialize with friends and relatives, exemplifying its role in creating a sense of place and community. Across age, but most especially among the younger females, respondents shared that the mobile device's primary use is for the maintenance of social ties, allowing them to share hobbies, interests, likes, photographs, videos, games, gossip, and news with relatives and friends. This supports previous findings, such as in Yoon' s ethnography in Seoul (2003), which showed how rituals of socialization and familial relationships were reenacted through practices of mobile phone use. This also runs parallel with Horst and Donner's (2006, p. 133-136) findings in the Caribbean wherein the ability to be connected and obtain a source of comfort through shared interests, hobbies, and even gossips becomes a source of comfort for women and a sense of fulfillment. Moreover, in Pertierra's research (2006, pp. 77-95) both males and females, married and unmarried, visit chat lounges using their mobile devices out of a sense of isolation or boredom.

Mobile Devices and Performance of Ritualized Gender Roles

Previous studies have argued that women have been found to feel emancipated by the enhanced connectivity facilitated by the mobile phone, while at the same time burdened by the responsibilities enabled by such perpetual connectivity (Lim & Soon, 2010). Findings from this study show that women do not seem to recognize the burden of connectivity. As their nurturing responsibilities are ingrained to their consciousness, the mobile phone is deemed an important tool to continually perform these responsibilities, despite physical or geographic barriers, while at the same time enabling them to insert opportunities for play, work, or socialization. For example, with three grown-up children, Connie, a 69 year old retired Professor, uses her old Nokia phone mainly for coordinating her family's day-to-day activities. One of Connie's sons has a major medical illness and lives with her, together with his daughter and wife in *Laguna*, a province close to Manila. Her other children, now grown-ups, live in separate apartments in Makati city. Connie spends the week moving from her family's home to her daughter's apartment. She stays with her at some days during the week and spends the rest in *Laguna*. "So the cellphone is very important because when I am in Makati, I would text and check on my son and my granddaughter. When I am in Laguna, I would check on my son and daughter in Makati." I probed what she normally texts them about:

Many things. Random things. Sometimes, scheduling my visits, sometimes about dishes that I can bring or about their clothes...because I prepare their clothes. Sometimes, I would make sure that the dogs are well-taken cared of when I am away. Sometimes, I text my other son about the medical routines or say hello to my granddaughter. Sometimes, I text our relatives to look after our old house, or I would exchange text messages with my old students and co-teachers.

Connie thinks that mobile phones have empowered women significantly. However, Connie's narrative shows that although her phone is deemed primarily for the performance of her nurturing role as mother, it also allows her to socialize with former students and friends, an experience shared by other adult women interviewed who use their mobile devices to connect with friends from church, old friends from college, or distant relatives despite primarily using the phone to support their role as domestic managers.

Mobile Devices and Reversal of Ritualized Gender Roles

In the context of families where wives work overseas, Tadeo-Pingol (1999, p. 22) argued that men are forced to negotiate their sense of masculinity and role as husbands or fathers amidst social and cultural expectations. In the case study below, Roland's narrative shows that the assumed reversal of traditional gender roles facilitated by the overseas labor migration of Filipina wives and mothers, in reality, coincides with the enactment of gender rituals.

Roland maintains a small family business and his wife is a care-giver in Dubai. Roland mentions that he is thankful that the mobile phone allows the cheap and constant communication between his family and his wife. Often, he and the kids would use the app *Viber* to message or call his wife over the phone, and this free application allows the family to discuss with her various issues from the mundane to the important ones. Roland narrated that during the initial months of his wife's placement overseas, she would text him every time to give household-management related instructions. "Yes, she would text everything, from reminding me about the children, to budgeting, to what is the good brand of laundry detergent." Roland explained that he was not really annoyed by such reminders. Instead, he is appreciative that he gets guidance on what to do because his wife knows the home-related matters best. "We have always been sharing household responsibilities, but I was also surprised at the many details that need to be considered. It can be very tiring. But I don't want her to be bothered too much because she also has many responsibilities there." But now, his wife's calls and messages are more about checking on the conditions of the family, the kids' school performance, and the family budget. Roland laughed when he said that he thinks his wife trusts him now about the household responsibilities. Contrary to previous findings (Cabanes & Acedera, 2012), Roland did not seem to have expressed a huge concern over his new responsibility as a home-maker and about being instructed by his wife to perform household responsibilities over the mobile phone. "Of course at first it was a bit uncomfortable, but my wife is lucky to have found a job there. It's not easy to get a good paying job these days."

The phenomenon of left-at-home fathers who perform the mother's role in the home (Cabanes & Acedera, 2012), is becoming common in light of increased migration of Philippine mothers. The study also affirmed earlier findings wherein wives use the phone to contact their husbands for guidance about domestic management. In this instance, it would seem that although the myth of the male breadwinner is challenged, the myth of the female homemaker is retained. This narrative shows that the re-enactment of women's roles can be seen side by side their capability to control, direct, question, or guide their husbands, especially in household management, which can be already considered as a significant challenge to traditional gender norms in the Philippine context. As in the case of Roland, he does not question his role in maintaining the household as a consequence of his wife's overseas work because he perceives manhood as a fulfillment of duties, regardless if these duties are gendered, and the wife seems to be able to support the husband in the performance of these "new duties" through constant communication. Such arrangements present opportunities for the challenging of what constitutes "mothering" and "fathering" and allows husbands and wives to mutually negotiate their roles given present-day realities.

Some respondents shared that the males of the house have become more active in searching for domestic-related information such as recipes, alternative medicine, or family illnesses though mobile applications. Other respondents shared that the task of grocery shopping may have become more distributed between couples since the arrival of the mobile device, as wives may sms the men or males may sms their wives to pick up grocery items depending on whoever is available. Although some married respondents narrated that their husbands "learned how to cook" through smartphone applications, this did not mean that the husband has taken over the full responsibility for cooking within the household and reversed particular role expectations in the household.

SOCIAL, CULTURAL, AND ECONOMIC IMPLICATIONS

Social Implications

There has been concern that the use of mobile technologies is increasingly facilitating individualism rather than strengthening of social ties (Turkle, 2011). As mobile devices with a multitude of applications continue to play a prominent role in people's lives, the sustained media consumption facilitated by mobile devices may alienate users and undermine important social relations. This critique of the social implications of technologies runs parallel to Putnam's concern about how technology operates to fragment social capital (Hjorth, 2009). In the Philippines, however, the mobile device is still actively utilized for maintaining and building social ties. For women in particular, the mobile device is used predominantly for socialization and "performing intimacy" with significant others so as to hold together the fabric of the family and the community by building and maintaining relationships (Wajcman, et.al., 2008, pp. 646-647).

The unique characteristic of mobile engagement emerging from the study, its personal yet 'shared' nature, facilitates the transgression of certain gender expectations and blurring of divides. Activities such as expressing oneself, enjoying the exploration of romantic partners, or exploring 'gender neutral' forms of play challenge gender norms and facilitates the performance of alternative gender subjectivities, thereby transforming gender into more fluid categories (Siapera, 2012, p.180). The mobile phone also seems to create a screen that provides women the confidence to express themselves, whether to a potential lover, a partner, or an aggressor. A female respondent mentioned that she considers the mobile phone as a security device, because based on what she sees in the news, women have been saved from dangerous situations using their mobile phones.

Mobile-mediated dating and social networking is also utilized by the LGBT community to identify each other stealthily, and this could prove helpful in the Philippines where LGBT relationships are still considered taboo or where LGBTs are in danger of being victims of hate crime. It also makes it easy for those 'in the closet' to interact with other LGBTs in their immediate localities. Such actions challenge the limits and strict adherence to traditional gender conventions and allow for more playful gender performances and subjectivities to take place. However, some have also raised concern about the kind of gay culture that applications such as *Grindr* might be shaping– hook ups vs. relationships, promiscuity, or apathy towards larger LGBT causes (Yiannopoulus, 2010), which may reproduce stereotypes and prejudices about the LGBT community. For women, on the other hand, while the capability of finding partners and expressing one's sexuality may be facilitated by the personal nature of the mobile phone, previous studies have shown how this can be used for prostitution or pornography which may also reproduce the misconstrued notion of women as primarily sexual objects (Gurumurthy, 2010).

Economic Implications

Mobile devices allow women to pursue the performance of their domestic obligations while expanding their opportunities for work or maintenance of a career. This includes women who migrate overseas to work, comforted by the fact that mobile phones can allow them to continually establish "perpetual contact" (Katz & Aakhus, 2002) with their families, while becoming economically productive actors. Studies found that women feel empowered by this economic capability, particularly as Filipinos overseas are able to transfer money directly to family back in the Philippines through their mobile devices (Madianou & Miller, 2011). However, others argue that although women are empowered economically, they retain a level of submission to their husbands and continually perform their domestic obligations, sometimes despite the availability of help from back home (Parreñas, 2005; Madianou & Miller, 2011). Thus, although economic roles are challenged, gender rituals remain where women are mothers or daughters first and need to perform their domestic obligations or keep the household together (Soriano, et al., forthcoming).

The uses of the mobile phone, and the perception of the value of the mobile phone to some of the respondents are heavily shaped by the roles traditionally expected of them. This is characterized by the elderly mother who uses the mobile phone to assert her role as the family's domestic manager, the overseas mother who attempts to perform her role as mother and wife through the mobile phone despite her physical absence, or testimonials of women that they are able to find more time for themselves because the mobile phone facilitates ease in the performance of their work and domestic obligations. These affirm the role of women as vanguards of the domestic front, despite the changing roles that they play in the economic front. Other studies have explored the dilemmas of working mothers who find mobile connectivity as both comforting yet burdensome as it stretches the performance of domestic duties beyond the home and even in the place of work (Madianou & Miller 2011; Cabanes & Acedera, 2012; Parreñas, 2001; Katz & Aakhus, 2002; Soriano et al., forthcoming). Yet, my interviews showed that Filipino women do not really recognize such "perpetual contact" as a bane, seeing their nurturing responsibility as natural and a given, they are relieved by the mobile's capability to allow them to perform their obligations, side by side the use of the mobile device for leisure, social networking, or other work obligations.

Moreover, the rise of small scale mobile enterprises such as mobile credit auto-loading or "mobile money services" which is very popular all around Metro Manila and in smaller provincial towns has also allowed many home-based women to earn a living. Such *auto-load* shops are incorporated into small village stores that sell "sachets" of everything, from shampoo to mayonnaise, to mobile phone credit (Mendes et al., 2007). Home-based mothers, who perceive the need to look after the home and children or who are unable to find formal employment, are thus aided by the mobile device as they become economically productive actors and participate in home-based business. Across all genders, the mobile device is used both for social and economic functions, such as looking for jobs and business opportunities, or finding cheap alternatives for particular products and services.

Political Implications

Equality in access to mobile devices pertains to equal opportunities for using the device for social or political expression. The role of mobile devices for political purposes also coincides with the political climate and the degree of tolerance for political expression (Hermanns, 2008). Since the election of former President Corazon Aquino in 1986, women have enjoyed a more active role in Philippine politics. Not only as direct politicians or candidates for government office, women are also active leaders of social movements, activist organizations, non-government organizations, and lobby groups. In all these activities, mobile devices equip women with tools to communicate their respective advocacies and perform their expected roles.

Mobile devices have also played an active role in coordinating the political campaign of the Philippine LGBT political party, Ladlad. In May 2010 and 2013, *Ladlad,* a national LGBT political party of lesbian, gays, bisexuals, and transgenders ran for the nationwide party list elections, which will give the community a seat in lawmaking. Although the community failed to obtain the required number of votes to secure a seat, leaders and members of the community have shared the active use of mobile devices and the Internet in the process of political mobilization

(Soriano, 2014). The value of mobile Internet for queer mobilization is congruent with the findings of Pullen (2010) in the United States, as well as Nip (2004) in the context of 'queer sisters' and electronic bulletin boards in Hongkong, and Lichterman's study among African-American queers (1999). Such spaces, "facilitate 'solidarity-building'... identity groups may well need a safe space to talk in angry unison about, to cry about, injustices perpetrated on one's 'own people,' without fearing rebuke..." (Lichterman, 1999, p. 135). Mobile connectivity also expanded the space for the sharing and discussion of queer issues, which allowed individual queers to act in solidarity with others of similar experiences. In the case of Ladlad, handy and powerful mobile devices allowed the leaders to organize the caravans and campaigns across the Philippine islands, while facilitating the members in joining in the conversation and helping in the coordination of campaign-related events (https://www.facebook.com/angLadlad). Members located in the various provinces also texted and posted updates of the campaigns in their *Facebook* and *Twitter* pages. With promotions from mobile and internet service providers for cheap unlimited access to social networking sites as well as calls and sms, LGBT organizers with limited funding were able to mobilize a collective force, including those who are uncomfortable with public expression of their identity and support to the cause.

CONCLUSION

The implications of technology on society may be understood through dichotomous oppositions: public/private; site of inequality / site of empowerment. Technology engagement, however, needs to be seen not just either as a site for domination or freedom but instead as containing patterns of relations and rituals that are nuanced and complex. This chapter examined the question of rituals and gender performance as mediated by the use of the mobile device. This approach to the understanding of the implication of mobile devices in society seeks to situate the relevance of mobile technologies in influencing everyday gendered practices, while emphasizing that its use is also heavily influenced by cultural rituals. Although the paper focuses on the Philippines as case study, both the approach and the findings can be related and compared to the mutual shaping of mobile devices and gender rituals globally.

This paper adopts the concept of gender as an identity that is constructed rather than one we are born with (Siapera, 2012, p. 179), generated and consolidated as a process of repetition and reiteration (Butler, 1993). Gender identities continue to acquire meaning through the various personal and social activities mediated by the mobile phone everyday as it helps the constitution of people as social subjects. As we establish ourselves through social networking, gaming, sms, or calls, our gender identities are constructed out of these individual interactions in everyday life. Such construction plays an important part in the ongoing self-shaping of particular gender subjectivities, and also in challenging gender norms.

The case studies showed that mobile devices are embedded in culturally valued relations, with the mobile device functioning as a site through which gender roles are maintained and negotiated. The maintenance and negotiation simultaneously occur through various ritualized interactions. Women are able to simultaneously perform their domestic roles while expanding their traditional social networks, performing leisure or work activities, or exploring their personal interests. Members of the LGBT community are also aided by their mobile devices for self-expression, finding partners, and forming social and political collectives. Mobile devices also facilitated the transgression of some gender boundaries in the rituals of leisure, romantic relationships, as well as the undermining of traditional lines that socially, economically, and politically limit the construction of gender identity. However, such "transgressive" activities are continually enacted within the context of continuing gender ideologies that have been constructed within gender rituals that foster ties, impose obligations and acceptable practices, and which work to moderate social order and relationships. The transgression of gender boundaries of play, for example, is performed in the context of the mobile phone's "shared" nature, a unique social arrangement influenced by strong family ties underlying Filipino culture. Secondly, although the phone helps expand the woman's economic possibilities, this is performed in the context of gender rituals -- where overseas Filipinas are still expected to perform or manage the same domestic obligations, and where they feel obliged to continually perform their ritualized nurturing roles for their families over the mobile device. Third, as more women use the mobile device to find space for leisure and social activities, this is negotiated by their roles as household manager or nurturer of children. Fourth, as a woman is seen emboldened by the mobile phone to pursue a potential partner or become an active, desiring individual, it still exists within the context of the need for the acceptance of such female persona by the male partner or the broader society.

An awareness of the ritualized scripting of mobile devices is a useful exercise in understanding the complex implications of mobile devices in society, as well as the changes, challenges, and chances for alternative gender performance. As we have seen from the findings, mobile communication helps to simultaneously erode and reinforce old gender categories and hierarchies, as the mobile device allows its subjects to experiment with socially, economically, or politically transformative relations of gender and sexuality.

REFERENCES

Aguilar, D. D. (1998). *Toward a nationalist feminism*. Quezon City, Philippines: Giraffe Books.

Angeles, L. C. (2001). The Filipino male as macho-Machunurin: Bringing Men and Masculinities into Gender and Development Studies. *Kasarinlan*, *16*(1), 9–30.

Aquiling-Dalisay, G., Nepomuceno-Van Heugten, M. L., & Sto. Domingo, M. R. (1995). Ang Pagkalalaki ayon sa mga lalaki: Pagaaral sa Tatlong Grupong Kultural sa Pilipinas. *Philippine Social Sciences Review*, *52*(1-4), 143–166.

Austria, F. (2007). Gays, the Internet, and Freedom. *Plaridel*, *4*(1), 47–76.

Berry, C., Martin, F., & Yue, A. (Eds.). (2003). *Mobile Cultures: New Media in Queer Asia*. Duke University Press. doi:10.1215/9780822384380

Brickell, C. (2012). Sexuality, power and the sociology of the internet. *Current Sociology*, *60*(28), 28–44. doi:10.1177/0011392111426646

Brown, B., Harper, R., & Green, N. (2002). *Wireless World: Social, Cultural, and Interactional Issues in Mobile Communications and Computing*. London: Springer Verlag. doi:10.1007/978-1-4471-0665-4

Burrell, J. (2010). Evaluating shared access: Social equality and the circulation of mobile phones in rural Uganda. *Journal of Computer-Mediated Communication*, *15*(2), 230–250. doi:10.1111/j.1083-6101.2010.01518.x

Butler, J. (1993). *Bodies that matter: On the discursive limits of sex*. London: Routledge.

Cabanes, J. V., & Acedera, K. (2012). Of mobile phones and mother-fathers: Calls, text messages, and conjugal power relations in mother-away Filipino families. *New Media & Society*, *14*(6), 916–930. doi:10.1177/1461444811435397

Cabrera-Balleza, M. (2005). Gendered, wired and globalized: Gender and globalization issues in the new information and communication technologies. *Review of Women's Studies*, *15*(2), 140–156.

Camacho, M. S. T. (2010). The public transcendence of intimacy: The social value of recogimiento. In *More Pinay Than We Admit: The Social Construction of the Filipina* (pp. 295–317). Manila: Vibal.

Campbell, S. W., & Kwak, N. (2010). Mobile communication and social capital: An analysis of geographically differentiated usage patterns. *New Media & Society*, *12*(3), 435–451. doi:10.1177/1461444809343307

Carling, J., Menjívar, C., & Schmalzbauer, L. (2012). Central themes in the study of transnational parenthood. *Journal of Ethnic and Migration Studies*, *38*(2), 191–217. doi:10.1080/1369183X.2012.646417

Caronia, L. (2005). Mobile Culture: An Ethnography of Cellular Phone Uses in Teenagers' Everyday Life. *Convergence*, *11*(3), 96–103. doi:10.1177/135485650501100307

Couldry, N. (2003). *Media Rituals: A Critical Approach*. London: Routledge.

Crampton, J. (2003). *The Political Mapping of Cyberspace*. Chicago: University of Chicago Press.

De Castro, L. (1995). Pagiging Lalaki, Pagkalalaki at Pagkamaginoo. *Philippine Social Science Review*, *52*(1-4), 127–142.

Dela Cruz, P. (1988). *Images of Women in Philippine Media: From Virgin to Vamp*. Manila: Asian Social Institute in cooperation with the World Association for Christian Communication.

Dobashi, S. (2005). The gendered use of *ketai* in domestic contexts. In M. Ito et al. (Eds.), *Personal, Portable, Pedestrian. Mobile Phones in Japanese Life* (pp. 219–236). MIT Press.

Doron, A. (2012). Mobile Persons: Cell phones, Gender and the Self in North India. *The Asia Pacific Journal of Anthropology*, *13*(5), 414–433. doi:10.1080/14442213.2012.726253

Dumdum, O. (2010). *Jerks without faces: The XTube spectacle and the modernity of the Filipino bakla*. Paper presented at the Annual Meeting of the International Communication Association. Suntec, Singapore.

Fortunati, L. (2009). Gender and the Mobile Phone. In G. Goggin, & L. Hjorth (Eds.), *Mobile Technologies: From Telecommunications to Media* (pp. 23–34). New York: Routledge.

Garcia, J. N. C. (2000). Performativity, the bakla and the orientalizing gaze. *Inter-Asia Cultural Studies*, *1*(2), 265–281. doi:10.1080/14649370050141140

Garcia, J. N. C. (2008). *Philippine gay culture: Binabae to bakla, silahis to MSM* (2nd ed.). Quezon City: University of the Philippines Press.

Garcia, L. (2008). *Manila beams with pride, despite debut of anti-gay protesters*. Retrieved from http://www.fridae.asia/newsfeatures/2008/12/08/2168.manila-beams-with-pride-despite-debut-of-anti-gay-protesters#sthash.rOWaDqSM.dpuf

(2010). Gender Rituals. In Salamone, F. (Ed.), *Routledge Encyclopedia of Religious Rites, Rituals and Festivals* (pp. 145–149). New York: Routledge.

Goffman, E. (1967). *Interaction Ritual: Essays on Face to Face Behavior*. New York: Pantheon.

Gonzales-Rosero, M. A. P. (2000). The Household as a Workplace: Articulation of Class and Gender in Filipino Middle Class Households. *Review of Women's Studies*, *10*(1-2), 41–68.

Green, E., & Singleton, C. (2007). Mobile Selves: Gender, ethnicity and mobile phones in the everyday lives of young Pakistani-British men and women. *Information. Cultura e Scuola*, *10*(4), 506–526.

Gross, L. (2003). The Gay Global Village in Cyberspace. In N. Couldry, & J. Curran (Eds.), *Contesting Media Power: Alternative Media in a Networked World* (pp. 259–272). New York: Routledge.

Gurumurthy, A. (2010). *Understanding gender in a digitally transformed world*. Retrieved from www.itforchange.net/

Hermanns, H. (2008). Mobile Democracy: Mobile Phones as Democratic Tools. *Politics*, *28*(2), 74–82. doi:10.1111/j.1467-9256.2008.00314.x

Hjorth, L. (2003). Pop and Ma: The Landscape of Japanese Commodity Characters and Subjectivity. In Mobile Cultures: New Media in Queer Asia (pp. 158-179). Duke University Press.

Hjorth, L. (2009). *Mobile Media in the Asia-Pacific: Gender and the art of being mobile*. New York: Routledge.

Hoflich, J., & Linke, C. (2011). Mobile Communication and Intimate Relationships. In *Mobile Communication: Bringing Us Together and Tearing Us Apart*. New Brunswick, NJ: Transaction Publishers.

Hondagneu-Sotelo, P., & Avila, E. (2003). I'm here, but I'm there: The meanings of Latina transnational motherhood. In P. Hondagneu-Sotelo (Ed.), *Gender and U.S. Immigration: Contemporary Trends* (pp. 317–340). Berkeley, CA: University of California Press. doi:10.1525/california/9780520225619.003.0015

Horst, H., & Miller, D. (2006). *The cell phone: An anthropology of communication*. Oxford, UK: Berg.

Huyer, S., Hafkin, N., Ertl, H., & Dryburgh, H. (2006). Women in the Information Society. In G. Sciadas (Ed.), *From the Digital Divide to Digital Opportunities: Measuring Infostates for Development*. Montreal, Canada: Orbicom.

International Telecommunication Union (ITU). (2013). *Key Global Telecom Indicators for the World Telecommunication Service Sector.* Retrieved from http://www.itu.int/ITU-D/ict/statistics/at_glance/KeyTelecom.html

Ito, M., Okabe, D., & Matsuda, M. (Eds.). (2005). *Personal, Portable, Pedestrian: Mobile Phones in Japanese Life*. Cambridge, MA: MIT Press.

Katz, J., & Aakhus, M. (Eds.). (2002). *Perpetual Contact: Mobile Communication, Private Talk, Public Performance*. Cambridge, UK: Cambridge University Press. doi:10.1017/CBO9780511489471

Komisar, L. (1987). *Corazon Aquino: The Story of a Revolution*. New York: George Braziller, Inc.

Lemish, P., & Cohen, A. (2005). On the gendered nature of mobile phone culture in Israel. *Sex Roles*, *52*(7/8), 511–521. doi:10.1007/s11199-005-3717-7

Levy, S., & Zaltman, G. (1975). *Marketing, Society, and Conflict*. Englewood Cliffs, NJ: Prentice Hall.

Libed, B. P. C. (2010). *Dekada '70 and activist mothers: A new look at mothering, militarism, and Philippine martial law*. (MA Thesis). San Diego State University, San Diego, CA.

Licoppe, C. (2004). 'Connected presence': The emergence of a new repertoire for managing social relationships in a changing communication technoscape. *Environment and Planning. D, Society & Space*, *22*(1), 135–156. doi:10.1068/d323t

Lim, S. S., & Soon, C. (2010). The influence of social and cultural factors on mothers' domestication of household ICTs –Experiences of Chinese and Korean women. *Telematics and Informatics*, *27*(3), 205–216. doi:10.1016/j.tele.2009.07.001

Ling, R. (2007). Mobile communication and mediated ritual. In K. Nyiri (Ed.), *Communications in the 21st Century*. Budapest: Academic Press.

Ling, R. (2008). The mediation of ritual interaction via the mobile telephone. In J. E. Katz (Ed.), *Handbook of Mobile Communication Studies* (pp. 165–176). Cambridge, MA: MIT Press. doi:10.7551/mitpress/9780262113120.003.0013

Ling, R., & Donner, J. (2010). *Mobile communication in everyday life: New choices, new challenges. Mobile Communication* (pp. 75–106). Malden, MA: Polity.

Ling, R., & Haddon, L. (2003). Mobile telephony, mobility, and the coordination of everyday life. In J. E. Katz (Ed.), *Machines that become us: The social context of personal communication technology* (pp. 245–265). New Brunswick, NJ: Transaction Publishers.

Madianou, M., & Miller, D. (2011). Mobile phone parenting: Reconfiguring relationships between Filipina migrant mothers and their left-behind children. *New Media & Society*, *13*(3), 457–470. doi:10.1177/1461444810393903

Matsuda, M. (2005). Mobile Communications and Selective Sociality. In M. Ito, D. Okabe, & M. Matsuda (Eds.), *Personal, Portable, Pedestrian: Mobile Phones in Japanese Life* (pp. 123–142). Cambridge, MA: MIT Press.

McVeigh, B. (2004). *Nationalisms of Japan: Managing and Mystifying Identity*. Oxford, UK: Rowman and Littlefield.

Mendes, S., Alampay, E., & Soriano, E. et al. (2007). *The Innovative Use of Mobile Applications in the Philippines: Lessons for Africa*. Stockholm: Swedish International Development Cooperation Agency.

Miles, M. B., & Huberman, A. M. (1994). *Qualitative data analysis*. Thousand Oaks, CA: Sage.

Mischler, E. (1986). *Research interviewing: Context and narrative*. Cambridge, MA: Harvard University Press.

Nagasaka, I. (2007). Cellphones in Rural Philippines. In *The Social Construction and Usage of Communication Technologies Asian and European Experiences* (pp. 100–125). University of the Philippines Press.

Nie, N. H. (2001). Sociability, interpersonal relations, and the Internet: Reconciling conflicting findings. *The American Behavioral Scientist*, *45*, 420–435. doi:10.1177/00027640121957277

Nip, J. Y. M. (2004). The Queer Sisters and its electronic bulletin board: A study of the Internet for social movement mobilization. *Information Communication and Society*, *7*(1), 23–49. doi:10.1080/1369118042000208889

O' Riordan, K. (2007). Queer Theories and Cybersubjects: Intersecting Figures. In K. O'Riordan, & D. Phillips (Eds.), *Queer Online* (pp. 13–30). New York: Peter Lang.

Parreñas, R. (2001). Mothering from a distance: Emotions, gender, and intergenerational relations in Filipino transnational families. *Feminist Studies*, *27*(2), 361–390. doi:10.2307/3178765

Parreñas, R. (2005). Long distance intimacy: Class, gender and intergenerational relations between mothers and children in Filipino transnational families. *Global Networks*, *5*(4), 317–336. doi:10.1111/j.1471-0374.2005.00122.x

Pertierra, R. (2006). *Transforming Technologies, Altered Selves. Mobile Phone and Internet Use in the Philippines*. Manila: DLSU Press.

Pertierra, R. Ugarte, E., Pingol, A., et al. (2012). Txt-ing selves: Cellphones and Philippine modernity. Manila: De La Salle University Press.

Philippine LGBT Crime Watch. (2012). *A Database of Killed Lesbian, Gay, Bisexual and Transgendered Filipinos*. Retrieved from http://thephilippinelgbthatecrimewatch.blogspot.com/

Philippine Overseas Employment Agency (POEA). (2012). *OFW Deployment Statistics.* Retrieved from http://www.poea.gov.ph

Pullen, C. (2010). The Murder of Lawrence King and LGBT Online Stimulations of Narrative Copresence. In C. Pullen, & M. Cooper (Eds.), *LGBT Identity and Online New Media* (pp. 17–36). New York: Routledge.

Quindoza-Santiago, L. (2010). Roots of Feminist Thought in the Philippines. In *More Pinay Than We Admit: The Social Construction of the Filipina* (pp. 105–119). Manila: Vibal.

Rakow, L. F. (1992). *Gender on the line.* Urbana, IL: University of Illinois Press.

Rivera, M., Walton, M., & Sreekumar, T. T. (2012). *ICTs, Gender, and Leisure: Soft Forms of Subversion.* Paper presented at the International Association of Media and Communications Research. Durban, South Africa.

Rook, D. (1985). The Ritual Dimension of Consumer Behavior. *The Journal of Consumer Research, 12*, 251–264. doi:10.1086/208514

Siapera, E. (2012). *Understanding New Media.* London: Sage.

Silverstone, R., & Hirsch, E. (Eds.). (1992). *Consuming Technologies: Media and Information in Domestic Spaces.* London: Routledge. doi:10.4324/9780203401491

Soriano, C. (2014). Constructing Collectivity in Diversity: Online Political Mobilization of a National LGBT Political Party. *Media Culture & Society, 36*(1), 20–36. doi:10.1177/0163443713507812

Soriano, C., Lim, S., & Rivera, M. (forthcoming). The Virgin Mother with a Mobile Phone: Ideologies of mothering and technology consumption in Philippine television advertisements. *Communication, Culture & Critique.*

Spradley, J. P. (1979). *The Ethnographic Interview.* New York: Holt, Rinehart & Winston.

Sreekumar, T. T. (2011). Mobile Phones and the Cultural Ecology of Fishing in Kerala, India. *The Information Society, 27*, 172–180. doi:10.1080/01972243.2011.566756

Tadeo-Pingol, A. (1999). Absentee Wives and Househusbands: Power, Identity & Family Dynamics. *Review of Women Studies, 9*(1-2).

Tenheunen, S. (2008). Mobile technology in the village: ICTs, culture, and social logistics in India. *The Journal of the Royal Anthropological Institute, 14*, 515–534. doi:10.1111/j.1467-9655.2008.00515.x

(2007). The Transformative Capacities of Technology: Computer Mediated Interactive Communications in the Philippines: Promises of the Present Future. InPertierra, R. (Ed.), *The Social Construction and Usage of Communication Technologies Asian and European Experiences* (pp. 189–226). University of the Philippines Press.

Turkle, S. (2011). *Alone together: Why we expect more from technology and less from each other.* New York: Basic Books.

Wajcman, J., Bittman, M., & Brown, J. (2008). Families without Borders: Mobile Phones, Connectedness, and Work-Home Divisions. *Sociology, 42*(4), 635–652. doi:10.1177/0038038508091620

Wearing, B. (1999). *Leisure and feminist theory.* London: Sage Publications Ltd.

Yiannopoulus, M. (2010, August 25). Grindr: Combatting loneliness or a cruising ground for gays? Gay social networks remain controversial and iPhone app. Grindr is no exception. *The Telegraph.* Retrieved from http://www.telegraph.co.uk/technology/social-media/7964000/Grindr-combatting-loneliness-or-a-cruising-ground-for-gays.html

Yoon, K. (2003). Retraditionalizing the Mobile Young People's Sociality and Mobile Phone Use in Seoul, South Korea. *European Journal of Cultural Studies*, *6*(3), 327–343. doi:10.1177/13675494030063004

Zainudeen, A., Iqbal, T., & Samarajiva, R. (2010). Who's got the phone? Gender and the use of the telephone at the bottom of the pyramid. *New Media & Society*, *12*(4), 549–566. doi:10.1177/1461444809346721

KEY TERMS AND DEFINITIONS

Gender: A social construction and representation, an attribute that is performed and shaped by social rules and the continued transformations, such as being male or female.

Gender Ritual: Refers to symbolic activities that in their repetitive form construct gender norms and ideologies within a particular locale.

Ladlad: A political party of lesbians, gays, bisexuals, and transgenders in the Philippines. The translation of 'ladlad' is to 'unfurl a cape used to cover one's body as a shield. It means to come out of the closet, to assert one's human rights as equal to that of other Filipinos' (Ladlad, website).

Prepaid: A payment scheme for mobile services for which credit is purchased (through prepaid cards) in advance of service use.

Ritual: Refers to a type of symbolic expressive activity constructed of multiple behaviors that occur in a fixed, episodic sequence and that tend to be repeated over time. Rituals shape the everyday sense of social order and identify the 'right way of doing things' in a locale.

Sachet Marketing: Refers to promotions that allow the purchase and exchange of items (including mobile credit) at small denominations, usually offered in small variety stores in rural and urban localities and villages.

Spanish Rulebooks: Pertain to documents written and directed to the mothers of the house during the Spanish period in the Philippines as scripts for how daughters and sons ought to be raised in accordance with particular values and role expectations.

Transnational Mothering: Pertains to the performance of mothers' nurturing responsibilities from geographically distant locations, normally performed by overseas working mothers.

Chapter 6
Mobile and Intimate Conflicts:
The Case of Young Female Adults in Nigeria

Gbenga Afolayan
Federal Polytechnic Ilaro, Nigeria

ABSTRACT

The recent trends in the developing world show that mobile telephony has signalled a significant milestone for the way people communicate. Even though there are still issues of unequal access to mobile phones, the fact remains that young people's access to mobile phones has become prevalent in Africa with its resultant impacts on virtual mobility, individual's technical capability, and personalised interaction. This chapter provides another local perspective on the transformative debate of mobile phones, the place of young adults in it, and its social consequences on intimate relationship in Nigeria. After reviewing the recent literature on mobile usage in Africa, and its social consequences on intimate conflicts, the chapter explores the ways in which the mobile phone has altered the modes of circulating information that is meant to remain secret. The young adults' usage indicate that the mobile phone, which is seen as an example of a globally imagined technological tool, is appropriated in localised ways in which the local slogan, "Wahala don sele" (problem has started), is articulated in intimate conflicts. The chapter concedes that mobile communication plays major roles in daily affairs of young female adults, but it has its social consequences. Therefore, it suggests that future research should look beyond the normative view of mobile development, which only concentrates on governance, education, business, and health.

INTRODUCTION

Mobile telephony has come of age 'with a big splash'—making history as one of the quickest diffusing communication technologies that reached almost six billion subscribers in 2011 (Wei, 2013). In the last decade, young people's access to mobile phones has quickly become prevalent in Nigeria just as in most other countries. The use of mobile phones has led to a new perspective of virtual mobility in terms of increase in the individual's technical capability, quick accessibility (to anybody, anywhere, and anytime), geographical extension and personalised social interaction. This growing perspective and use of mobile phones is an increasingly defining feature of contemporary

DOI: 10.4018/978-1-4666-6166-0.ch006

life in Nigeria, as well as elsewhere in Africa. Indeed, mobile phones are expanding at a faster rate in terms of their designs, contents and applications, with young people representing the greatest proportion of consumer market in the post-modern world.

In more than a decade, Nigeria has moved from a situation of limited and controlled media, marked mainly by government monopoly of communications, to a liberalised and pluralistic mobile communication environment by privately owned Global Satellite Mobile phone service providers. The story began in 2001 when ECONET Wireless and MTN Communications were issued a Global System for Mobile Communications (GSM) operating licence, while M-Tel (a national carrier) was given an automatic licence. By 2003, Globalcom was also granted operating licence as the second national carrier (Elegbeleye, 2005; Olatokun & Bodunwa, 2006). The introduction of mobile communication system by the Nigerian government was to increase the Nigerian teledensity and in extention to make telecommunication more cheaper and accessible to the common man. Just like in other developing countries, deregulation of telecommunications potentially contributed to the growth in mobile phone users in Nigeria (Elegbeleye, 2005). In 2004, there were almost two billion mobile users globally. Most developed countries have a penetration rate of 70 percent and most countries in the developing world follow close behind (Ling, 2004; Castells, Fernández-Ardèvol & Qiu, 2007).

In Nigeria today, mobile communication has transformed lives and generated new socio-economic opportunities (Pyramid Research, 2010). For example, daily activities such as banking transactions, shopping and trade have increasingly been transformed by new forms of mobile usage. Besides, the way mobile communication services have evolved across the country, and the speed at which they are being subscribed to have shown the potency of mobile communication services to both urban and rural Nigerians. A shining example of mobile innovation in Africa is M-Pesa—a mobile payment service in Kenya. It has provded opportunties to people without bank accounts to have access to easy-to-use, widely accessible and cheap money transfers. This mobile initiative has enabled potential customers to send mony quickly and securely to another mobile phone user (Hughes & Lonie, 2007; Etzo & Collender, 2010). In 2012, *Firstmonie*, a mobile financial service solution, was launched by the First Bank of Nigeria to enable subscribers to conveniently perform banking transactions through the use of mobile phone. In particular, the initiative helps to mitigate the challenges of banking services delivery to the vast unbanked market. Besides, in Nigeria, mobile media are also used as reliable tools for collection of medical information, supporting diagnosis and improving cancer care, and disseminating health education in a low resource setting (Odigie, Yusufu, Dawotola, Ejagwulu, Abur, Mai, Ukwenya, Garba, Rotibi & Odigie, 2012). Another example from Nigeria shows the new mobile health (mHealth) initiative. The organisation called 'mPedigree' has developed a service in Nigeria which enables consumers to check whether anti-malaria medicine produced by pharmaceutical companies are genuine or not. A code displayed on the anti-malaria medicine is sent via text message to a toll-free number, and within two seconds of texting the code, a message appears on the consumer's mobile phone with the word "YES"—a simple response meaning the drug is genuine (BBC, 2013).

Mobile media are also being used for political applications to enhance accountability and transparency (Etzo & Collender, 2010; Bratton, 2013). For example, during the Nigerian elections in 2007 and 2011, volunteers sent their observations about the poll via text message to a central database, and the reports were then relayed to other monitoring groups. Phenomena of these kinds demonstrate increasing world networking, however as Arminen (2007) notes, mobile communication is indexically tied to local circumstances and is underpinned by ways of life which tend to show persistent cultural differentiation. That means mobile media enable free-flow interaction between people and are integrated with everyday practices and identities (Ito, Okabe & Matsuda, 2005). In line with everyday practices and identities, attention to the phenomenon of young female adults and mobile usage typically focuses on its most transformational, incremental, sensational and titillating aspects. Ever since mobile media has been launched in 2001, the image of young female adult has become imprinted in the Nigerian mind as the symbol of concern. Yet the issues surrounding them in Nigeria are somewhat complex than this stereotypical image due to the way Nigeria (as elsewhere in Africa) is culturally structured (Afolayan, 2011). Mobile media tend to be rendering Nigerian society into some new form, and enabling different social patterns that have not been in existence in Nigeria for some time to evolve in socially significant ways.

Contemporary young people's communication and social interaction are mostly influenced by mobile communication technologies. In particular, the emergence of mobile phones into the world of intimacy, gender inequality, and interpersonal relationships has both facilitated and problematised the ways through which intimate and sexual relationships are constituted and managed in Nigeria . As mobile media start gaining prominence, they tend to fuel the incongruous debate about intimate relationships in Nigerian society, as elsewhere in developing countries, especially among the young ones. Likewise, it is quite challenging to ascertain the impacts of mobile media on the gender inequalities that support the social organisation of intimate and sexual relationships. On the one hand, demand for mobile phones might intensify the relationship between material inequality and sexual decision-making, putting young female adults at even greater risk of having sex with older men. On the other hand, as young female adults own mobile phones—often openly demanded as gifts from their male (older) lovers, they tend to use mobile media to control men's access to them and their access to men and their resurces (Smith, 2006).). All these demonstrate that mobile media expand both 'possibilities and paradoxes' of how men and women, boys and girls, men and young female adults relate to each other. One could then ask: In which ways can mobile phones influence social interaction or become a tool of intimate conflict? And do mobile media liberate young female adult?

Therefore, this chapter will focus on the ways in which the use of mobile phones has transformed information, not necessarily the often-prevailing 'useful information' considered by technological enthusiasts and proponents of the 'ICT for Development' (Robbins & Webster, 1999; Donner, 2008), but rather information that is expected to remain hidden. Following Archambault's work, this chapter put aside the techno-enthusiasm that often prevails to take a critical look into some instances when access to mobile communication backfires and when mobile phones are seen to fuel conflict. The chapter will argue that the prevention of incriminating phone calls and text messages is likely to open up new discursive-style spaces through which couples can 'navigate and negotiate' the terms of their social relationships. By so doing, this chapter attempts to provide another perspective on the transformative discourse of mobile communication in Nigerian context in order to enhance our understanding of the impacts of the spread of mobile communication. To do this, this chapter will set out the debate theoretically before presenting the landscape of mobile telephony in Nigeria and the place of young adults in it. Next, it will then take cursory look into the pivotal role that mobile phone communication plays in intimate conflicts, and its social consequences.

MOBILE IN AFRICA

Mobile telephony, a wireless communication system and hand-held platform for human communication, seems to have thoroughly transformed the daily affairs of people throughout the world. It has facilitated what communication scholars termed "the end of mass communication" (Maisel, 1973; Chaffee & Metzger, 2001). Mobile telephony strikes hard at the basic idea of the '*massness*' of communication in two ways: (a) people create and sustain social networks through the ubiquitous smartphone; and (b) individuals are connected via mobile phones that bring about what Licoppe (2004, p.135) called "connected presence" in mobile phone-saturated settings. For Ling (2004, 2008), he articulates this kind of mediated connectedness through the mobile phone as mobile connection. Globally, there are 68 mobile phone subscriptions and 27 Internet users per 100 people (ITU, 2010). Adoption rate of the mobile phone in Africa signals a significant milestone for the ways people communicate (Castells et al., 2007). From negotiations of gender discourse in both public and private to the development trajectories that are understood in daily struggles to communicate, it appears that mobile telephony has changed daily life of many people in sub-Saharan Africa (Archambault, 2011, 2013).

Indeed, things on the African continent have changed impressively since Castells (2000, p.92) talked about "Africa's technological apartheid at the dawn of the information age". For the vast majority of people in Africa, the spread of mobile phones has triggered puzzlement in the way people at the bottom of the income pyramid have access to mobile telephony. For example, electronically mediated communication gained popularity in Burkina Faso, "even when the economic environment seems to dictate quite different priorities" (Hahn & Kibora, 2008, p.94). In particular, some argue that mobile phones, as well as Information and Communication Technologies (ICTs), might even help 'save' Africa (Butler, 2005) because of its pivotal role in amplifying development efforts. Others demonstrate how users have developed different approaches to address the economic hurdles of mobile communication (Hahn & Kibora, 2008). Given the increased diffusion of mobile communication technologies in Africa, it was generally assumed that these technologies would help countries in the South "leapfrog" stages of development (Muchie & Baskaran, 2006). Proponents of this notion, typified as the 'ICT for Development' (ICT4D), hold that ICTs have the potential to empower people and enhance entrepreneurial activities by facilitating communication in contexts where other means of telecommunication are poorly developed (Afolayan, 2014).

However, in the last decade, several scholars have started considering unequal access to ICTs or what is known as 'digital divide'. And many observers have suggested that ICTs might even intensify new forms of exclusion (Muchie & Baskaran, 2006; Bridges, 2001, cited by Nielinger, 2006). Even though digital divide has been conceived as having wider consequences, it is often believed that ICTs have a leveraging potential. This point of view resulted in further mobilisation efforts towards universal access (Touré, 2008). Besides, there is also a shift from the transformation of technologies to the transformation of information (Robins & Webster, 1999; Osborn, 2008) which tend to permeate through policy and business lenses. This framework enriches mobile communication with the potential to boost development efforts by enabling the flow of 'useful information' which revolves around business, healthcare, education and governance-driven information (Melkote & Steeves, 2004; Donner, 2008). Taken together, the diffusion of mobile phones across African countries has triggered technological enthusiasm (Nyamba, 2000; Hahn, 2012). As a result, universal access is not only seen as an objective but also as a core issue to several problems confronting Sub-Saharan Africa.

Even though access is important, it is just only part of the ICT4D project. Users still need to use their mobile phones in development-inspiring ways. Concerted efforts to provide access tend to play down this perspective by assuming that once access is universally gained, people (users) will start accessing 'useful' information, and development will follow (Archambault, 2011). Several ethnographic studies on the use of mobile phones indicate, otherwise, that the relationship between ICTs and development usually rests more on individual thinking than empirical findings, perhaps because of young people's sociality (Yoon, 2003, 2006) and local realities of each environment. There is a great variation in the use of mobile phones. For example, Horst and Miller (2006) argued that mobile phones have the potency to contribute to the alleviation of poverty because of their role in the lubrication of coping strategies, but they are not always used to engage in entrepreneurial activities. In line with the foregoing, Burrell (2008) looks at the way in which internet users in Ghana tend to secure access to resources through misrepresentations of themselves and of Africa. Consequent upon its implications, however, she identified scamming as 'problematic empowerment' (ibid., p.27) which eventually problematize the gainful nature of information. Burrell (2008) also points out that inequalities are understood to reflect marginal access to wealth rather than limited access to information. All these, therefore, call for a new direction of the ICT4D discourse rather than a total rejection of what mobile communication can offer.

MOBILE PHONE COMMUNICATION, INTIMATE CONFLICTS, AND ITS SOCIAL CONSEQUENCES

Intimacy has various meanings to different people and cultures; however it is women that have been mostly implicated due to intimate and mobility footings (Hochschild, 2001, 2003). This is intensified by mobile media practices in which public and private distinctions are elusive. Mobile phone communication plays a crucial role (Wajcman, Brown & Bittman, 2009), particularly in what Gregg (2011) calls "presence bleed". As it is, debates on mobile phone communication and intimacy are undoubtedly beset with prevailing gendered practices. Contemporary couple intimacy in African societies revolves around diversity and interconnected transformations, with regards to the social context, gender relations, structures, family philosophies, practices, interactions and subjectivities (Ohiagu, 2010; Afolayan, 2011). Some prominent sociologists have investigated those practices and their social consequences for sexual intimacy in terms of a passage from traditional bonds to more consensual ones or 'pure relationships'. As Lasen and Casado (2012, p.550) note, "these assume or at least aspire to sexual and emotional equity, whose compromise can be undone". In a setting characterised by uncertainty and inequality, women are said to be unequal to men. Intimate conflicts then arise, when love is more essential than ever and also more intolerable (Beck and Beck-Gernsheim, 1995).

In this way, mobile communication practices play a pivotal role for intimate relationships, now that the characterization of both feminine and masculine subjects as equals open up many debatable issues around the world. Mobile media play a crucial part in the emergence, expression and management of these intimate conflicts and dynamics of power relations. Mobile media technologies have the potency to mediate disagreements, facilitate different strategies, and create platforms for arguing and making up. As Lasen and Casado (2012) observe, mobile media support the designing and defending of territories: either personal or collective, such as the couple or family ones. Analysing mobile phone communication within sexual relationships can be rewarding for an insightful investigation of the articulations between love, intimacy and conflicts and between the so-called 'traditional' and 'consensual' domains. This clarifies the fact that gendered inequality and subjectivities are not being increasingly dissolved, however they are frequently performed, mediated and remediated in ordinary practices of intimate relationships/couple intimacy, which now have to deal with more conflicts (Casado, 2002).

In the last decade, mobile media, social network sites and online dating webs usually play a part in love and intimate relationships until the breakup. These often happen in flirtation, adultery and other strategies of seduction (Wei, 2007). Amidst all these media, mobile phones are the most widespread and accessible, contributing largely to the reconfiguration of intimate relationships (Pertierra, 2005). The continuous contact and connected presence made possible mobile phones—'wireless leash' (Qiu, 2007), modifies communication patterns in the intimate relationship. This permanent availability of the partner's virtual presence plays a major role in setting the boundaries of the partners' or couple's territory. It can also bring about intimate conflicts between the partners as well, taking cognisant attention of the potential disagreements between one's own time, space, uncertainty, growing disparity, and private realm and the partner or couple sphere (Comaroff & Comaroff, 1999; Davidson, 2010). This intimate conflict can be experienced differently based on the gender positions, models and subjectivities. For example, as Archambault (2013, p.88) observes:

Mobile phones play a conspicuous role in the way young people in Inhambane, Mozambique, juggle visibility and invisibility in their everyday lives. By opening up new social spaces within which individuals can engage in various pursuits with some degree of discretion, mobile phone communication helps redress socioeconomic inequalities while preserving an unpleasant public secret about the workings of Mozambique's postwar economy: that young women are encouraged to exchange sexual favors for material gain.

Another different finding is that of Lasen and Casado (2012, p.552):

Most of the male participants in our research express their fears or unease toward a perceived or potential lack of autonomy within the couple's dynamics. They insist on how they need their "own space," using terms such as "invading" to reject the permanent contact that they attribute to their partners' wishes. This appears to be a reason to argue as well, when men feel that their partners' calls interfere with their work and answer them in what is perceived by women as an offensive or curt manner. At the same time, men acknowledge that this lack of autonomy is a normal feature of being in a couple. As a thirty-four-year-old participatn summarises it: "you are losing individuality and gaining as a couple".

These two findings show that balancing intimate relationship and autonomy is challenging due to the transformations engendered by gender and couple relations. As other scholars have argued (Horst & Miller, 2006; de Bruijn, Nyamnjoh & Brinkman, 2009), possessing a mobile phone is an evidence of being a member in a world that is ladden with paradoxes for most. While mobile phones are seen as a tool of liberation in some settings (Katz & Aakhus, 2002), their impact in other contexts is more ladden with ambiguities (Ito et al., 2005; Horst & Miller, 2006; Mazzarella, 2010). In a place like South-Western Nigeria, where most people have passed directly from no phone to mobile phones, mobile media open up new spaces of sociality, entrench and exacerbate unequal power dynamics between men and women, employers and employees, male and female partners and couples. The questions that guide this chapter are as follow:

- In spite of the fact that mobile media expand possibilities of how we relate to each other, do mobile media expand possibilities of how we intimately relate to each other, either as a partner or couple?
- In which ways can the mobile usage, often used by young people, influence the way intimacy and social interaction are constituted?

The aim of this chapter is to investigate how Nigerian people's mobile phone practices influence the relationship issues, and in particular to examine how intimate relationships are experienced and are acted out through mobile phone communication. The next section is based on the author's ongoing fieldwork in Ilaro town (one of the Yoruba communities in the South-Western Nigeria) between 2012 and 2013 on the usage of mobile phones. The author is opportuned to have lived in a peri-urban neighbourhood of Ilaro for more than 20 months. During this time, the author developed close relationships with a number of Ilaro indigenes and residents. This period was also complemented by short follow-up trips during the author's national service in 2008. Coupled with the author's direct observation, in-depth interviews were conducted with 50 young men and women aged between 15 and 33, concerning relationship issues, and how these closely aligned with the mobile phone practices. The author addressed the issue of mobile phone practices and other relevant discussions more directly in semi-structured group discussion sessions. Following Horst and Miller's (2006) and Archambault's (2011) work on phone use in Jamaica and Mozambique respectively, the author also used in-depth phone analyses with ten of the interviewed young adults via a collection of 'sent and received text messages'. Therefore, this chapter draws on these ongoing research findings.

LANDSCAPING MOBILE IN ILARO

Ilaro is one of the Yourba towns under Yewa South Local Government Area (LGEA) in Ogun State, Nigeria. It is a town that comprises mainly Christains, muslims and traditionalists. Before the advent of Christianity and Islam, it is a common saying that Yoruba believed in their own deities such as Sango (god of thunder), Ogun (god of iron), Yemoja and so many other gods which tend to change with each geographical location. Ilaro lies between the territory of the Yewas (formerly known as Egbados) and the international boundary with Benin Republic (Asiwaju, 2001, 2004). The Yorubas take their cultural values and traditions seriously because they form an important part of daily life. Ilaro town is approximately 50km from Abeokuta (the capital city of Ogun State), and about 100km from Lagos. Ilaro town was founded around 18th century amidst the proctracted communal wars from the neighbouring Benin Republic known then as Dahomey. By the mid 19th and early 20th century, Ilaro town benefited immensely from the Missionaries who introduced western education to the people of Yewaland (Asiwaju, 1976, 2001). Following the establishment of the Federal Polytechnic of Ilaro, Nigerian military barrack in Owode, cement factories (e.g. Dangote Cement Factory), many people living in rural areas fled to urbanised towns in Yewaland (Ilaro, Owode) in search of earning a living. This might have added pressure on urban towns in Yewaland where residents juggled limited job opportunities, the rising cost of living and the smuggling activities.

Ilaro town does not have the hustle and bustle associated with an urban environment because of its relatively small population. It is as rural as urban but not like Ibadan or Lagos (still in South-Western Nigeria). Most households rely on petty trade (e.g. okada riding), agriculture, teaching and a few government institutions to make ends meet, and some still face food shortages periodically. A number of households now have electricity and cement houses are being built here and there. The normal signs of economic progress are also increasingly visible in the town and its environs. In Ilaro town, shop windows display imported wares, shoes, alcoholic beverages/drinks, cars (popularly known as 'Tokunbo') and the latest mobile phones. Likewise, the spread of mobile phones has been remarkable in the last five years.

Since mobile communication technology has been introduced in Nigeria in 2001, the number of phone users rose to 1.6million within the first 16 months to 3.8 million within three years (Ndukwe, 2006). In 2008, the number of phone users has grown to over 62 million (ITU, 2008). By implication, it is evident that the number of users have grown in Ilaro. Looking at the specific segments of the population in Ilaro town, the figures are even more striking, although there is no authentic survey conducted yet. When one looks at the young adults who are at the centre of this research, some of them live in independent households while others are staying with their parents, usually in women-headed households. Also, some are still attending school, others have just recently finished schools, a few of them have a regular source of income from petty jobs, but most, even the educated ones, lacked the capital necessary to secure formal job, or to start their own businesses, and many are doing anything. All of them, however, aspired to a similar lifestyle in which the consumption of modern consumer goods figured prominently. Even though there is still lack of infrastructure in some areas, most have also passed from no phone to mobile phone.

Relevant to this chapter are the mobile phone practices and discourses from the perspective of young adults. This is not because they are the principal phone users in Ilaro town or Nigeria as a whole but because by standing at crossroads in terms of intimate relationships and childbearing, their aspect of lives offer interesting insights into the social consequences of the changing modes of access to information. Even as the demographic insights of African youths in itself calls for our attention, examining how young people create a platform for themselves through several options is important when it comes to determing development outcomes (Abbink, 2004)—whether ICT-driven or not. Generally, young adults in Ilaro use mobile communication technology in their everyday lives in different ways: searching for job opportunities and and coordinating illegitimate business activities like smuggling activities, contacting classmates, begging a relative for help with school fees and using pseudonames to ask for money from lovers, managing multiple relationships and controlling their partner's sexuality, communicating with distant relatives and insulting rivals. Inspite of the new openings created by mobile phone, many still express conflicting ideas towards it (Archambault, 2011). Listening to the narratives of young female adults, it became clear that the technology came with implicit costs that cannot be quantified. The most critical issue is that that they are concerned with the conflicts that emerge in their intimate affairs, following the interception of an incriminating phone call or text-message. This a key area for this chapter.

DEBATING INTIMATE CONFLICTS IN ILARO

Simply put, intimate relationships can likely be linked with perspectives such as 'daily' interactions, love, caring, sharing, close friendship, 'deep' and 'mutual' understanding and detailed knowledge. Having intense interaction with a person or sharing detailed knowledge about each other is required for trust and faith, which in turn are basic perspectives of intimacy and friendship. Thus, one of the distinguishing marks of an intimate relationship in this modern time is that friends are voluntarily chosen based on a personal predisposition, where mutual inclination and individual choices are required. This is in contrast with Yoruba kinship, which is not chosen, but obligatory and recognised. However, as Jamieson (1998) and Elegbeleye (2005) also point out, dimensions of intimate relationships and expectations of needs in terms of love and caring are culturally multifaceted, socially constructed and take on different meanings over time. Globally, the daily forms of social life of humans are changing. Upon witnessing the gradual processes—associated with ICTs, global forces and the overlay of a 'modern' upon traditional institutions—the contexts and vernacular aspects of how the meanings of intimate relationships seen are being less structured and scripted by Yoruba community norms, requiring abilities to navigate diverse social worlds. These and other historic transformations are altering the daily social experiences of adolescent and young female adults in Ilaro, as well as in other parts of Nigeria—even beyond the coast of Nigeria.

In particular, the average age of marriage has recently risen but the number of marriages in Ilaro has been decreasing. Instead of focusing on marriage, young adults engage in loosely intimate relationships (i.e. 'going out with someone'). This is a modern phenomenon which stems majorly from current socio-economic realities of Nigeria. In fact, communication is a significant perspective of this kind of relationship because young female adults see this characterisation as marking a break between themselves and previous Yoruba generations. There is, however, a double standard that tends to prevail and, unlike men for whom it is expected to have multiple partners based on social-cultural values of Yoruba people, a woman is not supposed to have more than one partner. Infidelity appears to be common and slightly expected, in particular from men but increasingly from young female adults in the last decade (including adult women) as well. Just like in some parts in Africa, jealousy appears to be one of the important themes of daily interaction. To some, jealousy is often seen, experienced and discussed as an evidence of love. For instance, although Adetoye was confident that his girlfriend did not cheat on him, he still constantly probed her fidelity, 'for her to know that I have feelings and caring thoughts for her', he expounded. In some situations, though, jealous feelings could turn to violence. Speaking from the perspective of women involved with married men Bola, another female student, explained:

Jealousy is part of cultural traits of men in Yoruba communities, although it may vary. It demonstrates the love such man has for his wife, although some are just jealous naturally to cause problem. As a person, I am even more jealous than my man friend.

In some peri-urban areas in Ilaro town (Oke ola, Oke ela, etc), daily life usually happens in the public sphere under the watch of neighbours, friends or passers-by, and rumours easily get spread through some gossipers in each neighbourhood. Without any evidence of dishonesty, rumours can be denied. As Adewale, a man in his early twenties, notes 'without evidence, I'll still love her—I mean my girlfriend, because people can be saying rubbish in order to destroy our relationships'. There is even a general belief that some people just take pleasure in destroying another people's relationships by tempting their partners or simply by spreading rumours about them. Likewise, as noted by Archambault (2011), some people can just be obsessed with neighbour's life in reference to:

Individual's acute interest in the private affairs of fellow residents. The dismissal of hearsay should, however, not be read as indicative of trust between partners; in many cases it rather reflects local notions of respect, wishful thinking or religious belief.

'WAHALA DON SE LE' (PROBLEM HAS STARTED)

The advent of mobile phones has altered the modes of circulation of information that is meant to remain secret. Mobile communication has, indeed, altered the way knowledge is disseminated in different ways; however some are more contentious than others. To start, technological mediation of interactions can facilitate attempts at falsification and misrepresentation (Burrell, 2008). To lie over the phone is quite easy, and as Archambault (2011, p.450) notes, "some have even designed complementary strategies in the hopes of turning the phone into a more efficient tool of control". For example, Samuel, a man in his early thirties who has just been transferred to Ilaro from Lagos, is kept away from home pending the time his family will join him and he really feels concerned about his wife's activities whenever he is not with them. He always calls his wife, even at midnight in order to confirm whether his wife is actually at home. When phoning her, he often asks his wife to wake up their daughter and put her on the phone. Aside this, phones act as a "repository of personal information" (Ling, 2008, p.97) and contain, in call logs, inboxes, sent messages, outbox, saved items, traces of interactions that provide material evidence of deceit. In particular, romantic messages serve as a compelling proof of infidelity than *'won ni won pe'* (hearsays) of a person who might have ulterior motive for spreading indicting information (cf. Paine, 1967). Consequent upon this, it cannot as easily be dismissed as rumours. Intercepted phone calls can, as shown below, also tell more of infidelity in exhilarating manners.

Noticeably, the ongoing survey showed that 34 per cent of male owners and 52 per cent of female phone owners reported having fought with their partners. In the words of Bayo, a young man in his late twenties:

Having a phone sometimes inspires one, however sometimes reverse be the case. This is because of relationship stuffs. For example, if you meet a girl and you exchange numbers with each other. Supposing, one day, she eventually phoned you late at night while you're with your girlfriend and your phone rings, what will you do? Some girls may say that such guy should put it on speaker. And if you talk....at least you can guess what will happen next, hahaha! Especially, if it happens that that girl keeps calling you, sweetheart, baby, my heart etc, wahala go sele that night!

Women can also get caught. As Nkechi, a young woman also in her late twenties, described:

I don put (have put) myself in this kind of trouble before. Sometimes last year, my boyfriend picked a call from my former boyfriend when I was taking my bath. That day, I received two dirty slaps of my life. He accused me that I might have been sleeping with him all this while.

To avoid that kind of trouble, Folasade usually empty her inbox. She explained further that 'even though my boyfriend suspects me, at least, he does not have a proof on which to base his suspicions'. Therefore, the interception of incriminating messages and calls is but one of the ways in which mobile phones can spark intimate conflict. 'When you try to reach your boyfriend and all you hear is 'pe̍ mi pada' (call me back), the first thing that comes to mind is that the person is with someone else,' explained Tofunmi. Archambault (2011) notes that reasons for suspicion that would never have existed in the past, are now appearing 'because of mobile phones'. Oluwakemi, a young woman who sells local drinks for drivers at the motor park, recounted the following 'phone story' which gives a sense of these dynamics. A particular night, while Oluwakemi was with her lover, her husband who works in Idogo (very close to Ilaro) and only comes to visit once or twice every two weeks, phoned her at around midnight, only to hear '*pe mi pada*' (call me back). When Oluwakemi reconnected her phone a couple of hours later, she received the call of an enraged husband who was very suspicious about finding his wife's phone disconnected at midnight. He promised to give her the beating of her life upon his next visit. Oluwakemi's friend (Zainab) suggested she should tell him that she was sleeping at the time, or that her phone battery just suddenly fell flat. Both discussed these options whether they sounded convincing. However, her husband later called and it further generated some arguments.

In some situations, 'wahala don sele' can be regarded as a common slogan when men and women are victims of false accusations. Within relationships that do not rest on exclusivity or at least the pretence of exclusivity, finding a suspicious message might not be such a problem, although both partners may still take offence if other relationships are revealed (Archambault, 2011).

It is another story when it comes to more serious relationships in which the woman is expected to be faithful and the man respectful, like married couples or those that have engaged as Yoruba custom demands. Rather than unfaithfulness, it is phone that is said to spawn break-ups of intimate relationships or conflict. When we examine the differences between suspicions of unfaithfulness fed by gossips and those fed through implicating messages and phone calls, it becomes clear that the phone's role in the circulation of information that is meant to remain secret has serious consequences (Ibid.). Mobile communication and the *'wahala'* (problem) it triggers opens up new conversational platforms within which couples can negotiate the terms of their intimate relationships.

By providing an 'evidence' of unfaithfulness, phones allow young female (married or unmarried) to talk with their husbands or boyfriends about matters that would otherwise be hurriedly rejected as unfounded. As Archambault notes, couples would not wait for mobile phones to start arguing but the phone is as "forensic science to crime scene investigation: it helps buttress accusations" (2011, p. 453). It is mainly for this reason that many people opine that mobile phone is a 'necessary evil' that one cannot do without.

CONCLUSION

Indeed, mobile telephony might have come of age 'with a big splash' as coined by Wei (2013, p.50), given the fact that young people's access to mobile phones has become prevalent in most African countries with resultant effects on virtual mobility, individual's technical capability, accessibility, personalised interaction, and so on. Still, this idea that mobile phones have the potency to contribute to socio-economic development has been critiqued by a number of scholars that highlight the importance of local appropriation of information that is facilitated (Ferguson, 2002; Escobar, 2005; Horst and Miller, 2006; Molony, 2008). These studies lay emphasis on the dialectical landscape of the link between the user and the technology, and explore the transformation of artefacts into socially meaningful and historically, socially as well as culturally constructed ones. According to Strathern (1992), 'the freedoms afforded by new technologies might feel new, but the tyrannies are likely to be all too familiar' (p. xi). Regarding Africa in particular, the phone has been described as the 'new talking drum,' thus highlighting continuity with older modes of communication. A salutary quote relevant to this claim reads that:

The mobile phone, through the social relationships it forges or entertains, and the economic possibilities it opens up, simultaneously challenges and reinforces the status quo, allowing for consolidation and renewal in ways not immediately obvious if treated in isolation or outside specific socio-economic contexts. In this way, the mobile phone reproduces social stratifications even as it is actively transforming them through the creativity and innovation that it provokes or condones (de Bruijn et al., 2009, p.15).

Views of mobile phones as change agents are, nonetheless, often closer to local interpretation than the appropriation reading (Sorensen, 2006). This is evident in the 'wahala don sele' slang that a young adults use to discuss conflicts that emerge when phone triggers conflict. Besides, just as in Ilaro, many Nigerians associate phone-related conflicts to their lack of experience or due to the fact that he/she is just a 'beginner'. Still, a recurring matter is that mobile phones were created with good intentions in mind (Robins & Webster, 1999). The resultant effects of the connection existing between users and ICTs frequently challenge expectations and problems that shadow several assumptions of the ICT for Development (ICT4D) discourse. The local discourses that are weaved into infidelity, as shown above are very critical. Indeed, the interception of information meant to remain secret alters, to some extent, infidelity across the lines of patriarchy and gender hierarchies that trigger it (Archambault, 2011; Afolayan, 2011). Therefore, information should be understood as relational (Burrell, 2008) and contextual (Afolayan, 2014) rather than as inherently beneficial (Castells, 2000).

In sum, it is clear that changes in the way information is accessed have a transformative possibilities as long as we know that what determines 'useful information' is contextual and varies across African countries, as well as elsewhere. As seen in the cases presented above, it also means it depends on where one stands in a falsehearted relationship. Unlike other ethnographic interpretations that critique the ICT4D paradigm, the ethnographic explanation highlighted here indicates that by facilitating the spread of information, mobile communication technology can liberate young female adult from the perspective of socio-economic transformations. Like Archambault (2011), this chapter concedes that by adding to the debate of redrawing of gender hierarchies and patriarchy in Nigeria as well as in West Africa, the pivotal role mobile communication technology plays in daily affairs of young female adults can likely have wider transformative social impacts. Future research should thus look beyond the normative view of socio-economic development which revolves around governance, education, business, and health.

REFERENCES

Abbink, J. (2004). Being young in Africa: The politics of despair and renewal. In *Vanguard or Vandals: Youth, Politics and Conflict in Africa* (pp. 1–34). Leiden: Brill Inc.

Afolayan, G. E. (2011). *Widowhood Practices and the Rights of Women: The Case of South-Western Nigeria*. (Unpublished MA dissertation). Erasmus University, Rotterdam, The Netherlands.

Afolayan, G. E. (2014). Critical Perspectives of E-government in developing World: Insights from Emerging Issues and Barriers. In *Technology Development and Platform Enhancements For Successful Global E-government Design* (pp. 395–414). Hershey, PA: IGI Global.

Archambault, J. S. (2011). Breaking up 'because of the phone' and the transformative potential of information in Southern Mozambique. *New Media & Society*, *13*(3), 444–456. doi:10.1177/1461444810393906

Archambault, J. S. (2013). Cruising through uncertainty: Cell Phones and the politics of display and disguise in Inhambane, Mozambique. *American Ethnologist*, *40*(1), 88–101. doi:10.1111/amet.12007

Asiwaju, A. I. (1976). *Western Yorubaland under European Rule, 1889–1945: A Comparative Analysis of French and British Colonialism*. London: Longman and Highland.

Asiwaju, A. I. (2001). *West African Transformations: Comparative Impacts of French and British Colonialism. Lagos*. Malthouse Press Ltd.

Asiwaju, A. I. (2004). Frontier in Egba [Yoruba] History: Abeokuta, Dahomey and Yewaland in the 19th Century. *Lagos Historical Review*, *4*(1), 18–48.

BBC. (2013). *Nigerian Texters to take on the drug counterfeiters*. Retrieved from http://www.bbc.com/news/world-africa-20976277

Beck, U., & Beck-Gernsheim, M. E. (1995). *The Normal Chaos of Love*. Cambridge, MA: Polity Press.

Bratton, M. (2013). Citizens and Cell Phones in Africa. *African Affairs*, *112*(447), 304–319. doi:10.1093/afraf/adt004

Burrell, J. (2008). Problematic empowerment: West African Internet scams as strategic misrepresentation. *Information Technologies and International Development, 4*(4), 15–30. doi:10.1162/itid.2008.00024

Butler, R. (2005). *Cell phones may help 'save' Africa*. Retrieved from http://newsmongabay.com/2005/0712-rhet_butler.html

Casado, E. (2002). *La construccio'n sociocognitiva de las identidades de ge'nero de las mujeres espan˜olas (1975–1995)*. Madrid: Universidad Complutense de Madrid.

Castells, M. (2000). *End of Millennium*. Oxford, UK: Blackwell.

Castells, M., Fernández-Ardèvol, M., Qiu, J. L., & Sey, A. (2007). *Mobile Communication and Society*. Cambridge, MA: The MIT Press.

Chaffee, S., & Metzger, M. (2001). The end of mass communication? *Mass Communication & Society, 4*(4), 365–379. doi:10.1207/S15327825MCS0404_3

Comaroff, J., & Comaroff, J. L. (1999). Occult Economies and the Violence of Abstraction: Notes from the South African Postcolony. *American Ethnologist, 26*(2), 279–303. doi:10.1525/ae.1999.26.2.279

Davidson, J. (2010). Cultivating Knowledge: Development, Dissemblance, and Discursive Contradictions among the Diola of Guinea-Bissau. *American Ethnologist, 37*(2), 212–226. doi:10.1111/j.1548-1425.2010.01251.x

de Bruijn, M., Nyamnjoh, F., & Brinkman, I. (2009). Introduction. In *Mobile Phones: The New Talking Drums of Africa* (pp. 11–22). Leiden: Langaa & African Studies Centre.

Donner, J. (2008). Research approaches to mobile use in the developing world: A review of the literature. *The Information Society, 24*(3), 140–159. doi:10.1080/01972240802019970

Elegbeleye, O. S. (2005). Prevalent Use of Global System of Mobile Phone (GSM) for Communication in Nigeria: A Breakthrough in Interactional Enhancement or a Drawback? *Nordic Journal of African Studies, 14*(2), 193–207.

Escobar, A. (2005). Imagining a post-development era. In *The Anthropology of Development and Globalization: From Classical Political Economy to Contemporary Neoliberalism* (pp. 341–351). Oxford, UK: Blackwell.

Etzo, S., & Collender, G. (2010). The Mobile Phone 'Revolution' in Africa: Rhetoric or Reality? *African Affairs, 109*(437), 659–668. doi:10.1093/afraf/adq045

Ferguson, J. (2002). Of mimicry and membership: Africans and the 'New World Society'. *Cultural Anthropology, 17*(4), 551–569. doi:10.1525/can.2002.17.4.551

Gregg, M. (2011). *Work's Intimacy*. Cambridge, MA: Polity Press.

Hahn, H. P. (2012). Mobile phones and the transformation of the society: Talking about criminality and the ambivalent perception of new ICT in Burkina Faso. *African Identities, 10*(2), 181–192. doi:10.1080/14725843.2012.657862

Hahn, H. P., & Kibora, L. (2008). The domestication of the mobile phone: Oral society and new ICT in Burkina Faso. *The Journal of Modern African Studies, 46*(1), 87–109. doi:10.1017/S0022278X07003084

Hochschild, A. (2001). *The Time Bind: When Work becomes Home and Home becomes Work*. Holt Press.

Hochschild, A. (2003). *The Commercialization of Intimate Life: Notes From Home and Work*. Berkeley, CA: University of California Press.

Horst, H. A., & Miller, D. (2006). *The Cell Phone: An Anthropology of Communication*. Oxford, UK: Berg.

Hughes, N., & Lonie, S. (2007). M-PESA: Mobile Money for the unbanked Turning Cellphones into 24-Hour Tellers in Kenya. *Innovations*, *2*(1–2), 63–81. doi:10.1162/itgg.2007.2.1-2.63

International Telecommunications Union (ITU). (2008). *Mobile cellular subscription*. Retrieved from http://www.itu.int/ITU-D/icteye/Reporting/ShowReportFrame.aspx?ReportName=/WTI/CellularSubscribersPublic&RP_intYear=2008&RP_intLanguageID=1

Ito, M., Okabe, D., & Matsuda, M. (Eds.). (2005). *Personal, Portable, Pedestrian: Mobile Phones in Japanese Life*. Cambridge, MA: MIT Press.

Jamieson, L. (1998). *Intimacy: Personal Relationship in Modern Societies*. Cambridge, MA: Polity Press.

Katz, J. E., & Aakhus, M. (2002). Conclusion: Making Meaning of Mobiles—A Theory of *Apparatgeist*. In *Perpetual Contact: Mobile Communication, Private Talk, Public Performance* (pp. 301–318). Cambridge, UK: Cambridge University Press. doi:10.1017/CBO9780511489471.023

Lasén, A., & Casado, E. (2012). Mobile Telephony and the Remediation of Couple Intimacy. *Feminist Media Studies*, *12*(4), 550–559. doi:10.1080/14680777.2012.741871

Licoppe, C. (2004). Connected presence: The emergence of a new repertoire for managing social relationships in a changing communication technoscape. *Environment and Planning: Society and Space*, *22*(1), 135–156. doi:10.1068/d323t

Ling, R. (2004). *The Mobile Connection: The Cell Phone's Impact on Society*. San Francisco: Morgan Kaufmann.

Ling, R. (2008). *New tech, new ties: How mobile communication is reshaping social cohesion*. Cambridge, MA: MIT Press.

Maisel, R. (1973). The decline of mass media. *Public Opinion Quarterly*, *37*(2), 159–170. doi:10.1086/268075

Mazzarella, W. (2010). Beautiful Balloon: The Digital Divide and the Charisma of New Media in India. *American Ethnologist*, *37*(4), 783–804. doi:10.1111/j.1548-1425.2010.01285.x

Melkote, D. R., & Steeves, H. L. (2004). Information and communication technologies for rural development. In *Development and Communication in Africa* (pp. 165–173). Oxford, UK: Rowman & Littlefield.

Molony, T. (2008). Non-developmental uses of mobile communication in Tanzania. In *Handbook of Mobile Communication Studies* (pp. 339–351). Cambridge, MA: The MIT Press. doi:10.7551/mitpress/9780262113120.003.0025

Muchie, M., & Baskaran, A. (2006). Introduction. In *Bridging the Digital Divide: Innovation Stems for ICT in Brazil, China, India, Thailand and Southern Africa* (pp. 23–50). London: Adonis & Abbey Publishers Ltd.

Ndukwe, E. C. (2006). *Three Years of GSM Revolution in Nigeria*. Retrieved from http://www.ncc.gov.ng/speeches_presentations/EVC's%20Presentation/GSM%20REVOLUTION%20IN%20NIGERIA%20%20-140504.pdf

Nielinger, O. (2006). *Information and Communication Technologies (ICT) for Development in Africa*. Frankfurt, Germany: Peter Lang.

Nyamba, A. (2000). La parole du téléphone: Significations sociales et individuelles du téléphone chez les Sanan du Burkina Faso. In *Enjeux des Technologies de la Communication en Afrique: Du Téléphone à Internet* (pp. 193–210). Paris: Karthala.

Odigie, V. I., Yusufu, L. M. D., Dawotola, D. A., Ejagwulu, F., Abur, P., & Mai, A. et al. (2012). The mobile phone as a tool in improving cancer care in Nigeria. *Psycho-Oncology, 21*(3), 332–335. doi:10.1002/pon.1894 PMID:22383275

Ohiagu, O. P. (2010). Influence of information & communication technologies on the Nigerian society and culture. In *Indigenous societies and cultural globalization in the 21st century*. Germany: Muller Aktiengesellschaft & Co.

Olatokun, M. W., & Bodunwa, I. O. (2006). GSM usage at the University of Ibadan. *The Electronic Library, 24*(4), 530–547. doi:10.1108/02640470610689214

Osborn, M. (2008). Fuelling the flames: Rumour and politics in Kibera. *Journal of Eastern African Studies, 2*(2), 315–327. doi:10.1080/17531050802094836

Paine, R. (1967). What is gossip about? An alternative hypothesis. *New Series, 2*(2), 278–285.

Pertierra, R. (2005). Mobile phones, identity and discursive intimacy. *Human Technology, 1*(1), 23–44.

Pyramid Research. (2010). *The Impact of Mobile Services in Nigeria: How Mobile Technologies are Transforming Economic and Social Activities*. Author.

Qiu, J. L. (2007). The wireless leash: Mobile messaging service as a means of control. *International Journal of Communication, 1*(1), 74–91.

Robins, K., & Webster, F. (1999). *Times of the Technoculture*. London: Routledge.

Smith, D. J. (2006). Cell Phones, Social Inequality, and Contemporary Culture in Nigeria. *Canadian Journal of African Studies, 40*(3), 96–523.

Strathern, M. (1992). Foreword: The mirror of technology. In *Consuming Technologies*. London: Routledge.

Touré, H. I. (2008). Welcome address. In *ICTs in Africa: A Continent on the Move*. ITU TELECOM Africa.

Wajcman, J., Brown, J., & Bittman, M. (2009). Intimate connections: The impact of the mobile phone on work life boundaries. In *Mobile technologies* (pp. 9–22). Routledge.

Wei, C. (2007). *Mobile Hybridity: Supporting Personal and Romantic Relationships with Mobile Phones in Digitally Emergent Spaces*. (Unpublished doctoral dissertation). University of Washington, Seattle, WA.

Wei, R. (2013). Mobile media: Coming of age with a big splash. *Mobile Media & Communication, 1*(1), 50–56. doi:10.1177/2050157912459494

Yoon, K. (2003). Retraditionalising the mobile: Young people's sociality and mobile phone use in Seoul, South Korea. *European Journal of Cultural Studies, 6*(3), 327–343. doi:10.1177/13675494030063004

Yoon, K. (2006). Local Sociality in Young People's Mobile Communications: A Korean case study. *Childhood, 13*(2), 155–174. doi:10.1177/0907568206062924

KEY TERMS AND DEFINITIONS

Digital Divide: It is simply considered as unevenly matched or unequal access to ICTs.

ICT4D: It is an acronym standing for 'ICT for Development'. It often connotes the initiatives designed through the utilisation of ICTs to achieve socio-economic development.

ICTs: It is an acronym that stands for 'Information and Communication Technologies'—detailing technological platforms used in the management of information through its 'Information Lifecyle'.

Intimacy: Means a sense of closeness and warmth that people feel towards someone they understand mutually and love deeply.

Intimate Conflicts: These are normal un-loving feelings, thoughts, interactions, tensions and verbal expressions that might run counter to intimacy and closeness that individuals have towards to someone they love.

Intimate Relationships: Intimate relationships are intense (daily) interactions between two or more persons (our emphasis is on two-person's intense interactions) who share detailed knowledge of love, caring, sharing, close friendship, deep and mutual understanding about each other.

Mobile Communication: This is a mobile media-enabled communication through mobile platform (or a wireless communication system) for the exchange of information between and among a small number of connected users, which can as well be identified.

Mobile Media: It is a personal, interactive, internet-supported and user-controlled portable platform that make the exchange of and sharing of personal and non-personal information among users who are inter-connected to be available (available in a range of hand-held devices from mobile phones, tablets, and e-readers to game consoles).

Social Network: It is a socially-enabled platform that is made up of social actors (individuals, organisations and governments), as well as a set of the dyadic ties between these actors.

Young Adults: Based on the location and context of this research, we technically refer to young adults as those people between the ages of 15 and 35 (they could be male, female, student, single, married), whoare capable to process, digest and exchange information—deliberating what to do, what to choose, what to think and who to relate with.

Section 3
Mobile Engagement and Movement

To engage people is to facilitate and foster civic conversation and participation, which has been part of the tradition of media, traditional or new. Switching from offline to online, media engagement does not change its function. To go mobile is another advance in media engagement. Instead of opening up a new space, mobile engagement extends what we already occupy, such as health, education, elections, debates, campaigns, journalism, banking, data collection, shopping, business, trade, gaming, entertainment, and narration. Speaking of narration, use of mobile is redefining narrative media engagement, which is investigated in Chapter 7. Besides being widely used in engagement, mobile has also been widely and increasingly used for civic movements not only in most liberal societies but also in most authoritative countries, which is unprecedented in history. For instance, as documented in Chapter 8, mobile communication has been used for seeking and sharing information, galvanizing support, and mobilizing action for political and social issues in Singapore. Mobile engagement and mobile movement go hand in hand in most cases. As examined in Chapter 9, in Malaysia, youth engagement is said to take the center stage of democracy, and young adults use mobile to mobilize members and resources for online and/or offline movements to seek social, economic, and political changes.

Chapter 7
Mobile Engagement:
Dynamics of Transmedia Pervasive Narratives

Elizabeth Evans
The University of Nottingham, UK

ABSTRACT

The evolution of mobile technologies with the capacity to act as conduits to narrative media has led to the emergence of increasingly transmedia and pervasive storytelling practices. This chapter interrogates a mobile phone-based pervasive drama, The Memory Dealer (Lander, 2010), and a focus group run with participants, focusing on two "shifts" within the performance: from internally focused narrative (via headphones) to externally focused narrative (via public speakers) and from an individual experience to a communal experience. By considering how The Memory Dealer's audience responded to such shifts, it becomes possible to consider the inherent nature of being "an audience" of narrative transmedia. It is argued that the experimental pervasive forms facilitated by mobile technologies raises the potential to critically re-examine the nature of media engagement, the relationship, and the technological, social, and cultural dynamics that shape and are shaped by that engagement.

INTRODUCTION

One of the most significant technological developments to occur in relation to mobile technology is its capacity to act as an access point for narrative media. Since the early 2000s, mobile telephones have not only become more complex in terms of computing power, but also in their capacity to act as a platform for film, television or gaming related texts. This technological convergence (Jenkins, 2006, p. 2) between mobile telephony and screen media has wider consequences for the social and cultural circulation and experience of media texts. This chapter will focus specifically on the experimental use of mobile technology in relation to fictional narratives and consider the implications of this experimentation on our understanding of the social dimensions of both mobile technology and narrative itself.

Engagement with narrative media has, until recently, been relatively singular and stable. Each narrative form has been associated with a specific

DOI: 10.4018/978-1-4666-6166-0.ch007

set of social and cultural factors that distinguish them from each other. Film and television involved moving images and sounds with little input from the viewer, watched in communal semi-private spaces and the home (see, for example, Morley, 1986; Carroll, 2003). Theatre brings the audience into closer contact with performers, privileging issues of liveness and presence (see Urian, 1998; Reason, 2004). Videogames prioritise 'interactivity', with the audience taking a direct role in moving the narrative forward (see, for example, Juul, 2004; Carr et. al. 2006). However the evolution of mobile technologies with the capacity to act as conduits to narrative media, has led to the emergence of increasingly transmedia and pervasive storytelling practices. By transmedia storytelling I refer to Henry Jenkins' (2006) model of 'a single narrative so large that it cannot be contained in a single medium' (p. 95; see also Evans, 2011). By pervasive I refer to a term beginning to gain currency within the study of human-computer-interfaces that refers to using mobile technologies to facilitate the bleeding of fictional narratives into real public spaces. With transmedia pervasive drama the established boundaries between narrative forms break down as different storytelling techniques, viewing contexts and modes of engagement are brought together into a single narrative experience.

This chapter will use *The Memory Dealer* (*TMD*), created by video artist Rik Lander, as a case study of experimental transmedia pervasive drama and a focus group run after its first public performance. *TMD*'s aim was to use mobile devices to create emotional engagement in a fictional narrative in public space. At the heart of the piece was a conscious attempt to explore the possibilities of mobile media technologies and the modes of engagement they encourage. Rather than act as an expositional drama, in which the audience are told events by a narrator or character, *TMD* acted instead as an 'experiential drama', in which the audience uncover the narrative as they take part in events themselves. This positioned the audience within the fictional diegesis in a more direct way than is the case in more traditional narrative forms. The focus group discussion explored audience members' responses to the piece and the mobile technology at its heart and raised issues concerning the ways in which social and technological factors played out in terms of their experience as 'an audience'.

Transmedia pervasive dramas offer a particularly useful way to interrogate the nature of engagement with both mobile technologies and fictional narratives. David Beer (2012) calls on Slavoj Žižek's theory of 'cogito' to examine how mobile media can reshape an individual's relationship to the world around them:

We might begin by thinking about mobile media in relation to this vision of cogito, the 'empty' in-between spaces of everyday life that exist between the more clearly defined spaces of the home [oikos] and public spaces of the city, town or even village [polis]...We might, perhaps more significantly, imagine that mobile media also allow space to be redefined, transforming the clearly defined spaces of the oikos and the polis into the empty and less definable space of the cogito (p. 362).

Transmedia pervasive dramas fit into this reshaping, transforming 'in between spaces' into narrative-enriched spaces, layered, as we shall see, with fictional memory and meaning. At the same time, the portable and individual nature of mobile devices creates narratives that exist everywhere, layered onto the material world, sitting amongst the social and cultural practices of daily life. They ultimately offer experiences of both mobile technology and narrative that are highly distinct from their well established uses. Single narratives are now found across multiple media platforms and within multiple public and semi-private spaces, with each platform and space facilitating different kinds of narrative engagement. The nature of these narratives is explicitly tied up with the nature of mobile technology and the way that it shapes (and is shaped by) its user's relationship to broader social and spatial structures. Examining this more experimental form of narrative allows us to consider the consequences of mobile technology's evolution into a platform for narrative media for our understanding of engagement.

Elsewhere I have discussed the difficulties in ensuring audience engagement with media when screen narrative is taken away from traditional technologies and spaces such as the television set and the home. In the case of mobile television drama, audiences were reluctant to engage with such content in public spaces because of a clash between expectations of the content to be emotionally immersive and the technology to be used in public spaces (Evans, 2011, pp. 133-139). The expectations of public space to be explicitly social, and its accompanying distractions and discomfort, can act as a barrier to the kind of immersion that the darkened spaces of the theatre or cinema, or the controllable space of the home facilitate. Despite these potential barriers, however, the practice of transmedia pervasive drama continues to grow. Such content remains predominantly within the field of experimental theatre, is consciously created for public spaces and does not have to fight with pre-existing expectations associated with more established forms such as theatre, film or television. At the same time, the focus of pervasive drama around *mobile* technologies allows greater movement for those engaging with it, both between technological platforms (and so the narrative forms best suited to them) and between spaces. By considering how audiences responded to this movement, which shifts were accepted and which became problematic, it becomes possible to consider the inherent nature of being 'an audience', rather than being a 'television audience', 'film audience', 'radio audience' or 'theatre audience'. Central to this is the relationship between the nature of mobile technology, the experience of audio-visual narrative and the social dimensions of 'audiencehood'.

THE MEMORY DEALER: CREATING A PERVASIVE TRANSMEDIA DRAMA

TMD was performed on 11th September 2010 at and around Broadway Cinema and Media Centre in Nottingham city centre and its experimental nature requires a certain amount of explanation to allow for a more in depth analysis. The drama brings together notions of pervasiveness and transmediality, combining textual and technological forms into a fictional narrative that is laid across real-world spaces. It used mobile audio devices, radio, projected film and theatre to create a science-fictional world where memories are recorded, downloaded and traded. Whilst large corporations seek to gain copyright over individual memories, a small guerrilla group, the XM, is beginning a black market trade in memories to prevent them from being identified with a specific individual. The drama focused around XM member Eve and each participant was positioned as her close, but frustrated, friend.

Act 1: Audioplay

Participants began by listening to a 15-minute audio play, written in the second person as if it were the participants' own thoughts. The recording directed them to leave Broadway and wander the surrounding streets; no specific directions were given so each participant followed their own, individual route. The soundtrack introduced them to the opening setting of the drama, that they are due to meet Eve at Broadway and are killing time until she arrives. Throughout Act 1 they are given suggestive instructions in which they are asked to look around their environment in terms that are ambiguous enough to remain relevant wherever they may happen to physically be:

How many times have you walked down this street without even seeing it? It might as well have been invisible. Your mind was on other things. But now you're starting to see a whole load of details that

Figure 1. Sylvia Robson as Eve in The Memory Dealer (© 2010, Bernard Zieja, used with permission)

you've never seen before. And although you've walked here many times the layers of memory are thin. Memories carelessly left to fade to nothing. How many layers of memory are hidden here?

This section also began to introduce elements of the wider diegesis including memory recording, the XM, Eve's possible involvement with them and the mysterious figures of the memory dealers who run a black market in memory trading. The section ends with a phone call from Eve, who tells the participant that she has been arrested and needs their help. She asks the participant to return to Broadway, locate a 'memory dealer' and acquire her memory from the previous day.

Act 2: Radio and Performance

In Act 2, participants became actors within the fictional diegesis. Upon returning to Broadway they were required to seek out an actress playing a memory dealer, who was hidden within the general public in the bar area. This actress then gave them a second audio device and instructed them to wait until they saw Eve appear in a large window overlooking the bar. At the same time, a fictional radio programme played over the bar's PA system, explaining that memories are now recordable, thanks to technology original intended to help Alzheimer's sufferers, and the ethical debates surrounding recording memories. Mocked up as an interview with a memory expert, the programme is interrupted by Eve calling in to argue against recording memories and supporting the XM in their mission to stop the trade. The radio section served to provide additional expositional material necessary for the participants' understanding of the drama, but without breaking the diegetic world.

Act 3: Theatre

Once all participants had collected their device from the dealer, the drama shifted into its third phase. At this point, the participants were introduced to the character of Eve in the flesh, played by actress Sylvia Robson (see Figure 1).

Eve then led the participants as a group back into the streets of Nottingham as they experienced her memory from the previous day by watching the actress and listening to the soundtrack on the device the dealer gave them. This soundtrack combined internal monologue from Eve, explaining more about the XM, and a narrator positioned as the participants' own thoughts. This recorded soundtrack was then intercut with live performance from Robson as she is called by an XM associate and asked to steal a bag from a nearby shop. The participants watched Eve steal the bag before following her back to Broadway for the concluding Act.

Figure 2. Filmed segment of TMD (© 2010, Rik Lander, used with permission)

Act 4: Film

TMD concluded back in Broadway's bar as the participants watched a film of Eve being interrogated by a policeman on the large public screen usually used for video art installation. In the film, Eve appears to have broken under interrogation and identified her collaborators within the XM. When the policeman tells her she may leave, she whispers to the camera that she hasn't, in fact, told him the correct names. Photographs of each participant getting a memory from the dealer (taken during Act 2) then appeared on screen and the policeman read out their names (taken from their registration details) from the file containing those he was planning on arresting. The drama ended with the actor playing the policeman arriving in the bar to arrest the participants.

INTERNAL/EXTERNAL, INDIVIDUAL/ COLLECTIVE: *THE MEMORY DEALER* AS EXPERIENTIAL DRAMA

Rather than utilising the third person mode of address normally employed in traditional time-based narrative forms such as film, television and theatre, *TMD* used the personal and directed nature of mobile media technology to explicitly address participants in the second person, as 'you'. This mode of address, most commonly found in videogames (see Buerkle, 2008), led creator Rik Lander to describe it as 'experiential drama'. Existing writing on the concept of 'experiential drama' has conceived it as an educational or political tool and focused on participants as performers (Boal, 1992; Krajewski, 1999; Tromski and Doston, 2003). *TMD* demonstrates the potential for considering 'experiential drama' as an entertainment form and its potential for opening up notions of audiencehood and engagement. Murray Smith (1995) has considered the form of emotional engagement generated through cinema, with a particular emphasis on a more indirect connection to the diegetic characters. He argues, in relation to engagement with horror films, that viewers 'experience vicariously the thoughts and feelings of the protagonist' (p. 77), He goes onto argue:

When the spectator Charles sees a fictional character faced by the Green Slime—to use the dramatis personae of Carroll's analysis—he does not experience an emotion identical to that of the character. Rather than experiencing fear of the Slime, Charles experiences anxiety for the character as she faces the slime. (p. 78, original emphasis)

In most older forms of narrative, the emotional engagement with the fictional world is partly detached from the audiences own emotions. The screen, proscenium arch and external fictional characters act as safety nets between themselves and the emotions the narrative explores.

In contrast to Smith's model, *TMD* focus group participants discussed feeling emotions more directly. Rather than anxiety for Eve as an external character, as in Smith's discussion above, they discussed the experience being more directly personal, of feeling the kind of emotions Eve was feeling themselves. Such a response was often directly connected to the mobile technology used, especially the headphones. The emotions alluded to through the character of Eve, primarily paranoia and apprehension of being caught as part of the illegal XM, were directly passed onto the audience. One participant, for instance, commented that:

the immersion is not necessarily about following the story, but in the anxiety of being slightly dislocated from the space you're in to a space in your headphones. And that is quite alienating and quite anxiety-inducing (male 1)[1]

One participant felt that the more direct form of address of Act 1, with instructions being hidden within the audio, was particularly immersive as he became part of the diegetic world:

The comments about looking around at things you hadn't noticed before were really relevant. Because you did find yourself slowing down, focusing with your head, it's very hypnotic so you're kind of [moves head slowly from side to side]. Saw some graffiti I've never seen which are huge and have been there for a long time. Looking at the tops of buildings as it suggests to you, that kind of thing. It was really interesting (male 2)

Pervasive drama of this kind has the potential to offer a more direct form of emotional immersion than that familiar from television and film, with the personal nature of the technology being a key part of this. Participants were encouraged to create their own, individual experiences that allow them to imprint themselves onto the narrative, and experience emotions first hand.

In many ways, this form of engagement echoes Gordon Calleja's (2011) 'incorporation' model of engagement with digital games. In this model, 'the virtual environment is incorporated into the player's mind as part of her immediate surroundings, within which she can navigate and interact' (p. 169). Simultaneously, 'the player is incorporated (in the sense of embodiment) in a single, systemically upheld location in the virtual environment at any single point' (p. 169). Although Calleja is discussing engagement with spaces that do not physically exist, rather than the layering of pervasive media, such a model remains useful in how it 'readily accounts for such instances of blending of stimuli from different environments' (p. 172). Such blending of fictional and real environments, and facilitating the audience's position within this blend, is integral to the kind of experience created by *TMD*. In fact, the blurring is to an even greater extent than in Calleja's model, where the computer or television screen acts as a barrier between the real and virtual environments. By using mobile technologies to allow participants to physically move through public spaces whilst 'within' the diegesis, *TMD* merges the virtual (or fictional) and real to a far greater extent. As one participant commented 'you felt, like, "Oh god I don't know who's acting and who isn't"' (female 2). *TMD* demonstrates the potential for exploiting mobile technologies' ability to blur perceptual boundaries, particularly in relation to experimental narrative forms. The narrative experience becomes far more direct, with ambiguity of its boundaries allowing it to shape not just the participant's understanding of the text, but also their perceptions of the world around them.

Whilst many participants enjoyed the more direct emotional engagement of *TMD*'s nature as experiential drama, the focus group did in a number of key shifts raise the technological and social dimensions of transmedia pervasive drama. The fact that *TMD* was transmedia and, as Henry Jenkins (2006) asserts, 'In the ideal form of transmedia storytelling, each medium does what it does best' (p. 96), resulted in the very different narrative forms (and so different forms of engagement) being brought together in close proximity. The fact that it was simultaneously pervasive, layering the narrative onto the real world, also contributed to these shifts in changing the audience's relationship to space and each other. Steve Benford et. al., (2009) in their discussion of multiplatform pervasive experiences similar to *TMD*, identify 'trajectories' or journeys which individuals follow. They identify the importance of managing moments when the nature of each trajectory is altered:

While trajectories through an experience are ideally continuous, maintaining continuity can raise significant challenges in practice. There are critical moments in an experience at which users must cross between spaces, rub up against schedules, take on new roles, or engage with interfaces, which need to be carefully designed if continuity and therefore coherence is to be maintained. (p. 6)

In the focus group, these transitions manifested in relation to two specific areas: the shift between internal and external narrative space, and the shift between individual and collective audience experience. Participants found both of these transitions potentially disruptive and in turn highlighted the specific nature of being a narrative audience. In doing so, they articulate key components of mobile media technologies' social dynamics.

FROM HEADPHONES TO SPEAKERS: EXTERNAL VS. INTERNAL SPACE

The first shift that participants particularly noticed related to the position of the narrative media in relation to the broader environment and the use of mobile technologies alongside static, contextual ones. This shift was discussed in the focus group primarily in reference to the transition between Act 1 and Act 2 and articulated as the difference between an internal engagement space and an external engagement space. In Act 1, the narrative is experienced via personal headphones that block out the surrounding soundscape. Consequently the sound manifests internally to the participant, inside his or her own head. It constructs itself, in effect, to replicate what Don Idhe (2003) calls 'auditory imagination' by pretending to be an 'inner speech' (pp. 64-65). Although the voice is naturally not that of the participants, the use of second person narrative and personal stereo equipment positions the information within the same, internal space as imaginary thought processes. In Act 2, the narrative becomes part of the broader environment, layered over the normal sounds and actions that make up Broadway's bar area. The meeting with the memory dealer takes place in public, and participants have to scan the faces of strangers to determine who to approach, with some approaching individuals who had nothing to do with the performance. The fictional radio programme is similarly integrated into the general ambience of the bar, played through loudspeakers rather than personal headphones; the audience's attention is not specifically drawn to it, it is merely part of the broader soundscape. One participant described the shift between the audio of Act 1 and the fictional radio show of Act 2 as follows: 'That was really interesting when you started with the audio in there [*indicates bar*] and you're listening to this in your head and you're sort of "this is outside my headphones"' (Male 2). This shift becomes physical, a shift in focus from the body to the external environment as the narrative shifts from mobile technology to environment-based technology.

The use of mobile technology facilitates the narrative in the first Act being directed and located in the body. This relates very closely to theories relating to mobile technologies, particularly Michael Bull's (2004) theories of the personal 'bubble':

historically, the construction of a "private bubble" of experience often required a level of silence, often institutionalized, as in prohibitions on talking loudly in library reading rooms or, more recently, in cinemas and concert halls...However, many are not seeking "silence" but their own, very personalized noise or soundscape. (p. 277)

Act 1 consciously operates to create such a private soundscape, blurring the individual's connection to the physical world around them and manipulating (and enhancing) their auditory imagination. In the second act, that bubble is broken, with the narrative leaking out into the space beyond the body as not only must the participant interact with another person, but the narrative also gets layered onto ambient sound via the use of Broadway bar's speaker system. The audience's primary attention is shifted from themselves as individuals to their wider environment. Margaret Morse (1990) discusses the nature of distraction when she argues that, 'it involves two or more close objects and levels of attention and the copresence of two or more different, even contradictory metapsychological effects' (p. 193). The relationship between audience members' multiple points of attention (the soundtrack playing in their head and the world around them) becomes unbalanced when the source of the narrative shifts from their personal headphones to a public address system.

In Act 3, when the audience follows Eve whilst simultaneously listening to a personal soundtrack, a second shift occurs, this time towards Jussi Parikka and Jaako Suominen's (2006) 'bi-psyche'. In its truest form, this bi-psyche involves an individual being involved with both the world of the entertainment on their mobile device and the real world, as in the following example concerning mobile gaming: 'the user...playing Sumea's *Extreme Air Snowboarding* (2003) while navigating through busy urban streets is an exemplary attentive subject, a true bi-psyche' ('The Bi-Psyche of the Online Gamer' para. 7). Alison Griffiths (2008) alludes to a similar sensation in her articulation of immersion: 'One feels enveloped in immersive spaces and strangely affected by a strong sense of the otherness of the virtual world one has entered, neither fully lost in the experience nor completely in the here and now' (p. 3). Whereas with traditional narrative forms, attention is directed towards a specific place (the stage, cinema screen or television set), the shifts that mobile technology allows within *TMD*, from personal bubble, to ambient attention, to bi-psyche, requires each audience members to shift their attention or risk missing the narrative.

Focus group responses to these shifts were generally positive, indicating that it generated a sense of unease, but one that added to, rather than detracted from, their enjoyment. One participant commented that:

the bit with the radio was a bit, kind of trippy, because you're sort of, you're trying to hear [indicated headphones in ears] and then you hear 'Eve, Eve' and it sort of punctures the kind of- your interior space in quite an interesting way" (male 1)

A second participant was similarly positive, indicating that the shift contributed to the experiential nature of the drama. He commented that:

But I think using the trick you did in [the bar], switching from the internal media to the external media and then doing the same thing with Eve, so you're listening to her memories and suddenly she's on the phone and she's shouting. When you do that, it gives you a paranoia that there's going to be other factors involved (Male 2).

These comments demonstrate how narrative information that is engaged with inside a sound bubble of mobile technology is experienced differently from narrative information that is layered onto a space external from the body, creating a slight moment of unease when an individual shifts from one to the other. However, they also indicate the potential to use this shift to create particular emotional responses, primarily those of paranoia and uncertainty. Suddenly changing the rules of where narrative information was coming from created a sense of paranoia that was fitting with the overall narrative arc concerning Eve's conflict with the authorities.

However, the group also indicated a potential problem for producers of transmedia pervasive drama that can be generated by requiring participants to shift their focus from internal to external media. The specific emotions discussed by the focus group (paranoia, anxiety and uncertainty) may not be successful within other narrative contexts. If one were to create a pervasive romantic comedy[2], for instance, the unease such shifts generates may be considerably more problematic. At the same time, any significant change in narrative source opens the potential for information to be lost. One participant was slightly unsure as to what she should be doing in the transition between Act 1 and Act 2, and worried that she had missed something. When discussing the narrative's shift to an external space she commented: "I wasn't sure what to expect...I was wondering if I was missing instructions" (Female 1). Henry Jenkins (2006) observes a similar issue with the emergence of mainstream transmedia texts, with *The Matrix* spreading key narrative information across multiple platforms, thereby raising the possibility of confusion for audience members who had not uncovered each component (pp. 93-95). A similar effect occurs here, with the requirement for audience members to shift their focus from narrative directed 'inside their head' to a narrative that plays out amongst the ambience of a public space potentially causing a lack of narrative clarity and disengagement.

The focus group's responses to the shift between Act 1 and Act 2 of *TMD*, both positive and negative, demonstrate that the kind of space a narrative occupies for its audience can be a key part of how they engage with it. The internal engagement space of the audio play and mobile device focuses the audience on their own bodies. The narrative plays out within a private bubble of sound; the boundaries of the engagement space are clear and end at the audience member's headphones. What attention they pay to the world around them is then down to them, it has little or no impact on how the narrative imparts its information. The integration of the radio play and performance mode of Act 2 into a busy public bar area creates the exact opposite form of engagement, with the narrative being layered onto real space and narrative information emerging from a number of un-signposted sources. Whereas one is concentrated, the other is dispersed, competing with other sounds and actions. The shift between these two modes of engagement created unease in that it required a sudden shift in attention, something that was unsettling or disruptive for some participants, but led to a heightened sense of experience for others.

FROM I TO US: COLLECTIVE VS. INDIVIDUAL EXPERIENCE

The second key shift that occurred throughout the piece related more directly to the *kind* of audience participants were positioned as, and in turn the ways mobile technologies function across communal and individual contexts. As Rowan Wilken (2010, p. 452) observes, much work around mobile technologies, including the work of Michael Bull discussed above, have discussed them as isolating. *TMD*, however, deliberately manipulated this assumption to combine individual and collective narrative formats. As Wilken goes on to argue, 'just as mobiles can be used to reinforce existing social networks (connecting known with known), they also have the potential to open up new social and interactive possibilities (perhaps through connecting known with unknown, and stranger with stranger)' (p. 452). Some parts of *TMD* used personal technologies (Act 1), whilst others used public ones (Act 2 and 4) and others both (Act 3, if an actor can be considered the 'technology' of theatre). On the surface, it may seem that a collective or individual mode of engagement is directly linked to whether that media is internal or external, with internal media creating an individual experience and external media creating a collective one. However this connection is far from straightforward. In Act 2 participants were listening to an external radio play, but were still positioned as individuals in their attempts to locate the memory dealer. In Act 3 they were listening to individual audio feeds, but were clearly positioned as a group, moving together through the streets of Nottingham and obviously distinct from the general shoppers around them. As one participant commented, there was no automatic connection between being physically with other audiences members and feeling part of a collective experience: 'Even though I walked around with [my partner] for the first bit, I still felt, because I had the personal headphones in, into my own space' (male 2).

In the focus group discussion, participants were uneasy with the shift between individual and collective experience in ways that again related to the drama's nature as experiential and the narrative focus on anxiety and paranoia. One participant commented that, in Act 3 the main cause of disengagement was, for him:

being on your own was much more interesting I think because you have that sense of isolation, so it's, like you say, you're walking around and you try not to look conspicuous because you're going round with headphones on but you're partly in your head and it's like, because I'm here with my wife, and we went off in different directions and when we came back together again and when you doing the exercise in here, I think you were much more, less immersed in it because you were much more conscious of what was going on around you. (male 1)

There was a direct correlation between how individual (and isolated) the experience was and how immersed in the experiential nature of the drama this participant became. The split in attention, discussed above in terms of bi-psyche, became problematic when the participant was made more aware of the real world co-existing with (and potentially overpowering) the fictional world.

Importantly, it was not the presence of the actress playing Eve that generated this lack of immersion, nor is it necessarily members of the general public out in Nottingham on a Saturday afternoon, but the presence of other audience members. He continued, in conversation with *TMD* creator Rik Lander, to say:

Rik Lander: Someone was making the comparison between following Eve and being on your own. So would you prefer being on your own to following her?

Male 1: Not so much the following her so much, or an actor in a space. It was being very conscious that we were a group of people [agreement from others]. So in dramatic terms, I think it works when you feel that you're on your own.

It is specifically the presence of bodies outside of the fictional world but who do not act like they 'naturally' belong in the real world that causes discomfort and disengagement. Participants became too conscious that they were part of a collective experience, that not only is the diegesis being layered over the real world is fictional, but they are visibly watching it. When their status as part of 'an audience' becomes overt, their immersion in the fictional experience becomes threatened.

A second participant commented that being conscious of other audience members affected *TMD*'s function as *experiential* drama, especially given the particular emotional focus of the narrative on Eve's paranoia about being watched. She commented that:

Female 2: I think if you had to do it as an individual it would be more immersive than doing it with a group

EE: Are you more conscious of what you're doing if you're a group – you got to recognise each other?

Female 2: You feel safe, because when you're with a group there's always someone else. It's like being in an audience. It's like you have the proscenium arch. Nothing too terrible can happen to me. But if you're on your own, you're kind of not safe.

EE: So you want the danger, paranoia?

Female 2: That's part of what you're trying to immerse into, her plight and so if you want to get that kind of experience, it would be more so if you were on your own.

For this participant it was not so much feeling conspicuous in a group, but more about losing the safety net of being with other people who, even though they are strangers, are sharing your experience. This is something that also appears in the quotes from the male participants above. The sense of isolation that being on your own in a busy city that the 'sound bubbles' of mobile technologies afford feeds into the emotional engagement with the piece. The isolation given to Eve within the narrative has the potential to be created in the audience, but only if any sense of collective experience is removed.

These discussions demonstrate a particular quality of mobile technologies: that they are also most affective as *personal* technologies. Engagement in a fictional pervasive narrative became lessened when a sense of collective experience was generated. This was either through embarrassment or through a sense of safety that undermined the experiential nature of the drama. Jennifer Radbourne, Katya Johanson and Hilary Glow (2010) argue that 'collective experience' is a key factor in audience evaluation of theatre, which they define as 'Ensuring expectations of social contact and inclusion are met, including shared experience…interaction or understanding between performers and audience' (online). They refer to Peter Eversmann's (2004) argument that:

[W]hile the emotional and perceptual dimensions are experienced individually, the cognitive analysis of a production is to a large extent a collective phenomenon, which may enhance the spectator's insight in a performance through communication with other audience members (p. 171).

Both of these arguments privilege a sense of communal experience with live performance. However in the kind of pervasive, mobile drama demonstrated by *TMD*, it is precisely the opposite that is desired by focus group participants. This opposition comes about precisely because of the dynamics that mobile technologies give such drama narratives. *TMD* exploits the ability of mobile technologies to remove their user from the wider, social structures around them but simultaneously bring them together with others in the same space. In doing so it reveals that whether a narrative experience is communal or personal is a defining characteristic of that experience.

THE SOCIAL AND TECHNOLOGICAL DIMENSIONS OF TRANSMEDIA PERVASIVE DRAMA

The development of mobile media technologies has allowed the emergence of new forms of drama, that integrate fictional narratives with real spaces and that bring together multiple technologies and forms. The key characteristic of such drama is the shift away from a narrative in which a story is told to a narrative that is experienced first-hand by the audience, putting them directly into the fictional world and generating emotional engagement that is experienced directly rather than vicariously. *TMD* is part of an early wave of such experimental pervasive drama and, although the focus group conducted after its performance was small, it generated some initial issues that are raised by this emerging form of drama about the relationship between narrative, technology and audience.

What became apparent during *TMD* was that the ability of mobile technologies to act as media platforms raises the potential to critically examine the nature of media engagement and the social and cultural dynamics that shape and are shaped by that engagement. Central to this is the exploitation (and understanding) of mobile technologies inherent mobility. This mobility allows pervasive drama creators to move their audiences through different kinds of space, take them away from other each or bring them together, and layer the external sounds of the material world over the internal sounds of a pair of headphones. In doing so, pervasive drama experiments with the boundaries between internal and external space, and individual and communal experience. As *TMD*'s audience indicates, these shifts between internal/external space and personal/collective experience create core changes in the nature of their engagement with the narrative. These changes sometimes heightened the drama's intended emotional immersion and sometimes endangered it.

However, the comments from *TMD*'s focus group are not limited to experiences of this specific form of narrative, but open up the opportunity to re-interrogate more traditional media forms and the nature of narrative engagement more broadly. What is particularly important about the *TMD* audiences' experience is that there were inherent differences in that experience that related to the technological and social dynamics of the piece, rather than its narrative form. As a result it is possible to consider the wider implications than simply those relating to the successful design of mobile-based transmedia pervasive drama. Film, television and videogames may equally be experienced individually or collectively and on communal or personal technology, the latter thanks to the broader adoption of mobile technology by the media industries. The internal/external and collective/individual dichotomies function as key characteristics of narrative engagement that sit across *all* forms of media. This offers a challenge to consider the characteristics of narrative engagement that shape an individual's engagement with it regardless of the specific form and reduce the tendency to separate out individual time-based narrative experiences such as theatre, film, television and radio that has so far defined scholarly work on audiences. The exploitation of the mobile phone as a platform for narrative forms is growing and, in many ways, is offering radically new forms of engagement. It is also, however, offering the chance to re-interrogate the more familiar social and technological dynamics of narrative engagement.

ACKNOWLEDGMENT

The author would like to especially thank Rik Lander for creating *The Memory Dealer* and those who took part in the performance and the focus group. Thanks also go to the Towards Pervasive Media Group at the University of Nottingham and the EPSRC, for funding the project, and to Broadway Cinema and Media Centre for hosting the event.

REFERENCES

Beer, D. (2012). The Comfort of Mobile Media: Uncovering Personal Attachments with Everyday Devices. *Convergence: The International Journal of New Media*, *18*(4), 361–367. doi:10.1177/1354856512449571

Benford, S., Giannachi, G., Koleva, B., & Rodden, T. (2009). From Interaction to Trajectories: Designing Coherent Journeys Through User Experiences. In *Proceedings of Conference on Human Factors in Computing Systems*. Retrieved from http://www.mrl.nott.ac.uk/~sdb/research/research-and-publications.htm

Boal, A. (1992). *Theatre of the Oppressed*. London: Pluto Press.

Buerkle, R. (2008). *Of Worlds and Avatars: A Playercentric Approach to Videogame Discourse*. (Unpublished PhD thesis). University of Southern California.

Calleja, G. (2011). *In-Game: From Immersion to Incorporation*. Cambridge, MA: MIT Press.

Carr, D., Buckingham, D., Burn, A., & Schott, G. (2006). *Computer Games: Text, Narrative and Play*. Cambridge, MA: Polity.

Carroll, N. (2003). *Engaging the Moving Image*. New Haven, CT: Yale University Press. doi:10.12987/yale/9780300091953.001.0001

Evans, E. (2011). *Transmedia Television: Audiences, New Media and Daily Life*. London: Routledge.

Eversmann, P. (2004). The Experience of the Theatrical Event. In V. A. Cremona et al. (Eds.), *Theatrical Events: Borders Dynamics Frames* (pp. 139–174). Amsterdam: Rodopi.

Gauntlett, D., & Hill, A. (1999). *TV Living: Television, Culture and Everyday Life*. London: Routledge.

Griffiths, A. (2008). *Shivers Down Your Spine: Cinemas, Museums and the Immersive View*. New York: Columbia University Press.

Idhe, D. (2003). Auditory Imagination. In *The Auditory Culture Reader*. Oxford, UK: Berg.

Jenkins, H. (2006). *Convergence Culture: When New and Old Media Collide*. New York: New York University Press.

Juul, J. (2004). *Half-Real: Video Games between Real Rules and Fictional Worlds*. Cambridge, MA: MIT Press.

Krajewski, B. (1999, October). Enhancing Character Education through Experiential Drama and Dialogue. *NASSP Bulletin*, 40–45. doi:10.1177/019263659908360906

Morley, D. (1986). *Family Television: Cultural Power and Domestic Leisure*. London: Routledge.

Morse, M. (1990). An Ontology of Everyday Distraction: The Freeway, the Mall, and Television. In *Logics of Television: Essays in Cultural Criticism*. London: BFI.

Parikka, J., & Suominen, J. (2006). Victorian Snake? Towards a Cultural History of Mobile Gaming and the Experience of Movement. *Game Studies, 6*(1). Retrieved from http://gamestudies.org/0601/articles/parikka_suominen

Radbourne, J., Johanson, K., & Glow, H. (2010). Empowering Audiences to measure Quality. *Participations: International Journal of Audience Research, 7*(2).

Reason, M. (2004). Theatre Audiences and Perceptions of Liveness in Performance. *Participations: International Journal of Audience Research, 1*(2).

Smith, M. (1995). *Engaging Characters: Fiction, Emotion and the Cinema*. Oxford, UK: Clarendon Press.

Tromski, D., & Doston, G. (2003). Interactive Drama: A Method for Experiential Multicultural Training. *Journal of Multicultural Counseling and Development, 31*(1), 52–62. doi:10.1002/j.2161-1912.2003.tb00531.x

Urian, D. (1998). On Being an Audience: A Spectator's Guide. In *On the Subject of Drama* (pp. 133–150). (N. Paz, Trans.). London: Routledge.

Wilken, R. (2010). A Community of Strangers? Mobile Media, Art and Urban Encounters with the Other. *Mobilities*, *5*(4), 449–468. doi:10.1080/17450101.2010.510330

KEY TERMS AND DEFINITIONS

Audiences: Individuals that take part in narrative events, either individually or as a group.

Engagement: The emotional, intellectual and physical experience that an individual or group has with a narrative.

Experiential Drama: Storytelling technique in which the narrative is experienced by the audience by taking on performative or interactive roles; contrast to 'exponential' drama in which the narrative is told via a narrator or observing actors.

Medium Specificity: Theory that each form of storytelling in unique, for example the idea that film is inherently different from television.

Mobile Media: Small technologies that are carried on the body and are capable of carrying media content such as video and audio files.

Narrative: The plot, characters and story world constructed within an entertainment experience.

Pervasive Drama: Storytelling form that places narratives within, through and across real, public spaces, blurring boundaries between reality and fiction.

Sound Narrative: The use of aural technologies to convey a story through single voice narrative, dialogue, sound effects or music.

Transmedia Drama: Storytelling technique that bring together multiple technological platforms into a single narrative.

ENDNOTES

1 Many of the quotes discussed in this chapter come from a small number of focus group participants. This is mainly as the group itself was dominated by small percentage of participants. However, their comments were consistently backed up either by general agreement (both verbal and non-verbal) or by other participants alluding to the same issues in less direct ways. The limited range in participants is more to do with how they articulated their opinions, rather than their opinions being unrepresentative of the rest of the group.

2 Pervasive dramas (and related content such as Alternate Reality Games) have tended to focus on fantasy, science fiction, or action based genres. Examples of such work include *The Beast* (Microsoft, 2001), *Uncle Roy All Around You* (Blast Theory, 2003) and *The Malthusian Paradox* (Urban Angel, 2012).

Chapter 8
Mobile Media and Youth Engagement in Malaysia

Joanne B. Y. Lim
The University of Nottingham, Malaysia Campus, Malaysia

ABSTRACT

Youth engagement is often said to be at the heart of democracy, though the extent of such efforts in affecting policymaking remains highly debatable. Nonetheless, there have been heightened attempts by "ordinary citizens" to reform the country's state of politics and to improve society's living conditions. Malaysia's authoritarian democracy has been a crucial motivation for young adults to "have their say" in challenging the current regime. This chapter highlights the various ways in which young adults use mobile media to activate and participate in civic, community, and political engagement whilst taking into account the many restrictions that are set up by the ruling government to monitor and control such engagements. Discussed alongside youth definitions of nationalism, citizenship, and activism that are embedded within the interviews, the findings are juxtaposed with present post-election discourses taking place within the country. The relationship between mobile media and youth engagement further affirms the idea of a new generation of mobile users that are not just technologically savvy but are using their knowledge to affect significant societal changes.

INTRODUCTION

In May 2013, more than 100,000 Malaysians, mobilized and coordinated by waves of text messages, assembled at the Kelana Jaya Stadium – the first of five Black 505 rallies held nationwide to protest electoral fraud. During the previous year, following the Bersih 3.0 rally which garnered over 250,000 people at the Merdeka Square, Parliamentary member Fong Po Kuan's (DAP – Batu Gajah) queried regarding jammed telecommunication lines during the rally. In a written reply, Information, Communications and Cultural Minister Datuk Seri Dr Rais Yatim explained how the system could not handle the high usage:

Cellular telecommunication systems generally do not have enough capacity to support such high numbers of telephone conversations and transfer of data, as what occurred during the Bersih 3.0

DOI: 10.4018/978-1-4666-6166-0.ch008

Figure 1. Undi PRU13 mobile app screen capture

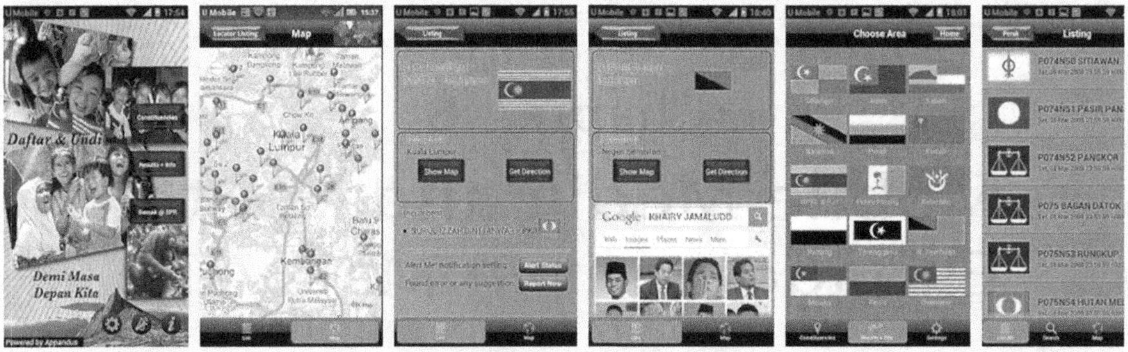

gathering...when such cells are full, the system will reject calls and this results in the user being unable to make calls. (Yuen, 2012)

Both the *Black 505* and *Bersih* rallies offer clear examples of how online social networks contribute to the mobilization of individuals leading to more civic engagement in the physical world. The response on telecommunication jamming presents us with two indications pertaining to mobile media in Malaysia. First, the current condition of cellular telecommunication system in Malaysia leaves much to be desired, and second, mobile phones are the main form of communication during such rallies. A counter argument arising from the query was that authorities had jammed phone services to hinder further mobilization of people. If this was true, then the rationale for doing so would be based on the idea that mobile phones are indeed the main source of information and method of communication among rally supporters, thus allowing crowd engagement via mobile phone updates. This further affirms the significance of mobile media in influencing participation among youths in Malaysia. However, as put forward by Cherian George (2005), in his well-known thesis on Internet penetration, it is necessary for the Internet to be studied alongside political contexts – whereby high Internet penetration does not imply higher participation. The extent to which mobile penetration affects physical engagement in civic, community and political activities must be challenged and probed.

While social media platforms such as Facebook and Twitter continue to serve as the main communication tools accessed via mobile phones, another important function gaining increasing popularity are mobile applications (or mobile apps), developed for 3G phones including the iPhone, BlackBerry, Android and Windows phones. In May 2012, a study reported that more mobile subscribers used apps than browsed the web on their devices (Perez, 2012).

During the recent 13th General Elections, several mobile apps were launched, including the most frequently installed *UndiPRU13* (created by a local app developer, Appandus Sdn Bhd – see Figure 1), to enable users to get information on the 727 constituencies, including their locations, and a flexible search function for key words relating to a (previous) member of Parliament, State Assemblymen and election results since March 2008 (The Malaysian Insider, 2013).[1] The apps, downloadable for free via Google Playstore and Apple iTunes, could even set a reminder on the various *ceramahs* (campaign talks). The opposition, Parti Keadilan Rakyat (PKR) on the other hand, launched a mobile app called 'PKR' as a campaign tool to disseminate information to the public. Various features included providing the latest updates on election-related issues, voters' registration check, info graphics, videos of *cera-*

Figure 2. A sample from over 40 mobile apps related to the GE13

mahs, press conferences, the party's campaign activities and schedule, and profiles of the party's candidates (see Figure 2). The app was linked to "social media platforms such as Facebook and Twitter for a wider outreach" (Amin, 2013).

In addition, Virtual Logic Sdn Bhd developed a mobile app game, *JustUndi GE13*[2] (see Figure 3) with the aim to:

> *...educate all users [on the] election...and also enable the users to participate in election (decision-making) process of voting. The polling process is in real time...this is just an education game and it would not reflect or affect the actual result of election (Virtual Logic Sdn Bhd, 2013).*

Figure 3. JustUndi GE13

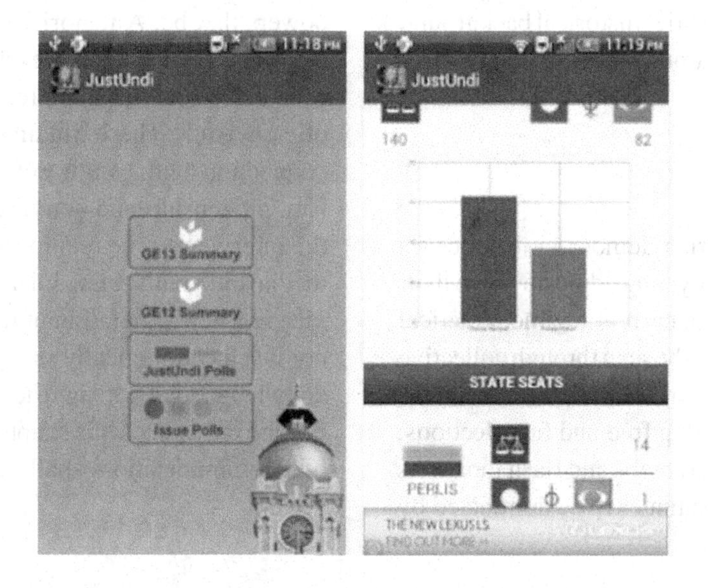

In reflecting on the use of mobile phones and the various apps developed to encourage 'ordinary citizens' to reform the country's state of politics and to improve society's living conditions, it is imperative we first consider the historical background of Malaysia's mobile industry, including regulations in place to 'protect' the current regime from being subjected to threats by civil society. Following this, it is useful to examine the types of mobile media/communications (e.g., short message service) young adults are using to engage in the political process and to what effect. This chapter also seeks to discuss the role of mobile devices in changing relationships between young citizens' and the political institutions (e.g., the government, NGOs, etc); and, examine how the usage of mobile devices is transforming the nature of young citizens' political/civic/community engagement. While the target users of mobile phones in Malaysia are not restricted to young adults, it is accurate to claim the majority of mobile users are between the ages of 17 and 26 (64.9%)[3]. Hence, this chapter focuses on the rise of mobile media among Malaysian youth and highlights the various discourses affecting youth participation, especially in relation to civic and political engagement. In order to better understand the significance of mobile communication in Malaysia, the following sections provide an overview of the fieldwork and a brief historical background of its telecommunications/mobile industry.

AN OVERVIEW

Malaysia's authoritarian democracy has been a crucial motivation for young adults to 'have their say' in challenging the current regime. Practical dissent in Malaysia can be seen through collective efforts such as the *Occupy Dataran* movement, the *Bersih* rallies advocating free and fair elections, and various other smart mobs and flash mobs. As mainstream media continues to be dominated by the ruling party, young adults are using alternative platforms such as mobile and social media to encourage participation in civic, community and political activities. Increasingly, ideas of nationalism and citizenship do not necessarily conform to state definitions when developed and championed by various groups of youth activists.

Based upon interviews and focus group sessions conducted between 2011 and 2012 with 80 young Malaysian activists from a recently concluded research project on Youth, ICTs and political engagements in Asia, it was found that one of the main tools used to organize such activities is mobile phones. It was also noted that the introduction of more affordable smart phones and large screen tablets in 2012 increased the use of mobile phones to access the Internet and to update social media sites such as *Facebook, Twitter*, *YouTube* and *Instagram,* among others. In general, young adults claim mobile phones (including mobile apps) are used to gather resources, attract members/ followers/ fans, to organize and subsequently mobilize members to participate in offline events.

Many young Malaysian activists who participated in the interviews, especially those who are more vocal /visible online, or whose social media posts have become viral (widely circulated online) expressed an awareness of being monitored by the powers that be. A majority exercise self-control over their own content (despite varying levels of censorship and disparate reactions to the notion of activism). These findings are discussed in connection with youth definitions of nationalism, citizenship and activism embedded within the interviews. The relationship between mobile media and youth engagement further affirms the idea of a new generation of mobile users who are not just technologically savvy but are using their knowledge to affect significant societal changes. For the purpose of this chapter, the scope of 'mobile communication' shall be limited to mobile/

Table 1. Percentage of users who download mobile apps (adapted from the MCMC 2012 handphone users survey (Malaysian Communications and Multimedia Commission, 2012)

Downloaded Mobile Apps	% Smartphone Users
10-20	34.6
20-30	16.3
30-40	7.4
40-50	4.2
50-60	1.5
>100	1.4

smartphones, excluding tablets, iPads and other 'mobile' devices not carrying 3G functions.

MOBILE INDUSTRY IN MALAYSIA AND THE SMARTPHONE PHENOMENON

Since the 1980s, Malaysia's telecommunication services has been dominated by Telekom Malaysia Berhad – which became a publicly listed company in 1990 with the government being a major shareholder as well as retaining a 'special share' to ensure "Telekom Malaysia's important operational decisions are consistent with Government policy" (Bursa Malaysia, p. 159). Though started as a monopolistic regime,[4] the industry now has seven licensed domestic network operators including Maxis Broadband®, Telekom Malaysia Berhad, TT dotcom®, Celcom Transmission (M) Sdn Bhd®, DiGi®, Fiberail Sdn Bhd® and Prismanet (M) Sdn Bhd®. Nonetheless, the 'liberalisation' of the industry in 1999 took place following a new regulatory regime introduced by the government in November 1998 to govern telecommunications, broadcasting and Internet industries. The Malaysian Communications and Multimedia Act 1998 (CMA) was enacted to address the convergence of the said industries and to "promote greater transparency and clarity as well as industry self-regulation." (Bursa Malaysia, p. 164). Mobile regulation in Malaysia comes under the perusal of the Malaysian Communications and Multimedia Commission (MCMC), established under the CMA, the same government division that proposed an amendment to the Section 114A Evidence Act, then passed in parliament (without debate) in April 2012. The Act legislates for 'presumption of fact in publication' on the Internet with further proposals in 2013 to censor Internet content. Among its functions, the Commission is responsible for supervising and monitoring communications and multimedia activities; enforcing provisions and recommending reforms to the communications and multimedia laws. The Commission also has the power to implement policies, issue directions to licensees, to make determinations, to hold public inquiries and to conduct investigations.

Malaysia's telecommunication industry continues to be transformed due to the emergence of new technologies, requiring service providers to differentiate their products and services in order to stay ahead of the competition. According to the Nielson 2012 Smartphone Insights study, Malaysia's smartphone penetration rate is at 27% with more than one-third (35%) of users claiming "availability of a wide choice of applications carried the most weight in their choice of model", with apps centered on social network, location-based and games being most popular (Wong, 2012). A separate study conducted by GfK Asia revealed smartphone penetration levels in Malaysia have "reached a high of 88%, translating to almost nine out of every ten in the general population being a smartphone user;" 58% of the phones are equipped with a camera of at least 8 megapixel (GfK Retail and Technology Asia, 2012). GfK's findings have been challenged by media analysts who claim the actual figure should be between 30-35% (Singh, 2012).[5]

Nonetheless, the increase in smartphone users is inline with various government incentives to increase smartphone penetration especially among young adults – when tabling the 2013 Budget, Prime Minister Najib Razak announced a one-off rebate of RM200 for 3G smartphone purchases by the first 1.5 million eligible youths through the Youth Communication Package "in efforts to give youths access to the information highway" (The Borneo Post, 2012).[6] MCMC's Handphone Users Survey (Malaysian Communications and Multimedia Commission, 2012) found as many as 68.8% of smartphones users accessed the Internet through their smartphones – 27.9% of whom used more 3G than WiFi, 26% used more WiFi than 3G. In their report, MCMC claimed that: "...mobile apps make the smartphone more than just a phone and Malaysian smartphone users have a good appetite for mobile apps." (Malaysian Communications and Multimedia Commission, 2012, p. 20)

In addition, the findings suggest that 42.3% of smartphone users leveraged on mobile apps such as instant messenger or over-the-top (OTT) messaging apps to avoid voice call/SMS charges. Such activities would have adverse affects on traditional telecom services, many of which are also wireless carriers. In dealing with the larger issue of mobile content, the Malaysian Mobile Content Providers Association (MMCP) was set up in 2007 to represent "the voice of mobile content industry...[and encourage] alliances with the industry stakeholders which include regulators and mobile operators" while at the same time prioritizing "consumer protection and privacy...to ensure a positive mobile experience" (Malaysian Mobile Content Provider Association, 2012). The extent to which it serves to "represent voices" or impose guidelines set by the MCMC[7] remains debatable, nonetheless, the Association claims to encourage active participation in discussions/ forums/ task forces with the MCMC, the Consumer Forum of Malaysia (CFM), and the Communications and Multimedia Content Forum Malaysia (CMCF).

With the regulating body having such a strong presence in the activities of the mobile industry, it is not surprising that while Malaysia's mobile market is growing exponentially, the development of mobile content, the significance of usage, and the level of penetration are still rather dismal. Nonetheless, young adults active users of mobile communication, have begun to realize the possibilities with mobile technology - to connect with various pockets of society, to engage with people sharing similar interests and ambitions via social media applications, and to organize large scale events such as flashmobs and political rallies using mobile devices. The following sections draw upon the author's research findings from the IDRC project to discuss the process of mobilization and/ or youth engagement via mobile communication in Malaysia.

THE RISE OF MOBILE MEDIA AMONG MALAYSIAN YOUTH

Vast amount of research has been conducted to explore and discuss the 'culture of participation' made possible via the Internet (see Jenkins, 2006) advocating social media as a tool for political engagement. To an extent, this chapter critiques Jenkins' notion of participatory culture within the context of engagement instrumented via mobile phones. The arguments in this chapter largely stem from George's (2005) seminal comparative study of Malaysia and Singapore, pointing to the paradox of Internet penetration and participation, which then challenges us to further scrutinize the context of mobile communication.

As discussed earlier, the rise of mobile media among young adults in Malaysia takes place alongside attempts to regulate content and monitor youth (political) engagements. Nonetheless, before we fall into the trap of generalizing all forms of engagement as 'meaningful participation' (see Lim, 2013a), it is important we recognize how young adults 'engage' – what does it mean to 'like' a Facebook post/page; does forwarding politically-inclined SMSes make one an activist; what is actually being created, co-created, disseminated and shared; and how do young adults see their role as agents of change in an authoritarian democracy like Malaysia? In addressing this question, we are once again confronted by the need to reconsider notions of personal and public spaces – and to question how mobile media engender the personal within public spaces (see following discussion on 'Politically-motivated SMS-blasting'). Balakrisnan & Shamim (2013) reported, Malaysians have the highest number of Facebook friends and risk addiction to the social platform. This is greatly implicated by the use of mobile phones to access social media sites (see Leong et al., 2011). Nonetheless, the purpose for engaging in such online activities bear different connotations even among young activists. Responding to a question on activism, interviewees offer alternative terms (see Lim, 2013b) to explain their work either due to an attempt to avoid negative connotations associated with 'activism' in Malaysia, or because they feel unworthy to be called activists compared to those who have "dedicated their whole time to that" (youth respondent). Young adults call themselves 'advocates', 'responsible *rakyats* (citizens)', and 'social commentators'.

...a lesser form of an activist...to kind of tone down the thing...I think that the term activist cannot, should not, be used too loosely. (community youth leader)

When people mention activist, I think of someone who has dedicated his or her life to a certain issue or an event. And goes to extreme length and measures to achieve their objectives in that issue. (Sarawak youth leader)

Activists...the ones who go against the government. (university student leader)

Regardless of how young adults define activism or whether they see themselves as activists, many interviewees decide to post about their activities only after careful consideration of the implications on their family and other online friends.

...[because of] my family members, I have to censor some things...I don't want them to be worried about me. So sometimes my work as an activist puts me in a dangerous position. Like when I attended a rally...here is the fear of being beaten up, or caught, or arrested by police, or [being] tear gassed or whatever. So, some things I don't put online. As much as I want people to know what I go through, but because my family is there, I don't want to worry them. (non-governmental organization coordinator)

If there is something offensive, that someone shared with me I wouldn't share it or post it on my wall, because I know I've got some friends from that community and I don't want to be offensive to them. (youth leader)

Yes, by everything that I say online as social commentary, I know that a chunk of them are are being read...they are watching what I'm saying. I say it because I want it to be read. (social activist and 'mob' organizer)

It's dangerous to spread political messages via mobile phones because they can track you down. (opposition party youth volunteer)

Nonetheless, several interviewees feel that while Facebook 'likes' are important, the actual value in the numbers cannot account for anything significant as many use the 'like' function as a 'bookmark' in order to get future notifications on a page.

It depends on people. Some people are just plain lazy. They see many likes on this page and they automatically think it's a good site it's a good page. (youth activist)

One activist shared that despite having people 'like' his page either meaningfully or just in passing, the number of 'likes' serves as a form of motivation:

Sometimes what I feel is that every time we post something on the page and people 'Like' it and comment on it, I think it is a good response and it can also encourage us to do some more. For example, we do an event and [we put up] all the pictures on Facebook [for people to see], they like and comment on them. After we see all these comments, it can encourage us to do more events and give them what they like. (leader of a youth skating community)

Generally, we know that the more likes the page has, it's obviously better...it draws many others to like the page as well. (youth activist)

Joining a 'group' requires more purposeful reasoning, often with an intent to contribute to the group due to the condition that group memberships require approval from the administrator(s) and one feels the need to actively add to ongoing discussions.

Joining a group is being part of something that you share together and you (would) want to be part of the process, part of the group in doing something... when I join a group and I am approved, I can give my comments and start a discussion. Somehow, if you join a group, it shows that you are interested in knowing what is going on in the group. Liking a page, to me, just shows that you just know what they are doing. (university student leader)

Between liking a page and joining a group, I would say the latter one shows more interest and commitment. (youth microblogger)

Returning to George's penetration/ participation paradox, it is clear there are various motivations and different levels of engagement among young adults, from disseminating/ contributing to information via the often taken-for-granted actions of liking a page, to joining a group and following a Twitter handle as being among the most popular activities of smartphone users. It may be argued, the more conventional SMS-blasting or forwarding of text messages require even more commitment by the sender (due to the telco charges imposed). *Penang Youths* have garnered up to 15,000 members online and have since organized the largest Penang Youth Merdeka Carnival – Youth Jam 2009, attracting over 10,000 Penang youths solely through SMS blasting:

...one message, many recipients. We blast to update the youths on minutes of our meetings, after event news, etc. (Penang Youths co-founder)

Another popular smartphone function among Malaysian youth is the sharing of content via instant messaging clients such as MSN, WhatsApp and TalkBox (see Osman et al., 2012). In particular, interviewees acknowledged the use of WhatsApp to organize group chats, coordinate meetings, update events and send photos and videos.

It's a very good platform to brainstorm about the slogan, brainstorm about the design. Just snap a photo and put it on the Whatsapp or Facebook group. Then we discussthe votes and we decide to use this design. Discussion-wise is very good. And when we come to a certain point in the online discussion online whereby we know it cannot be continued, then we say 'Okay, let's have a meeting tonight', to come out and meet. (Polling Agent & Counting Agent – PACA – volunteer)

With free messages and calls (via Skype and Viber platforms) available to smartphone users, more young adults are finding it easier to organize and mobilize for various purposes.

MOBILIZING THE ONLINE COMMUNITY

Mainly organized online via mobile technology, flashmobs and smartmobs are gaining popularity among young adults in Malaysia. These offline events depend largely (if not solely, in the case of the Bersih rallies) on social/mobile media to garner hundreds to thousands of 'followers', invited to congregate in physical spaces for a particular cause. Be it for political or entertainment purposes, these mobs can be deemed as an extended form of participation, transferring online communities onto offline platforms thereby further enhancing online social connections. Online discussions and forums (also informed by mobile apps) result in smart mobs that ultimately lead to more civic engagement executed in the physical world.

Using methods like guerrilla, flashmobs and other forms of participatory art, a group of youth who founded RandomAlphabets.com put together various projects aimed at "bringing human beings together" (RandomAlphabets.com, 2010). A single flashmob[8] event, for example, requires the group to create a micro-site for uploading video tutorials of choreographed movements/dance steps via YouTube, then downloaded by 'random' people visiting their site with the intent to participate in the mob. While Nicholson (2005), Molná (2010) and Duran (2006) have rather optimistically highlighted the possibilities of mobile connectivity and the reframing of public space through digital mobilization, such collective movements are more problematic when organized in a country like Malaysia. An interview respondent from RandomAlphabet states:

Difficulties…they thought it was illegal but I was convinced it was not illegal. I felt very frustrated in the sense that, if in the Constitution states you don't have the freedom of association…even though that's unfair, if it is the rule of law, it's okay… but if the constitution and the local law of the land states that you can gather, that you can meet up, you can…a lot of people come together, but people believe that they can't…that's what I call self censorship. So the frustrating part about that [flashmob] was that a lot of people wanted to come to the event but…they were a bit apprehensive because they thought it was illegal. To address this problem, we need a minor mindset change to tell them that look, this is not illegal. I think that is my primary target, to change the mindset. (Founder, RandomAlphabets)

While this may have been the case as recently as 2010 (at the time of the interview), the 'culture of fear' (see Wong, 2004; Azizuddin, 2008) seemed to have been overcome by the youth advocates, to an extent, through an increased sense of empowerment to claim their rights to freedom of expression and to make a difference in society. Thus, participation in flash/smartmobs in Malaysia increased from 2011 onwards with the more politically motivated mobs organized prior to the 13th General Elections and post-elections. Regarded as a form of collective expression to demonstrate the massive power of community, the 'trend' is for a group of strangers to come together, make an impact, and then to continue living their lives. As these flashmobs may be regarded as a form of performing arts, more young adults are less concerned about the repercussions of 'illegal gatherings'.[9] The table below highlights the 10 most significant flash/ smartmobs in Malaysia.

Contrary to the more common connotations associated with mobs, smartmobs (see Cheverst et al., 2004) encourage social coordination whereby the community behaves 'efficiently' due to increasing network links, enabling people to connect to information and others. There are arguably distinct differences between flashmobs and smartmobs (some perceive flashmobs as a subset of smartmobs) whereby smartmobs usually serve either practical functions or political ends. Nonetheless, both provide evidence of network power via SMS (short messaging system)-blasting. The process usually involves one person sending

Table 2. Flash/smart mobs organized by young adults in Malaysia

Flashmobs	Links
Glee Flashmob Dance, in The Gardens, by Random Alphabets, (key person: Zain HD, supported by Astro Star World), 2010 - one of the bigger, popular flashmobs **KL Freeze in Unison**, in Pavilion, by Random Alphabets, (key person: Zain HD), 2008 - one of the first few successful flashmobs **Keretapi Sarong**, aka Wear Sarong Day, by Random Alphabets (key person: Zain HD) - 2012, 2013 **untitled**, in KLCC, flashmob for press freedom, reading the newspaper upside down, 2010 - only 30, but a creative and powerful gesture which caught the attention of many **Bersih**, in KLCC (no explicit organiser), 2011 - wearing yellow **Gangnam Style flash mob**, in Ipoh (organised by Astro and Sunway - a performance for marketing purposes), 2012 **V for Merdeka: V-mask flash mob**, in Dataran Merdeka (organised by #OccupyDataran), New Year's Eve, 2011	http://randomalphabets.com/2010/05/glee-flashmob-dance-2010/ http://randomalphabets.com/2008/04/official-video-kl-freeze-in-unison/ http://randomalphabets.com/2012/02/keretapi-sarong/ http://randomalphabets.com/2013/03/keretapi-sarong-2013-2/ http://helpvictor.blogspot.com/2010/05/flash-mob-in-klcc.html http://www.malaysiakini.com/news/183757 http://www.nst.com.my/streets/northern/gangnam-style-flash-mob-sets-new-record-1.212421 http://www.youtube.com/watch?v=OqewXj1NWjM https://www.facebook.com/events/180134328752210/ http://www.pakistaniyan.com/2013/08/what-is-meaning-of-raabiaraabar4bia.html http://samidoun.ca/2012/05/facebook-profiles-share-solidarity-with-striking-palestinian-prisoners/
Smartmobs	
Facebook profile photo changed to **Green** ("Anti-Lynas" and "Save Merdeka Park"), **Black** ("Stop 114a"), **Yellow** ("Bersih"). Solidarity (especially among the Malaysian Muslims) with Egypt's anti-coup movement ("the **Rabia** sign"), 2013 Solidarity with Palestinian prisoners (among Malaysian Muslims and human rights activists), 2013	

a text message to certain people to kick-start a chain reaction, often being able to gather a large crowd within minutes. What began as a struggle against having society form sites for communal isolation as a result of an over-connected age (Muse, 2010), has, instead, led communities to recognize the power of physical collectivism to fight for urgent societal causes and political change (such as in the cases of the *Bersih* and *Black 505* rallies advocating free and fair elections). However, there is a need to consider the possibility of manipulation by those who are in a position to control the 'mobbing system' (often by people with the means to forward instant messages to a group) and what implications this may have on the individuals subscribed to the contact list(s).

POLITICALLY MOTIVATED SMS-BLASTING

In recognizing the speed of mobile messages to penetrate the everyday lives of young voters, the ruling party engaged in a series of SMS-blasting to new registered voters during the election campaign period (see Figure 4). The extent to which these messages 'changed' the mindset of the voters needs to be further explored. However, the outreach of such messages went beyond the individual's mobile phones when young adults who were either 'disgusted' or 'proud' to receive the messages chose to share screenshots of the SMSes on their social media pages. What can we make of political leaders misusing (or overusing) their control over mobile communications (or public funds) to solicit electoral votes?[10]

Translations:

1. Nationwide gathering with the Prime Minister on 12 Jan (Saturday) at Dataran TESCO, 9am. Come add to the festivities. Love Selangor. Trust BN.
2. Salam 1Malaysia. Received BR1M RM500, assistance for school children, licensing assistance for youth. Do not forget to REPAY

Figure 4. SMS sent particularly to young voters by the ruling coalition Barisan Nasional (BN) prior to the GE13

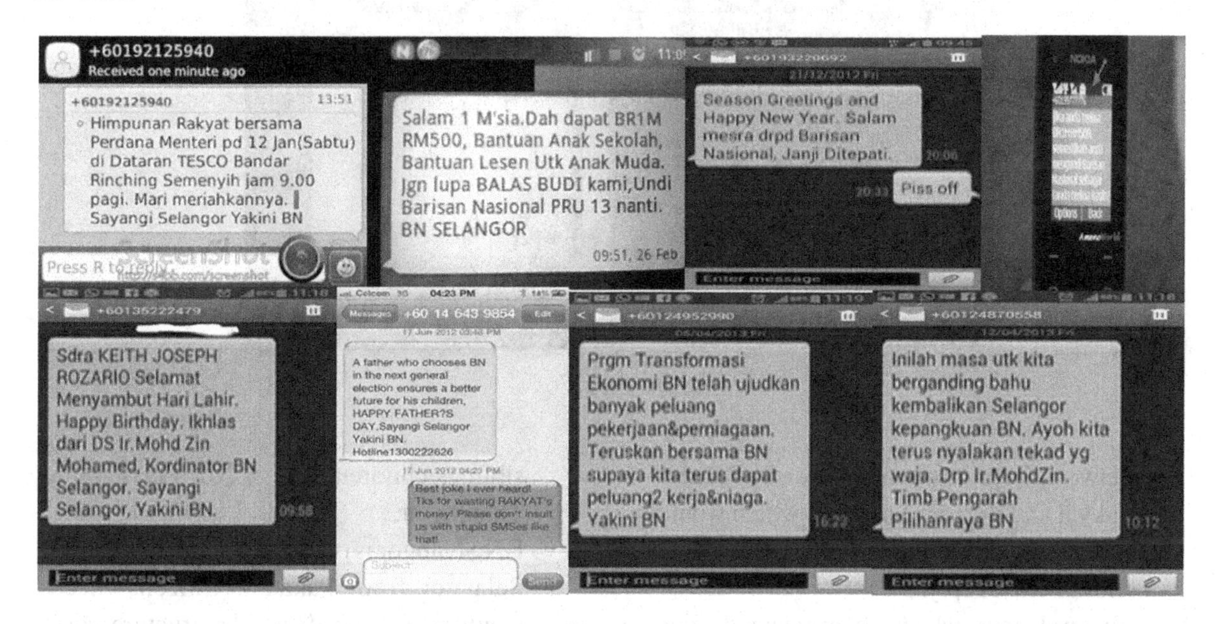

OUR KINDNESS. Vote for BN at the GE13. BN Selangor.

3. Season Greetings and Happy New Year. Kind wishes from BN. Promises kept.
4. If you received BR1M RM500, shouldn't you vote BN as a sign of gratitude?
5. Brother Keith Joseph Rozario. Happy Birthday. Sincerely from Datuk Seri Ir. Mohd Zin, Selangor BN Coordinator. Love Selangor. Trust BN.
6. A father who chooses BN in the next GE ensures a better future for his children. HAPPY FATHER'S DAY. Love Selangor. Trust BN. Hotline.
7. BN's Economic Transformation Programme has created many jobs and businesses. Continue to be with BN so that we can continue to have jobs and business opportunities. Trust BN.
8. This is the time for us to work together to return Selangor to BN. Let's continue to persevere. From Ir. Mohd Zin, Deputy Director BN Elections.

Such efforts to engage the youth to carry out their 'civic responsibilities' by casting their votes for BN were widely criticized by many recipients. Furthermore, young voters were unhappy about the 'intrusion of privacy' upon receiving SMSes including birthday greetings from BN, in addition to campaign materials. Complaints lodged with the service providers returned replies that the identity of the sender could not be revealed without prior approvals (Woon, 2013). Malaysian blogger Keith Rozario received over 20 political 'spam' SMSes from the BN 'Yakini BN' (Trust BN) campaign, various political surveys (see Figure 5) and two messages from his local PAS (Opposition) member of parliament in the span of six months: "they know my FULL name, my contact number and even my place to vote. They know more about me than I've ever given out to ANYONE from any survey, and I consider this an invasion of my privacy" (emphasis in original – Rozario, 2013).

Translations:

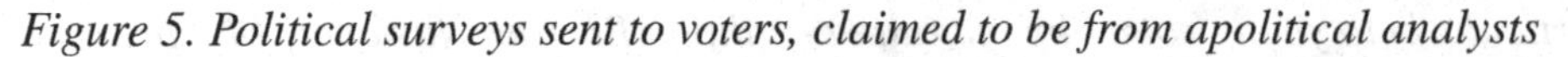

Figure 5. Political surveys sent to voters, claimed to be from apolitical analysts

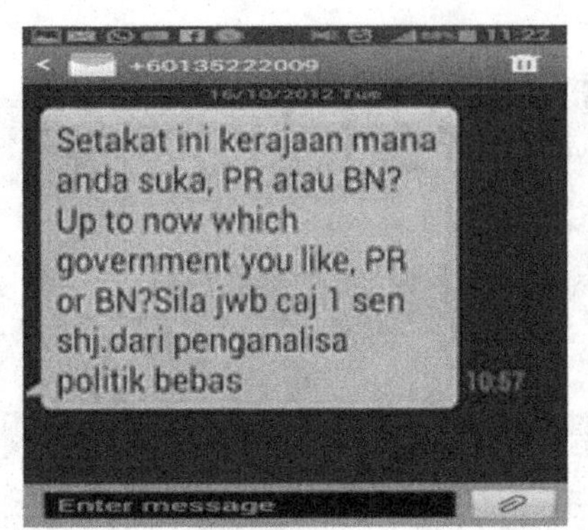

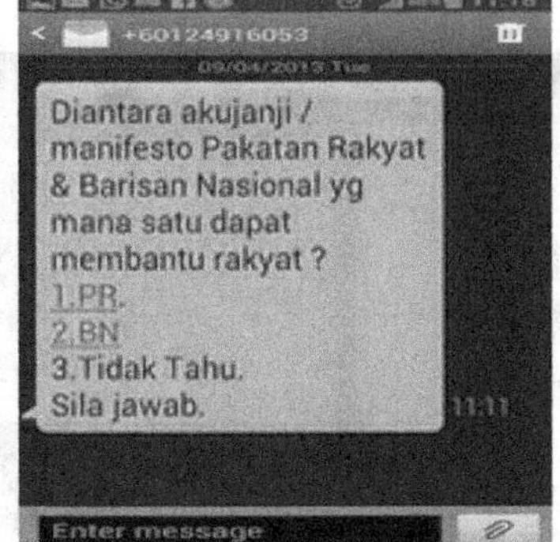

1. Between the promises/manifestos by Pakatan Rakyat and Barisan Nasional, which would better serve the nation? 1 PR; 2 BN; 3 Don't know. Please reply.
2. Up until now, which government do you like, PR or BN? Please reply. Only 1 cent charge. From apolitical analyst.

In response to the questions, Selangor BN Coordinator Mohd Zin Mohamed explained that: "the details of the SMS recipients were gathered by experts from Selangor BN's machinery"; adding, they had also "engaged professionals to record the details of Selangor folks by going house-to-house"; Mohd Zin had referred to the cost as a "trade secret" (Malaysian Digest, 2013). The content of such messages included provocations about race, education, healthcare, infrastructure and government subsidies (see Figure 6). A Malay resident received an SMS (in Malay), which read: "Congratulations! You are a registered voter for the Subang parliamentary constituency. Please fulfill your responsibility by ensuring that Malays continue to rule. BN IS THE CHOICE OF THE YOUTH [emphasis in original]"

Translations:

1. Hello. Budget 2013 - Assistance of RM100 for all primary and secondary school students by the BN government able to help your siblings/children? Yes or No? Please reply. Thank you.
2. BN Selangor is optimistic about taking over Selangor from Pakatan. Citizens' needs would be well cared for. Infrastructure would be better maintained. Love Selangor. Trust BN.
3. 1Malaysia Clinic brings medical services closer to the citizens. 90 clinics are in operation and another 70 will be opened from 10am to 10pm with the fee of only RM1. Trust BN.

Such efforts to capture the votes of the new electorates (between the ages of 21 and 35) are symptomatic of the mobile media outreach attempted by the ruling party. It would seem that mobile communication proliferated by the government are often criticized because they are deemed propagandist in nature (in addition to the assumption that the government's machinery is operated with public funds). On the other hand, the idea that individual/communities' public and private domains should be protected continues to dominate how young adults perceive their roles as responsible citizens. Thus, it is not surprising, along with the mushrooming of various alternative/ independent Internet platforms, that the mobile industry (including mobile content) is often considered an alternative space for the sharing of

Figure 6. SMS sent as part of the Yakini BN (Trust BN) campaign

information and the organization of events, not otherwise discussed or shared in physical locations. Nonetheless, we recognize the increased use of mobile technology to engage young adults in various civic, community and political activities offline.

YOUTH ENGAGEMENT: TOWARDS A BETTER MALAYSIA

The numerous flash/smartmobs and political rallies taking place in Malaysia within the past year are testaments to the rising levels of nationalism among young Malaysians. It was clear in the interviews that young adults felt they were better citizens compared to their parents, partly due to technological advancement:[11]

I'd like to think that we Malaysians are improving with each generation. I think I am better because I read and share a lot and take initiative to create action...the idea of 'networking'...especially social, online. I think I am more politically active than my parents, but in a proactive way. One thing's for sure: I'll be registering myself as a voter when I'm 21. Something they both did not do. (digital youth activist)

The trick is to find a common ground and get both [generations] to work together. Inspire and you earn their trust. (youth entrepreneur)

I am definitely more vocal...less afraid to speak out and have access to platforms that they are not familiar with such as social media. (social media advocate)

Nonetheless, views on the infrastructure of ICT in Malaysia were hardly optimistic, again citing problems with accessibility: "they only let you have what they can control" (youth environmental activist). Almost all interviewees highlighted the need for quicker and better enhancements of the country's mobile technology. Contrary to the call by Home Minister Ahmad Zahid Hamidi for Malaysians who are unhappy with the country's political system to "leave the country" (The Edge Malaysia, 2013), young adults are determined to stay in Malaysia in order for them to make a difference:...we are trying to change people's mindset...we need not go out of Malaysia...we can make Malaysia a better place. (youth leader)

Young adults who were asked about what they thought make a good citizen expressed the importance of having a sense of responsibility to contribute to the community/ country: "or at least work on self-development in order to improve oneself and indirectly improve the country... to increase the level or standard of living" (co-founder of Penang Youths).

A good citizen is one who doesn't make excuses, doesn't blame others, and makes the best of what he has, no matter who he is or where he is. (digital youth activist)

Knowing what's happening in your country, knowing the political views. (TedEx KL youth organizer)

...constantly tries to help his country achieve greatness because of his high spirit of patriotism. (member of a Student Representative Council)

Such responses tend to challenge previous ideas on political engagement among youth in Asia as 'declining' (Takahashi & Hatano, 1999). On the contrary, there is a strong indication of the extent to which youth engagement in civic, community and political activities has become a necessary platform for youth to be fully incorporated into society. Rather than seeing agency or engagement as a result of bitter resentment with political systems (Urdal, 2006), it is necessary to view their motivations as stemming from a process of negotiation between expression and regulation. Young adults are increasingly seeing their role as agents of change in an authoritarian democracy and are constantly struggling to establish new and/or alternative localized arenas for public debate and cultural/political expression in order, more importantly, to function in meaningful unrestrictive public spheres.

CONCLUSION

The state of mobile communication in Malaysia opens up both opportunities and constraints – regulatory powers have developed vast ICT infrastructure while at the same time imposing numerous (at times contradictory) sets of rules deeply enforced by various parties. For many youth advocates, such circumstances have become the very purpose for civic/political engagements. The use of mobile technology has undeniably brought about great potential for innovation by youth activists developing their capacity to use various apps and mobile functions to communicate, organize and mobilize across very different cultural, social and economic backgrounds. In many ways, these tools serve as a bridge between urban/ rural youth, youth and organizations, youth and policies, thus making collective action more possible and meaningful.

Elements of networking and collaboration are functioning deeply within the mobile sphere and, to an extent, the present mechanism adopted by young adults to engage with each other and with the rest of society compensates for the failures of a nation seeking to celebrate multiculturalism while heading towards further segregation. Attention to youth as 'citizens' in this developing nation expands the meaning of and motivation behind civic and political action. Many citizens long regarded as insignificant members of society are becoming increasingly visible, while trying not to be mistaken with being violent. Young adults are winning majority votes and exercising their rights to represent the people at parliamentary debates. While this is taking place, young adults continue to congregate – to demand change, to advocate unity, or to participate at any level deemed suitable.

This chapter discussed the rise of mobile media among youth by reflecting upon the various networking tools used to put forward their own agency. The findings in this chapter give us reason for cautious optimism. They highlight the persistence of today's youth in challenging the existing social and political structures, while portraying youth as active participants striving for social change, constructing their own identities. However, while youth may be able to overcome great obstacles and make significant contributions, we cannot overlook the role of society (especially the State) in providing them with useful resources and support, which will enable them to claim their proper position in society and to take the nation to greater heights.

ACKNOWLEDGMENT

The author would like to thank both the International Development Research Centre, Canada and the Ministry of Higher Education, Malaysia for funding two separate research projects (Youth, ICTs, and Political Engagement in Asia; Social Media and The Agency of Youth in Malaysia) from which data for this chapter were obtained.

REFERENCES

Amin, S. (2013, March 20). PKR Launches Mobile App. As Election Campaign Tool. *Malaysian Digest*.

Attorney-General's Chambers. (2012). *Laws of Malaysia, Peaceful Assembly Act 2012*. Retrieved August 28, 2013, from http://www.federalgazette.agc.gov.my

Azizuddin Mohd Sani, M. (2008). Freedom of speech and democracy in Malaysia. *Asian Journal of Political Science*, *16*(1), 85–104. doi:10.1080/02185370801962440

Balakrishnan, V., & Shamim, A. (2013). Malaysian Facebookers: Motives and addictive behaviours unraveled. *Computers in Human Behavior*, *29*(4), 1342–1349. doi:10.1016/j.chb.2013.01.010

Borneo Post. (2012, December 28). *Registration for 3G smartphone rebate start from next Tuesday –MCMC*. Retrieved September 27, 2013, from http://www.skmm.gov.my/Media/Press-Clippings/Registration-for-3G-smartphone-rebate-start-from-n.aspx

Bursa Malaysia. (n.d.). Retrieved September 21, 2013, from http://announcements.bursamalaysia.com/edms/hsubweb.nsf/1c0706d8c060912d48256c6f0017b41c/48256aaf0027302c48256bcd001175b4/$FILE/Maxis-IndustryOverview-Material%20Indebtedness-Conditions-RPT-Additional%20Info%20(780KB).pdf

Cheverst, K., Fitton, D., Rouncefield, M., & Graham, C. (2004). *Smart mobs and technology probes: Evaluating texting at work*. Academic Press.

Chye, K. T. (2013, February 15). *Gross Encounters with the Zin Kind*. Retrieved August 17, 2013, from http://www.malaysiandigest.com/opinion/256182-gross-encounters-of-the-zin-kind.html

Coleman, S. (2007). How democracies have disengaged from young people. In B. D. Loader (Ed.), *Young citizens in the digital age: Political engagement, younger people and new media*. London: Routledge.

Duran, A. (2006). Flash mobs: Social influence in the 21st century. *Social Influence*, *1*(4), 301–315. doi:10.1080/15534510601046569

Edge Malaysia. (2013, May 16). *New home minister tells unhappy Malaysians to emigrate*. Retrieved August 5, 2013, from http://www.theedgemalaysia.com/political-news/239111-new-home-minister-tells-unhappy-malaysians-to-emigrate.html

GfK Retail and Technology Asia. (2012, April 26). *Boom Times Continue as Southeast Asia's Smartphone Market Value Expands by 62 Percent in Quarter One 2012: GfK Asia*. Retrieved September 27, 2013, from http://www.gfkrt.com/asia/news_events/news/news_single/009744/index.en.html

Leong, L. Y., Hew, T. S., Ooi, K. B., & Lin, B. (2011). Influence of gender and English proficiency on Facebook mobile adoption. *International Journal of Mobile Communications*, *9*(5), 495–521. doi:10.1504/IJMC.2011.042456

Lim, J. B. Y. (2013a). *Youth Participation: Social Media and East Asian Cultures in Malaysia*. Paper presented at the International Symposium - The Korean Wave in Southeast Asia: Consumption and Cultural Production. Kuala Lumpur, Malaysia.

Lim, J. B. Y. (2013b). Videoblogging and Youth Activism in Malaysia. *International Communication Gazette*, *75*(3), 300–321. doi:10.1177/1748048512472947

Malaysian Communications and Multimedia Commission. (2012). *Hand Phone Users Survey 2012*. Retrieved September 20, 2013, from http://www.skmm.gov.my/skmmgovmy/media/General/pdf/130717_HPUS2012.pdf

Malaysian Digest. (2013, February 19). *Mohd Zin Defends Selangor BN's SMS Campaign*. Retrieved September 5, 2013, from http://www.malaysiandigest.com/sports/259902-mohd-zin-defends-selangor-bns-sms-campaign.html

Malaysian Insider. (2013, April 8). *Get 'Undi PRU13' apps for free on your Androids or Apple smartphones*. Retrieved August 10, 2013, from http://www.themalaysianinsider.com/malaysia/article/get-undi-pru13-apps-for-free-on-your-androids-or-apple-smartphones/

Malaysian Mobile Content Provider Association. (2012). *Home*. Retrieved September 4, 2013, from http://www.mmcp.org.my/

Malaysian Mobile Content Providers Association. (n.d.). Retrieved September 4, 2013, from http://www.mmcp.org.my/

Mohd Azam Osman, A. Z.-Y. (2012). A Study of the Trend of Smartphone and its Usage Behavior in Malaysia. *International Journal on New Computer Architectures and Their Applications*, *2*(1), 274–285.

Molnár, V. (2010). Reframing public space through digital mobilization: Flash mobs and the futility (?) of contemporary urban youth culture. *Theory, Culture & Society*.

Muniandy, P. (2012). Malaysias Coming Out! Critical Cosmopolitans, Religious Politics and Democracy. *Asian Journal of Social Science*, *40*(5-6), 5–6. doi:10.1163/15685314-12341263

Muse, J. H. (2010). Flash Mobs and the Diffusion of Audience. *Theater*, *40*(3), 9–23. doi:10.1215/01610775-2010-005

Nicholson, J. A. (2005). Flash! mobs in the age of mobile connectivity. *Fibreculture Journal, 6*.

Osman, M. A., Talib, A. Z., Sanusi, Z. A., Shiang-Yen, T., & Alwi, A. S. (2012). A Study of the Trend of Smartphone and its Usage Behavior in Malaysia. *International Journal of New Computer Architectures and their Applications, 2*(1), 274-285.

Perez, S. (2012, July 2). *comScore: In U.S. Mobile Market, Samsung, Android Top The Charts, Apps Overtake Web Browsing*. Retrieved from http://www.techcrunch.com

RandomAlphabets.com. (2010, May 15). *Glee Flashmob Dance 2010*. Retrieved September 2, 2013, from http://randomalphabets.com/2010/05/glee-flashmob-dance-2010/

Rozario, K. (2013, April 14). *BN Spam SMS: Why it's such a bad idea.* Retrieved September 4, 2013, from http://www.keithrozario.com/2013/04/barisan-nasional-bn-spam-sms.html

Schlozman, K. L. (2002). Citizen participation in America: What do we know? Why do we care? In I. Katznelson, & H. V. Milner (Eds.), *Political science: The state of the discipline* (pp. 433–461). New York: W. W. Norton & Company, Inc.

Singh, K. (2012, May 15). *Interesting insights into smartphone behavior.* Retrieved September 27, 2013, from http://www.digitalnewsasia.com/node/107

Takahashi, K., & Hatano, G. (1999). Recent Trends in Civic Engagement among Japanese Youth. In *Roots of civic identity: International perspectives on community service and activism in youth*. Academic Press.

Urdal, H. (2006). A clash of generations? Youth Bulges and Political Violence. *International Studies Quarterly*, *50*, 607–629. doi:10.1111/j.1468-2478.2006.00416.x

Virtual Logic Sdn Bhd. (2013, March 4). *JustUndi GE13*. Retrieved September 19, 2013, from https://play.google.com/store/apps/details?id=com.etoff.undilah&hl=en

Wong, E. L. (2012, May 8). *More Malaysians Using Smartphone*. Retrieved September 27, 2013, from http://www.marketing-interactive.com/news/32749

Wong, K. (2004). Asian-Based Development Journalism and Political Elections Press Coverage of the 1999 General Elections in Malaysia. *Gazette*, *66*(1), 25–40. doi:10.1177/0016549204039940

Woon, L. (2013, February 22). *DAP man sees red over BN SMS*. Retrieved September 5, 2013, from http://www.freemalaysiatoday.com/category/nation/2013/02/22/dap-man-sees-red-over-bn-sms/

Yuen, M. K. (2012, June 14). *Govt Denies Jamming Calls during Bersih Rally*. Kuala Lumpur, Malaysia: Star Publications (M) Bhd. Retrieved August 3, 2013, from http://www.thestar.com.my/News/Nation/2012/06/14/Govt-denies-jamming-calls-during-Bersih-rally.aspx

KEY TERMS AND DEFINITIONS

Civic Engagement: Efforts to encourage awareness and participation among members of society in social, political and economic issues.

Cyber-Urban Connections: The relationship between online-offline participation which leads to mobilization of the masses in both virtual and physical spaces. Also includes the blurring of the two spaces especially where technology is readily available and has become part of one's everyday life.

Mobile Activism: The use of mobile phones, smart phones and tablets to disseminate information, garner followers and encourage mobilization.

National Identity: A sense of belonging or the spirit of unity and nationalism experienced by citizens. The notion of national identity may differ between what is put forward by the ruling government and by members of its society.

Participatory Culture: The involvement of individuals and/or communities in activist movements both online and offline. Participatory culture may be arguably influenced by mob mentality or meaningful engagement with the actual causes.

Political Mobilization: Society (individuals, governmental and non-governmental organizations) actively involved in advocating change within the electoral system, though not necessarily related to partisan politics.

Social Media Users: Members of society who use social media platforms actively, including Facebook, Twitter, YouTube, Instagram and other online networks.

Youth Activism: Young adults between the ages of 18 and 35 who are engaged in political/ civic/ community activities for social change. Their focus encompasses issues relating to human rights, freedom of expression, sexuality, and electoral politics, among others.

ENDNOTES

1. Google Playstore recorded user 50,000-100,000 installations, receiving 4.2 stars from 237 reviewers.
2. Attracted 10,000-50,000 user installations, rated 4.4 stars by 83 reviewers.
3. See Mohd Azam Osman, A. et al. (2012). A Study of the Trend of Smartphone and its Usage Behavior in Malaysia.
4. Telekom Malaysia monopolized mobile, fixed line and international telephony services until 1999 when the government introduced its "equal access" policy to allow for a more transparent, liberalized structure to promote increased competition and allow Malaysian consumers to choose their telephone carriers/network operators.
5. Maxis Communications Bhd vice president and head of Product, Device, Innovation and Roaming, T. Kugan states, an internal survey done by Maxis last year revealed 31% of its mobile subscribers are smartphone users. An analyst report out in *The Edge Financial Daily* (2012) indicated that around 20% of DiGi Telecommunications Bhd customers are smartphone users.
6. Youths qualifying for the RM200 rebate would be able to purchase selected 3G smartphones costing up to RM500 from selected dealers and agents appointed by the providers.
7. The Association seeks to educate mobile content providers on various guidelines including SMS marketing, provision of mobile and content services, and other "mandatory guideline(s)" set up by the MCMC.
8. Duran (2006) define flashmobs as strangers who come together in a public place, perform an unusual behavior, and randomly disperse.
9. Under the Peaceful Assembly Act 2012, authorities have the right to refuse approval of certain assemblies that are deemed to threaten public order. While some believe that the Act encourages healthy participation in assemblies, others feel this is yet another move by the powers that be to impose restrictions upon human rights/ freedom (see Muniandy, 2012).
10. Political writer Kee Thuan Chye wrote: "… they (the SMSes) have become such an annoyance that the recipients invariably express nothing but disgust each time they receive such messages." – Chye, K. T. (2013, February 15). *Gross Encounters with the Zin Kind.* Retrieved August 17, 2013, from Malaysian Digest: http://www.malaysiandigest.com/opinion/256182-gross-encounters-of-the-zin-kind.html
11. Youth is found to be less actively engaged than the elders in almost every established form of political activities in Western countries (Coleman, 2007; Schlozman, 2002).

Chapter 9
Mobile Communication and Bottom-Up Movements in Singapore

Carol Soon
National University of Singapore, Singapore

Cheong Kah Shin
The Straits Times, Singapore

ABSTRACT

At Hong Lim Park in February 2013, the "No to 6.9 Million Population" protest saw 5,000 Singaporeans expressing their unhappiness with the government's Population White Paper. Touted to be the largest demonstration since Singapore's independence, it bore witness to digital technologies' mobilization effects. Personal and organization websites, discussion forums, blogs, and social media provide viable spaces for individuals and marginalized groups to circumvent offline media regulations and participate in counter-hegemonic discourse. Surveys indicate that Singaporeans are increasingly leveraging mobile communication for utility purposes—seeking and sharing information—and for networking. This chapter identifies digital bottom-up movements that took place in recent years, the anatomy of these movements, and how digital technologies were used. What is evident is that groups championing different causes are using a wide range of digital platforms to galvanize support and mobilize action for political and social issues. However, the link between that and mobile communication remains unclear. This chapter concludes by presenting recommendations for future studies on mobile communication and bottom-up movements.

INTRODUCTION

On 16 February 2013, a few thousand Singaporeans gathered at the Speakers' Corner. They were taking part in the "No to 6.9 Million People" protest - expressing their unhappiness with the government's population policy. Touted to be the largest demonstration since Singapore's independence in 1965, the protest showcased digital technologies' mobilization qualities as protest organizers relied on social media for information dissemination and mobilization. Post-General

DOI: 10.4018/978-1-4666-6166-0.ch009

Elections 2011, Singapore appears to have arrived at a new dawn of politics with the emergence of frequent collective criticisms of policies and calls for change. In other parts of the world, the Occupy Wall Street movement, uprisings in the Middle East and Jasmine Revolution in China are proof of how ordinary citizens use digital technologies to challenge the incumbent leadership and establish a new social order.

Although cautioning against technological determinism, social movement scholars Diani (2000), Tarrow (1998) and Tilly (2004) acknowledged that information communication technologies are fast becoming an integral part of the social movement repertoire. The rhetoric surrounding technology's impact on civic engagement is not new. Analyses pertaining to its democratizing effects have assumed a dominant focus from the 1990s till now, particularly in the fields of political science, sociology and media studies (see for example, Bimber, 1999; Ibrahim, 2009; Johnson, 2003; Schumate & Lipp, 2007; Soon & Kluver, forthcoming). Mattoni's (2013) study on the mobilizations against precarity in Italy established that activists and social movement groups developed their own communication repertoires and were adept at using a variety of media for different purposes.

A large body of work scrutinizes the contributions of mobile communication to political participation and collective action. Although revolutions have been part and parcel of the Arab world's tumultuous history, the uprisings in 2011 have been linked to mobile communication. Mobile devices catalysed the organization of mass demonstrations by allowing organizers and participants to circumvent traditional media that are compliant with despotic rulers, share "sticky" information (in the form of text, images and videos) and establish alliances with one another (Abdelhay, 2012; Hussain & Howard, 2013; Khondker, 2011; Paradiso, 2013). In 2005, mobile videos uploaded to YouTube and other video-sharing sites disseminated actual footage of vote counting and rigging during the Egyptian elections where Mubarak's party won 89% of the vote, fuelling protests against acts of injustice by the ruling elites and repression by the military (Hussain & Howard, 2013). Paradiso (2013) advocated that "technologies such as cell phones, YouTube, Facebook, Twitter, and satellite television clearly enhanced the diffusion of indignation and anger, and facilitated the coordination of demonstrations" (p.180).

In Castell's "mobile civil society", mobile phones create "a powerful platform for political autonomy on the basis of independent channels of autonomous communication, from person to person, and from group to group" (Castells, 2005/2006, p. 112). According to Castells, the distributed network of mobile communication provides the foundation for the formation of new public spaces. Such a distributed network facilitates high-volume communication, personalization and interactivity, and allows users to circumvent traditional media. In developing countries where access issues, low literacy and high costs pose as barriers to the adoption of computing devices (such as personal computers, laptops and tablets), mobile devices provide a viable alternative to the populace in communicating with many others almost instantaneously (Latham, 2007; Skuse & Cousins, 2008).

While there has been burgeoning scholarship concerning how Internet technologies enhance civic engagement and political participation in Singapore, little is known about the role of mobile communication in the broader picture of collective action. During the campaigning period leading to the general election on 7 May 2011, social media catalysed the exchange of political quips, photographs and videos of election rallies, with online discourse largely dominated by those opposing the ruling elite (George, 2011; Ng, 2011).

However, research on citizen-driven movements, in particular their deployment of digital platforms, is at a nascent stage. We examine current scholarship on mobile communication in the international and Singapore contexts, and analyse bottom-up movements in Singapore. While the analysis does not provide any empirical evidence on the use of mobile communication in these movements, it is nonetheless instructive to identify the range of technological platforms used in ground-up movements. In so doing, the anatomical perspective to bottom-up movements used in this chapter helps identify gaps in studying the relationship between mobile communication and civic engagement in a well-connected society as Singapore's. We conclude by proposing research approaches for future evaluations of mobile communication and collective action in connected societies.

MOBILE INDIVIDUALS IN MOBILE SOCIETIES

Communication scholars have lauded mobile communication for transforming people's lives. The proliferation of social media and mobile communication has given rise to bigger and rapidly-formed social networks of knowledge construction (Hofheinz, 2011). Touted to be social levellers, mobile phones accord to individuals greater networking opportunities and provide them with low-cost means for building relationships, thereby expanding their social capital. This is a particularly significant contribution in countries with an unequal distribution of technology adoption.

Hampton, Sessions and Her (2011) compared the 2008 US General Social Surveys with the 1985 and 2004 surveys. They found, similar to Internet use, a positive correlation between mobile phone use and one's network size and diversity, hence challenging conventional wisdom concerning the displacement of core networks by geographically dispersed weak ties. Not only did social isolation not increase in the past twenty years, Hampton et. al. found that mobile phone users and those who use social media (specifically for Instant Messaging and uploading photos to share online) had larger core networks. These core networks are important valuable sources of social support and emotional aid. Their conclusion is reiterated by Campbell and Kwak (2010) whose study informed the role of text messaging within mobile communication. Text messaging was a conduit for highly personal form of interaction which cultivated personal relationships for individuals who used mobile phones to contact people within the local community and those who used the device to communicate with people living in another geographic area.

A strong link has been drawn between mobile communication's social capital-building capacities and civic engagement. Through different approaches, ranging from anecdotal to structural analyses, scholars have identified specific ways through which mobile communication affects the political landscape. While some scholars study how mobile communication has changed (or not changed) political participation, others have adopted a more macro-perspective in determining how mobile communication fits into the larger ecology of information communication technologies. A few others adopted a comparative approach to understand the consequences of mobile phone use in civil society. This body of work addresses how mobile communication liberates repressed citizens in authoritarian regimes and makes it possible for them to counter hegemonic discourse, and factors that influence the use of mobile communication for civic and political engagement.

Mobilizing Repressed Polities

Scholars who studied autocratic regimes with undeveloped Internet infrastructure confirmed that mobile communication plays a critical role in challenging existing power structures. More than half of Arab countries have mobile penetration well over 100%, including several of the countries where Arab Spring protests were most successful (Hussain & Howard, 2013). In South Africa, while Internet access is largely restricted to businesses and the urban elite, mobile penetration reached

83.33 subscribers per hundred as early as in the mid-2000s (UNSTATS 2008).

Besides the low cost of mobile phones, the short-messaging system (SMS) holds another attraction for mobile phone users due to the low cost of sending SMS, and how text messages "can be sent, stored, and retrieved in a delayed fashion, thus ensuring communication whether the recipients' phones are off, busy, malfunctioning, or roaming" (Yu, 2004, p.185). Innovative and flexible "pay-as-you-go" pricing models and extensive network coverage made mobile phones part of the lives of many who were otherwise disconnected (Skuse & Cousins 2008; Tenhunen, 2011). Such a pattern was also evident in China where mobile phones are used far more widely than the Internet and by a much more diverse population (Latham, 2007).

Khondker (2011), Hussain and Howard (2013) posited that the portability of mobile phones was key in their role as a mediating instrument bridging communication gaps among protesters in the Arab uprisings. Thus, "mobile telephony, in the form of small consumer-based communication devices, has allowed regular citizens to bear witness, record, and disseminate acts of injustice and repression" (Hussain & Howard, p.61). This concept of the public empowered by technology to monitor those in power was reiterated by Cammaerts (2012) who described how protesters used handheld cameras for "sousveillance tactics", which he defined as "bottom-up surveillance by the citizen/activist on the state or public figures" (p.127).

In addition, mobile technologies enable individuals to overcome barriers to Internet access and connect to blogs, photo-sharing sites and social networking sites (Abdelhay, 2012). The power of the mobile phone both as a recorder and disseminator of information was demonstrated in the case of the Tunisian uprising. The harassment of a Tunisian fruit vendor who subsequently committed suicide brought simmering public unhappiness to a boiling point in December 2010. His suicide was captured on cell phone video and posted to Facebook. The visceral image was shared by large numbers of people who accessed the social networking site via smartphones (Paradiso, 2011).

The bridging effects of mobile communication led scholars such as Loudon (2010) to conclude that mobile devices are used to reach previously disconnected majorities, strengthen their involvement in existing processes and extend a movement's reach. By facilitating horizontal communication, mobile communication enables individuals to challenge traditional top-down communication and strengthen their political agency, thereby creating what Rheingold (2002) described as "smart mobs".

Engendering New Communicative Practices

Besides providing accessible and affordable means for information dissemination and the organization of activities, mobile devices are also creating alternative spaces where individuals participate in counter-hegemonic discourses (Yu, 2004). Latham (2007) advocated that SMS messaging, which became popular in the late 1990s and early 2000s, became the next most widely-used tool for information-seeking after fax machines, pagers, fixed-line phones and the email. The "disorderly" content present in telecommunications is "unregulated, contingent, unpredictable, and largely unknowable" and thus makes it difficult for the Communist Party in China to control as they have done with traditional media, a state apparatus for building ideological consensus (p.303). The individualized and small-group nature of mobile communication further challenges the state's ability to monitor interpersonal exchanges as "they cannot easily be switched off, controlled, or made safe by centralized supervisory authorities" (*ibid*).

In a separate study, He (2008) uncovered "two distinct discourse universes" emerging in China, with one being "the official universe, characterized by vagueness, abstractness, ambiguity, and indoctrination" and the other being "the private universe, characterized by non-hegemonic expressions ranging from radical nationalism to liberalism, materialism, and extreme cynicism" (p.182). He described the wide spectrum of non-official discourse in the following excerpt:

On the extreme side are messages that are politically subversive and morally "decadent," and on the mild side are humour and jokes that can be found on any other platform. In between are smart pieces that are politically or morally satirical and defiant of the communist political and moral norms (p.186).

Supporting Latham's thesis, He argued that while the "official universe" monopolized public spaces of expression, especially in the state-controlled mass media, the "private universe" thrived in the oral sphere which spilled over to the Internet and SMS messages. The "semiotic power" experienced by people who say almost anything about politics through SMS "grants them a sense of being equal in political discourse" (p.187).

Decomposing Mobile's Mobilizing Qualities

While research on mobile communication in non-democratic regimes typically adopts the case study approach to determine the impact of the medium, other researchers have used a correlational approach to decompose various forms of mobile phone usage and their effects on political participation. Rojas and Puig-i-Abril (2009) analysed how blog and mobile technology use is linked to civic and political participation. Their study confirmed that blog use and informational uses of cellular phones are positively related to expressive political engagement – "a form of political participation that entails the public expression of political orientations" (p.906). The significance of their study lies in its confirmation that mobile technologies' democratizing potential is comparable to that of web-based platforms.

The link between mobile phone use for information-seeking purposes and political participation received further support from Kwak, Campbell, Choi and Bae's (2011) study. They found that when individuals used their mobile phones to seek out public affairs information and discuss politics, they were more likely to participate in civic and political activities. A separate analysis by Campbell and Kwak (2010) nuanced the impact of mobile communication on political participation and posited that individual competency was a moderating factor. Individuals who reported higher levels of comfort with mobile phones were more civically and politically engaged than those who reported less comfort with the technology.

Other scholars have attempted to identify how groups leverage the medium for organizing and mobilizing their target constituencies (Loudon, 2010; Qiu, 2008). Tenhunen (2011) advocated that mobile communication facilitates both horizontal (among activists and with external organizations) and vertical information transmission (between leaders and subordinates). Loudon's study of Treatment Action Campaign (a South African social movement campaigning for the rights of people living with HIV/AIDS) uncovered how non-governmental organizations leveraged SMS to send mass reminders to members and partner organizations about events and meetings, follow-up messages on door-to-door advocacy campaign visits, and daily SMS messages containing inspirational messages.

Mobile communication also provided access to a "wiki-type encyclopaedia of treatment literacy information" that was searchable via a cell phone web browser. Qiu's (2008) comparison of the People Power 2 movement in The Philippines 2001 and Nosamo movement in South Korea in 2002 revealed that the organisational forms of movements were shaped by unique contextual and structural factors as well as the common historical and institutional conditions they shared. More importantly, he found that while mobile communication contributed significantly to political mobilisation, coordination, and identity-formation among members of the civil society, mobile phones did not work in silos but with the Internet and other media as part of an "enlarged communication ecology" instead (p.53).

MOBILE COMMUNICATION IN SINGAPORE

On the whole, the revenue for the information communications (comprising hardware, software, IT services, telecom services and content services) industry saw a year-on-year growth of 19%, reaching $83.42 bill in 2011 (Infocomm Development Authority of Singapore, Annual Survey on Inforcomm Industry for 2011). Within the growing sector, domestic revenue grew 18% to reach $24.72 bill in 2011 with telecommunications services being the highest contributor. On the level of consumer adoption, upward trends were observed for both Internet and broadband penetration within the Singapore populace. Approximately 85% of households had home Internet access and almost all of such households connected to the Internet via broadband (Infocomm Development Authority of Singapore, Annual Survey on Inforcomm Usage in Households and by Individuals for 2011).

Regarding mobile communication, the same survey found that although computers remain as the main equipment used to access the Internet at home, Internet-based mobile access has increased from 34% in 2010 to 54% in 2011 (*ibid*). In 2011, more than half of the residents used a smartphone, with usage highest among those aged 25 to 34 years old. A study on smartphone penetration and usage released by market research company Blackbox in 2012 indicated that the distribution of smartphones is relatively consistent even among older netizens (Blackbox, Smartphones in Singapore, A Whitepaper Release).

A separate survey conducted by Ericsson ConsumerLab provided a more fine-grain analysis of the activities mobile phone users engage in using their phones. Using social networks, sending and receiving email, getting information and general web browsing, and reading online news were ranked the top four activities, with most users aged 15 to 34 years old. These activities were followed by playing or downloading computer/mobile/video games, getting information about goods and services, using instant messaging, and downloading movies short films and music (Networks Asia, Jun 18, 2013). Latest figures from the report point to an increase in smartphone penetration from 74% in 2012 to 78% in 2013, and from 31% in 2012 to 42% in 2013 for tablet penetration.

These trends show that Singaporeans are increasingly leveraging mobile communication for utility purposes – seeking and sharing information, and for networking. However, with the exception of a few studies that addressed digital divide issues in mobile phone use and the social effects of specific uses of mobile communication (e.g. sexting) on adolescents and determinants of the adoption of multi-media system messaging (Lim, 2009; Ji & Skoric, 2013; Wei & Lo, 2013), research on the impact of mobile communication on political participation and its connection to web-based technologies is extremely thin.

Existing literature on mobile communication in Singapore, though growing, is typically centred on usage patterns and adoption factors in areas of health management and education (see for example, Lim et.al., 2011; Menkhoff & Bengtsson, 2010, Sha et.al, 2012a, 2012b; So, Tan & Tay, 2012; Xue et.al., 2012; Ng et.al., 2012). These studies examine the roles mobile communication play in ageing women's information-seeking for health information, enhancing self-management of health issues, for healthcare delivery, student engagement in learning and knowledge building. The next section provides a contextualized analysis of Singapore's political and historical trajectories and the significance of digital technologies in engendering collective action in Singapore. In so doing, we explore the identify trajectories for future work.

COLLECTIVE ACTION IN SINGAPORE

Up until 1959 when Singapore achieved self-rule from its British colonial masters, individuals and groups were given high civil autonomy by the British to pursue their own political and social agenda (Gillis, 2005). However, the landslide victory of the People's Action Party (PAP) in the 1959 Legislative Assembly election marked a new era in the governance of Singapore. Although the conditions in post-colonial Singapore posed many challenges for the new government, the delivery of economic affluence by the state soon affirmed the polity's belief that PAP's mode of pragmatic governance was effective and cultivated "co-option and political discipline" among the citizenry (Rodan, 1998, p.67).

Threats to nation-building efforts in the myriad forms of trade unions, student unions and vocal media were eradicated. The regulation of mainstream media was justified on the grounds of building a national identity and social cohesion among Singapore's richly diverse polity and was implemented through a complex set of laws (Banerjee, 2002; Kuo, 1995). The growth of civil society was in part stymied by the PAP-led government's success in supplanting many of the social and economic functions that were traditionally performed by private individuals and organizations during colonial times (Tan, 2007). Any potential growth of a civil society was further constrained through the Societies Act which granted the state discretionary power to deny permit to groups that are "likely to be used for unlawful purposes or purposes that may be prejudicial to public peace, welfare, and good order or against national interest" (Koh & Ooi, 2004, p.181).

A turning point came in the 1990s when the Singapore economy underwent a major shift - when the government embarked on transforming the economy into one that is driven by innovation rather than manufacturing. Within several years, Singapore was ranked among the top 10 economies in the world for active-mobile broadband subscriptions (International Telecommunication Union, The World in 2011, ICT Facts and Figures). Despite liberalizing the telecommunications sector, the state regulated online discourse through a complex set of rules and regulations including the Internet Code of Practice and the Class License Scheme. Several incidents demonstrated the state's resolve in curbing threats to the country's political and social stability. However, the government's attempt to strike a balance between "illiberal political interventions with market-oriented strategies for economic growth", coupled with the architecture of the Internet soon created loopholes that are exploited by marginalized groups and individuals (George, 2003). Developments in the cyberspace since late 1990s suggest that the Internet has opened up spaces for marginalized individuals and groups to connect with like-minded others, overcome regulatory constraints and contest hegemonic discourse (George, 2003; Ho, Baber & Khondker, 2002; Ibrahim, 2009).

Emergence of Online Collectives

Soon and Cho (*2014*) found that although formalized or traditional civil society groups continued to play a critical role in civil society, the Internet has added diversity in both actors and issues. By according the comfort of anonymity and enhancing ease of participation, Internet technologies lower barriers to participation in political and social activism. With their instantaneity, reach and interactivity, computer-mediated communication made it possible for political bloggers to locate and identify those who share a similar ideology or grievances with ease and speed, leading to quick formation of issue-based groups (Soon & Cho, *2014*, Soon & Kluver, *2014*).

Based on interviews conducted with forty over political bloggers, Soon and Cho found that digital technologies played an important role in encouraging collective action by eradicating barriers posed by structural proximity. They established that digital technologies made significant contributions in two ways. First, they connected activist bloggers to others who shared similar goals; second, the flourishing of online-based movement groups reduced participation costs and inducted fledging activists.

The mainstream media reported that civic engagement among Singaporeans has been on the rise in recent years. Singaporeans are using blogs, social networking sites, forums and online videos to raise awareness and garner support for myriad causes (Tan, 2008). Some of these endeavours include the "No to Rape" campaign, which lobbied for the criminalization of marital rape, started by three youths who leveraged new media technologies such as blogs, Facebook and Twitter (Chew, 2009). In spite of the growing trend of digital bottom-up movements such as "No to 6.9 Million People" protest, Pink Dot, Repeal 377A and Anti-Mandatory Death Penalty campaign, little is known on how the public uses web-based technologies for political and social advocacy.

We focus on digital bottom-up movements, and not how non-governmental and voluntary groups use digital technologies to advance their agenda. Bottom-up movements are started by individuals who do not belong to any organizations and are not distinctly linked to ant non-governmental organizations. Members in bottom-up movements are brought together by shared concerns and motivation to push for change. The following section identifies digital bottom-up movements which took place in recent years and the web-based technologies used in these campaigns.

Bottom-Up Movements in Singapore

To recap, we defined bottom-up movements as campaigns as those that originate in cyberspace (as opposed to campaigns launched by established non-governmental or voluntary organizations) and use digital technologies as the main mode of organizing. By web-based technologies, we refer to the wide array of online platforms that includes websites, blogs, file-sharing platforms and social media. We identified 11 bottom-up movements through keyword searches conducted in February 2013 via commercial search engines.

Some of these movements took place recently between 2011 and February 2013 (e.g. Internal Security Act campaign, Save Bukit Brown, Slut Walk, Sticker Lady Petition and Slut Walk)[1] while other campaigns were included in our study due to their continuity (No to Rape), prominence (Pink Dot.sg) and a resurfacing of their campaign issues (Repeal 377A). Appendix A provides a brief description of each campaign. We then coded the textual content and structural features found on the sites to determine the following:

1. Campaign issue and organizer(s);
2. The online platforms used – to identify an exhaustive list of online platforms adopted for the campaign (e.g. blog, website, Facebook)
3. The actions mobilized – mobilization of online and offline actions (see Table 1[2])
4. The outcome(s) – depending on the action called, the outcome can be the number of participants who turned up for an event or the number of people who signed a petition. Anticipated outcomes such as a meeting between activists and government officials are also included in our analysis.

To identify the web-based platforms used for online campaigns, we performed a Google search using campaign names as keywords. Facebook pages, YouTube videos, Twitter feeds and online petition pages emerged from this search.

The Heart of the Matter: Campaign Issues

The issues advocated by digital bottom-up movements could be categorized into three types. The first type focused on *human-rights issues*. These movements advanced the rights of marginalized communities (e.g. lesbians, gays, bisexuals and transsexuals [LGBT]) as in the case of Pink Dot. sg. Others sought to restore human dignity and

Table 1. Coding scheme for technology deployment

Usage	Definition
Information provision	Provide information pertaining to campaign activities.
Mobilizing online action	Encourage direct action online though online petitions and providing action alerts etc.
Mobilize offline action	Coordinate, organize or plan offline actions through calendar of events, describing campaign actions or volunteer opportunities.
Promote dialogue and discussion	Promote discussions via the comment and/or email functions.
Connect with local individuals and groups	Hyperlink to other individuals or groups that share similar interests/goals.
Connect with international organizations	Hyperlink to organizations or groups outside Singapore that share similar interest/goals.
Fundraising	Solicit donations, sell merchandize, sell subscriptions or carry advertising.

freedom for individuals perceived to be unjustly treated by the law, e.g. the Internal Security Act (ISA) campaign and the Anti-Mandatory Death Penalty campaign. The second category of campaigns dealt with *bread and butter issues* such as rising costs of living, transportation issues and income inequality.

The first instalment of the "No to 6.9 Million People" was attended by almost 5,000 citizens[3] from different walks of life who expressed their rejection of the government's Population White Paper which they felt would aggravate existing infrastructural problems and competition for jobs. The Lehmann Brothers minibonds rally was organized to facilitate collective action among those purportedly misled by financial institutions and lost their money in Lehmann-linked structured financial products. The movements in the third category are *incident-specific*, organized to galvanize a collective response to specific incidents (e.g. Sticker Lady petition and Save Bukit Brown). These campaigns are more ad hoc and transient in nature. They also tend to attract a narrower segment of the population (e.g. the academic community, nature lovers, and the art community).

Using Locher's (2002) typology, we classified the movements based on their aims. The purpose and scope of each campaign are detailed in Table 2. Digital bottom-up movements in Singapore are alternative, redemptive and reformative in nature. None of the campaigns were revolutionary in nature. Two possible reasons explain this. First, despite unhappiness with specific policies, Singaporeans by and large are content with the existing system and recognize the capabilities of the one-party dominated government. Second, laws such as the Internal Security Act and Sedition Act serve as a major deterrent for individuals who are concerned with being viewed as subversive. With the exception of the Pink Dot.sg and Slut Walk, the targets of claims were institutions of authority such as the government.

Facebook pages surpassed websites/blogs to be the most popular platform among campaign organisers, with 9 out of 11 campaigns using them. For campaigns that did not have a website/blog (i.e. Free Sticker Lady and Occupy Raffles Place), Facebook played a key role in getting the word out as it allows users to "push" material to their friends via Facebook feeds. Hence, they might be preferred over static website/blogs, which depend on a "pull" mechanism to get users to their site. Following Facebook, websites/blogs were the second most popular platform, alongside e-petitions, then Twitter and YouTube. The nascence of Singapore's Twitter culture could account for the slightly more modest use of Twitter compared to high-profile foreign movements like the Obama campaign or the Arab Spring. As for YouTube, making videos takes arguably more time and technical expertise than starting a blog or Facebook page. These barriers to entry could

Table 2. Aims of bottom-up movements

Alternative Change opinions pertaining to a specific issue among a target group	Redemptive Bring about a dramatic change to the lives of individuals belonging to a specific group
Occupy Raffles Place Call for greater transparency and accountability for investments made by Temasek Holdings and Government of Singapore Investment Corporation **Pink Dot.sg** Eradicate prejudice against the LGBT community **Slut Walk** Raise awareness for the problem of sexual harassment of women and eradicate victim-blaming	**Lehmann Brothers and Minibonds** Seek compensation for investors who lost money in Lehmann-linked structured products **Internal Security Act (ISA) Campaign** Increase awareness for ISA detainees and their experiences; and call for the abolition of the Internal Security Act
Reformative Change an entire community or society in a specific way	**Revolutionary Eradicate an old social order and replace it with a new one**
"No to 6.9 Million People" Petition the government to reconsider population policies and curbing immigration **No to Rape** Advocate the total abolition of marital rape immunity in Sections 375(4) and 376(5) **Repeal 377A** Seek the repeal of Section 377A of the Penal Code which criminalizes sex between mutually consenting adult men **Save Bukit Brown** Petition against the government's decision to build a highway through Bukit Brown cemetery, a heritage site **Sticker Lady petition** Petition the Ministry of Home Affairs to reduce Sticker Lady's charge and recognise her work as art, not vandalism. **Anti-Mandatory Death Penalty campaign** Seek legislative changes to the death penalty for drug-related offenses	Nil

be plausibly explain why the platform is not first choice for average citizen campaigners.

To analyse how digital platforms were used in these campaigns, we classified data according to Stein's (2007) typology of communication in Internet campaigns (see Table 3). The campaigns organizers used their websites and Facebook pages extensively to disseminate information on the campaign. Sites also often contained critiques of laws and policies that campaigners were campaigning against. Most sites provided contact information so that readers could contact campaign administrators for more information. Many of these campaign sites, especially those on Facebook, were not static troves of information. Instead, they were fertile grounds for on-going conversations about the various campaign issues. Some such as Save Bukit Brown and Pink Dot.sg had a comprehensive blogroll to individuals or groups that shared a common interest. Other campaigns such as the Anti-Mandatory Death Campaign and the No to Rape campaign were promoted on a host of civil society blogs.

Communication scholars have suggested that hyperlinking and cross-advertising identifies blogs that webmasters perceive as credible (Evans & Wurster, 1999; Hagel & Armstrong, 1997; Park, 2002; Shapiro & Varian, 1999). Cross-advertising and hyperlinking allied blogs adds legitimacy to a campaign: it is not a lone voice in the wind, but a collective of mutually trusted websites, championing the cause. E-petitions were common with eight out of 11 campaigns mobilising action online through this tool. Notably, dedicated petition websites like change.org allow users to articulate their reasons for signing the petition. Such voices demonstrate conviction to the cause, thus making the e-petition more than a numbers game. When it came to mobilising offline action, nine out of 11 campaigns did so via digital means, most commonly through a Facebook event page such as the "No to 6.9 Million People" page. Several were promoted on civil society blogs (e.g. "No to 6.9 Million People" on TR Emeritus, "We Remember" on civil society film maker Martyn See's blog, "Pink Dot" on fridae.com). Evidently, allied alternative media platforms function as a precious support network to get the word out.

Last but not least, fundraising was a moderately popular feature with online campaigners. This was most commonly done through the online sale of merchandise or adding a "donate" button to the blog. Some campaigns like "Save Bukit Brown", "No to Rape" and "Free Sticker Lady" were purely civic responses to policies or specific legislation, and no donations were solicited.

FUTURE RESEARCH

As presented earlier, the target of claims for the majority of bottom-up movements in Singapore is the government. While some campaigns are targeted at getting the government to reconsider specific policies (Save Bukit Brown and "No to 6.9 Million People"), most of the bottom-up campaigns advocate constitutional or legislative changes. Campaigns pushing for amendments to the Singapore constitution include amending the death penalty for drug-related offenses (Anti-Mandatory Death Penalty campaign), repealing Section 377A of the Penal Code (Repeal 377A), abolishing marital rape immunity in Sections 375(4) and 376(5) (No to Rape), and abolishing the Internal Security Act (ISA campaign). An evaluation of the effectiveness of these campaigns is not possible at this juncture as with the exception of the Anti-Mandatory Death Penalty campaign, other campaigns are still on-going. Even in the case of the death penalty campaign, the Minister for Law stated that legislative changes made to the death penalty in 2012 were the result of the government's review of laws, not from activist campaigning[4]. An assessment of campaign outcomes should thus take into account other forms of support or endorsement such as the turn out for publicly-held and closed-door events, and the number of signatures for petitions. See Appendix B for the outcomes of the bottom-up movements.

In terms of event turn-out, Pink Dot.sg and "No to 6.9 Million People" protest saw large numbers of participants, with the formal garnering a steady increase for its annual gathering held in June. The original target for the "No to 6.9 Million People" protest was 1,000 participants but post-event, organizers reported an approximate turn-out of 5,000 people[5], with the Associated Press reporting a number of 3,000 people[6]. However, organized by anonymous individuals, the "Occupy Raffles Place" protest saw less than 20 attendees, comprising mainly of members of the media and foreigners[7].

Some digital bottom-up movements led to private meetings. One of which was the meeting between a small group of activists and Mr Tan Chuan-Jin (Minister for Manpower) to discuss the government's plans to build a highway over Bukit Brown. However, as explained by the Minister, the meeting was to explain the government's policy pertaining to development plans, and not to consult activists on which course of action to take. A large proportion of the digital bottom-up movements galvanized support for causes through online petitions. The Singapore Anti-Mandatory Death Penalty campaign garnered the highest number of signatures, followed by the Sticker Lady and "No to 6.9 Million Population" petitions. Although one may gauge the popularity of a campaign by the number of "Likes" it receives on Facebook, the level of support (based on number of Facebook "Likes") may not be a reliable indication of the campaign outcome. Despite garnering a large

Table 3. Functions of campaign sites

	Name of Campaign	Information provision	Promote dialogue and discussion	Hyperlink to other individuals or groups that share similar interests	Hyperlink to national or international social movements organizations	Mobilise online action	Mobilise offline action	Fundraising
1	Anti-Mandatory Death Penalty Campaign	✓	✓	✓	✓	✓	✓	✓
2	Internal Security Act Campaign	✓	✓	✓		✓	✓	✓ (sale of books and T-shirts)
3	Free Sticker Lady	✓	✓			✓		
4	Save Bukit Brown	✓	✓	✓		✓	✓	
5	Repeal 377A (website hacked)	NA	NA	NA	NA	✓	NA	NA
6	Lehmann Brothers mini-bonds protest	✓	✓	No but blog documents minibonds saga in Hong Kong	No but blog documents minibonds saga in Hong Kong	✓	✓	
7	Pink Dot	✓	✓	✓			✓	✓(sale of decals and plushies)
8	Occupy Raffles Place	✓	✓				✓	
9	Slut Walk Singapore	✓	✓	✓ (Hyperlink on Facebook page)			✓	✓
10	No To Rape	✓	✓	✓		✓	✓	
11	No to 6.9 Million People	✓	✓			✓	✓	✓(sale of T-shirts)

number of "Likes" on Facebook, "Occupy Raffles Place" (12,405 "Likes") drew a dismal turn-out at the protest.

In Singapore, digital technologies play an important role in facilitating civic engagement among ordinary citizens as they provide alternative means of assembly and organizing to those who want to advance specific causes. Prior to the Internet, individuals who wanted to form groups and organize activities to further their goals have to register with the Registrar of Societies. Under the Societies Act, the state has discretionary power to deny permit to groups and upon successful registration, civil society organizations are closely monitored by the authorities to ensure that they keep to the agenda stated in their constitution and mission.

Although the same regulation exists today, digital technologies have enabled individuals to circumvent regulatory constraints. Now on the

Internet, individuals can converge with ease and advocate causes which they believe in, and the anonymity afforded by the Internet lowers risks of participation. This chapter has established that bottom-up collectives use a wide array of digital platforms to carry out offline and online actions. These platforms are used to increase awareness, organize and encourage participation among their target constituencies. In addition to providing campaign details, bottom-up groups leverage the mobilization capacity of the Internet to distribute action alerts, plan and coordinate offline activities, and execute online actions.

With the exception of Pink Dot.sg, Slut Walk and the Lehmann Brothers and minibonds petition, majority of the bottom-up movements target the government and legislation. The generalizability of findings from this study must be considered, relative to other new media practices and regime types in other countries. The nature of bottom-up movements is conditioned by the historical and political variables in Singapore context within which activism developed. As Cammaerts (2012) hypothesised, the "mediation opportunity structure for protest movements and activists cannot be separated from the wider political and economic opportunity structures – they are clearly enmeshed with each other" (p.129). The authoritative stance adopted by the government in regulating societal discourse could have led to the burgeoning of anti-establishment voices and aggregations in the cyberspace. Anonymity and low barriers to publication and participation cultivate a conducive environment and to a large extent safe haven for the expression of dissenting views. Although set in the context of Singapore, this study nevertheless reflects a critical extension in a literature that is typically North American centric and calls for comparative analyses of how new media technologies interact with collective action in different regimes.

However, the role mobile communication plays in advancing bottom-up movements begets further analysis. We propose the following considerations for future studies.

First, there is a need to *question underpinning assumptions* and contextualize the impact of mobile communication against information communication technology adoption. Much of the libertarian rhetoric surrounding the citizen empowerment and the democratization effects of mobile communication are influenced by political upheavals in Arab countries, regimes where access to Internet in home and public settings is low, compared to mobile phone penetration (Abdelhay, 2012; Loudon, 2010; Skuse & Cousins, 2008). Furthermore, computing devices (desktop and portable) remain priced out of reach for the majority in these countries. The high Internet and broadband penetration in countries such as Singapore could possibly downplay the contribution of mobile communication and render its role as a mediating instrument that bridges communication gaps less vital.

Second, *the uses of mobile communication have to be decomposed.* In Singapore, surveys conducted by the Infocomm Development Authority and the industry indicate that the public is leveraging mobile devices for information-seeking and social networking. As mentioned earlier, a survey conducted by Blackbox points to a somewhat equal distribution of smartphones across age groups. A more fine-grained approach is required to decompose the types of connections, communication activities and information-seeking motivations in order to derive valid conclusions pertaining to mobile communication's contribution to one's social capital and civic engagement. *To what extent are text and multi-media messaging used to communicate with others on political and civic activities? To what extent do individuals seek political news and information when they browse the web on their mobile phones? To what extent do individuals use their mobile devices to access digital platforms (e.g. social networking sites) that enable them to seek out others who share similar political and civic interests?* Asking such questions will help researchers determine if and how mobile devices complement or substitute other devices in mobilizing and galvanizing collective action.

Third, *mobile communication's contribution within the new media ecology should be isolated* in order to draw reliable conclusions on mobile communication's roles. Future studies would have to adopt a more parsimonious approach to ascertain how usage of mobile device for civic engagement is similar to or different from that of other technologies. Although addressing a different context, Lim's (2009) study involving Singapore school-going children showed that access point (home, school or mobile access) affects usage type and proficiency. In so doing, movement actors can then develop their own communication repertoires and exercise options strategically when engaging with internal and external target constituencies (Mattoni, 2013). Several issues should therefore be considered: *To what extent are myriad mobilizing platforms accessed through mobile devices? How does mobile communication rank in usage for civic engagement when compared to other devices such desktop computers, laptops and tablet computers? What problems and limitations are associated with access through mobile devices?*

Finally, more work needs to be done *to identify determinants of mobile adoption for civic engagement.* Although limited in number, existing studies have shown that individuals' adoption and usage of mobile communication is shaped by factors beyond access issues. While the extent to which people use mobile phones for professional and social connections has been linked to one's gender and social resources (Ji & Marko, 2013), social influence, perceived usefulness and perceived ease of use were found to affect people intention to adopt 3rd Generation mobile services (Cho, 2011).

Thus, future research on individual-cognitive factors, technological convergence and interface issues, and fine-grained studies of uses and gratifications behind mobile use will yield a more cogent analysis of mobile communication's effects on civic engagement. In terms of methodology, surveys involving a representative segment of the population will be instrumental for baseline studies; in-depth interviews would yield insights into the factors and conditions that drive political and civic uses of mobile communication.

REFERENCES

Abdelhay, N. (2012). The Arab uprising 2011: New media in the hands of a new generation in North Africa. *Aslib Proceedings: New Information Perspectives*, *64*(5), 529–539. doi:10.1108/00012531211263148

Banerjee, I. (2002). The locals strike back? *Gazette: The International Journal for Communications Studies*, *64*(6), 517–535. doi:10.1177/17480485020640060101

Biddix, P. J., & Park, H. W.Han Woo Park. (2008). Online networks of student protest: The case of the living wage campaign. *New Media & Society*, *10*(6), 871–891. doi:10.1177/1461444808096249

Bimber, B. (1999). The Internet and citizen communication with government: Does the medium matter? *Political Communication*, *16*(4), 409–429. doi:10.1080/105846099198569

Blackbox. (2012, May). *Smartphones in Singapore – A whitepaper release*. Retrieved from http://www.blackbox.com.sg/wp/wp-content/uploads/2012/05/Blackbox-YKA-Whitepaper-Smartphones.pdf

Cammaerts, B. (2012). Protest logics and the mediation opportunity structure. *European Journal of Communication*, *27*(2), 117–134. doi:10.1177/0267323112441007

Campbell, S. W., & Kwak, N. (2010). Mobile communication and civic life: Linking patterns of use to civic and political engagement. *The Journal of Communication*, *60*(3), 536–555. doi:10.1111/j.1460-2466.2010.01496.x

Castells, M. (1997). *The information age: Economy, society and culture: The power of identity* (Vol. 2). Oxford, UK: Blackwell Publishing.

Castells, M., et al. (2005/6). Electronic communication and socio-political mobilization. *Global Civil Society,* 266-285.

Chew, C. (2009, August 15). Young crusaders. *The Straits Times*, pp. D1-9.

Cho, H. (2011). Theoretical intersections among social influences, beliefs, and intentions in the context of 3G mobile services in Singapore: Decomposing perceived critical mass and subjective norms. *The Journal of Communication, 61*(2), 283–306. doi:10.1111/j.1460-2466.2010.01532.x

Diani, M. (2000). Social movement networks virtual and real. *Information Communication and Society, 3*(3), 386–401. doi:10.1080/13691180051033333

Evans, P., & Wurster, T. S. (1999). Getting real about virtual commerce: Strategies for success in electronic and physical retail commerce. *Harvard Business Review, 77*(6), 84–94. PMID:10662007

Flanagin, A. J., Stohl, C., & Bimber, B. (2006). Modeling the structure of collective action. *Communication Monographs, 73*(1), 29–54. doi:10.1080/03637750600557099

George, C. (2003). The Internet and the narrowing tailoring dilemma for Asian democracies. *Communication Review, 6*(3), 247–268. doi:10.1080/10714420390226270.002

George, C. (2011, June 25). Alternative media: New era in ties. *The Straits Times*, p. A32.

Gillis, E. K. (2005). *Singapore civil society and British power*. Singapore: Talisman.

Hagel, J., & Armstrong, A. G. (1997). *Net gain: Expanding markets through virtual communities*. Boston: Harvard Business School Press.

Hampton, K. N., Sessions, L. F., & Her, E. J. (2011). Core networks, social isolation, and new media. *Information Communication and Society, 14*(1), 130–155. doi:10.1080/1369118X.2010.513417

He, Z. (2008). SMS in China: A major carrier of the nonofficial discourse universe. *The Information Society, 24*(3), 182–190. doi:10.1080/01972240802020101

Ho, K. C., Baber, Z., & Khondker, H. (2002). 'Sites' of resistance: Alternative web sites and state-society relations. *The British Journal of Sociology, 53*(1), 127–148. doi:10.1080/00071310120109366 PMID:11958682

Hofheinz, A. (2011). Nextopia? Beyond revolution 2.0. *International Journal of Communication, 5*, 1417–1434.

Hussain, M. M., & Howard, P. N. (2013). What best explains successful protest cascades? ICTs and the fuzzy causes of the Arab Spring. *International Studies Review, 15*(1), 48–66. doi:10.1111/misr.12020

Ibrahim, Y. (2009). Textual and symbolic resistance: Re-mediating politics through the blogosphere in Singapore. In A. Russell, & N. Echchaibi (Eds.), *International blogging* (pp. 173–198). New York, NY: Peter Lang Publishing.

Infocomm Development Authority of Singapore. (2010). *Annual survey on infocomm usage in households and by individuals for 2010*. Retrieved on September 23, 2012 from http://www.ida.gov.sg/doc/Publications/Publications_Level3/Survey2010/HH2010ES.pdf

Infocomm Development Authority of Singapore. (2013). *Annual survey on infocomm industry for 2011*. Retrieved from http://www.ida.gov.sg/~/media/Files/Infocomm%20Landscape/Facts%20and%20Figures/SurveyReport/2011/ASInfocommIndustry11pdf.pdf

Infocomm Development Authority of Singapore. (2013). *Annual survey on infocomm usage in households and by individuals for 2011*. Retrieved from http://www.ida.gov.sg/~/media/Files/Infocomm%20Landscape/Facts%20and%20Figures/SurveyReport/2011/2011%20HH%20mgt%20rpt%20public%20final.pdf

International Telecommunication Union. (2009). *The World in 2009, ICT Facts and Figures*. Retrieved from http://www.itu.int/ITU-D/ict/material/Telecom09_flyer.pdf

Ji, P., & Skoric, M. M. (2013). Gender and social resources: Digital divides of social network sites and mobile phone use in Singapore. *Chinese Journal of Communication*, *6*(2), 221–239. doi: 10.1080/17544750.2013.785673

Johnson, D. G. (2003). Reflections on campaign politics, the Internet and ethics. In D. M. Anderson, & M. Cornfield Johnson (Eds.), *The civic web: Online politics and democratic values* (pp. 19–34). Lanham, MD: Rowman & Littlefield.

Khondker, H. H. (2011). Role of the new media in the Arab Spring. *Globalizations*, *8*(5), 675–679. doi:10.1080/14747731.2011.621287

Koh, G., & Ooi, G. L. (2004). Relationship between state and civil society in Singapore: Clarifying the concept, assessing the ground. In *Civil society in Southeast Asia* (pp. 167–197). Singapore: ISEAS Publications.

Kreimer, S. F. (2001). Technologies of protest: Insurgent social movements and the First Amendment in the era of the Internet. *University of Pennsylvania Law Review*, *150*(1), 119–171. doi:10.2307/3312914

Kuo, C. Y. (1995). The making of a new nation: Cultural construction and national identity. In B. H. Chua (Ed.), *Communitarian ideology and democracy in Singapore* (pp. 101–123). London: Routledge.

Kwak, N., Campbell, S. W., Choi, J., & Bae, S. Y. (2011). Mobile communication and public affairs engagement in Korea: An examination of non-linear relationships between mobile phone use and engagement across age groups. *Asian Journal of Communication*, *21*(5), 485–503. doi:10.1080/01292986.2011.587016

Langman, L. (2005). From virtual public spheres to global justice: A critical theory of internetworked social movements. *Sociological Theory*, *23*(1), 42–74. doi:10.1111/j.0735-2751.2005.00242.x

Latham, K. (2007). SMS, communication, and citizenship in China's information society. *Critical Asian Studies*, *39*(2), 295–314. doi:10.1080/14672710701339493

Lim, S., Xue, L. S., Yen, C. C., Chang, L., Chan, H. C., & Tai, B. C. et al. (2011). A study on Singaporean women's acceptance of using mobile phones to seek health information. *International Journal of Medical Informatics*, *80*(12), E189–E202. doi:10.1016/j.ijmedinf.2011.08.007 PMID:21956003

Lim, S. S. (2009). Home, school, borrowed, public or mobile: Variations in young Singaporeans' Internet access and their implications. *Journal of Computer-Mediated Communication*, *14*(4), 1228–1256. doi:10.1111/j.1083-6101.2009.01488.x

Locher, D. A. (2002). *Collective behaviour*. Upper Saddle River, NJ: Prentice Hall.

Loudon, M. (2010). ICTs as an opportunity structure in southern social movements. *Information Communication and Society*, *13*(8), 1069–1098. doi:10.1080/13691180903468947

Mattoni, A. (2013). Repertoires of communication in social movement processes. In B. Cammaerts, A. Mattoni, & P. McCurdy (Eds.), *Mediation and protest movements* (pp. 39–56). Bristol: Intellect.

McAdam, D., Tarrow, S., & Tilly, C. (2001). *Dynamics of contention*. New York, NY: Cambridge University Press. doi:10.1017/CBO9780511805431

Menkhoff, T., & Bengtsson, M. L. (2010). Engaging students in higher education through mobile learning. *Communications in Computer and Information Science*, *112*, 471–487. doi:10.1007/978-3-642-16324-1_56

Networks Asia. (2013, June 18). *Singapore smartphone and tablet penetration on the rise, app. usage increasing*. Retrieved from http://www.networksasia.net/content/singapore-smartphone-and-tablet-penetration-rise-app-usage-increasing?page=0%2C0

Ng, T. P., Lim, M. L., Niti, M., & Collinson, S. (2012). Long-term digital mobile phone use and cognitive decline in the elderly. *Bioelectromagnetics*, *33*(2), 176–185. doi:10.1002/bem.20698

Ng, T.Y. (2011, May 17). Facebook trolls can be the most valuable fans. *The Straits Times*, p. A2.

Norris, P. (2002). *Democratic phoenix*. Cambridge, UK: Cambridge University Press. doi:10.1017/CBO9780511610073

Paradiso, M. (2013). The role of information and communications technologies in migrants from Tunisia's Jasmine Revolution. *Growth and Change*, *44*(1), 168–182. doi:10.1111/j.1468-2257.2012.00603.x

Park, H. W. (2002). Examining the determinants of who is hyperlinked to whom: A survey of webmasters in Korea. *First Monday*, *7*(11). doi:10.5210/fm.v7i11.1005

Qiu, J.L. (2008). Mobile civil society in Asia: A comparative study of People Power II and the Nosamo Movement. *Javnost – The Public, 15*(3), 39-58.

Rodan, G. (1998). The Internet and Political Control in Singapore. *Political Science Quarterly*, *113*(1), 63–89. doi:10.2307/2657651

Rojas, H., & Puig-i-Abril, E. (2009). Mobilizers mobilized: Information, expression, mobilization and participation in the digital age. *Journal of Computer-Mediated Communication*, *14*(4), 902–927. doi:10.1111/j.1083-6101.2009.01475.x

Schumate, M., & Lipp, J. (2007, May). *Connective collective action online: An examination of the network structure of the English speaking Islamic Resistance Movement*. Paper presented at International Communication Association Conference. San Francisco, CA.

Sha, L., Looi, C. K., Chen, W., Seow, P., & Wong, L. H. (2012). Recognizing and measuring self-regulated learning in a mobile learning environment. *Computers in Human Behavior*, *28*(2), 718–728. doi:10.1016/j.chb.2011.11.019

Sha, L., Looi, C. K., Chen, W., & Zhang, B. H. (2012). Understanding mobile learning from the perspective of self-regulated learning. *Journal of Computer Assisted Learning*, *28*(4), 366–378. doi:10.1111/j.1365-2729.2011.00461.x

Shapiro, C., & Varian, H. R. (1999). *Information rules: A strategic guide to the network economy*. Boston: Harvard Business School Press.

Skuse, A., & Cousins, T. (2008). Getting connected: The social dynamics of urban telecommunications access and use in Khayelitsha, Cape Town. *New Media & Society*, *10*(1), 9–26. doi:10.1177/1461444807085319

So, H. J., Tan, E., & Tay, J. (2012). Collaborative mobile learning in situ from knowledge building perspectives. *Asia-Pacific Education Researcher*, *21*(1), 51–62.

Soon, C., & Cho, H. (2014). OMGs! Offline-based movement organizations, online-based movement organizations and network mobilization: A case study of political bloggers in Singapore. *Information Communication and Society*, *17*(5), 537–559. doi:10.1080/1369118X.2013.808256

Soon, C., & Kluver, R. (2014). Uniting political bloggers in diversity: Collective identity and web activism. *Journal of Computer-Mediated-Communication. Journal of Computer-Mediated Communication*, *19*(3), 500–515. doi:10.1111/jcc4.12079

Stein, L. (2007, May). *National social movement organizations and the World Wide Web: A survey of web-based activities and attributes*. Paper presented at International Communication Association Conference. San Francisco, CA.

Tan, K. P. (2007). New politics for a renaissance city? In K. P. Tan (Ed.), *Renaissance Singapore?* (pp. 17–36). Singapore: NUS Press.

Tan, W. (2008, August 9). Rise of online activists. *The Straits Times*, p. B11.

Tarrow, S. (1998). *Power in movement: social movements and contentious politics*. New York, NY: Cambridge University Press. doi:10.1017/CBO9780511813245

Tenhunen, S. (2011). Culture, conflict, and translocal communication: Mobile technology and politics in Rural West Bengal, India. *Ethnos*, *76*(3), 398–420. doi:10.1080/00141844.2011.580356

Tilly, C. (2004). *Social movements, 1768-2004*. Boulder, CO: Paradigm Publishers.

UNSTATS. (2008). *Millennium Development Goals Indicators*. Retrieved from http://mdgs.un.org/unsd/mdg/Data.aspx

Van Laer, J., & Van Aelst, P. (2010). Internet and social movement action repertoires. *Information Communication and Society*, *13*(8), 1146–1171. doi:10.1080/13691181003628307

Wei, C., & Lo, V.-H. (2013). Examining sexting's effects among adolescent mobile phone users. *International Journal of Communication*, *11*(2), 176–193.

Xue, L. S., Yen, C. C., Chang, L., Chan, H. C., Tai, B. C., & Tan, S. B. et al. (2012). An exploratory study of ageing women's perception on access to health informatics via a mobile phone-based intervention. *International Journal of Medical Informatics*, *81*(9), 637–648. doi:10.1016/j.ijmedinf.2012.04.008 PMID:22658778

Yu, H. (2004). The power of thumbs: The politics of SMS in urban China. *Graduate Journal of Asia-Pacific Studies*, *2*, 30–43.

KEY TERMS AND DEFINITIONS

Bottom-Up Movement: Collective action that originates with the common people (e.g. members of the public) or people of a lower rank.

Civic Engagement: Individuals' participation in collective action to address issues of public concern.

Collective Action: Individuals working in concert towards a common goal, i.e. to engender social, political or economic change.

Digital Technologies: Hardware or software systems that make use of binary or digital logic.

Mobile Communication: A communication network which does not rely on physical connections between two communication entities (individuals, groups, organizations). Users are mobile during communication.

Political Participation: Taking part in an activity or series of activities to influence policymaking and the selection of those who are involved in the crafting and implementation of policies. Political participation assumes different forms, such as voting, signing a petition and attending a rally.

Social Capital: The benefits that one gain from being part of a social network and the cooperation and sharing that goes on in that network. Besides economic benefits, social capital also includes reciprocity, information and trust which enhance the well-being of a community.

ENDNOTES

1. We had initially included the campaign launched by Toh Yi residents to garner support for their petition against the government's plan to build a home for the aged. The campaign attracted media coverage in mainstream press. However, we had to exclude the campaign as the original campaign site (hosted on Facebook) is no longer in use.
2. We adopted Stein's (2007) instrument to analyse how various web-based technologies were used.
3. As per figure given by organizers of the event.
4. "Death penalty change came from review, not activists: Shanmugam" (Nov 5, 2012) http://sg.news.yahoo.com/death-penalty-change-came-from-review--not-activists--shanmugam.html
5. "S'poreans hold protest against White Paper on Population" (Feb 16, 2013) http://www.channelnewsasia.com/stories/singaporelocalnews/view/1254631/1/.html
6. "Rare Singapore protest against population plan" (February 17, 2013) http://www.thejakartapost.com/news/2013/02/17/rare-singapore-protest-against-population-plan.html
7. "Less than 20 turn up for Raffles Place rally" (Oct 16, 2011) http://www.asiaone.com/News/AsiaOne+News/Singapore/Story/A1Story20111016-305317.html
8. We were not able to ascertain details on the Repeal 377A campaign as the website was hacked and is now defunct.

APPENDIX A: DESCRIPTIONS OF DIGITAL BOTTOM-UP MOVEMENTS

Table 4.

Movement	Description
1. Singapore Anti-Mandatory Death Penalty	The Singapore Anti-Mandatory Death Penalty campaign protests against Singapore's mandatory death penalty sentence for murder and the trafficking of drugs. In some cases, the drug mules are unaware that they are carrying illicit substances. Thus, activists argue that sentencing them to death will not solve problems of greater kingpins at work in the drug trade. Instead, leniency should be provided based on the defendant's co-operation with the police. Singapore human right activists like Kirsten Han, M Ravi, Rachel Zeng and Andrew Loh were central to the campaign. In November 2012, the mandatory death penalty was amended so that judges now have discretion on whether to impose a death sentence or lighter punishment on drug-traffickers and murder convicts.
2. Campaign against the Internal Security Act	The campaign advocates revoking the Internal Security Act, a piece of legislation that allows authorities to detain individuals without trial. Though the Internal Security Act has been used to detain terrorists, it was, however, used in the eighties to allegedly detain peaceful activists under an alleged Marxist conspiracy. Detainees were seized in a 1987 home security operation called "Operation Spectrum". The ISA was also invoked to detain political activists and journalists in Singapore's formative years. To date, detainees' accounts have largely been left out of polished mainstream accounts of Singapore history. The campaign surfaces these alternative histories, it aims to repeal the law to prevent further injustice and encourage the young to be unafraid of a political past.
3. Free Sticker Lady	25-year-old Samantha Lo (also known as Sticker Lady) was arrested for pasting stickers and spray painting public property with tongue-in-cheek messages. The penalty for graffiti in Singapore includes fines and/or jail and corporal punishment. The campaigners advocated that Lo's graffiti was not wanton destruction, instead it had aesthetic merits and it was well-liked by a considerable section of the online population. Thus, the petition received widespread support.
4. Save Bukit Brown	The Save Bukit Brown movement hopes to preserve the Bukit Brown cemetery after officials announced plans to build a major road through it. Bukit Brown activists say that the place is a precious heritage site and a sanctuary for diverse flora and fauna. Activists propose for the cemetery to be turned into a heritage park, where Singaporeans can simultaneously get to know their own roots, while enjoying nature. The activists are a loosely assembled group including those from the Nature Society, SOS Bukit Brown, All Things Bukit Brown, Green Drinks and Asia Paranormal Investigators.
5. Repeal 377A	Repeal 377A is a movement to repeal section 377A of Singapore's penal code which criminalises sex between mutually consenting adult men. The issue was first raised in parliament in 2007 and it led to a high-profile debate on social norms and public morality. Then-Nominated Member of Parliament (NMP) Siew Kum Hong was seen to champion the repeal 377A camp while NMP Thio Li-ann was seen to champion the keep 377A camp. There were online petitions for both movements. The issue resurfaced in 2013, after two men challenged the constitutionality of section 377A. A pro-377A Christian pastor also posted his views online, provoking online rejoinders from more conservative pastors. However, since the constitutionality of the law was challenged in 2013, people were told not to speak about it publicly. Doing so would be sub-judice.
6. SlutWalk	SlutWalk is a protest against excusing rape by using a woman's appearance or her dress choices as an excuse. The international movement first started in Toronto and has since spread to Asian cities including India, Malaysia and Singapore. The Singapore chapter started in 2011 and is sponsored by AWARE.
7. Lehmann Brothers Minibonds Protest	American bank Lehmann Brothers declared bankruptcy in 2008 and about 10,000 retail investors in Singapore lost all or a large part of their investments totalling over S$500 million in products linked to Lehmann Brothers. The financial institutions that distributed the products were accused of having mis-sold these relatively high-risk products to the investors, many of whom were the elderly and the less educated. Ordinary citizen Tan Kin Lian (later a presidential candidate hopeful in 2011) campaigned for investors' money to be returned to them.

continued on following page

Table 4. Continued

Movement	Description
8. Pink Dot	Pink Dot SG is an annual, non-profit movement, free-for-all-event which started in 2009, in support of the Lesbian, Gay, Bi-sexual and Transexual community in Singapore. Attendees of the Pink Dot SG events gather to form a giant pink dot in a show of support for inclusiveness, diversity and the freedom to love. Pink Dot, which originated in Singapore, has spread to London, Montreal and Hong Kong.
9. Occupy Raffles Place	"Occupy" is an international movement against political and social inequality. It started first with the Occupy Wall Street movement in 2011 where Americans protested against the huge income gap between the working classes and the top earners in the financial sector, or the "99% vs. 1%". The Occupy movement spread to 951 cities over 82 countries, including Singapore. However, Occupy Singapore where organisers called for protesters to "Occupy Raffles", was largely unsuccessful as nobody turned up for the event.
10. No to Rape	The campaign advocates that sexual violence by any person, against any person, is criminal violence. The premise for the campaign is: regardless of whether the victim and perpetrator are married to each other, non-consensual sexual penetration should be treated as rape. The online petition is coordinated by a team of concerned Singaporeans who came together to promote change on this issue.
11. "Say No to 6.9 Million People"	In Feb 2013, the Singapore government tabled a population white paper that targeted a 6.9 million population for Singapore by 2030. The release of the white paper was set against concerns about high costs of living, rising housing prices, lack of space and infrastructure that Singaporeans had been experiencing due to an influx of foreigners since the early 2000s. The "6.9 million" figure provoked more anger on the ground. Citizen Gilbert Goh held a protest at Hong Lim Park, where ordinary Singaporeans and opposition politicians spoke voiced their opposition to the plan. Gilbert Goh stressed several times that the protest targeted government policies and it was non-xenophobic in nature.

APPENDIX B: OUTCOMES OF BOTTOM-UP MOVEMENTS

Table 5.

Campaign	Outcome
Internal Security Act campaign	Over 400 people attended 'We remember' rally on 2 February 2013.
Lehmann Brothers and minibonds	Over 500 people participated in a rally on 11 October 2011.
"No to 6.9 Million People" protest	Attracted about 5000 participants 3295 signatures for Tan Kin Lian's petition 452 signatures for Transitioning.org's petition
No to Rape	3618 signatures for online petition
Occupy Raffles Place	Less than 20 people turned up Facebook likes: 12,405 likes as of 20/2/13 Twitter: 1,525 followers as of 20/2/13
Pink Dot.sg	2009: 1,000 to 2,500 attendees 2010: 4,000 attendees 2011: 10,000 attendees 2012: 15,000 attendees Facebook: 14,750 likes as of 20/2/13
Repeal 377A[8]	2007 petition: unknown as link does not work 2013: 55 likes for Facebook page (bear in mind sub judice clause)
Save Bukit Brown	1347 signatures for "Save Bukit Brown" e-petition (goal was 5000 signatures) Dialogue session with Minister Tan Chuan-Jin in March 2012 1024 likes for "SOS Bukit Brown" on Facebook 756 likes for "Save Bukit Brown Cemetery, The Roots Of Our Nation" on Facebook
Slut Walk	600 attendees at the 2011 Slut Walk 1319 likes on Facebook as of 21/2/13 372 followers on Twitter as of 21/2/13
Singapore Anti-Mandatory Death Penalty	42099 signatures for "Save Yong Vui Kong" petition 80 turned up at Speaker's Corner to on Yong Vui Kong's 24th birthday 150 turned up at Speakers' Corner for "Give Vui Kong a second chance protest";
Sticker Lady	More than 70 artists and members of public attended a townhall meeting to discuss the issue 15221 signatures for online petition Sticker Lady has been released on police bail pending "further investigations"

Section 4
Mobile Children and Parenting

Mobile children are those who are using mobile as an indispensable part of their daily lives. Besides enabling children to stay connected with peers and parents, mobile has also ushered in new forms of interaction and coordination among children for flexible and fluid social networking. Mobile has become a tool for children for peer identification, interaction, engagement, and communication. Earlier studies regarding use of mobile among children largely involved feature phones. With the advent of smartphones, more and more children are replacing their feature phones with smartphones. Instead of a mere change of device, that shift opens up a whole new world for children. Beyond phone calls and text messaging, they use smartphones for, among others, chatting, gaming, surfing the Internet, music, taking and sharing photos, and/or staying social. Equipped with advances in mobile technologies such as mobile location-based services, smartphones have brought children not only new opportunities but also new risks. Besides identifying both new opportunities and new risks, Chapter 10 examines major implications of the use of smartphones among children and sheds light on new areas for research on smartphones for children. Closely related to mobile children is naturally mobile parenting. Using mobile as a tool for parenting, parents extend their attention, care, and surveillance when they are away from their children. With an increasing number of migrant workers leaving their children in their home countries, how to parent them has become a bigger challenge for migrant parents. Chapter 11 investigates the practice of mobile parenting and its related demographic, social, economic, and technological challenges and implications through a case study of Filipino migrant mothers in Singapore.

Chapter 10
Mobile Communication and Children

Mascheroni Giovanna
Università Cattolica del Sacro Cuore, Italy

ABSTRACT

The shift from mobile phones to smartphones, which integrate mobile communication, social media, and geo-locative services, expands the scope of mobile communication and opens up new opportunities for young people, as well as new risks. However, research on the adoption and use of smartphones among young people is still sparse. Therefore, the consequences of smartphone use on children's communicative cultures and relational practices remain yet to be explored. Drawing on the consistent literature on children and mobile communication and on the findings of preliminary research on younger users' domestication of smart mobile devices, the present chapter discusses the implications of mobile media use amongst young people in the light of the changes associated with smartphones and the mobile Internet.

INTRODUCTION

Since the widespread adoption of mobile phones in the second half of the Nineties, childhood and adolescence have been central categories within mobile communication research: some of the earliest and seminal work in the field focused on mobile phone use amongst teenagers, teens' mobile communicative practices and the meanings of mobile communication in the transition between childhood to adulthood (Green & Haddon, 2009). These studies - predominantly conducted in Europe, though significant contributions came also from the Asia-Pacific and the U.S. - pointed to an intimate connection between mobiles and young people: this group of users arguably played a key role in pioneering specific communicative practices such as texting, and thoroughly incorporated mobile technologies in their everyday lives. Indeed, certain aspects of mobile cultures were strongly associated with youth and their distinctive cultures of communication globally (Goggin, 2006; Caron & Caronia, 2007; Castells, Fernandez-Ardevol, Sey & Qiu, 2007; Green & Haddon, 2009; Goggin, 2013).

More recently, while the mobile phone has acquired the status of a taken for granted condition of our social ecology (Ling, 2012) - with the

DOI: 10.4018/978-1-4666-6166-0.ch010

majority of children and adolescents in developed and, increasingly, developing countries growing up with mobile phones (Ling & Bertel, 2013) – scholars have turned their attention to smartphones, or "the mobile as an online, networked *media* device" (Hjorth, Burgess & Richardson, 2012, p. 1) which integrates mobile communication, social media and geo-locative services. This convergence expands the scope of mobile communication and opens up new opportunities for young people. Preliminary research has started to examine emerging practices and analyse changes in practices and meanings of mobile communication. However, the consequences of smartphones' use on young people's communicative cultures and relational practices remain yet to be explored and identified, for a number of reasons. On the one hand, research on children's adoption and use of smartphones is still sparse and at an earlier stage compared to the rich body of empirical studies on young people and mobiles, which has predominantly focused on texting and voice calls. On the other hand, the mobile internet is currently being domesticated (Silverstone, Hirsch & Morley, 1992; Haddon, 2004) and it is not completely taken for granted, at least among the youngest: habits and meanings, expectations and norms, are currently being negotiated. The question whether and to what extent the anywhere, anytime opportunity to access the internet is altering the meaning of, or adding new meanings to mobile phones cannot yet be fully answered, while being a key issue on the research agenda.

The present chapter aims to provide a brief overview of the literature on children and mobile communication, and discuss the implications of mobile media use amongst young people in the light of the changes associated with smartphones and the mobile internet, drawing on the findings of preliminary research on younger users' domestication of smart mobile devices.

MOBILE COMMUNICATION AND CHILDREN

Earlier and notable contributions provided a review of the numerous studies on youth and mobile communications (Ling & Haddon, 2008; Green & Haddon, 2009; Ling & Bertel, 2013). Drawing on this body of writing, continuing themes in the field from the late Nineties can be identified in the following: the investigation of particular communicative practices, such as texting; the analysis of the social functions of mobile phones, in terms of micro- and hyper-coordination; and the understanding of its social consequences, namely connectivity and social cohesion, emancipation, social exclusion, the disturbance of the public sphere, the extension of one's personal sense of safety. Moreover, while the majority of the studies focused on how mobile communication fitted into teenagers' peers relations, the issue of parent-child interactions in relation to this technology also attracted a good deal of attention.

In a seminal study of Norwegian teenagers, Ling and Yttri (2002) analysed the adoption of mobile telephony among Norwegian teens and concluded that mobile communication yielded new forms of interaction and coordination, called micro- and hyper-coordination. The concept of micro-coordination points to the role of mobile phones in facilitating the coordination of face to face encounters: indeed, mobile communication allows for a more flexible, fluid and personalised planning of if, where and when to meet (Ling & Haddon, 2008). In this sense, the mobile phone "is both a substitute for and a supplement to time as a basis of coordination" (Ling, 2004, p. 80). This augmented temporal and spatial flexibility in social arrangements is particularly common among teens and is deemed to have altered the nature of teenage meetings: indeed, the mobile phone is also used to "anticipate and to summarize physical encounters" (Ling & Haddon, 2008, p. 145).

Therefore, the consequent blurring of the boundaries of a meeting is the product not only of instrumental uses of mobile communication, as it happens in micro-coordination and the iterative planning of face-to-face encounters: rather, it is also engendered by more expressive uses of mobile telephony, through which young people fill in the gaps between face to face encounters and "extend the interaction beyond the period of co-presence" (Ling, 2008, p. 128). This kind of hyper-coordination through texting, voice calls and beeping provides teenagers with a sense of "perpetual contact" (Katz & Aakhus, 2002), or "connected presence" (Licoppe, 2004): frequent and continuous communication exchanges, the content of which is often secondary or irrelevant, provide young people with the opportunity to strengthen their sense of belongingness to the peer group, while simultaneously reassuring friends about their own engagement in the relationship. Therefore, expressive uses of mobile communication respond to a phatic function, that is, to confirm friendship ties and check out on the status of the relationship: text messages are incorporated within a "gift economy" (Johnsen, 2003), whereby the act of giving and reciprocity are markers of friendship. However, scholars have argued that this "anywhere, anytime" accessibility is only potential: phatic communicative practices "maintain the illusion of a constant connection" (Licoppe, 2003, p. 172). We can thus conclude, as does Ribak, that "the mobile is a tool for potential, rather than actual communication; its significance lies in its availability rather than in its actual use" (Ribak, 2009, p. 188).

The material reviewed here suggests a second way in which the mobile phone is used to tie the group together and, at the same time, manage the fragile balance between identification and individuation in interpersonal relationships with peers: the mobile is, indeed, a significant repository of personal information and an identity-making device (Caron & Caronia, 2007, p. 105). Mobile phones have been domesticated by youth and turned into tools for co-present interaction: communication practices such as calls and SMS, multimedia content such as personal photos and videos, as well as the mobile phones themselves are shared with friends (Weilenmann & Larsson, 2001; Scifo, 2005) as symbols of trust and engagement, but also as a way to claim one's status within the group. The model of the phone, the mobile phone address book – especially the number and status of the individuals in the list – as well as the history of text messages and their call logs are all marker of status amongst peers (Ling, 2004; Green & Haddon, 2009).

Youth, therefore, have domesticated mobile phones as both tools providing a sense of connected presence with peers and symbolic resources for identity formation. One of the main and more relevant social consequences of these expressive uses of mobile communication has been identified in social cohesion (Ling, 2004, 2008; Ling & Haddon, 2008): more specifically, for teenagers hyper-coordination and conversational uses of texting are an important way to build cohesion with their peers and reinforce pre-existing ties. The continual access to social circles provided by mobile communication is usually restricted to a limited number of strong ties: even teenagers, who send and receive more texts than any other age group and exchange SMS with a wider variety of people, communicate on a regular basis only with a select group of around 5-8 individuals (Ling, Bertel & Sundsøy, 2012). This is an important distinctive feature of mobile communication, that differentiates it from other communicative practices that have become highly popular with young people, such as social networking: research has shown that, although to keep in touch with pre-existing friends is the "the dominant and normative [...] usage pattern" of social networking sites (boyd, 2009, p. 89; see also Livingstone, 2008; Livingstone, Haddon, Görzig & Ólafsson, 2011), social network platforms enable youth to connect with their "extended social networks" (boyd & Ellison, 2007) and to activate so-called "latent ties" (Haythornthwaite, 2005), while mobile phones are the tool for "the full time intimate sphere" (Matsuda, 2005; Ling, 2008). With mobile phones, thence, the "community of interactions" becomes "a mobile phenomenon" (Meyrowitz, 2005, p. 26): the link to the circle of closest friends and family can be activated anywhere, anytime. This constant connectivity has two sides: while it facilitates teens' social inclusion in their peer groups, this kind of sociability may also be risky since it can exclude others. As Ling noted (2004), mobile phone use may lead to the formation of "walled communities" and to a withdrawal from the public into the private sphere.

A second outcome of perpetual or quasi-perpetual connectivity with peer networks can be found in the process of teenage emancipation: by providing teens with a direct and personal link to peers, and by supporting their development of a self-identity, mobile communication has altered the way both children and parents perceive the process of emancipation (Ling & Haddon, 2008; Clark, 2013). Mobile telephony provides adolescents with an unprecedented independence since it grants privacy in making and receiving calls and texts, out of parental surveillance. Nonetheless, this growing autonomy can be understood as "a conditional freedom" (Haddon, 2000; Ling & Bertel, 2013) and a source of tensions in family relationships.

While for children the primary motivation for adopting mobile communication is social access to peers (Lenhart et al., 2010; Ling & Bertel, 2013), mobile phones enter the parent-child relationship usually as a gift from parents (Caron & Caronia, 2007), whose main motivation is safety – the possibility to reach their children regardless of space and time constraints, as well as the opportunity for children to reach their parents whenever they need. Therefore, as pointed out by a consistent body of literature (see, among others, Fortunati & Manganelli, 2002; Caron & Caronia, 2007; Green & Haddon, 2009; Ling, 2012; Clark, 2013), parents and children tend to mobilise different, at times contrasting, understandings and expectations around mobile phones. Parents invest mobile communication with a paradoxical value: on one side, parents want to foster children's autonomy and emancipation, and therefore employ mobile phones as an opportunity for children to grow up and take on responsibilities – for the device itself and for its costs; on the other, however, they understand and use mobile phones as means for extending parental control and parental care outside the domestic environment, to keep track of children's movements as much as to enable a perpetual contact. This extended surveillance has been variously termed as "remote parenting" (Haddon & Ling, 2008; Clark, 2013), "remote motherhood" (Rakow & Navarro, 1993), or "electronic" or "digital leash" (Ling, 2004; Caron & Caronia, 2007). The resulting paradox is that children experience more freedom as well as more control in their everyday mobilities: they are more free to go out, but need to report on their movements more often than ever. Moreover, through mobile phones children bring their families with them when out and about, and are never alone. On these bases, therefore, scholars have questioned the role of mobile communication in supporting children's and teenagers transition to adulthood. Again, the answer is not straightforward: while empirical evidence shows that children engage in a variety of "parent management strategies" (Green & Haddon, 2009) - such as turning off the phone and later giving the excuse that the battery was dead – in order to avoid parental surveillance and maximise the greater freedom provided by mobile phones, other research argues that this safety link may actually reduce children's autonomy and increase children's dependence on their parents (Green & Haddon, 2009; Ling, 2012). Indeed, the mobile phone is a "safety link" and, as such, provides youth with a sense of protection and security that arises from perpetual contact with parents. As a consequence, young people tend to rely more and more on their parents in a crisis situation, as when they get in trouble with their peers (Clark,

2013), At the same time, young people tend to incorporate this notion of the "safety link", showing a "perceived responsibility to be available all the time to communications from their parents" (Green & Haddon, 2009, p. 103). Other studies suggested that young people share and anticipate their parents' concerns by calling in advance to reassure them – a practice which is legitimised as part of being a responsible teenager (Ribak, 2009). Thus, the meanings and uses of the mobile phone in the context of the parent-child relationship reveal all its ambiguity: while it is turned into an "anxiety- reducing device" by parents (Ribak, 2009, p. 191), it may actually cause new anxieties to children (Bond, 2010).

THE DOMESTICATION OF SMARTPHONES AMONG CHILDREN: CONTINUITIES OR CHANGES?

Recent technological developments – namely the integration of functions and features typical of personal computing with those already supported by mobile phones in new smart convergent devices – are expanding the set social and communicative practices developed around mobile communication. Smartphones are complex digital environments based on online, convergent, networked technologies which remediate prior generation of mobile phones and mark the shift to the so-called "post-desktop" ecology, whereby mobile convergent media are fast overtaking the desktop computer as a prioritised means of internet access.

As a consequence, mobile telephony is being re-domesticated, calling for new social legitimations and posing new challenges for how we study mobile communication. In a study on the domestication of smartphones among young Danes, Bertel and Stald (2013) identify three features of the new devices, in which emerging uses are grounded: smartphones are defined as devices "that 1) have the computing power and technical platform to run application and access Internet content, 2) have – in principle – a persistent data connection to the Internet, and 3) are typically equipped with positioning technology, often GPS" (Bertel & Stald, 2013, p. 200). Thence, providing anywhere, anytime connectivity and internet access, smartphones may actually reconfigure both mobile communication and online practices among young people. Before examining the emerging communicative practices and modes of sociability that smartphones enable, however, it is worth asking which children and teenagers go online using smartphones and what they actually do on their mobile devices.

Research on the adoption and use of smartphones among this age groups is sparse and not easily comparable across countries; nonetheless, a few studies help us make some reflections on which children use mobile convergent media to go online and how. One of the first studies pointing at the growing diffusion of smartphones among the younger population is an international comparison published by GSMA & NTT Docomo in 2012. The research investigates mobile phone use among children aged 10-18 in Japan, India, Indonesia, Egypt and Chile, and shows that 21% of children with a mobile phone own a smartphone, while 63% have a feature phone[1], and that smartphone ownership increases with age. Similarly, mobile internet use increases with age and is higher among smartphone owners. While a small amount of children use primarily a mobile phone to go online, the percentage of children accessing the internet from a mobile device rises up to 32% among smartphone owners, and is higher in Japan and Indonesia, where 62% and 46% of children using a smartphone use it as their primary means of accessing the internet.

Similarly, a report by the Pew Research Center (Madden, Lenhart, Duggan, Cortesi & Gasser, 2013) shows that smartphones have become pervasive among American teens aged 12-17: nearly half (47%) of the sample own a smartphone and nearly three in four teens access the internet on their phones or tablets computer at least occasionally.

For a smaller proportion of teenagers, however, mobile internet use is not an occasional experience but, rather, an ordinary presence in their everyday lives: indeed, one in four teens report using the internet mostly from their smartphones. The Pew report also highlights divides in the use of the internet and smartphones amongst American teens based on socio-economic status: teens from less advantaged families are less likely to use the internet overall; however, teenagers living in lower-income and lower-educated households are just as likely or in some cases more likely than peers from higher socio-economic status to use smartphones as their primary means of internet access.

The Net Children Go Mobile project[2] (Mascheroni & Ólafsson, 2014) maps the use of the internet and mobile convergent media among children aged 9-16 years old in 7 European countries (Belgium, Denmark, Ireland, Italy, Portugal, Romania and the UK). The findings are consistent with those of the projects briefly outlined above, and indicate that smartphones are increasingly popular among youth: 46% of the interviewees own a smartphone or have it for private use, though smartphone ownership is strongly differentiated by age – just 20% of children aged 9-10 have a smartphone while 64% of teenagers aged 15-16 possess one - and country – with higher penetration in Denmark (84%) and lower in Romania (26%). The average age at which children receive a smartphone is 10 years old; however, younger children are more likely to be given a smartphone when they are only 9, while older teenagers were 14 when they got their first smartphone. One in six children were given a smartphone as their first mobile phone.

The data also indicate that smartphones have become well integrated in children's daily online practices: 41% of the sample report using a smartphone to go online several times a day or at least once a day. Despite smartphones are the device most likely to be used on the move – 18% say they access the internet from a smartphone at least once a day when out and about, or on the way to school and other places - smartphones also top the lists of devices that children use to go online at home – 32% use smartphones in their own bedroom on a daily basis and 33% in another room at home[3]. This finding confirms the increasing privatisation of internet access and use and the pervasiveness of "bedroom culture" (Livingstone & Bovill, 2001) among European children, while suggesting that the role of the personal computer is being challenged by the highly portable mobile phone. However, the preference for smartphones in the domestic context may also represent a continuity with practices developed around mobile phones in youth cultures, rather then a dramatic novelty. As Caron and Caronia (2007) already noted in their ethnographic studies of young people's domestication of mobile communication, teens develop a repertoire of cultural reasons that favour the choice of mobile phones over the fixed phone: the mobile is preferred because it is more accessible and always at hand. Thus mobile phone fits better in teenagers' everyday lives because it responds to "teenage laziness" (Caron & Caronia, 2007, p. 144): going online from a smartphone is easier than leaving the sofa and moving in another room and starting a computer. Similarly, Bertel and Stald (2013, p. 202) note that "mobile devices afford easier handling than, for instance, a laptop computer in some situations".

Therefore, smartphones provide children with the opportunity to access online services and functionalities typically associated with computers from a personal, highly portable and almost invisible device. Whether going online from a smartphone is "an *extension of* or a task-specific *substitute for* the personal computer" (Bertel & Stald, 2013, p. 208) is not clear so far. While some findings suggest that at least a small proportion of children use a smartphone as their primary means of internet access, other data support the hypothesis that children use a variety of devices to go online, and that smartphones supplement rather than replace desktop or laptop computers.

A further question is the issue of social and communicative practices that young people develop around smartphones, that are likely to remediate the meanings and uses of mobile communication in children's everyday life. The GSMA & NTT Docomo (2012) study identifies entertainment as the type of internet content that children are more likely to access from their mobile phones: despite countries differences, the most popular activities include playing games, watching videos and listening to music. The second most common activity is social networking, with around half of the children who access their profiles on a social network site from mobile phones. Entertainment and communication apps are also the most popular apps downloaded and used by children.

The findings of the Net Children Go Mobile project confirm the popularity of communicative practices and entertainment among the activities children perform online on a daily basis. Moreover, it shows that smartphone users engage more in each of the online activities asked about: the most noticeable growth concerns communication practices - visiting a profile on a social network site is practised several times a day or at least every day by 70% of smartphone users and 39% of children who don't use a smartphone to go online; instant messaging such as WhatsApp and Skype by 58% of smartphone users and 28% of non users - and, in second place, entertainment activities (listening to music and watching video clips). However, children who use a smartphone are also more likely to use the internet for schoolwork on a daily basis (38% of smartphone users vs. 21% of the whole sample).

Therefore, the empirical evidence so far available on social uses of smartphones amongst young people points to the importance of being connected and to potential reconfiguration of communicative practices: texting practices are changing – with instant messaging services that are starting to erode the volume of SMS[4] – while, at the same time, new modes of sociability emerge that enable users to engage with different social circles. The nature of the mobile phone as the means of the intimate sphere par excellence seems, however, unaltered. Indeed, when asked about their communication repertoire, children indicate their parents and friends as the contacts they are in touch with on a daily basis through voice calls, SMS, and social network sites: more specifically, 71% and 57% of the respondents report talking on the mobile to their friends and parents respectively at least once a day; 66% and 41% use SMS and other forms of message as a daily means of keeping in touch with friends and parents; while social network sites are primarily used to communicate with friends (74% use it on a daily basis, vs. just 10% who use it with their parents daily). Moreover, just 16% of respondents communicate with the "extended" network - namely, friends of friends and online contacts - on social network sites on a daily basis (Mascheroni & Ólafsson, 2014). Therefore, while smartphones expand the range of communicative practices and the type of audiences children are now able to engage with, they are still the means for regular contact with family and friends.

A further issue that the shift from mobile phones to smartphones is likely to reconfigure, at least partially, is the understanding of mobile technologies within the parent-child relationship: from a safety link and a tool for interpersonal monitoring, in parents' perceptions the smartphone may be turned into a risky opportunity, since may facilitate greater exposure to a variety of online risks. Indeed, anxieties over children's mobile phone use – such as concerns on screen-time and potential withdrawal from co-present interaction due to absorption in mediated communication - are not new (Haddon, 2012). However, the diffusion of smartphones among youth has raised new parental worries, also due to media panics.

The GSMA & NTT Docomo's report (2012) on children use of smartphones investigated also parental perceptions and concluded that over 62% of parents in Chile, Egypt, India, Indonesia and Japan are "very concerned" with the risk of their children being exposed to inappropriate content (such as sexual content) on their smartphones; nearly half express considerable concern for excessive use, and around 40% are very worried for privacy issues.

These findings might not scale to other countries characterised by different media and childhood cultures. However, a recent qualitative study conducted in Italy on the domestication of smartphones by parents and children (Mascheroni, 2013) has equally shown that smartphones raise contradictory feelings and create new dilemmas for carers. On the one hand, parents who decide to provide their children with the latest technological equipment reaffirm the dominant framework on mobile telephony and build new social legitimations for the mobile internet: in their view, smartphones combine the safety link already provided by mobile phones - and help them manage the potential risks encountered by their children through the perpetual contact thus activated – with the opportunities offered by the internet – namely always on access to educational resources, cultural and civic opportunities. On the other side, however, parents tend to project on smartphones their concerns for inappropriate online situations: they worry for sexual content, contact with strangers, sexting and bullying. Moreover, they lament that smartphones guarantee only a limited and impoverished access to online opportunities: indeed, their children seem to favour gaming and social networking.

Of course, alternative framings of smartphones within the parent-child relationship are possible: indeed, new mobile devices my well be perceived as less risky than the usual desktop-based experience of the internet, in so much as mobile phones are seamlessly integrated in the everyday life contexts and thus invisible. What this ambivalent attitude towards children's smartphones use suggests, as did negative accounts of mobile telephony in the past, is that there's no settled agreement on the legitimate users and uses of smart portable devices, which are currently being domesticated (Ling, 2012).

IMPLICATIONS OF SMARTPHONE USE: NEW ISSUES ON THE RESEARCH, POLICY, AND PUBLIC AGENDAS

As we have seen, the social uses of mobile telephony among children and teens have been associated with some social effects and problematic issues (Ling & Bertel, 2013). Associating smartphones with a clear set of social consequences, however, may be more problematic for a number of reasons. First, because the domestication of smartphones is an ongoing process and, therefore, meanings and practices around mobile convergent media are still negotiated, invented and resisted. Second, the consequences of ICTs are never straightforward nor easily predictable: as the domestication approach (Silverstone, Hirsch & Morley, 1992; Haddon, 2004) has shown, by virtue of their "double articulation" media devices are appropriated, made meaningful and contextualised within a complex set of relationships between culture and technology, the symbolic and the material, which emerge in the everyday practices of consumption; in a word, technologies "are both shaped and shaping" (Silverstone, Hirsch & Morley, 1992, p. 27). So, while mobile convergent media are shaping the conditions in which childhood and adolescence are played out, smartphones are also appropriated and shaped by children teenagers to fit into their everyday lives. Finally, I agree with Bertel and Stald who conclude that the social consequences of smartphones may be "less striking than those of the mobile phone and the internet – evolutionary rather than revolutionary" (2013, p. 200) because they combine features, meanings and practices of mobile phones and the internet.

Indeed, among the issues concerning the social implications of smartphones that are particularly debated on the policy as well as research agendas, those that have migrated from internet studies to the field of mobile communication research have attracted considerable attention and are worth mentioning.

First is the question whether the diffusion of the mobile internet may bridge inequalities in internet access and use, or, rather, result in new inequalities (Castells et al., 2007; Hargittai & Kim, 2011). Global statistics (ITU, 2013) show that the number of mobile phone subscriptions has approached the global population number, though the average global rate of mobile broadband subscriptions is 30%. So, while mobile phones provide almost every individual with connectivity, just a small proportion of mobile devices worldwide enable a broadband connection to the internet. Moreover, prior research indicates that different platforms do not lead to the 'same' internet experience (Donner et al., 2011). The Net Children Go Mobile data we have summarised above do not provide evidence that children do a fewer range of online activities, and mainly communication and entertainment, when they access the internet from their smartphones. However, despite the fact that smartphones are more likely to be used everywhere on a daily basis than any other device, children who participated in the study have usually access to the internet from a variety of devices. Research on less advantaged children, who are more likely to access the internet mostly if not exclusively from a smartphone will help understand whether mobile internet use while reducing inequalities in access and, therefore, the so-called first level digital divide, may actually produce new divides and lead to a more restricted online experience.

A further issue that arises from the convergence of mobile communication and online media is the concern over internet risks, and, more specifically, the question whether the intensification in space and time of the practices related to computing, internet and social media use provided by smartphones may extend the exposure to a range of online risks, while also posing new specific challenges to children's online safety. These concerns are part of the media panics that arise wherever a new medium emerges, when its use by children is considered. Media anxieties over cyberbullying and sexting, however, were paralleled by concerns on the policy agenda, that, at least at the European level, have been central in shaping the climate for research in the area of children and the internet.

So far, empirical evidence that mobile access to the internet is accompanied by increased exposure to online risks or to new emerging risks is limited. The 25 country EU Kids Online 2 survey showed that in 2010 around a third of 9-16 year old internet users said that they used a mobile phone or a handheld device to go online (Livingstone et al., 2011). The findings show that there is little difference in the likelihood of experiencing risks on the internet, between those children who use a mobile phone to go online and those who do not have mobile access. It is the integrations of various platforms which results in more time spent online; the more intense use of the internet, in turn, is related to increased exposure to risk (Stald & Ólafsson, 2012). However, though smartphones were not very popular among children at the time of the EU Kids Online survey - and initial uses may differ from uses that become socially accepted once a new technology is more embedded in society - the empirical evidence provided by this study shows that the introduction of smartphones into everyday life is not without consequences in terms of online experiences: we find slightly increased exposure to risks related to using a smartphone compared to going online from a mobile phone (Stald & Ólafsson, 2012).

The EU Kids Online findings do not provide empirical evidence of specific mobile internet risks. However, the greater diffusion of these devices among youth may have changed the picture: public and policy concerns point to the potential new risks associated with geo-positioning and near-field communication technologies that are able to locate one's position in space, connect the user with content, services and other users located nearby and access contextual information (Gordon & De Souza, 2011). However, a recent report from the Pew Research Center (Madden, Lenhart, Cortesi & Gasser, 2013) shows that around half (46%) of the teenagers surveyed have turned location-tracking functions off on their mobile phones or specific apps, and more than half have decided to avoid certain apps because of privacy concerns.

The Net Children Go Mobile data are consistent with those of the EU Kids Online in pointing to a greater exposure to risks among smartphone users: while 17% of the respondents reported having been bothered or upset by something online, this rises up to 24% of children using a smartphone, while it decreases to 12% of children who don't use nor a smartphone nor a tablet computer to go online (Mascheroni & Ólafsson, 2014). The findings, however, warn against assuming any straightforward or causal relationship: rather than indicating that the conditions of mobile internet use are risky *per se*, the data suggest that smartphone users are generally older and are likely to take up more activities on the internet. As a consequence, they develop more digital and safety skills but, at the same time, are exposed to more online risky experiences.

CONCLUSION

The present chapter has addressed the issue of the domestication of mobile phones and smartphones among children and teenagers, in order to understand whether the use of smartphones implies a radical remediation of mobile communication, and its associated practices and meanings. Drawing on the consistent body of writing in the field of mobile communication and children, it has summarised the main social uses of mobile telephony among young people, and its social consequences.

It has then explored the sparse and emerging research on children and smartphones, in order to advance some potential empirically-based implications of smartphones and mobile convergent media among young people. As other scholars have argued before (Bertel & Stald, 2013), rather than enabling radically new social and communication practices, smartphones are more likely to continue and partially remediate functions and meanings associated with both the internet and the mobile phone. While expanding the scope of communicative practices, the opportunity to access the internet from mobile devices does not seem to alter the meaning of the mobile phone as the tool of the intimate sphere. However, smartphones clearly offer access to a variety of online activities - from entertainment to communication, to more capital-enhancing activities - that are equally being re-mediated by virtue of the very nature of the devices, which are portable, always at hand and personal. Therefore, new issues are emerging on both the research and policy agenda that are new to the field of mobile media studies and rather migrate from the study of the internet and childhood. Issues of digital inequalities and online risks are among the clearest manifestation of the emerging hybrid research agenda at the crossroads of internet and mobile studies.

REFERENCES

Bertel, T., & Stald, G. (2013). From SMS to SNS: The use of the internet on the mobile phone among young Danes. In K. Cumiskey, & L. Hjorth (Eds.), *Mobile media practices, presence and politics: The challenge of being seamlessly mobile* (pp. 198–213). New York: Routledge.

Bond, E. (2010). Managing mobile relationships: children's perceptions of the impact of the mobile phone on relationships in their everyday lives. *Childhood*, *17*(4), 514–529. doi:10.1177/0907568210364421

boyd, d., & Ellison, N. (2007). Social Network Sites: Definition, History, and Scholarship. *Journal of Computer-Mediated Communication, 13* (1), 210-230.

boyd, d. (2009). Friendship. In M. Ito, et al. (Eds.), *Hanging Out, Messing Around, and Geeking Out: Kids Living and Learning with New Media* (pp. 79-115). Cambridge, MA: MIT Press.

Caron, A., & Caronia, L. (2007). *Moving cultures*. Montreal, Canada: McGill-Queen's University Press.

Castells, M., Fernandez-Ardevol, M., Sey, A., & Qiu, J. L. (2007). *Mobile communication and society: A global perspective*. Cambridge, MA: MIT Press.

Clark, L. S. (2013). *The parent app*. Oxford, UK: Oxford University Press.

Cotten, S. R., Anderson, W. A., & Tufekci, Z. (2009). Old wine in a new technology, or a different type of digital divide? *New Media & Society*, *11*(7), 1163–1186. doi:10.1177/1461444809342056

Donner, J., Gitau, S., & Marsden, G. (2011). Exploring mobile-only internet use: Results of a training study in urban South Africa. *International Journal of Communication*, *5*, 574–597.

Dürager, A., & Livingstone, S. (2012). *How can parents support children's Internet Safety?* London: EU Kids Online. Retrieved December 15, 2013 from http://www.eukidsonline.net

Fortunati, L., & Manganelli, A. (2002). Young people and the mobile telephone. *Estudios de Juventud*, *57*(2), 59–78.

Goggin, G. (2006). *Cell phone culture: Mobile technology in everyday life*. London: Routledge.

Goggin, G. (2013). Youth culture and mobiles. *Mobile Media & Communication*, *1*(1), 83–88. doi:10.1177/2050157912464489

Goggin, G., & Hjorth, L. (Eds.). (2009). *Mobile technologies: From telecommunications to media*. London: Routledge.

Gordon, E., & de Souza e Silva, A. (2011). *Net Locality: Why Location Matters in a Networked World*. Boston: Blackwell-Wiley.

Green, N., & Haddon, L. (2009). *Mobile communications: an introduction to new media*. Oxford, UK: Berg.

GSMA & NTT Docomo. (2012). *Children's use of mobile phones: An international comparison 2012*. Retrieved December 15, 2013 from http://www.gsma.com/publicpolicy/wp-content/uploads/2012/03/GSMA-ChildrenES_English-2012WEB.pdf

Haddon, L. (2000). The social consequences of mobile telephony: Framing questions. In R. Ling & K. Thrane (Eds.), *Sosiale Konsekvenser av mobiltelefoni: Proceedings fra et seminar om samfunn, barn og mobiltelefoni* (pp. 2-7). Kjeller: Telenor FoU N 38.

Haddon, L. (2004). *Information and Communication Technologies in Everyday Life*. Oxford, UK: Berg.

Haddon, L., & Vincent, J. (2009). Children's broadening use of mobile phones. In G. Goggin, & L. Hjorth (Eds.), *Mobile technologies: from telecommunications to media* (pp. 37–49). London: Routledge.

Hargittai, E., & Kim, S. J. (2011). *The Prevalence of Smartphone Use Among a Wired Group of Young Adults*. Northwestern University. Retrieved December 15, 2013 from http://www.ipr.northwestern.edu/publications/docs/workingpapers/2011/IPR-WP-11-01.pdf

Haythornthwaite, C. (2005). Social networks and Internet connectivity effects. *Information Communication and Society*, *8*(2), 125–147. doi:10.1080/13691180500146185

Hjorth, L., Burgess, J., & Richardson, I. (2012). Studying the Mobile. Locating the Field. In L. Hjorth, J. Burgess, & I. Richardson (Eds.), *Studying mobile Media: Cultural technologies, Mobile Communication and the iPhone* (pp. 1–7). London: Routledge.

Ito, M., Okabe, D., & Matsuda, M. (Eds.). Personal, portable, pedestrian: Mobile phones in Japanese life. Cambridge, MA: MIT Press.

ITU. (2013). *The World in 2013: ICT Facts and Figures*. Retrieved December 15, 2013 from http://www.itu.int/en/ITU-D/Statistics/Documents/facts/ICTFactsFigures2013.pdf

Johnsen, E. (2003). The Social Context of the Mobile Phone Use of Norwegian Teens. In J. Katz (Ed.), *Machines that Become Us: The Social Context of Personal Communication Technology* (pp. 161–169). New Brunswick, NJ: Transaction Publishers.

Katz, J. (Ed.). (2003). *Machines that Become Us: The Social Context of Personal Communication Technology*. New Brunswick, NJ: Transaction Publishers.

Katz, J., & Aakhus, M. (2002). *Perpetual Contact: Mobile Communication, Private Talk, Public Performance*. Cambridge, UK: Cambridge University Press. doi:10.1017/CBO9780511489471

Lenhart, A. (2009). *Teens and Sexting: How and why minor teens are sending sexually suggestive nude or nearly nude images via text messaging*. Washington, DC: Pew Research Center. Retrieved December 15, 2013 from http://pewresearch.org/assets/pdf/teens-and-sexting.pdf

Lenhart, A., Ling, R., Campbell, S., & Purcell, K. (2010). *Teens and Mobile Phones*. Washington, DC: Pew Research Center.

Licoppe, C. (2003). The Modes of Maintaining Interpersonal Relations Through the Telephone: From the Domestic to the Mobile Phone. In J. Katz (Ed.), *Machines that Become Us: The Social Context of Personal Communication Technology* (pp. 171–185). New Brunswick, NJ: Transactions.

Licoppe, C. (2004). 'Connected' presence: The emergence of a new repertoire for managing social relationships in a changing communication technoscape. *Environment and Planning. D, Society & Space*, *22*(1), 135–156. doi:10.1068/d323t

Ling, R. (2004). *The Mobile Connection: The Cell Phone's Impact on Society*. San Francisco: Morgan Kaufmann.

Ling, R. (2008). *New Tech, New Ties. How Mobile Communication is Reshaping Social Cohesion*. Cambridge, MA: MIT Press.

Ling, R. (2012). *Taken for grantedness: The embedding of mobile communication into society*. Cambridge, MA: MIT Press.

Ling, R., & Bertel, T. (2013). Mobile communication culture among children and adolescents. In D. Lemish (Ed.), *The Routledge international handbook of children, adolescents and media* (pp. 127–133). London: Routledge.

Ling, R., Bertel, T., & Sundsøy, P. R. (2012). The socio-demographics of texting: An analysis of traffic data. *New Media & Society*, *14*(2), 281–298. doi:10.1177/1461444811412711

Ling, R., & Haddon, L. (2008). Children, youth and the mobile phone. In K. Drotner, & S. Livingstone (Eds.), *The international handbook of children, media and culture* (pp. 137–151). London: Sage. doi:10.4135/9781848608436.n9

Ling, R., & Yttri, B. (2002). Hyper-coordination via mobile phones in Norway. In J. Katz, & M. Aakhus (Eds.), *Perpetual Contact: Mobile Communication, Private Talk, Public Performance* (pp. 139–169). Cambridge, UK: Cambridge University Press.

Livingstone, S., & Bovill, M. (2001). *Children and their changing media environment: A European comparative study*. Mahwah, NJ: Lawrence Erlbaum.

Livingstone, S., Haddon, L., Görzig, A., & Ólafsson, K. (2011). *Risks and safety on the internet: The perspective of European children: Full findings*. London: LSE. Retrieved December 15, 2013 from http://lse.ac.uk/EUKidsOnlineReports

Madden, M., Lenhart, A., Cortesi, S., & Gasser, U. (2013). *Teens and Mobile Apps Privacy*. Washington, DC: Pew Research Center.

Madden, M., Lenhart, A., Duggan, M., Cortesi, S., & Gasser, U. (2013). *Teens and Technology 2013*. Washington, DC: Pew Research Center.

Mascheroni, G. (2013). Parenting the mobile internet in Italian households: Parents' and children's discourses. *Journal of Children and Media*. DOI: 10.1080/17482798.2013.830978

Mascheroni, G., & Ólafsson, K. (2013). *Mobile internet access and use among European children: Initial findings of the Net Children Go Mobile project*. Milan, Italy: Educatt. Retrieved December 15, 2013 from www.netchildrengombile.eu

Mascheroni, G. and Ólafsson, K. (2014). *Net Children Go Mobile: Risks and Opportunities*. Second Edition. Milano: Educatt. Retrieved on July 18, 2014

Matsuda, M. (2005). Mobile communication and selective sociality. In M. Ito, D. Okabe, & M. Matsuda (Eds.), *Personal, portable, pedestrian: Mobile phones in Japanese life* (pp. 123–142). Cambridge, MA: MIT Press.

Meyrowitz, J. (2005). The Rise of Glocality: New Senses of Place and Identity in the Global Village. In K. Nyìri (Ed.), *The Global and the Local in Mobile Communication* (pp. 21–30). Wien: Passagen Verlag.

Ofcom. (2013). *Communications market report: United Kingdom*. London: Ofcom. Retrieved from http://stakeholders.ofcom.org.uk/market-data-research/market-data/communications-market-reports/cmr13/uk/

Rakow, L., & Navarro, V. (1993). Remote mothering and the parallel shift: Women meet the cellular telephone. *Critical Studies in Mass Communication*, *10*(2), 144–157. doi:10.1080/15295039309366856

Scifo, B. (2005). The Domestication of Camera-Phone and MMS Communication: Early Experiences of Young Italians. In K. Nyìri (Ed.), *The Global and the Local in Mobile Communication* (pp. 363–373). Wien: Passagen Verlag.

Silverstone, R., Hirsch, E., & Morley, D. (1992). Information and communication technologies and the moral economy of the household. In R. Silverstone, & E. Hirsch (Eds.), *Consuming Technologies: Media and Information in Domestic Space* (pp. 15–31). London: Routledge. doi:10.4324/9780203401491_chapter_1

Stald, G., & Ólafsson, K. (2012). Mobile access – Different users, different risks, different consequences? In S. Livingstone, L. Haddon, & A. Görzig (Eds.), *Children, risk and safety online: Research and policy challenges in comparative perspective* (pp. 285–295). Bristol, MA: Policy Press. doi:10.1332/policypress/9781847428837.003.0022

Weilenmann, A., & Larsson, C. (2001). Local Use and Sharing of the Mobile Phone. In *Wireless World: Social, Cultural and Interactional Issues in Mobile Communications and Computing* (pp. 92–107). London: Springer.

KEY TERMS AND DEFINITIONS

Cyberbullying: Relates to aggressive and mean behaviour directed to harm others, which is conducted through ICTs such as social network sites, mobile phones etc. Compared to face-to-face bullying, cyberbullying can be anonymous, is more persistent - due to the persistence and replicability of digital communication - and involves wider audiences, including "invisible audiences".

Electronic Leash: This expression is used to describe the meaning of mobile phones for parents, as a tool for extending parental control outside the domestic environment and for keeping track of children's movements (Caron & Caronia, 2007; Ling, 2004).

Hyper-Coordination: Indicates expressive uses of mobile communication among teenagers, to mark belonging to the peer group as well distinction. Forms of hyper-coordination include ritual communicative practices, the use of the group's argot, as well as personalisation of the device.

Micro-Coordination: Refers to the use of mobile phones to coordinate face-to-face interactions. Compared to prior means of communication, mobile phones render micro-coordination more flexible, allowing for a continuous planning on the fly.

Perpetual Contact: Refers to the particular sense provided by mobile communication that the tie with the intimate sphere can be activated anytime, anywhere. This sense of "connected presence" (Licoppe, 2004) is often maintained through frequent, short and continuous communication.

Safety Link: The anywhere, anytime connectivity provided by the mobile phone promoted its domestication as a "safety link" (Ling, 2012), especially in the parent-child relationship. The expression refers to the sense of personal safety and protection that arises from being able to dispense parental care at a distance - from the parents' side - and to reach parents in case of problems - from children's side.

Sexting: Relates to the exchange of sexually explicit content online or via mobile phones and smartphones. It has been defined as the "exchange of sexual messages or images" (Livingstone et al., 2011) and "the creating, sharing and forwarding of sexually suggestive nude or nearly nude images" (Lenhart, 2009). It may overlap with cyberbullying, e.g. when someone shares sexually explicit photos of their ex once they break up.

ENDNOTES

1. A feature phone is a mobile phone that includes some functions of a smartphones, such as a camera and internet access, but usually does not support Apps.
2. The Net Children Go Mobile is a research project funded by the Safer Internet Programme of the European Commission (SI-2012-KEP-411201) in order to investigate through quantitative and qualitative methods how the changing conditions of internet access and use - namely, mobile internet and mobile convergent media - bring greater, lesser or newer risks to children's online safety. Participating countries include Denmark, Italy, Romania, the UK. Belgium, Ireland and Portugal joined the project on a self-funding basis. For more information on the project visit: www.netchildrengomobile.eu
3. The second most common device is a laptop, used to go online from their bedroom or elsewhere at home by 30% of children in both locations.
4. For example, the recent Communications market report by Ofcom, indicates that the volume of SMS has been decreasing since 2011 due to the increased availability of alternatives offered by services such as WhatsApp and by messages capabilities of social network services like Facebook.

Chapter 11
Mobile Parenting and Global Mobility:
The Case of Filipino Migrant Mothers

Ma. Rosel S. San Pascual
University of the Philippines, Philippines

ABSTRACT

This chapter focuses its attention on Filipino mothers in diaspora and their mediated parenting as it tackles the centrality of mobile communication in transnational parenting. Apart from describing Filipino migrant mothers' mobile parenting practice in terms of the occurrence and content of their mediated parenting efforts, this chapter discusses mobile parenting by looking into the intersections of the socio-demographic, socio-economic, and socio-political landscapes of Filipino migrant mothers' mobile parenting with the social and technological dimensions of mobile parenting. In describing the landscapes, dimensions, and practice of Filipino migrant mothers' mobile parenting, evidence from interviews of 32 Singapore-based Filipino migrant mothers were supplemented by evidence from the literature on Filipino migrant parents and their transnational parenting efforts, their children left behind, and their children's caregivers.

THE GLOBAL MOBILITY OF FILIPINOS

Global diaspora is progressively becoming a familiar circumstance among the world's population. Based on the 2013 United Nations (UN) global migration statistics, there is an estimated 232 million international migrants worldwide such that 3.2% of the world's population are migrant residents with half of them concentrated in the United States of America, Russian Federation, Germany, Saudi Arabia, United Arab Emirates, United Kingdom, France, Canada, Australia, and Spain (UN Website, accessed 13 September 2013). The latest global migration statistics also reveal that Asians comprise the largest block of international migrants followed by Latin Americans (UN Website, accessed 13 September 2013).

Present-day international migration is marked by the worldwide influx of itinerant labor from developing countries to more developed nations. Among the world's active labor-exporting coun-

DOI: 10.4018/978-1-4666-6166-0.ch011

tries is the Philippines and official statistics show an increasing number of Filipinos being deployed abroad, reaching over a million annually since 2006 (Philippine Overseas Employment Administration [POEA], 2006; POEA, 2011; POEA, 2012). In 2012, an average number of 4,937 Filipinos leave the country daily for overseas employment (POEA, 2012). The International Organization for Migration (IOM) even noted that, based on UN statistics, Filipinos make up the second largest population of migrants living abroad (IOM Website, accessed 13 September 2013). The popular host countries of Filipino migrant workers include Saudi Arabia, United Arab Emirates, Singapore, Hong Kong, Qatar, Kuwait, Taiwan, Malaysia, Italy, and Bahrain (POEA, 2012).

Parents who are leaving their home and family in order to seek better economic opportunities abroad comprise a considerable portion of Filipino migrant workers. According to Mirca Madianou and Daniel Miller (2012), it is estimated that over 10 million Filipino children are left behind by their migrant parents. Given that family reunification is not always possible or immediately feasible (Asis, 2000; Asis, 2008), the occurrence of transnational families is steadily becoming commonplace in contemporary Philippine society (Madianou & Miller, 2012; Parreñas, 2005a).

The consequent separation of parents from their children has been raising concerns on the psychological and social effects of transnational migration on family well-being and family dynamics (for instance, Alunan-Melgar & Borromeo, 2002; Añonuevo, 2002; Asis, 2000; Asis, 2008; Battistella & Conaco, 1998; Beltran, Samonte, & Walker, 1996; Madianou & Miller, 2012; Parreñas, 2001; Parreñas, 2005a; Parreñas, 2005b; Parreñas, 2008; Pernia, Pernia, Ubias, & San Pascual, 2013; Sobritchea, 2007; Uy-Tioco, 2007). Admittedly, literature offers mixed observations regarding the psychological and social adjustments of transnational family members (for instance, Alunan-Melgar & Borromeo, 2002; Añonuevo, 2002; Battistella & Conaco, 1996; Betran, Samonte, & Walker, 1998; Madianou & Miller, 2012; Parreñas, 2001; Parreñas, 2005a; Parreñas, 2005b; Parreñas, 2008). Nonetheless, literature documents that, for Filipino transnational families, physical distance does not necessarily mean affective separation (Madianou & Miller, 2012; Parreñas, 2001; Parreñas, 2005a; Parreñas, 2005b; Sobritchea, 2007; Uy-Tioco, 2007).

In fact, literature chronicles the tremendous role of mediated communication in managing transnational family life (Madianou & Miller, 2012; Paragas, 2005; Paragas, 2008; Parreñas, 2001; Parreñas, 2005a; Parreñas, 2005b; Sobritchea, 2007; Uy-Tioco, 2007). With the recent advancements in communication media and technologies, the capacity of transnational families to remain connected regardless of space and time has vastly improved (Aguilar, 2009; Paragas, 2005; Paragas, 2008). Indeed, connection mobility complements global mobility. James Katz and Mark Aakhus (2002b) describe this as being "physically mobile, but socially 'in touch'" (p. 301).

While providing for a better life persists to be the predominant motivation for Filipino parent migration, long-distance communication offers migrant parents immense opportunity to extend their relationship with their children beyond economic provision. While transnational migration undeniably poses significant challenges to migrant parents and on their parenting, transnational communication gives them the chance to address the parenting challenges imposed by their migration-led separation from their children. Thus, communication allows migrant fathers and mothers to parent their children even across borders.

Filipino mothers make up a significant portion of Filipino parents in diaspora (Madianou & Miller, 2011). Hence, this chapter focuses its attention on Filipino migrant mothers and their mediated parenting efforts. This chapter tackles the centrality of mobile communication in transnational parenting and refers to Filipino migrant mothers' use of mobile communication for transnational parenting as "mobile parenting."

According to Madianou and Miller (2011), "The possession of mobile phones by migrant mothers represented a true catalyst for new developments in family communication" (p. 462).

As can be expected, this chapter describes Filipino migrant mothers' mobile parenting practice in terms of the occurrence and content of their mediated parenting ventures. However, to further understand Filipino migrant mothers' mobile parenting, aside from inspecting their practice of mobile parenting, it is also important to look into the intersections of the socio-demographic, socio-economic, and socio-political landscapes of mobile parenting with the social and technological dimensions of mobile parenting. As such, this chapter discusses mobile parenting by describing the landscapes of Filipino migrant mothers' mobile parenting, the dimensions of their mobile parenting, and their practice of mobile parenting.

This chapter presents evidence primarily from interviews of 32 Singapore-based Filipino migrant mothers. Findings from the said interviews, which were conducted from October 2010 to March 2011 for the author's graduate thesis entitled Communicated Parenting: Singapore-Based Filipino Working Mothers and Their Long-Distance Parenting of Their Teenage Children in the Philippines (National University of Singapore, 2012), were supplemented by evidence from the literature on Filipino migrant parents and their transnational parenting efforts, their children left behind, and their children's caregivers.

THE LANDSCAPES OF FILIPINO MOBILE PARENTING

In order to understand Filipino migrant mothers' mobile parenting, it is important to look into the socio-demographic, socio-economic, and socio-political landscapes where their mobile parenting is situated. The socio-demographic landscape locates Filipino migrant mothers' mobile parenting in the context of the parenting needs of their children who are left at home. Meanwhile, the socio-economic landscape positions Filipino migrant mothers' mobile parenting in the context of access to mobile communication technology, which limits or expands the occasions for mobile parenting. Last, but not the least, the socio-political landscape stations Filipino migrant mothers' mobile parenting in the context of gendered norms and expectations, which potentially extends to gendered mobile parenting.

The Socio-Demographic Landscape of Filipino Mobile Parenting

The socio-demographic landscape situates the parenting needs of children who are living back home while one or both of their parents have migrated to work abroad. Children's parenting needs may be characterized by "the basic needs of children as they grow up" (Herbert, 2004, p. 57) such as "physical care and protection," "affection and approval," "stimulation and teaching," "discipline and controls that are consistent and appropriate to the child's age and development," and "opportunity and encouragement to acquire gradual autonomy, so the child takes gradual control of his or her own life" (Herbert, 2004, p. 58). Parents attempt to address these basic needs of their children through their parenting efforts and activities (Herbert, 2004; Hoghughi, 2004).

Parenting may be broadly defined as "activities that are specifically aimed at promoting the child's welfare" (Hoghughi, 2004, p. 6) through "the promotion of the positive and anything that might help the child" to "the prevention of adversity and anything that might harm the child" (Hoghughi, 2004, p. 7). As such, it is easy to understand how parenting activities are easier performed when parents and children are within close proximity of each other.

The interviews of Singapore-based Filipino migrant mothers reveal their regard of parenting as both a visual and tactile activity. Thus, they admitted that performing parenting functions while they are apart from their children poses significant challenges.

As a visual activity, on the one hand, the migrant mothers interviewed shared that they feel more at ease if they could "see" that their children are fine and they are comforted if they could personally do something if they "see" that their children are not well. A migrant mother expressed,

Parenting is literally a visual activity. As parents, we want to monitor our children in person. We want to personally see what their needs are, what they are doing, how they are doing, how they are feeling. That is why living apart from them is so difficult for me as a parent.

Another mother elucidated,

I feel more unperturbed when I can personally monitor my children and when I can personally address what is bothering them.

According to these mothers, they are reassured of their children's well-being when they can actually see that their children are thriving and can personally do something when they see that they are not.

As a tactile activity, on the other hand, the migrant mothers interviewed shared that they are more comfortable when they could personally take care of their children's daily nurturing needs and when they could physically express their affection. A migrant mother said,

It is natural for me and my kids to express our affection through hugs, kisses, and caresses. Besides, who doesn't want to be hugged? I myself love getting hugs from my family and friends.

Hence, as a tactile activity, some migrant mothers narrated feeling frustrated during stressful circumstances, especially when they know that they could do more for their children if they were spatially closer to them. For instance, a mother who is a registered nurse shared,

I am a nurse, yet, I could not personally attend to my daughter when she is sick because I am here and she is there.

Because of the visual and tactile aspects of parenting, the migrant mothers interviewed admitted that their physical distance from their children poses significant challenges to their practice of parenting and in addressing the needs of their children. Hence, they heavily depend on communication to take up the absence that their migration-led separation brings. While they recognize the extent of parenting that communication enables, these mothers acknowledge that communication is still the best response given their situation.

The Socio-Economic Landscape of Filipino Mobile Parenting

The socio-economic landscape situates migrant mothers' access to mobile communication technology that enables mobile parenting. Socio-economic variables, such as migrant mothers' occupation and income, could heavily influence opportunities for "perpetual contact" (Katz & Aakhus, 2002a). Occupation, which sets the employment conditions of migrant mothers, and income, which sets the parameters of affordable mobile communication technology, may expand or limit occasions for mobile parenting.

In fact, there is a disparity in the array of easily accessible communication media and technologies among the Singapore-based Filipino migrant mothers interviewed. The mothers who are employed in higher-skilled occupations (i.e., mothers who are professionals, associate professionals, managers) and their families have easy access to both mobile phone and Internet-connected computer so they are able to readily partake of mobile-based and Internet-based communication technologies such as SMS, mobile voice call, Internet chat, e-mail, Internet voice/video call, and social media. In contrast, the mothers who are employed in lower-skilled occupations (i.e., mothers who are household service workers) and their families only have easy access to mobile phone so they are only able to readily avail of SMS and mobile voice call. Thus, the interviews reveal that the mothers who are employed in higher-skilled occupations and their families are the ones who are truly operating in what Madianou and Miller (2012) refer to as a "polymedia" environment because they can choose from a wider array of communication media and technologies for their transnational contact.

Income is a significant factor for the wider range of easily accessible communication technologies for the mothers in higher-skilled occupations and their families. They can afford to purchase mobile phone and personal computer and they can manage to pay for mobile and Internet services. On the contrary, the mothers in lower-skilled occupations and their families generally do not have the funds to purchase a computer and, usually, they do not have enough money to cover the cost of Internet subscription as well.

Some of the mothers interviewed who are working as household service workers reported that their employers encourage them to borrow their employers' Internet-connected computer. Some of them even shared that their employers tutor them on using the computer for overseas contact. However, they disclosed that they are uncomfortable in using media that they do not personally own so they are wary of using their employers' computer and even their employers' land-based phone for overseas contact. As such, these mothers predominantly depend on their mobile phone for their mobile parenting.

Overseas contact is not free of charge. Thus, while the cost of long-distance communication has significantly gone down with the advent of newer forms of communication media and technologies (Aguilar, 2009; Madianou & Miller, 2011; Paragas, 2005), it could still make a considerable dent in the financial resources of migrant parents, especially among those who are employed in lower-skilled jobs (Paragas, 2005; Uy-Tioco, 2007). Hence, the migrant mothers interviewed employ various strategies to manage their overseas communication expenses.

For their routine overseas contact, the mothers employed in lower-skilled occupations exchange SMS with their families for their daily communication and they initiate mobile voice call for their weekly conversations. In this manner, the mothers in lower-skilled occupations manage to remain connected with their families while keeping within their overseas communication budget. In contrast, the mothers employed in higher-skilled occupations depend on Internet voice/video call for their daily communication with their children. While they also regularly send SMS, habitual and lengthy conversations with their families are typically conducted through Internet voice/video calls.

Furthermore, applying Richard Daft and Robert Lengel's (1984) Information and Media Richness Theory, Internet voice/video call offers the most nonverbal cues among the array of currently available communication media and technologies, thereby allowing richer communication experience among transnational family members. Hence, the mothers with access to Internet-connected computer and smartphone are able to engage in frequent, longer, and richer conversations with their families at no extra cost. This is ironic given that the mothers in higher-skilled occupations earn comparatively more but they spend relatively less considering the frequency, length, and richness of their long-distance communication experience. As Anastasia Panagakos and Heather Horst (2006) put it, "Internet penetration into the most isolated corners of the world would allow transnational migrants in industrialized countries unprecedented access to their families and networks back home and vice versa" (p. 120).

Thus, it would be interesting to monitor how expanding smartphone penetration among migrant parents and their families and how increasing mobile Internet affordability could revolutionize transnational contact among transnational migrant family members and how it could transform mobile parenting. During the time of the author's data gathering, only the migrant mothers in higher-skilled occupations have access to smartphones and have availed of mobile Internet services. Moreover, only these mothers reported that their children have access to smartphones.

The interviews of Singapore-based Filipino migrant mothers, especially those who are employed as household service workers, reveal that they typically own a prepaid Singapore sim card as well as a prepaid Philippine sim card activated for international roaming. The prepaid nature of their subscription permits them to operate within their overseas communication budget. Furthermore, they use their Singapore sim card for exchanging local SMS and local voice calls while they use their Philippine sim card for maintaining contact with their families back home. However, opinions are mixed whether it is cheaper to initiate overseas contact through their Singapore sim or Philippine sim. Some of them use their Singapore sim for initiating overseas contact because they avail of international SMS and international voice call promotions that Singapore telecommunications networks offer.

While these mothers' opinions vary with respect to the more economical sim card to use when they are the ones initiating contact, these mothers agree that owning a Philippine sim activated for international roaming allows their families back home to affordably initiate contact through SMS. The cost of sending SMS from the Philippines to a Philippine sim activated for international roaming is only PhP1/SMS (roughly around $0.02 USD). Thus, when migrant mothers, especially those who are employed in lower-skilled occupations, own a Philippine sim activated for international roaming, they allow their families back home to send them SMS at an affordable price.

However, these migrant mothers dissuade their families to initiate long-distance voice calls because it is cheaper for the migrant mothers from initiating such calls. While the regular cost for voice call applies to the family when they call the migrant mother's Philippine sim activated for international roaming, the migrant mother would have to shoulder the international roaming charges of an overseas incoming call. Then again, if the family calls the migrant mother through her Singapore sim, they would then shoulder the additional charges associated with calling an international number. Hence, it is more cost-effective if it is the migrant mother who would initiate overseas voice calls because she is the only one who would incur long-distance charges, as incoming calls in the Philippines is still free of charge. Thus, if their families need to contact them through voice call, their families usually just send them an SMS to request for an overseas phone call.

Occupation sets the employment condition of migrant parents and this could affect their opportunities for engaging in transnational communication. Based on the interviews, some of the migrant mothers employed as household service workers multitask as they use the hands-free capability of their mobile phone while doing their work. One such migrant mother narrated,

I avail of international voice call promos so I can have extended conversations with my family back home. I even talk to them while working. I simply plug my earphones, tuck my phone in my pocket, and I talk and work at the same time. There were even instances when my employer thinks that I was talking to her only to find out that I am actually in a voice call conversation with my family.

However, some of them disclosed that they are uncomfortable in using their mobile phone during work hours because they do not want to offend their employers and compromise their job. Furthermore, some of them are not permitted to use their mobile phone during work hours. Ironically, the mothers employed in higher-skilled occupations are the ones who are more constrained in using their mobile phones for voice conversations during work hours. Thus, they depend on asynchronous forms of contact like SMS, Internet chat, e-mail, and social media post.

Then again, the migrant mothers in higher-skilled occupations have predictable blocks of free time that they can devote for transnational communication with their families. Moreover, they also have vacation leaves that they can use to visit home and their family, which they can also afford. In contrast, the free time of the migrant mothers who are employed as household service workers more or less depends on the demands of their household chores. Their live-in employment arrangement also contributes to the unpredictability of their free time as they can be called to serve their employers' needs anytime. Additionally, their employment contract typically allows them infrequent vacation leaves. Besides, these migrant mothers are also constrained by the cost of going home, which includes airfare, the cost of gifts that they are expected to dole out to family members, kin, and neighbors, as well as the cost of parties and gatherings that they are expected to host during their return visit.

The Socio-Political Landscape of Filipino Mobile Parenting

The socio-political landscape situates mobile parenting in the context of gendered norms and expectations. As the Filipino family is striving to adapt to social changes, it is also concurrently attempting to hold on to traditional values and norms (Medina, 2001). In particular, contemporary Filipino families are still observing conventional gender roles while, at the same time, relaxing traditional gender boundaries to respond to the evolving social tide. According to Belen Medina (2001), while women are increasingly becoming "co-breadwinners for the family" and while men are becoming more "personally and emotionally involved in day-to-day child rearing," women are still "mainly and ultimately responsible for domestic tasks and child-care" (p. 280). As such, even with the changes happening in contemporary Philippine society, men and women are still performing traditional gender roles and responsibilities.

Filipino transnational families break the norm of conventional Filipino family type, where spouses and unmarried children cohabit the same household and jointly experience family life (Parreñas, 2001). Nonetheless, even though Filipino transnational families diverge from the traditional mold of a Filipino family, transnational family members still more or less perform traditional gender roles and responsibilities (Cabanes & Acedera, 2012; Parreñas, 2001; Parreñas, 2005a; Parreñas, 2008). Hence, it is not surprising for Filipino migrant mothers to observe traditional and gendered parenting norms in their mobile parenting (Parreñas, 2005a; Parreñas, 2008).

To a considerable extent, gendered parenting norms set higher transnational parenting expectations from migrant mothers. For instance, Graziano Battistella and Ma. Cecilia Conaco (1998) found in their study of children left behind that "the absence of the mother has the most disruptive effect in the life of the children" (p. 237). However, Rhacel Parreñas (2005a) responded that, "it is the continued nurturing of mothers that sets apart children who find less dissatisfaction in the transnational family" (p. 107). Thus, even if migrant mothers cross the role of being providers for their families as they engage in overseas employment, as mothers, they are still regarded as nurturers of the family. As nurturers, they are expected to support their children's physical, social, emotional, and intellectual health (Le Poire, 2006) and are nonetheless expected to perform distant caring apart from providing for their families.

In contrast, because fathers are principally regarded as the provider of the family, when migrant fathers are able to provide, they are deemed to have fulfilled their principal responsibility (Parreñas, 2008). As providers, fathers are expected to make food, clothing, shelter-related items, and money available for the family (Le Poire, 2006). Nonetheless, this does not mean that migrant fathers could get away from establishing affective ties with their children as Parreñas' (2008) interview of children left behind reveal that they also desire to form a more intimate bond with their migrant father.

In spite of this, based on their interviews of children of migrant mothers, Madianou and Miller (2011) wrote, "But if, in general, mothers see new media as to some degree enabling them to reconstitute their role as mothers and thereby ameliorating their situation of absence, the children's assessment of transnational communication seems to be divided into two broad patterns. Although several children were largely positive about the fact that they could keep in touch with their mothers, as many were very critical about their mother's ability to successfully reconstitute their role" (p. 465). Hence, while communication and new media have the capacity to help migrant parents to expand their role beyond economic provision, there are still children who are ambivalent regarding the reconfigured role of their migrant mothers.

As Parreñas (2005a) pointed out, societal definition of the "right kind of family" (p. 30) and the gendered expectations of family roles significantly burden transnational family members in accepting, rationalizing, and defending the consequent changes in family structure and dynamics brought about by transnational movements. Thus, the church, school, media, and other institutions of social learning would greatly contribute in reconfiguring and expanding the definition of family to embrace transnational families (Parreñas, 2005a).

THE DIMENSIONS OF MOBILE PARENTING

Mobile parenting, as a mediated communication activity, has social and technological dimensions. The social dimension of mobile parenting pertains to the long-distance interactions of migrant mothers with their children and their children's caregivers. Meanwhile, the technological dimension of mobile parenting refers to the mobile communication technologies accessed by migrant mothers, their children, and their children's caregivers in their transnational interactions.

The Social Dimension of Mobile Parenting

Because of their transnational distance from their children, migrant parents cannot conventionally perform parenting activities. Through long-distance communication, however, migrant parents are given the chance to address the visual and tactile aspects of parenting that physical distance challenges. As noted by the joint study of the Episcopal Commission for Pastoral Care of Migrants and Itinerant People-Catholic Bishops Conference of the Philippines/Apostleship of the Sea-Manila, Scalabrini Migration Center, and Overseas Workers Welfare Administration (2004), "...communication... made it possible for fathers and mothers to continue their parenting role" (p. 43).

Communication is a central component of mobile parenting. Mobile parenting permits migrant parents to address the visual and tactile aspects of parenting wherein the visual aspect is relieved by periodic updates from their children and their children's caregivers and the tactile aspect is supplanted by verbal exchanges of care and affection. Thus, while migrant parents primarily leave their home country so that they could better provide for their family, mobile parenting allows these migrant parents to function beyond economic provision by giving them the opportunity to care for and guide their children even across borders.

In their mobile parenting efforts, parents must engage their children to talk to them so that they could get updates directly from their children. It is comparatively effortless to engage children in conversations when parents and children have a good relationship. For instance, one migrant mother recounted,

It is easy for me to talk to my daughter about anything and everything because we are friends.

In addition, engaging in meaningful conversations with their children is also relatively easier as children get older. Besides, the level of maturity of children would also facilitate easier discussion of personal and family matters. However, it could be arduous to deal with children when they exhibit obstinate attitudes. For example, one migrant mother talked about her daughter's rebellious tendencies and emotional volatility. She said,

It is difficult to reprimand my daughter because I don't know how she will react. Dealing with her is quite a challenge.

Indeed, Carolyn Sobritchea (2007) also found from her interviews of migrant mothers that communicating with their children is not always easy. Sobritchea (2007) wrote, "The effort to communicate their love for their children was often hamstrung by their own inability to 'find the right words' and 'the right time' to do it" (p. 186).

In her interviews of migrant parents' children, Parreñas (2005b) wrote that, "The children who receive constant communication from migrant parents are less likely to feel a gap in intergenerational relations. Moreover, they are also more likely to experience 'family time' in spatial and temporal distance" (p. 328). However, there is also evidence that children view communication as "intrusive and unwanted" (Madianou & Miller, 2012, p. 2). Nevertheless, communication is one method for migrant parents to convey their presence in their children's life despite the distance.

Still, Parreñas (2008) uncovered that, children of migrant mothers feel close to their migrant parent while children of migrant fathers feel an emotional gap with their migrant parent. The closer relationship between migrant mothers and their children may be attributed to the frequency of communication between them. According to Parreñas (2008), "In contrast to the sporadic communication that embodies relations between migrant men and their children in the Philippines, migrant mothers are ever-present in the lives of their children" (p. 1067).

The intimacy of relationship between migrant parents and their children may also be attributed to gendered parenting norms (Parreñas, 2008). It is evident from the interviews of Singapore-based migrant mothers that they attempt to live up to their culturally expected role as the nurturer of the family, and in doing so, nurturing acts abound during their habitual overseas communication with their children and their children's caregivers. Meanwhile, Parreñas (2008) noted that migrant fathers also strive to carry out their culturally expected role as the economic provider of the family and the enforcer of discipline among the children. Practicing their role as economic provider requires migrant fathers to communicate with their children's caregivers who will then convert their financial remittance to material provision. Hence, practicing their role as economic provider does not necessarily require migrant fathers to converse with their children unlike when they are performing their role as enforcer of discipline, which would then necessitate direct communication with their kids (Parreñas, 2008). Unfortunately, such disciplining acts engender children to feel an emotional gap with their migrant father (Parreñas, 2008). Therefore, Parreñas (2008) suggested that "Efforts of migrant fathers to nurture from a distance, communicate regularly with their children, and reconstitute fathering to not centre on disciplining could enhance intergenerational relationships in transnational families" (p. 1070).

Aside from communicating with their children, migrant parents must also be able to communicate with their children's caregivers. Thus, it is not only important for migrant parents to leave their children in the care of family members whom they trust, but they should also be able to discuss co-parenting matters with them. When kids are younger, migrant parents are also more dependent on caregivers to give them updates about their children since they are not yet old enough to personally give their parents updates. Furthermore, caregivers could inform migrant parents about matters that have been intentionally or unintentionally left out by their children during their conversations.

Based on literature, the designated caregivers vary depending on the gender of the migrant parent (Episcopal Commission for Pastoral Care of Migrants and Itinerant People-Catholic Bishops Conference of the Philippines/Apostleship of the Sea-Manila, Scalabrini Migration Center, and Overseas Workers Welfare Administration, 2004; Parreñas, 2005a; Parreñas, 2008). In father-away families, mothers are still the ones administering child-care and, as such, father-away transnational families resemble nuclear households (Parreñas, 2005a; Parreñas, 2008). On the contrary, in mother-away families as in the case of the Singapore-based migrant mothers interviewed, fathers usually share child-care responsibilities with other family members, most especially with women relatives. One of these migrant mothers expressed how she feels comfortable that her mother is primarily in-charge of taking care of her daughter. She said,

I completely trust my mother as my daughter's caregiver. I am confident that she will never compromise my child's welfare. I know that even if I am not with my daughter, my mother can take care of her very well.

Thus, mother-away transnational families are more dependent on a network of extended kin and not just on the children's father (Parreñas, 2005a). For instance, a mother interviewed shared,

My sister-in-law helped my daughter when she got her first menstrual period. Otherwise, it would have been awkward and uncomfortable for my daughter if my husband, father, and brother were the ones to help.

Another mother said,

Of course I worry when my child gets sick. But I also feel at ease because I know that my mother, father, and sibling are around to take care of my child.

Another mother shared,

I have a special arrangement with my brother – that he could be strict with my children and that he should not be persuaded by their charm and allow them to do what they want to do.

The migrant mothers interviewed admitted that they experience easier co-parenting when they have smooth interpersonal relationship with their children's caregivers and when they can comfortably discuss co-parenting matters with them. One migrant mother interviewed narrated that even though she is separated from her daughter's father, she has good communication with him so it is easy to consult co-parenting matters with each other. She said,

My daughter and I are very close and she tells me about the guys courting her. Since I am not around to monitor her initiation to romance, I talked to my daughter's father and enlisted his help. So we share monitoring responsibility: I monitor and guide my daughter through our frequent and open conversations and her father monitors on-site.

Conversely, another mother shared that she does not have a good relationship with her estranged husband:

My children's father is now their primary caregiver. However, our parenting situation is very difficult because we do not have a good relationship. In

fact, I prefer not to talk to him. As a consequence, we could not synchronize our parenting.

Caregivers could also nurture parent-child relationship and help migrant parents have a closer bond with their children. For instance, one migrant mother, who has been working in Singapore for 13 years, narrated that she left her family when her eldest daughter was only 4 years old. She explained her husband's instrumental role for her to have a close relationship with her two children, who are now 17 and 13 years old. She expressed,

I am very close with my eldest daughter. She told me that I am her best friend, isn't that great?!!

She also quipped that her daughter sends her SMS to greet her,

Hello, my mommy-best friend!

The migrant mothers interviewed shared that their children's caregivers support them in making their children understand and appreciate their efforts to provide for them, help them in reminding their children of their love and concern, and facilitate their habitual communication with their children. One mother explained that caregivers could assist in making their children recognize that their mothers are "real" and not just some fictional character doling out money and gifts.

Aside from conversing with their children's primary caregivers, some of the migrant mothers interviewed reported that they also talk to their children's teachers as part of their mediated parenting efforts. One mother shared,

I always request my children's teachers for their mobile number so I could get in touch with them if needs arise. I also give them my mobile number so that they could SMS me if they need to confer matters with me. I don't hesitate to call overseas if there is a need to discuss matters with them.

Moreover, the mother who is a registered nurse also recounted,

When my daughter gets sick and she needs to visit her doctor, I personally call her doctor so that I could be directly informed about my daughter's condition. I even offer medical inputs and suggestions. So sometimes, I think my child's doctor feels like I'm the one issuing the doctor's order!

The Technological Dimension of Mobile Parenting

Mobile parenting also depends on the ability of migrant parents, their children, and their children's caregivers to use mobile communication technology. In fact, the mobile phone is the most accessible medium for migrant parents to use in their mobile parenting efforts because they can easily get in touch and be in touch with their children and their children's caregivers through it. For one, mobile phone penetration is high in the Philippines compared to other communication media and technologies. Second, mobile signal reaches even the obscure corners of the Philippine archipelago. Third, Filipinos are generally proficient in the use of basic mobile-based technologies, such as SMS and voice call. Fourth, these basic mobile-based technologies are increasingly becoming affordable with the various promotions and packages offered by telecommunications networks in the Philippines and in various host countries. In the case of Singapore-based Filipino migrant mothers, they avail of promotions and packages offered by Singapore telecommunications networks to make their overseas contact more affordable.

The interviews of these mothers reveal their preference for newer forms of communication media and technologies in their overseas communication with their families. Hence, these migrant mothers aim to own newer forms of media for maintaining contact with their families.

Ownership of mobile phone is universal among the migrant mothers interviewed while ownership of Internet-connected computer and smartphone is common only among the mothers employed in higher-skilled occupations. Similarly, access to mobile phone is universal among the families of the migrant mothers interviewed while access to Internet-connected computer and smartphone is only common among the families of mothers employed in higher-skilled occupations. As technological symmetry is necessary, the mothers in higher-skilled occupations are able to choose from a wider array of communication media and technologies, from mobile-based to Internet-based technologies, for their overseas contact with their families since their families back home also have access to a similarly wider array of communication media and technologies.

Then again, data from the interviews of migrant mothers reveal that their choice and use of communication media and technologies depends on the situational context of overseas contact. Since the migrant mothers in lower-skilled occupations principally depend on their mobile phone for their overseas communication and since they do not own a smartphone, their choice is only limited between SMS and mobile voice call. During emergency situations and moments of distress when synchronous exchange is crucial, these mothers automatically resort to mobile voice call. Similarly, even if the migrant mothers in higher-skilled occupations can choose from mobile-based and Internet-based technologies, they automatically choose mobile voice call during emergency situations and moments of distress when synchronous exchange is necessary.

One migrant mother interviewed recounted,

My 17-year old daughter's boyfriend sent me an SMS asking for my daughter's hand in marriage. Of course, this SMS alarmed me because my daughter is still young. Besides, she has been repeatedly assuring me that she is prioritizing her studies and she is determined to graduate on time. So, I responded by calling my daughter's boyfriend right away to express my negative reaction. However, he did not answer my repeated calls; perhaps he was worried of my reaction? So in order to reach him, I opted to send him SMS instead. This way, I was still able to get my sentiments across. Since he was not picking up, I called my daughter to clarify matters and she herself was shocked about her boyfriend's SMS. She assured me that she would keep her promise to finish her studies and to secure a stable job before getting married. She also assured me that she would immediately contact her boyfriend to clarify things.

Fortunately, the Singapore-based migrant mothers related that emergency situations and moments of distress are uncommon occurrences. Hence, these mothers rarely experience the need for urgent overseas contact. Given that routine communication occurs more frequently, it also means that their technology of choice during routine conversations is the one frequently used.

If the choice were only between SMS and mobile voice call, choosing one over the other generally depends on the need for synchronous interaction. If the circumstance requires immediate exchange, the migrant mothers reported that they would automatically resort to mobile voice call. Unlike SMS that depends on keyboard encoding, mobile voice call uses voice format which makes it ideal for immediate exchange. Aside from this, the migrant mothers reasoned out that exchanging SMS would depend on both parties having prepaid load credits. One mother explained,

If my children do not have prepaid load, then they won't be able to respond to my SMS. Thus, it is more expedient if I call them.

Furthermore, the choice would also depend on the need for communicating more nonverbal cues. Again, applying Daft and Lengel's (1984) Information and Media Richness Theory, mobile voice call offers more nonverbal cues than SMS. This is echoed by Madianou and Miller (2011) when they wrote, "Interestingly, all mothers, even those who have access to a wider range of communication platforms, report that their preferred medium for keeping in touch with their children is the mobile phone partly because of the emotionality of voice communication (p. 467)."

THE PRACTICE OF MOBILE PARENTING

The practice of mobile parenting incorporates the occurrence and content of transnational contact. The occurrence of contact refers to the frequency of long-distance communication that opens the channel for mobile parenting. Meanwhile, the content of contact pertains to the topics of conversations exchanged that address the visual and tactile aspects of parenting challenged by transnational labor migration.

The Occurrence of Mobile Parenting

With respect to occurrence of contact, the interviews reveal that Singapore-based Filipino migrant mothers usually talk to their children at least once a week. Thus, compared to migrant fathers who generally engage in infrequent long-distance communication with their children (Parreñas, 2008), the migrant mothers interviewed reported a more habitual overseas contact.

While these mothers and their children routinely observe this once-a-week minimum contact, they no longer adhere to a strict schedule for their overseas communication. This is in sharp contrast with how transnational families maintained contact decades before the widened coverage of land-based phone lines and before the advent of mobile phones (Paragas, 2005). Filomeno Aguilar (2009) chronicled that phone call during those years "was a prearranged and synchronized event" and that "the recipient needed to be at the designated place at the appointed time" as delays "could be costly" (p. 208). As such, it was common among migrant parents and their families during that period to schedule their overseas contact (Aguilar, 2009; Paragas, 2005). Nowadays, the emergence of mobile communication technologies enables migrant parents and their families to be in "perpetual contact" (Katz & Aakhus, 2002a) so they need not worry about setting a strict schedule for their overseas communication.

Contact between the migrant mothers interviewed and their children also occurs during special occasions and events. Birthday and holiday greetings are expressed during special occasions while words of encouragement and congratulations are articulated during special events. A migrant mother interviewed explained,

Calling my children during their birthday makes the greeting extra special. For one, the long-distance call is made especially for the birthday celebrant, which is done apart from my regular overseas contact with my family. Besides, there is an impression that overseas call is more expensive than SMS so by making an overseas call, I also express that the occasion merits spending.

Another interviewed mother narrated,

If the birthday falls during a school day, I would usually send a birthday greeting through SMS at the stroke of midnight so the birthday celebrant would wake up with a birthday greeting from me. Then I will call the birthday celebrant sometime during the day. I don't just leave an SMS greeting. I make it a point to call because I value being able to personally extend my love and affection to my children during their birthday.

These mothers also shared the same sentiments during holidays and other special occasions. With respect to special events, a mother shared,

When my child joined a school competition, my husband gave me periodic updates through SMS on what was happening so I felt that I was also there.

During special occasions and events, the migrant mothers interviewed reported that they make sure that they have enough prepaid load credits in their mobile phone so they could engage in overseas conversation.

Only a number of migrant mothers interviewed reported that they experienced dealing with stressful circumstances in their transnational parenting. For those who did, they dealt with adolescent concerns, school matters, financial issues, health alarms, caregiver relationships, romantic engagements, and children misbehaviors.

For instance, one mother shared,

I have been separated from my children's father for a long time. However, my teenage daughter was still wishing for us to get back together. So, when she found out that her father already has another family, she got really upset and she refused to talk to me. I tried reaching her through voice call but she kept on ignoring my calls. So I just sent her SMS. This way, I was able to reach her without exactly talking to her.

She further narrated,

Since my daughter won't personally talk to me, I asked my mother and my siblings to help me reach out to her. I also requested them to help me explain to her the circumstance of my relationship with her father.

This experience further supports the claim that mobile parenting also includes contact between migrant parents and their children's caregivers. The aforementioned mother added,

Through a series of SMS explaining my side, through reassurances of my love and support, and through the help of my family, we were able to surpass this challenge.

Some of the mothers interviewed shared that, as much as possible, they avoid having disagreements with their children because their distance makes patching things up challenging. They also admitted that they try hard to stretch their patience and become more understanding of their children. Others are careful when reprimanding their kids. One mother explained,

Since we are already dealing with physical distance, I don't want to further complicate it with emotional distance.

Another mother shared,

I try to explain to my daughter where I am coming from when I am reprimanding her to circumvent any untoward misunderstandings and ill-feelings.

Only a few mothers experienced exchanging heated conversations with their children. A mother who experienced such expressed,

While I usually look forward to and enjoy speaking with my kids, arguing with them is so emotionally stressful.

Another mother narrated,

When there are strong disagreements, I sometimes opt to cool down first before talking to them. This way, we are calmer and we avoid saying hurtful words.

The Content of Mobile Parenting

With respect to content of mobile parenting, Brett Laursen and W. Andrew Collins (2004) explained, "As families navigate the transition from childhood into adulthood, the frequency and content of their interactions change" (p. 333). The interviews of Singapore-based Filipino migrant mothers show that the content of routine communication between them and their children typically revolves around updates on how their children are, matters related to school, talks about their children's well-being and safety, and guidance on their teenage children's initiation to romance. Some of them also shared that they tutor their kids during their overseas communication with them. A number of them also disclosed that they reprimand their children when necessary. Aside from these, the migrant mothers also reported that they habitually exchange affective expressions such as "*I love you*", "*I miss you*", "*Good morning*", "*Goodnight*", "*God bless*", and "*Ingat*" (an expression popularly used by Filipinos which means "take care"). These topics of conversation and statements of affection between migrant mothers and their children reflect the traditional gendered role of mothers as the nurturer and giver of care in the family. Hence, these mothers still carry out traditional gendered parenting role even in their mobile parenting.

In contrast with the assurances of love and concern which abound during these Singapore-based migrant mothers' conversations with their children, Parreñas (2008) discovered that when migrant fathers talk to their children, they do not "allay the fears and insecurities of children" (p. 1068). In addition, while the migrant mothers interviewed attempt to avoid reprimanding their children as much as possible, Parreñas (2008) uncovered that migrant fathers "often communicate to discipline their children" (p. 1068). In this manner, migrant fathers still practice their traditional role as the disciplinarian in the family (Medina, 2001; Parreñas, 2008). Unfortunately, children in father-away families observe that their infrequent communication with their father is even dominated by disciplining acts (Parreñas, 2008). Nonetheless, even if they only communicate to impose discipline, doing so is also an act of mobile parenting.

The specific content of discussion might pose a communication challenge between parents and children. For instance, as Filipino parents are generally inclined to be more permissive of their sons and more restrictive of their daughters when it comes to engaging in romantic relationship, daughters could be wary of opening up to their parents about their romantic interest. In order to avoid this from happening, one migrant mother interviewed narrated,

I encourage my daughter to talk to me about the boys courting her. I try to keep an open mind about it so she would feel comfortable to talk about such things with me. This would help me monitor the situation and this would also enable me to guide her accordingly. In contrast, she is afraid to talk about boys with her father because he is so strict.

Moreover, in conservative Philippine society, discussion about sensitive matters, such as premarital sex, could also be awkward. The migrant mothers interviewed tend to talk about the consequences of early pregnancy instead of focusing on premarital sex per se because they feel uncomfortable discussing it.

In addition, children might be hesitant in opening up about topics that they perceive to be worrisome. One mother recounted,

When my son had a motorcycle accident, I was not immediately informed about it, even though I regularly talk to my children. When I learned about it later on, I asked them why they did not inform me right away. They said that the accident was not serious, that they capably took care of it, and that they did not want me to worry.

To avoid this, another mother shared,

I told my son that he could open up about anything to me, even if he thinks that the matter is unpleasant or even if he thinks that I might consider the matter unpleasant. I told him, 'Don't worry about me worrying about you. It's my role as a mother to worry about you.'

As co-parents in the context of transnational parent migration, the migrant mothers interviewed also engage in long-distance communication with their children's caregivers. Co-parenting discussions revolve around the caregivers' co-parenting responsibilities, which may range from taking care of the daily needs of children, monitoring children's welfare, imposing discipline, and handling remittances. One migrant mother narrated,

My husband and I frequently consult each other. One time, my 13-year old daughter asked us for permission to go out with her friends. My husband was initially hesitant but after we consulted with each other, we eventually allowed our daughter.

MOBILE PARENTING AND GLOBAL MOBILITY

The occurrence of transnational families is a manifestation of how transnational family members creatively and strategically respond to the challenges of family separation brought about by international labor migration (Parreñas, 2001). Maruja Milagros Asis, Shirlena Huang, and Brenda Yeoh (2004) noted that, "While transnational migration is reshaping the contours of the Filipino family, it has in no way diminished the importance of being, or the desire to be, 'family'" (p. 204).

Mobile parenting hinges on the ability of migrant parents, their children, and their children's caregivers to communicate and to use mobile communication technology in their transnational interactions. Mobile parenting is contingent on the ability of migrant parents to harness the power of mediated communication in parenting their children across borders. Thus, skills in interpersonal communication and easy access to mediated forms of contact are progressively becoming vital among transnational family members given that transnational family is gradually becoming a common type of Filipino family. Society should recognize the increased indispensability of communication and communication media and technologies among families in diaspora given the heightened dependence of transnational families on mediated communication in experiencing family life.

In the case of migrant parents, especially those who are largely dependent on mobile phone, enhancing their mobile parenting competency is the most natural, innate, and logical response to aid parents and children who are coping with the separation brought about by transnational labor migration. Equipping these migrant parents with necessary mobile parenting skills would help them better navigate through the vicissitudes of transnational family life. Furthermore, increasing smartphone penetration and expanding affordable mobile Internet services would enable migrant parents and their families to enjoy Internet-based technologies, which are cheaper than conventional SMS and regular mobile voice calls.

Findings from both interviews of Singapore-based Filipino migrant mothers and extant literature (for instance, Parreñas, 2001) uncover migrant parents' recognition of the extent of parenting that long-distance communication enables. Madianou and Miller (2011) wrote, "Although mobile phones are empowering for female migrants and present a number of opportunities for intimacy and care at a distance, our evidence suggests that we need to be cautious with regard to the celebratory discourse about the potential of the mobile phone to overcome problems of family separation" (p. 467). Then again, findings from the interviews of Singapore-based migrant mothers also reveal their appreciation of the value of communication in helping them ease the transnational distance. As an interviewed migrant mother succinctly puts it,

It's hard, but through communication, I can still be a mother to them.

Thus, even though mediated communication cannot perfectly replace migrant parents' on-site presence, their personal administration of care, and their physical expressions of affection, mediated parenting offers them the best strategy for adapting to the parenting challenges imposed by their transnational separation from their children. Therefore, in this era of global mobility, parenting through mediated communication is increasingly imperative for the survival and growth of transnational families.

REFERENCES

Aguilar, F. V. Jr. (2009). *Maalwang buhay: Family, overseas migration, and cultures of relatedness in Barangay Paraiso. F.V. Aguilar, Jr. with J.E.Z. Peñalosa, T.B.T. Liwanag, R.S. Cruz I, & J.M. Melendrez*. Quezon City: Ateneo de Manila University Press.

Alunan-Melgar, G., & Borromeo, R. (2002). The plight of children of OFWs. In E. Dizon-Añonuevo & A. Añonuevo (Eds.), Coming home: Women, migration, and reintegration (pp. 106 - 114). Balikbayani Foundation, Inc., & Atikha Overseas Workers and Communities Initiatives, Inc.

Añonuevo, A. (2002). Migrant women's dream for a better life: At what cost? In E. Dizon-Añonuevo, & A. Añonuevo (Eds.), *Coming home: Women, migration, and reintegration* (pp. 73 - 83). Balikbayani Foundation, Inc., & Atikha Overseas Workers and Communities Initiatives, Inc.

Asis, M. M. B. (2000). Imagining the future of migration and families in Asia. *Asian and Pacific Migration Journal, 9*(3), 255–272.

Asis, M. M. B. (2008). The social dimensions of international migration in the Philippines: Findings from research. In M. M. B. Asis, & F. Baggio (Eds.), *Moving out, back and up: International migration and development prospects in the Philippines* (pp. 77–108). Quezon City: Scalabrini Migration Center.

Asis, M. M. B., Huang, S., & Yeoh, B. S. A. (2004). When the light of the home is abroad: Unskilled female migration and the Filipino family. *Singapore Journal of Tropical Geography, 25*(2), 198–215. doi:10.1111/j.0129-7619.2004.00182.x

Battistella, G., & Conaco, M. C. G. (1998). The impact of labour migration on the children left behind: A study of elementary school children in the Philippines. *Sojourn: Journal of Social Issues in Southeast Asia, 13*(2), 220–241. doi:10.1355/SJ13-2C

Beltran, R. P., Samonte, E. L., & Walker, L. (1996). Filipino women migrant workers: Effects on family life and challenges for intervention. In R. P. Beltran, & G. F. Rodriguez (Eds.), *Filipino women migrant workers: At the crossroads and beyond Beijing* (pp. 15–45). Quezon City: Giraffe Books.

Cabanes, J. V. A., & Acedera, K. A. F. (2012). Of mobile phones and mother-fathers: Calls, text messages, and conjugal power relations in mother-away Filipino families. *New Media & Society, 14*(6), 916–930. doi:10.1177/1461444811435397

Daft, R. L., & Lengel, R. H. (1984). Information richness: A new approach to managerial behavior and organizational design. In B. M. Staw, & L. L. Cummings (Eds.), *Research in organizational behavior: An annual series of analytical essays and critical reviews* (Vol. 6, pp. 191–233). Greenwich, CT: JAI Press Inc.

Episcopal Commission for Pastoral Care of Migrants and Itinerant People-Catholic Bishops Conference of the Philippines/Apostleship of the Sea-Manila, Scalabrini Migration Center, and Overseas Workers Welfare Administration. (2004). *Hearts apart: Migration in the eyes of Filipino children*. Manila.

Herbert, M. (2004). Parenting across the lifespan. In M. Hoghughi & N. Long (Eds.), Handbook of parenting: Theory and research for practice (pp. 55-71). London; Thousand Oaks, California; & New Delhi: Sage Publications.

Hoghughi, M. (2004). Parenting: An introduction. In M. Hoghughi & N. Long (Eds.), Handbook of parenting: Theory and research for practice (pp. 1-18). London; Thousand Oaks, California; & New Delhi: Sage Publications.

International Organization for Migration (2013). *Facts and figures infographics*. Accessed on 13 September 2013 from International Organization for Migration Website.

Katz, J. E., & Aakhus, M. (Eds.). (2002a). *Perpetual contact: Mobile communication, private talk, public performance*. Cambridge, UK: Cambridge University Press. doi:10.1017/CBO9780511489471

Katz, J. E., & Aakhus, M. A. (2002b). Conclusion: Making meaning of mobiles – A theory of Apparatgeist. In J. E. Katz, & M. Aakhus (Eds.), *Perpetual contact: Mobile communication, private talk, public performance* (pp. 301–318). Cambridge, UK: Cambridge University Press. doi:10.1017/CBO9780511489471.023

Laursen, B., & Collins, W. A. (2004). Parent-child communication during adolescence. In A. L. Vangelisti (Ed.), *Handbook of Family Communication* (pp. 333–348). Mahwah, NJ: Lawrence, Erlbaum, Associates, Inc.

Le Poire, B. A. (2006). *Family communication: Nurturing and control in a changing world. Thousand Oaks, California; London*. New Delhi: Sage Publications, Inc.

Madianou, M., & Miller, D. (2011). Mobile phone parenting: Reconfiguring relationships between Filipina migrant mothers and their left-behind children. *New Media & Society*, *13*(3), 457–470. doi:10.1177/1461444810393903

Madianou, M., & Miller, D. (2012). *Migration and new media: Transnational families and polymedia*. Oxon, UK: Routledge.

Medina, B. T. G. (2001). *The Filipino family* (2nd ed.). Quezon City: The University of the Philippines Press.

Panagakos, A. N., & Horst, H. A. (2006). Return to cyberia: Technology and the social worlds of transnational migrants. *Global Networks*, *6*(2), 109–124. doi:10.1111/j.1471-0374.2006.00136.x

Paragas, F. (2005). Migrant mobiles: Cellular telephony, transnational spaces, and the Filipino diaspora. In K. Nyiri (Ed.), *A sense of place: The global and the local in mobile communication* (pp. 241–249). Vienna: Die Deutsche Bibliothek.

Paragas, F. (2008). Migrant workers and mobile phones: Technological, temporal, and spatial simultaneity. In R. Ling, & S. W. Campbell (Eds.), *The reconstruction of space and time: Mobile communications practices* (pp. 39–65). New Brunswick, NJ: Transaction Publishers.

Parreñas, R. S. (2001). *Servants of globalization: Women, migration, and domestic work*. Stanford, CA: Stanford University Press.

Parreñas, R. S. (2005a). *Children of global migration: Transnational families and gendered woes*. Stanford, CA: Stanford University Press.

Parreñas, R. S. (2005b). Long distance intimacy: Class, gender and intergenerational relations between mothers and children in Filipino transnational families. *Global Networks*, *5*(4), 317–336. doi:10.1111/j.1471-0374.2005.00122.x

Parreñas, R. S. (2008). Transnational fathering: Gendered conflicts, distant disciplining and emotional gaps. *Journal of Ethnic and Migration Studies*, *34*(7), 1057–1072. doi:10.1080/13691830802230356

Pernia, E. M., Pernia, E. E., Ubias, J. L., & San Pascual, M. R. S. (2013). *International Migration, Remittances, and Economic Development in the Philippines*. National Research Council of the Philippines. Unpublished.

Philippine Overseas Employment Administration. (2006). *OFW global presence: A compendium of overseas employment statistics 2006*. Retrieved from the Philippine Overseas Employment Administration Website.

Philippine Overseas Employment Administration. (2011). *2007-2011 Overseas employment statistics*. Retrieved from the Philippine Overseas Employment Administration Website.

Philippine Overseas Employment Administration. (2012). *2008-2012 Overseas employment statistics*. Retrieved from the Philippine Overseas Employment Administration Website.

San Pascual, M. R. S. (2012). *Communicated Parenting: Singapore-Based Filipino Migrant Mothers and Their Long-Distance Parenting of Their Teenage Children in the Philippines*. (Unpublished Master's Thesis). National University of Singapore, Singapore.

Sobritchea, C. I. (2007). Constructions of mothering: The experience of female Filipino overseas workers. In T. W. Devasahayam, & B. S. A. Yeoh (Eds.), *Working and mothering in Asia: Images, ideologies and identities* (pp. 173–194). Singapore: NUS Press.

United Nations, Department of Economics and Social Affairs, Population Division. (2013). *UN Global Migration Statistics 2013*. Retrieved on 13 September 2013 from http://esa.un.org/unmigration/wallchart2013.htm

Uy-Tioco, C. (2007). Overseas Filipino workers and text messaging: Reinventing transnational mothering. *Continuum (Perth)*, *21*(2), 253–265. doi:10.1080/10304310701269081

KEY TERMS AND DEFINITIONS

Communication Media and Technologies: Include older forms of media and technologies such as postal mail and land-based phone call as well as newer forms of media and technologies such as SMS, mobile voice call, e-mail, Internet chat, Internet voice/video call, and social media.

Content of Contact: Pertains to the topics of conversations exchanged that address the visual and tactile aspects of parenting that transnational labor migration challenges.

Mobile Parenting: Refers to the use of mobile phone and mobile communication technology for transnational parenting.

Occurrence of Contact: Refers to the frequency of long-distance communication that opens the channel for mobile parenting.

Socio-Demographic Landscape: Locates Filipino migrant mothers' mobile parenting in the context of the parenting needs of their children who are left at home.

Socio-Economic Landscape: Positions Filipino migrant mothers' mobile parenting in the context of access to mobile communication technology, which limits or expands the occasions for mobile parenting.

Socio-Political Landscape: Stations Filipino migrant mothers' mobile parenting in the context of gendered norms and expectations, which potentially extends to gendered mobile parenting.

Social Dimension: Pertains to the long-distance interactions of migrant mothers with their children and their children's caregivers.

Technological Dimension: Refers to the communication media and technologies accessed by migrant mothers, their children, and their children's caregivers in their transnational interactions.

Transnational Parenting: Pertains to migrant parents' long-distance parenting as they engage in overseas communication with their children and their children's caregivers. Transnational parenting is facilitated by the use of communication media and technologies.

Section 5

Mobile Marketing and Advertising

Both marketing and advertising have gone mobile, be it "mobile too," "mobile first," or "mobile only." While mass media are declining, mobile is increasing in popularity as a platform to catch attention, increase awareness, instill desires, and increase sales. Equipped with advanced technologies and connected with positioning technologies, mobile enables users to access Mobile Location-Based Services (MLBS). While enjoying MLBS benefits, mobile users are exposed to privacy issues, which are examined socially, economically, and politically in Chapter 12. Although mobile location-based advertising has all the benefits of location awareness, sociability, and spatiality, it does have constraints, as investigated in Chapter 13, which also discusses potentials of MLBA-enabled advertising in China. For marketing, mobile location-based services have also been employed in a game to create hybrid spaces, in which mobile users can be more engaged and empowered through brand experience, productive play, and value co-creation. Although it may be accompanied by some unintended consequences, mobile location-based games can prove to be effective in marketing, as demonstrated in Chapter 14.

Chapter 12
Mobile Location Based Services:
Implications on Privacy

Hee Jhee Jiow
National University of Singapore, Singapore

ABSTRACT

Mobile Location Based Services (MLBS) have been in operation since the 1970s. Conceived initially for military use, the Global Positioning System technology was later released to the world for other applications. As usage of the technology increased, mobile network points, developed by mobile service operators, supplemented its usage in various applications of MLBS. This chapter charts the trajectory of MLBS applications in the mass market, afforded by the evolution of technology, digital, and mobility cultures. Assimilating various MLBS classifications, it then situates examples into four quadrants according to the measures of user-position or device-position focus, and alert-aware or active-aware applications. The privacy implications of MLBS are captured on the economic, social, and political fronts, and its future is discussed.

DEFINITION AND HISTORY OF MOBILE LOCATION BASED SERVICES

Location and positioning technology has been in operation since the 1970s. Conceived for the United States' (US) military, the Global Positioning System (GPS), a satellite-based technology, was used to locate people, places and objects via Cartesian coordinates (Brimicombe & Li, 2009; Chen, 2012; Crato, 2010; Giaglis, Kouroutha-nassis, & Tsamakos, 2003; Spiekermann, 2004). Eventually, in the 1980s, this technology was made available to the world by the US government. This meant that commercial and government entities were free to incorporate GPS technology into their existing infrastructures. Many proclaimed this development as the birth of Mobile Location Based Services (MLBS). However, MLBS was then at a nascent stage, as the GPS technology was typically found only in mobile devices with dedicated functions, and was not widely incorporated into mobile devices, such as mobile phones, together with other integrated technologies. Moreover, the GPS technology, which required mobile devices to be connected to at least three satellite stations in order to provide positional data, was rendered ineffective in built-up places or indoors due to

DOI: 10.4018/978-1-4666-6166-0.ch012

poor connections with GPS satellites; signals are weakened upon passing through various media (e.g. atmosphere, walls, trees), or messed up when bouncing off buildings. In general, the technology was lacking in location accuracy and costly (Junglas & Watson, 2008). While this limitation did not greatly affect its usage in the maritime industry for tracking ships and cargo, due to their movements in open spaces, it has significant negative impact on the accuracy, timeliness of information and computing cost for mass consumer adoption (Rao & Minakakis, 2003). As such, its usage then was predominantly in the navigational systems of vehicles, such as cars, ships and planes, and for freight tracking.

In the mid 1990s, mobile operators, through their mobile network cell stations, enabled more reliable uses of MLBS. Emergency 911 services in the US first deployed this technology afforded by mobile operators to plug a life-threatening gap (Gow, 2005). There were many cases in which distressed callers were not able to verbally inform 911 service staff of their locations. As such, the US Federal Communications Commission established Wireless E9-1-1, a service concept requiring mobile phone operators to have systems in place to acquire location data of 911 callers, and route the information to the nearest emergency call station (Spiekermann, 2004). This demands that the "location accuracy must be within 50 to 100 meters for 67% of all calls and within 150 to 300 meters for 95% of all calls" (Junglas & Watson, 2008, p. 66).

Because the mobile phone industry was mandated to have accurate positioning technologies, it prompted them to venture into incorporating MLBS into their services for the masses. As a result, MLBS, defined as "services that integrates a mobile device's location or position with other information so as to provide added value to a user" (Spiekermann, 2004, p. 10), saw tremendous growth and adoption. This is broadly due to technological advancements that improved the accuracy, timeliness and reliability of the location information, and certain key factors, such as the formation of digital and mobility cultures, that promoted its mass adoption.

TECHNOLOGICAL ADVANCEMENTS

Technological advances in three distinct areas, namely Satellite-Based Technologies, Network-Based Technologies and Sensor-Based Technologies, have unlocked vast opportunities for MLBS. These developments afforded improved accuracy, reliability and timeliness of positioning information. Moreover, the integration of these technologies further enriched the location information. The following sections chart these developments.

Satellite-Based Technologies

Global Navigation Satellite Systems (GNSS) is a collective term referring to the many standalone satellite systems hovering above the earth. GPS is owned by the US and it is "the only one fully in operation and still in upgrading" (Chen, 2012, p. 203), with 31 satellites providing global coverage. The GPS system ensures that, at any one time, there are 5 to 8 satellites visible from any point on the earth for triangulation purposes (Brimicombe & Li, 2009). The Russians own GLObalnaja NAvigatsion-naja Sputnikovaja Sistema (GLONASS), which began development in 1976, but was only fully functional with total global coverage by its 24 satellites in 2011. The European Union's Galileo will be completed in 2018 and will boast over 30 satellites. The Chinese BeiDou Satellite Navigation System (BDS), currently meant for regional positioning purposes, will be expanded to a global COMPASS (or BeiDou-2) navigational system around the same time. India is developing an independent regional satellite navigation system called Indian Regional Navigational Satellite System, while Japan is proposing to build its own version called Quasi-Zenith Satellite System. While GNSS is widely available for use, individual countries can regulate the signal strength received at various regions and limit its usage, giving them control of access to this location technology in times of conflicts.

While GNSS-enabled devices enable the capture of satellite signals, these data need to be decoded and integrated with other parameters, such as satellite clock data, to produce positioning information. It could take minutes, or even tens of minutes, to generate information on the mobile device's location, especially in challenging signal settings such as urban built-up areas or indoors. To reduce the processing time, Assisted-GNSS (A-GNSS) technology was developed; it works by sending complementary data (instructions) to the mobile device to look for specific satellites, instead of combing the whole space, thereby off-loading the mobile device's computing burden (Giaglis et al., 2003). This is achieved through connection to the Internet via Network-Based technologies. Besides reducing the time required to generate positioning information, A-GNSS technology significantly improves positioning accuracy to a range of 1 to 10 meters. Currently, Assisted-GPS (A-GPS) is the predominant technology used by mobile devices.

Network-Based Technologies

Though A-GNSS and A-GPS technology employ the existing networks to function, they primarily serve the interpretation of GNSS data into location information. This is arguably different from Network-Based technology, which predominantly caters to communication of data such as voice, data packets, texts and pictures (Brimicombe & Li, 2009). Yet, Network-Based technology unintentionally unlocked the potential for MLBS to improve in its accuracy and timeliness of information.

Cell Station Identification

When a mobile device is in use, it is always connected to a cell station. As such, the mobile networks or service providers are able to get an approximate location of the mobile device based on the location of the fixed cell station [also known as Base Transceiver Station, Base Station or mobile network cell station (Brimicombe & Li, 2009)]. However, this approximation is only accurate to a range of 200 meters to tens of kilometers. This accuracy decreases with a reduced density of cell stations in a particular geographic location, and with weaker signal strength of cell stations. As such, it is expected that urban areas would experience higher location accuracy compared to rural expanses. Despite its shortcomings, this method is touted as the "most basic manifestation of the ability to provide location services" (Giaglis et al., 2003, p. 70), and used by all mobile communication devices.

Cell Station Triangulation

This technology involves measuring the time taken for mobile device signals to reach three different cell stations. Together with the locations of these fixed cell stations, the data will be computed to triangulate the position of the mobile device. Using this method, the Cartesian coordinates of the mobile device can be approximated due to the known Cartesian coordinates of the three different fixed cell stations. While this method is obviously more accurate than Cell Station Identification, the technological requirements for measuring and computing the data is costly. Giaglis et al. (2003) differentiate this method into two practices – Time Of Arrival and Observed Time Difference. Time Of Arrival places the location-measuring and time-synchronization-with-GPS tasks on the cell stations, which are both costly processes. In contrast, Observed Time Difference shifts the location-measuring burdens onto the mobile device, thereby lowering the cost to the mobile network operators as fewer cell stations are required to be equipped. However, this translates to significant computational cost to the mobile device users.

Bluetooth

Meant for low-cost close-proximity data connections and transfers, such as wireless connections between personal computers, printers and fax machines within an office, Bluetooth technology uses short-range Radio Frequency. The "nominal link range [for a Bluetooth connection] is 10cm to 10m, but it can be extended to more than 100m by increasing the transmit power" (Electronic Times, 1998, p. 46), and it provides robust point-to-point or point-to-multi-point linkages. Due to the relatively small coverage of Bluetooth signals, location data provided by this technology has high accuracy.

Wireless Local Area Networks

Using radio frequency to connect enabled devices together, Wireless Local Area Networks replaces the cabled Local Area Network to provide these devices greater mobility over a certain area. The Access Point connects the wireless devices (e.g. computers, laptops, printers and fax machines) to the wired network (e.g. the Internet). Compared to Bluetooth technology, it has a larger coverage range of approximately 30 meters in a walled area, to hundreds of meters when there is direct line of sight for receiving and transmitting signals. This makes location data less accurate, which is further exacerbated in a large network with multiple Access Points which has the ability to seamlessly hand off device connection from one Access Point to another, as the user moves around. Though its benefit to MLBS is limited, its cost-effective setup and high speeds makes Wireless Local Area Networks "a promising technology for the future data communications market" (Garg, 2007, p. 713).

Radio Frequency Identification (RFID)

RFID is a technology that allows for tracking of RF transponders or tags through a localized area by means of a reader, which may be mobile or fixed. Conceived primarily for the identification of objects in logistics industries, the RFID also affords positioning of the object by computing the signals received. The use of RF boosts its signals' ability to penetrate obstacles and to cover huge distances of up to hundreds of meters when unblocked. Its other advantages include "simplicity of the system, low cost of the device, high portability, [and] ease of maintenance" (Chen, 2012, p. 71). It was widely used in the 1990s for warehouse management, tracking of goods and containers, and protection of items from theft in shops. Passive RFID tags have a simplistic function - carrying only minimal information for identification purposes, these tags normally lack a power source and draws its energy from the RF reader device through inductance. As such, it has to be in close proximity, of approximately a few millimeters to 10 meters, to the reader for information transfer. Active RFID tags can afford a greater gap (over 15 meters) from the RF reader device, as they carry their own power source. The usage of RFID tags can be distinguished into two approaches. In the first approach, the RFID receivers are placed in fixed locations and the RFID tags are attached to the mobile object. In close proximity, the information will be transferred via triggering a response. An example of this practice is the tagging of goods in a shop, and the placement of the receivers at its entrances and the exits; this is for theft deterrence. An alarm is activated when the goods, with the tags attached, is out of the receiver's signal range. In the second approach, the tag is placed at known positions, and the receiver given mobility instead. An example of this practice is the tagging of checkpoints along the route of security guards patrolling an area. The GPS information is thus relayed to the mobile receiver when it is in the tag's vicinity. Typically, the lowest-cost approach, in which the tags outnumber the receiver in a system, is adopted.

Sensor-Based Technologies

Nowadays, mobile devices (predominantly mobile smartphones) come equipped with accelerometer, magnetometer and gyroscope to measure the devices' speed, direction of magnetic field (performing the functions of a compass), and orientation or rotation. These sensors help to determine an object's position when the GNSS signal is weak or when Network-Based connections are absent.

Technological advancements have also made possible the integration of Satellite-Based, Network-Based and Sensor-Based technologies to produce a smooth and continuous positioning tracking when moving across indoor and outdoor spaces (de Souza e Silva, 2013). Known as Hybrid Positioning (Chen, 2012), this ubiquitous positioning solution boasts substantial improvements in the availability, accuracy and reliability of location information over individual technologies (Arikawa, Konomi, & Ohnishi, 2007; Kaupins & Minch, 2005). For example, Google Maps for Mobile integrates Satellite-Based technologies and Network-Based technologies in its MLBS called 'My Location'.

MASS ADOPTION

While technological advancements have made location data more efficient and effective, other major factors have contributed to MLBS's increased penetration and adoption in the mass consumer market. This chapter has identified 4 factors: the portability of mobile devices, the enriched information provided by the integration with other value-added data, the users' mobility patterns and the contemporary consumer trends.

First, mobile devices capable of acquiring and processing location information have become more portable (Brimicombe & Li, 2009; Park, 2005). Though miniaturization of mobile devices is attributed to developments in technology, it does not serve to improve on the quality of location information. Instead, making mobile devices smaller and more convenient to carry have led to their increased adoption by the masses. Moreover, it has become cheaper to acquire these mobile devices and incorporate location technology in them. Despite the presence of dedicated devices such as GPS navigational systems in ships and cars, mobile phones are currently the most pervasive mobile communication device affording MLBS usage. Mobile phone subscribers have outpaced landline subscribers since 2005 – there are 2.2 billion mobile subscribers compared to 1.3 billion landline subscribers in 2005 (Brimicombe & Li, 2009). Latest data shows worldwide mobile phone subscription totaling 6.8 billion, almost equaling the world's population, and representing a near 100% penetration rate (Telecommunication Development Bureau, 2013). However, this does not mean that everyone on the planet possesses a mobile phone, as there are many people with more than one mobile phone subscription. In reality, the number of unique subscribers is less than half of the total subscription figures (GSMA Intelligence, 2013).

Second, the improved interactivity of location data with other data has released a floodgate of opportunities for MLBS, and is often touted as the critical success factor for mass adoption (Business Wire, 2012; de Souza e Silva, 2013; Giaglis et al., 2003). Early mobile phone models could only receive and transmit voice and text messaging data (SMS). They then evolved to accommodate other data streams such as video, music, photos and Internet data. The integration of these data produces higher-value information to the users of mobile devices, especially mobile phones and tablets. An example would be the use of location data in the tagging of photos or videos in Facebook and Foursquare to provide more "personal, spontaneous and interactive" (Brimicombe & Li, 2009, p.

17) information in the socialization process (Raper, Gartner, Karimi, & Rizos, 2007). Botfighters, a very popular game in Russia, which utilizes location data and integrates it into its gameplay, is another example (de Souza e Silva, 2013; Soh & Tan, 2008). Social event organizers use location data to recommend social events within the vicinity of the device (Calabrese, Crowcroft, Di, Lathia, & Quercia, 2010). While the smartphones afford more options for integration of data, the normal mobile phones are still conduits of MLBS. For example, mobile phone operators typically provide paid location service which sends the location of the child's mobile phone to the parents' when requested (Marmasse & Schmandt, 2003; Singtel, 2013). Current US data suggest a mobile phone ownership rate of 88%, of which at least 56% are smartphones (Smith, 2012, 2013).

Third, people are geographically more mobile – they travel more often and over larger distances. With increased economic competition, air travel, including international trips, has become more affordable. Additionally, with current mobile phone technology that allows constant connection with people at home when traveling, the 'out of touch' feeling is significantly reduced. With such prevalent travel patterns, demand for navigational services, mobile guides and proximity-based advertising services have surged, likewise usage of location tagging on social media platforms. Arguably, this is a symbiotic relationship – MLBS spurs travel patterns and vice versa.

Fourth, consumer mobility has been enhanced (Brimicombe & Li, 2009). Using the Internet and innovative payment options (such as use of credit and debit cards), consumers can now purchase products and services from abroad. Likewise, consumers can track the location of their products en route to their destination or locate services, boosting the use of MLBS. While consumers are spoiled for choice and no longer confined to local consumption sources, markets are experiencing intense competition, resulting in cheaper products and services. As such, many have incorporated MLBS as a way to differentiate their products or services. Thus, consumer mobility has driven the utilization of MLBS.

Thus far, the chapter has reviewed the beginnings of MLBS and how technological advancements in generating accurate, reliable and timely location information have improved the effectiveness and efficiency of MLBS. The portability of mobile devices and interactivity of data to produce high-value customized information, and enable sharing on consumers' travel and purchasing patterns, have also aided and hastened the adoption of MLBS by the masses (Shiode, Li, Batty, Longley, & Maguire, 2002). The next section of the chapter will situate various MLBS applications based on their basic operations.

CLASSIFICATION OF MOBILE LOCATION BASED SERVICES

Many attempts have been made to classify the range of MLBS available. Raper et al. (2007) categorized the major MLBS into Mobile Guides, Intelligent Transport Systems, up-and-coming Location Based Gaming, Assistive Technology and Location Based Health, based on the key functional architecture of the respective services. The authors have also broken down each major category into smaller ones. Giaglis et al. (2003) classified MLBS according to their main uses in Emergency, Navigation, Information, Advertising, Billing and Tracking services. Barkuss and Dey (2003) differentiated MLBS as location-tracking and location-aware. Their classification is based on the relative positioning between two static or mobile entities, where one of the entities' location data is recorded and used (Junglas & Watson, 2008). Tracking the location of an entity, such as the child-tracking services provided by mobile service providers, is categorized as location-tracking. Car navigational systems would be termed as location-aware services. Junglas and Watson (2008) found that location-tracking services were more valuable and useful to the user than location-aware services. Spiekermann (2004) differentiated MLBS into four quadrants. The quadrant which an application falls into is determined by two axes, whether it is device or person-oriented, and whether it is a push or pull service. However, advances in technology and usage have raised several issues and call for further refinement to these classifications. This chapter will adapt these classifications and discuss their social, economic and political implications.

Person-oriented services are user based, which means the use of the person's (user) position to enrich the service. An example of MLBS in this category is navigational systems. As such, "user-position focus" would capture the MLBS available and robustly serve the concept. Device-oriented services are device based, which means the use of the position of the device to enhance the service. The device may also represent another person, as in the child-location services provided by some mobile service providers. Again, as the object can mean either a person or a device, an object-position focus would work to capture this dimension of the MLBS more aptly. While closer examination of these two categories shows a glaring resemblance to location-aware and location-tracking terms, many experts have added further a distinction to MLBS in the measure it represents a push or pull service (Raper et al., 2007). Push services "imply that the user receives information as a result of his or her whereabouts without having to actively request it" (Spiekermann, 2004, p. 14). Pull services imply that the user uses an application by 'pulling' the location data off the network. As MLBS evolves, the blurring of the line between push and pull services is emerging. First, many applications allow both push and pull service options. Friend finder services, for example, allow notifications to be pushed to the user, or passively wait for the user to request that information (Spiekermann, 2004). Second, as a pedestrian who uses the popular Navitime system 'pulls' on the service, his/her mobile navigational device will have to go through extended monitoring, resembling the network resources and processes involved in 'push' technology, until he/she arrives at the destination, which could take a couple of hours (Arikawa et al., 2007). As technology enables constant tracking of mobile devices, it would be more apt to use the terms Alert-Aware Services, suggesting that the user becomes aware of positioning applications at work when he/she is alerted. In this way, the Active-Aware Services term is used to define applications that require the active engagement and continued awareness of its usage. Table 1 (below) situates some examples of MLBS within the proposed framework.

IMPLICATIONS OF MOBILE LOCATION BASED SERVICES

Economic Implications

The network demands for alert-aware and active-aware services are drastically different, and therefore have distinctive economic implications for service providers. MLBS that operates on an active-aware basis are reactive in nature. As and when information is required, the user activates the network and initiates the process of generating location information and supplying it to the user's mobile device. Being reactive does not require much resources compared to the active nature of MLBS that employ alert-aware technology. In the case of alert-aware services, the network has to constantly be in conversation with the user's mobile device to pinpoint its location, process, update and monitor its location, and initiate a response by supplying information to the user's mobile device when the certain criteria are met. For example, for a store's coupon service to alert-aware its promotions to customers in its proximity, network resources are required to continuously scan the vicinity for every mobile device (phone) that passes through it, and send coupons to those mobile devices that meet the application's crite-

Table 1. MLBS examples in categories

	Alert-Aware Services	Active-Aware Services
User-Position Focus	1. User is alerted when he/she enters a friend's vicinity. Eg. Friend finder services such as FindMyFriends for iOS devices. 2. User is alerted when he/she enters the vicinity of certain establishments such as restaurants and stores. Eg. Infotainment services. 3. An alert is sent to the user in the form of a bill due to his/her use of services in a particular location. Eg. Location-sensitive billing of call rates, which are typically reduced when called from the subscriber's home, or when a car passes through a gantry for toll charge purposes.	1. User views a mobile map to find out where he/she is positioned. Eg. GPS navigational systems in cars. 2. User calls for assistance, but is not able to communicate his/her location. Eg. Automotive assistance services or emergency services. 3. Players use their mobile devices to locate and "kill" game opponents. Eg. Botfighters game
Object-Position Focus	1. User is alerted when a purchased item has reached certain destinations. Eg. Amazon's shipment tracking system. 2. User is alerted when his/her child leaves a certain zone. Eg. Safe & Sound	1. User requests information on where his cars or ships are. Eg. Fleet monitoring services by transport companies. 2. User requests information on where his/her child is by sending a text message to the mobile service provider. Eg. Location services offered by mobile service providers.

ria. This activity would require huge resources which adversely impacts the profitability of such a MLBS. However, increasing usage demands of active-aware applications and technological advances may, in future, reduce this resource gap.

Yet there are potential economic benefits to commercial entities that employ MLBS. Object-position focus applications have been a crucial service feature to postal service customers for tracking of registered parcels, and to consumers of online merchants for tracking of purchases. However, for some postal services, a premium has to be paid by customers if they want to be able to track their packages – this is termed as user-charged services (Giaglis et al., 2003). An example is the United States Postal Service, which charges a higher premium for point-by-point tracking of articles (United States Postal Service, 2013). However, Amazon provides similar tracking services free of charge, which makes for a better customer experience and enhances customer loyalty. Many have also claimed that the prevalent use of MLBS has boosted international trade.

The incorporation of positioning technologies into the mass-market mobile phones have lowered the cost of using MLBS by individuals to almost nothing (Bellavista, Kupper, & Helal, 2008; Kaupins & Minch, 2005). For example, car navigational services are available on mobile smartphone applications, such as the widely popular and free Google Maps; costly, dedicated GPS navigational devices are no longer needed. As such, usage of MLBS by mobile subscribers are expected to hit about 35%, equivalent to 255 million users by 2017 (Business Wire, 2012).

Social Implications

The advent of location technology has boosted many applications used for safety purposes. As mentioned earlier, its first foray into the societal sector involves the use of positioning data for emergency rescue purposes. It has since developed and is widely used for other safety applications. While tracking of an object can provide a sense of security, society as a whole has benefited greatly from safety applications that protect people. Examples include provision of navigation through unfamiliar areas and child-tracking applications. Here, we will specifically highlight child-tracking applications as a crucial social benefit of MLBS (de Souza e Silva, 2013; Haddon, 2013; Jiow & Lin, 2013).

Communication technologies such as mobile phones have been used by parents to exert some control over their children's travel boundaries and be more aware of their whereabouts (Ledbetter et al., 2010). In the past, this typically involved contacting their children via phone calls or SMS messages to solicit information about their activities and whereabouts. However, recent advancements in mobile phone technology have provided supervision applications to parents (Barreras & Mathur, 2007). There are claims that these may deter a child from misbehaving by providing a sense of the parent's "psychological presence" (Stattin & Kerr, 2000, p. 1084). This technology has allowed "remote parenting" to take place (Boesen, Rode, & Mancini, 2010, p. 68). Examples of child-tracking applications include Safe & Sound (Marmasse & Schmandt, 2003), which is alert-aware in nature, and Locator which is active-aware in nature (Singtel, 2013).

As commercial entities strive to provide better MLBS to consumers, they have resorted to consumer profiling (Shiode et al., 2002). Examples include a theatre ticket reservation being made, at the destination country, in preparation for an individual who travels abroad due to the profiling and processing of the individual's lifestyle patterns. However, this has an impact on the individual's privacy – a key societal concern of MLBS usage (Barkuss & Dey, 2003; Barreras & Mathur, 2007; Beresford & Stanjano, 2003; Boesen et al., 2010; Consolvo et al., 2005; Gow, 2005; Junglas & Watson, 2008; Spiekermann, 2004).

To maintain the contextual integrity of this chapter, the discussion will focus on location privacy, which is defined as "the ability to prevent other parties from learning one's current or past location" (Beresford & Stanjano, 2003, p. 46). As such, location privacy concerns are typically centered on how sensitive the information (nature of information) is and how it is managed (management of information) (Duckham & Kulik, 2006). Identifiable information, such as one's full name and social security number, are considered sensitive information. Location information tied with identifiable information would be deemed as highly intrusive. Pervasive information, or large amounts of data, would also be considered sensitive information – characteristic of many Alert-Aware applications. For example, an individual may not care if others have information on his location at a specific time, but would be very concerned if others have information on his/her every movement at every second. As such, the generation of large amounts of information would inevitably surface some potentially sensitive ones. Information showing an individual's frequent visits to an AIDS clinic is an example of sensitive information generated by pervasive tracking (Duckham & Kulik, 2006). Moreover, the more accurate the location information, the more intrusive the positioning application would be.

The manner in which the information is managed is also a concern, specifically, in two areas – how information is used, and who has access to it. The sending of unsolicited messages (spam), in the form of promotional materials, by stores that employ proximity technologies is typically frowned upon and considered a nuisance by the user who has entered the vicinity. While unconsented use of the location data clearly raises privacy concerns, information managers have been shrewd to circumvent this by including its data usage practices in wordy privacy policies, which are often difficult to understand and which most users do not actually read before they agreed to them. The entity that has access to location information is also a concern. While revealing one's location to a friend or family member may be favorable, revealing it to others may arguably not be, especially for sensitive information. Locating a friend during lunch time may be acceptable, but locating an employee during his/her lunch break may come across as intrusive (Kaupins & Minch, 2005). As such, the person accessing the information is of primary importance (Fusco, Michael, Michael, & Abbas, 2010). A study showed that youths, despite being of an age demanding individuation and autonomy, generally do not hold negative views when being tracked by their parents through their mobile phones (Jiow & Lin, 2013). Pure GPS technology, typically used in powerful and dedicated navigational devices, does not require mobile devices to relay back information to GPS satellites. As such, the mobile devices would possess and process the location data, giving total ownership to the user and allaying privacy concerns (Duckham & Kulik, 2006).

However, with the adoption of hybrid positioning technologies, and integration of other information to provide quality and relevant MLBS to users, there will inevitably be trade-offs because "some goals are clearly mutually exclusive and cannot be simultaneously satisfied" (Beresford & Stanjano, 2003, p. 46). For example, a person cannot expect his friends to locate him while keeping his position and his identifiable information, such as his name, secret. A parent cannot track the location of his child accurately, through the use of hybrid positioning technologies, without allowing mobile service operators to have that information. Such trade-offs seem to be accepted by society. A study showed that despite people finding MLBS useful, the object-position focus is perceived as less intrusive than user-position focus (Barkuss & Dey, 2003). Yet it is crucial for government to intervene with policy formulations so as to strike a balance in these trade-offs.

Political Implications

Because of such concerns, several governments have made attempts to enact new privacy laws. Fueled by privacy invasions by many child-location application developers, the US Congress enacted the Location Privacy Protection Act in 2012 (Kaiser, 2012; Kaupins & Minch, 2005). The Act covers the management of location information in terms of how the information is collected and used. It also stipulates adequate consent processes, but does not mention how long the data can be kept. Korea has also proposed regulations to ban all MLBS providers access to location information (Korea Times, 2004). While the US has specific legislations for location data, other countries have relied on existing general privacy laws. The Norwegian Personal Data Act and the Finnish Personal Information Law and Law about Privacy and Security of Telecommunications both claim to provide adequate coverage on privacy and consent issues associated with location data (Kaupins & Minch, 2005). Consistent with many privacy laws, regulatory approaches need to consider the issues of notice and transparency to the user, consent to the use of information and its limitations, access and participation afforded to users, integrity and security of the personal data, and enforcement and accountability of the collectors of information (Duckham & Kulik, 2006). While governments impose regulations to protect location privacy, they have also been accused of abusing it. In the wake of Snowden's leaks on US surveillance activities, many are calling for transparency in how the US government makes use of location tracking devices (The International Herald Tribune, 2013). Yet MLBS has positive implications for crime prevention and prosecution. For example, a thief was tracked via a GPS-enabled phone that he stole (de Souza e Silva, 2013). Law enforcement has used location data to place the criminal at the scene of the crime too.

LIMITATIONS, RECOMMENDATIONS, AND FUTURE RESEARCH DIRECTIONS

Rapid technological advancements have made MLBS more accurate, reliable and interactive. Coupled with consumers' digital and mobility cultures, MLBS is postured for ubiquitous usage. As MLBS evolves with innovative features, its use will encroach more aggressively into the mass market, and inadvertently pose a challenge to privacy rights and protection. Though there are technological tools developed to protect users' privacy, such as anonymisation tools and third party storage of identifiable information, developers of MLBS have to adopt integration of location data with other data (identifiable ones being the most attractive) in order to provide more novel services and increase its market share, thereby winning the race against privacy. As such, while the balance between users' perceived usefulness of location features (in the use of MLBS) and privacy is struck differently among people, there will be a shift in favour of usefulness.

While policy development by the government sector would have some impact on privacy, enforcement would be challenging. Moreover, legal formulation process typically loses out to the speed of innovation of MLBS. However, in the light of Snowden's revelation about government surveillance activities, location privacy protection from the state's intelligence agencies is highly questionable. While extreme, the way forward for users could be summed up by Sun CEO Scott McNealy's comment: "You have zero privacy anyway, get over it!" (Sprenger, 1999, p. 1).

Future research would do well to explore magnitude of users' privacy concerns afforded by the different segments of the framework. It is conceivable that MLBS would elicit the most privacy concerns for Alert-Aware services that are User-Position Focus, and the least privacy concerns for Active-Aware services that are Object-Position Focus. Yet, the MLBS that would be perceived as most useful, providing users with convenience, would arguably be the Alert-Aware services. It would be interesting if future research addresses these issues and inform policy-making.

REFERENCES

Arikawa, M., Konomi, S., & Ohnishi, K. (2007). Navitime: Supporting Pedestrian Navigation in the Real World. *IEEE Pervasive Computing / IEEE Computer Society [and] IEEE Communications Society*, *6*(3), 21–29. doi:10.1109/MPRV.2007.61

Barkuss, L., & Dey, A. (2003, September). *Location-Based Services for Mobile Telephony: A study of user's privacy concerns.* Paper presented at the 9th IFIP TC13 International Conference on Human-Computer Interaction. Zurich, Switzerland.

Barreras, A., & Mathur, A. (2007). Wireless Location Tracking. In K. R. Larson, & Z. A. Voronovich (Eds.), *Convenient or Invasive - The Information Age* (pp. 176–186). Boulder, CO: Ethica Publishing.

Bellavista, P., Kupper, A., & Helal, S. (2008). Location-Based Services: Back to the Future. *IEEE Pervasive Computing / IEEE Computer Society [and] IEEE Communications Society*, *7*(2), 85–89. doi:10.1109/MPRV.2008.34

Beresford, A., & Stanjano, F. (2003). Location Privacy in Pervasive Computing. *IEEE Pervasive Computing / IEEE Computer Society [and] IEEE Communications Society*, *2*(1), 45–55. doi:10.1109/MPRV.2003.1186725

Boesen, J., Rode, J. A., & Mancini, C. (2010, September). *The domestic panopticon: Location tracking in families.* Paper presented at the 12th ACM International Conference on Ubiquitous Computing. Copenhagen, Denmark.

Brimicombe, A., & Li, C. (2009). *Location-based services and geo-information engineering*. Chichester, UK: Wiley-Blackwell.

Business Wire. (2012). Research and markets: Mobile LBS - for 2017 in the EU27, mobile LBS penetration of mobile subscribers will reach 35% - or 255 million users. *Business Wire*. Retrieved September 5, 2013, from http://search.proquest.com/docview/1015995647?accountid=13876

Calabrese, F., Crowcroft, J., Di, L. G., Lathia, N., & Quercia, D. (2010, December). *Recommending Social Events from Mobile Phone Location Data*. Paper presented at the IEEE International Conference on Data Mining. Sydney, Australia.

Chen, R. (2012). *Ubiquitous Positioning and Mobile Location-Based Services in Smartphones*. Hershey, PA: Information Science Reference. doi:10.4018/978-1-4666-1827-5

Consolvo, S., Smith, I. E., Matthews, T., LaMarca, A., Tabert, J., & Powledge, P. (2005, April). *Location disclosure to social relations: Why, when, & what people want to share*. Paper presented at the Human Factors in Computing Systems. Portland, OR.

Crato, N. (2010). How GPS Works. In N. Crato (Ed.), *Figuring It Out: Entertaining Encounters with Everyday Math* (pp. 49–52). Berlin: Springer. doi:10.1007/978-3-642-04833-3_12

de Souza e Silva, A. (2013). Location-aware mobile technologies: Historical, social and spatial approaches. *Mobile Media & Communication*, *1*(1), 116–121. doi:10.1177/2050157912459492

Duckham, M., & Kulik, L. (2006). Location Privacy and Location-Aware Computing. In J. Drummond, R. Billen, E. Joao, & D. Forrest (Eds.), *Dynamic & Mobile GIS: Investigating Change in Space and Time*. Boca Raton, FL: CRC Press.

Electronic Times. (1998). How Bluetooth works. *Electonics Times*, 46.

Fusco, S. J., Michael, K., Michael, M. G., & Abbas, R. (2010). *Exploring the Social Implications of Location Based Social Networking: An Inquiry into the Perceived Positive and Negative Impacts of Using LBSN between Friends*. Paper presented at the Ninth International Conference on Mobile Business/2010 Ninth Global Mobility Roundtable. Piscataway, NJ.

Garg, V. K. (2007). Wireless Local Area Networks. In V. K. Garg (Ed.), *Wireless Communications & Networking* (pp. 713–776). Burlington, MA: Morgan Kaufmann. doi:10.1016/B978-012373580-5/50055-7

Giaglis, G. M., Kourouthanassis, P., & Tsamakos, A. (2003). Towards a Classification Framework for Mobile Location Services. In B. E. Mennecke & T. J. Strader (Eds.), Mobile Commerce: Technology, Theory and Applications (pp. 64-81). Hershey, PA: Idea Group Inc (IGI).

Gow, G. A. (2005). Information Privacy and Mobile Phones. *Convergence: The International Journal of Research into New Media Technologies*, *11*(2), 76–87. doi:10.1177/135485650501100208

Haddon, L. (2013). Mobile media and children. *Mobile Media & Communication*, *1*(1), 89–95. doi:10.1177/2050157912459504

Intelligence, G. S. M. A. (2013). *Global mobile penetration — Subscribers versus connections*. Retrieved September 3, 2013, from https://gsmaintelligence.com/analysis/2012/10/global-mobile-penetration-subscribers-versus-connections/354/

International Herald Tribune. (2013, August 19). GPS and the law. *The International Herald Tribune*.

Jiow, H. J., & Lin, J. (2013). The influence of parental factors on children's receptiveness towards mobile phone location disclosure services. *First Monday*, *18*(1). doi:10.5210/fm.v18i1.4284

Junglas, I. A., & Watson, R. T. (2008). Location-based services. *Communications of the ACM*, *51*, 65–69. doi:10.1145/1325555.1325568

Kaiser, T. (2012). *Location Privacy Protection Act Passed by Senate Committee*. Retrieved August 27, 2013, from http://www.dailytech.com/Location+Privacy+Protection+Act+Passed+by+Senate+Committee/article29428.htm

Kaupins, G., & Minch, R. (2005). *Legal and Ethical Implications of Employee Location Monitoring*. Paper presented at the 38th Annual Hawaii International Conference on System Sciences. Hawaii, HI.

Korea Times. (2004, July 3). Location-Based Information Service Due Next Year. *Korea Times*. Retrieved from http://www.koreatimes.co.kr

Ledbetter, A. M., Heiss, S., Sibal, K., Lev, E., Battle-Fisher, M., & Shubert, N. (2010). Parental Invasive and Children's Defensive Behaviors at Home and Away at College: Mediated Communication and Boundary Management. *Communication Studies*, *61*(2), 184–204. doi:10.1080/10510971003603960

Marmasse, N., & Schmandt, C. (2003). *Safe & sound: A wireless leash*. Paper presented at the Human Factors in Computing Systems. Ft. Lauderdale, FL.

Park, W. (2005). Mobile Phone Addiction. In R. Ling, & P. E. Pederson (Eds.), *Mobile Communications* (Vol. 31, pp. 253–272). London: Springer. doi:10.1007/1-84628-248-9_17

Rao, B., & Minakakis, L. (2003). Evolution of mobile location-based services. *Communications of the ACM*, *46*(12), 61–65. doi:10.1145/953460.953490

Raper, J., Gartner, G., Karimi, H., & Rizos, C. (2007). Applications of location-based services: A selected review. *Journal of Location Based Services*, *1*(2), 89–111. doi:10.1080/17489720701862184

Shiode, N., Li, C., Batty, M., Longley, P., & Maguire, D. (2002). *The Impact and Penetration of Location-Based Services*. London: University College London.

Singtel. (2013). *Locator Plus*. Retrieved September 5, 2013, from http://info.singtel.com/personal/phones-plans/mobile/vas/locator-plus/detail

Smith, A. (2012). *17% of cell phone owners do most of their online browsing on their phone, rather than a computer or other device*. Retrieved September 5, 2013, from http://pewinternet.org/Reports/2012/Cell-Internet-Use-2012/Key-Findings.aspx

Smith, A. (2013). *Smartphone Ownership - 2013 Update*. Retrieved September 5, 2013, from http://pewinternet.org/Reports/2013/Smartphone-Ownership-2013/Findings.aspx

Soh, J., & Tan, B. (2008). Mobile gaming. *Communications of the ACM*, *51*(3), 35–39. doi:10.1145/1325555.1325563

Spiekermann, S. (2004). General Aspects of Location-Based Services. In J. Schiller, & A. Voisard (Eds.), *Location-Based Services* (pp. 9–26). San Francisco, CA: Morgan Kaufmann Publishers. doi:10.1016/B978-155860929-7/50002-9

Sprenger, P. (1999). *Sun on Privacy: 'Get Over It!'*. Retrieved September 2, 2013, from http://www.wired.com/politics/law/news/1999/01/17538

Stattin, H., & Kerr, M. (2000). Parental Monitoring: A Reinterpretation. *Child Development*, *71*(4), 1072–1085. doi:10.1111/1467-8624.00210 PMID:11016567

Telecommunication Development Bureau. (2013). *The World in 2013 - ICT Facts and Figures*. Retrieved September 5, 2013, from http://www.itu.int/en/ITU-D/Statistics/Pages/default.aspx

United States Postal Service. (2013). *Track & Confirm*. Retrieved August 26, 2013, from https://tools.usps.com/go/TrackConfirmAction!input.action

KEY TERMS AND DEFINITIONS

Active-Aware Services: Applications that require the user's active engagement and continued awareness of its usage.

Alert-Aware Services: Applications where user becomes aware of positioning applications at work when he/she is alerted.

Location Information: Information on the user's or device's location.

Location Privacy: "The ability to prevent other parties from learning one's current or past location" (Beresford & Stanjano, 2003, p. 46).

Mobile Location Based Services: "Services that integrates a mobile device's location or position with other information so as to provide added value to a user" (Spiekermann, 2004, p. 10).

Object-Position Focus: The focus on the device's position to enrich the service.

User-Position Focus: The focus on the person's (user) position to enrich the service.

Chapter 13
Location-Aware Mobile Media and Advertising: A Chinese Case

Mei Wu
University of Macau, China

Qi Yao
University of Macau, China

ABSTRACT

Location-Based Services (LBS) that are combined with ubiquitous smartphones usher in a new form of information propagation: Location-Based Advertising (LBA). Modern technologies enable mobile devices to generate and update location information automatically, which facilitates marketers to launch various types of location-aware advertising and promotional services to users who are in the vicinity. This chapter conceptualizes location-aware mobile communication as the locative and mobile media with a McLuhan's notion of retrieve of "locality" in the "networked" space of information flows, and examines the current dilemma faced by LBA in China through a case study. It first defines location-aware mobile technologies and influences such media afford for location-aware advertising and information propagation. It then provides an overview of the development of LBS and LBA in China. A case study of the LBA app "SBK" further offers a detailed examination how new models of advertising are developed with the technical affordances of location awareness, sociability, and spatiality. The chapter concludes with a discussion on the constraints and potential of LBA in China.

INTRODUCTION

With the arrival of mobile era, wireless communication devices such as mobile phones, which enable people to communicate anytime and anywhere, have transformed our lives tremendously. Specifically, the advent of smartphones has brought forth a new form of advertising --mobile advertising, which has become a rapid growing business providing brands and companies a novel

DOI: 10.4018/978-1-4666-6166-0.ch013

channel to connect with consumers directly and interactively beyond traditional methods of advertisement.

The history of LBS could trace back to 1993. On November 13, 1993, an 18-year girl Jennifer Koon was kidnapped and murdered. Before her death she called 911 for help; however the technology at that time was not allowed to locate her position. This event resulted in E911 by U.S. Federal Communications Commission (FCC), which enables emergency services to identify the caller's geographic location. LBS was not popular until the release of smart phone and related operating system such as Apple's iPhone 3 and Google's Android system. The growth of mobile phone and related application software (apps) enables users to explore and extend the functions of mobile phones; one of the new uses is location-based service (LBS). LBSs, combined with ubiquitous smartphones, usher in a new form of information propagation: location-based advertising (LBA). Modern technologies enable mobile devices to generate and update location information automatically, which facilitates marketers to launch various types of location-aware advertising and promotional services to users who are in the vicinity. This article investigates location-based mobile applications with a specific attention given to the context of location-based advertising in China. The research questions are concerned with such issues as how to conceptualize locative and mobile media or location-aware apps? How such technologies combined with smartphones and Internet-based social media provide affordance for the creation of a new form of commercial information dissemination? As a mediated tool, in what pattern and to what extent does LBS offer a unique communicative environment for persuasive communication like LBA to take place? What are the influence implications and potentials of location-aware mobile media as ubiquitousness of LBSs increases?

This article conceptualizes location-aware mobile communication as the locative and mobile media with a McLuhan's notion of retrieve of "locality" in the "networked" space of information flows and examines the current dilemma faced by LBA in China through a case study. It first defines location-aware mobile technologies and influences such media afford for location-aware advertising and information propagation. It then provides an overview of the development of LBS and LBA in China. A case study of the LBA app "SBK" further offers a detailed examination how new models of advertising are developed with the technical affordances of location awareness, sociability and spatiality. The article concludes with discussion on constraints and possible potentials of LBA in China.

DEFINING LOCATION-AWARE MEDIA

Location-aware media is defined here as any form of networked service via wireless communication technologies to mobile terminals which enable users to be aware of the location of themselves and/or others in the vicinity (Licoppe, 2013). Location-aware technologies and applications may include GPS-based geo-location or cell phone triangulation (Licoppe & Inada, 2006), but most distinctively the self-declarative location-sharing applications such as with Gowalla, Foursquare, Facebook place, Grindr and Digu. In this manner, the user indicates his/her location in way of "check-ins". The location notification may be made public or restrict to his/her circle of "friends" (Tang, Lin, Hong, Siewiorek, & Sadeh, 2010; Licoppe, 2013). This is based on a variety of wireless communication technologies such as Wi-Fi and Bluetooth, through which most smartphones and other mobile devices are capable of recognizing one another within a range of certain distance. Users to whom such connection and service are available can receive notifications of their "friends" or other users in proximity (Licoppe, 2013).

Locative media arouses a lot of research interest because it brings forth a new form of media experience in the age of "networked individualism" as in the words of Rainie & Wellman (Rainie & Wellman, 2012) and "virtual communities"(Rheingold, 2002) of social networks. To follow the thesis of Marshall McLuhan about any new media, this novel experience is reflective of the retrieve of "locality". It is the "connectivity that paradoxically provides both location awareness ("where are you?") and facilitates communication without regard to location" (Wellman & Rainie, 2013). When much of public discourse and popular rhetoric focus on the Internet and social media being emancipated from spatial constraints and overcoming the barrier of physical place, location-aware media revives the "locality" back into our environment of communication.

However, this recovered realm of "locality" differs from the default "locality" we used to know. It is first the locality that connects the "real place" with the "virtual space". The communicative flow of information that takes place in location-aware media has both components of binding to a physical location and unrestraint in global cyber connectivity. Second, it is "networked locality". Our awareness of location is connected to the social network media such as Facebook, Twitter, Renren or QQ. Gordon and de Souza e Silva (2011) defines it as "net locality" which is practiced hybrid space, developed by the constant enfolding of digital information and networked connections into local spaces. Net locality is based on the recognition that we are networked, but still connected to local spaces, and that belonging to a global network strengthens local connections. This concept directly challenges traditional views of mobile communication, which emphasize users' disconnections from local spaces. In net locality, remote connections are still present, but become part of the space in which the mobile user is, instead of removing users from it. (de Souza e Silva, 2013)

Location-Aware Media for Advertising

Location-aware media has potential to offer a new type of brand and commercial message production, transmission and delivery mobile terminal users such as smartphones, tablets, and eReaders. Much of the academic interest in locative media derives from the notion that location awareness might be supplementary to existing social media platforms or even transform them into location-based social networks (or LBSN) (de Souza eSilva & Frith, 2010).

The commercial use of locative media has been the stimulator for the LBS development. LBS refers to the retrieval of data directly based on the location of the mobile phone user when making the request (Dibdin, 2001). In the early stage, LBSs required the users to manually input their location data such as a street name or zip code. In the early stage, users had more control of the location-aware service and hence would evoke less concern about privacy leakage. However the weakness was also obvious: Users might not know their exact location, and the position provided by users restricted marketers to offer various personalized services (Teri, 2000). Later, a variety of new technologies were developed to facilitate marketers to wage location-based advertising, promotion and other types of services when the users were in a certain pre-defined region (Unni & Harmon, 2003). These technologies include wireless application protocol (WAP), Bluetooth, radio frequency identification (RFID), GPS and augmented reality (Leppäniemi, 2006).

The market of LBS is classified into various categories involving coordinating social events, reserving nearby business or service, navigation, finding or locating people, receiving alerts, location-based billing, etc. Its applications are growing with innovations and the expansion of mobile phone market. Five types of social use of LBS are listed:

1. **Emergency:** It provides concurrent aid to the mobile phone users by locating their precise locations in case of any emergency situations.
2. **Navigation:** When referring to navigation, LBS sends various data directly to the users such as location of destination and even provides the best route for journey based on the user's current location.
3. **Information:** LBS users can obtain different types of information such as traffic status, route in a unfamiliar place, maps, transportation services, restaurants and even can perform as a tourist or shopping guide and etc.
4. **Tracking and Management:** Vehicle tracking is critical for fleet management especially for cargo service companies. Not only the location of vehicles, companies can also locate the positions of the employees and then inform the latest information to the customer about the arrival time. Meanwhile the vehicle tracking can be utilized to find the nearest ambulance and send emergency messages.
5. **Location-Based Billing:** This function enables the service providers to charge the users for using their services in different locations (Gordon C. & Anand Kumar, 2010).

Location-based advertising (LBA) is a new type of advertising method that combines mobile advertising with LBS. LBA was defined as direct advertising information sent to mobile device users by an identified service provider who is specific to the location of the user (Ramaprasad Unni & Robert Harmon, 2007). There are basically two types in delivering LBA - pull and push advertising (Paavilainen & Jouni, 2002). Pull LBA refers to sending advertising messages to the users' devices in a certain location only when users explicitly requested; whereas push LBA is delivered to users' mobile phones on the basis of the users' location and previously known product preferences (Paavilainen & Jouni, 2002). In a research by Ramaprasad Unni (2007), the authors stated that the advertising effect of pull LBA is better than push LBA. However, perceived benefits of LBA and the intention to use the service seem to be very low.

Furthermore privacy protection is the most concerned issue when using of location-aware mobile technologies. Because privacy policies are so obscure that users often feel vulnerable when they fail to know whether their locational information is shared or not, since they either do not understand the complicated policies or do not know how their privacy is leaked to ulterior business entities (de Souza e Silva & Frith, 2012). Some scholars proposed solutions on privacy protection. Sharad and Animesh (2010) introduced the "matching service" which exchanges encrypted data from various entities, so that each part could only have limited access to the users' data. Xu (2012) examined the effects of three privacy protection methods, namely government regulations, user self-protection and industry self-regulation, and suggested that perceived control over the context of users' locational data is the key in handling LBS privacy issues.

Rao (2011) indicated how far consumers will go to purchase a good deal. When considering a $100 product, 55% of consumers would travel up to 15 minutes. When given 10% discount, 45% would go 30 minutes. And if a 50% discount is given, 40% consumers would travel an hour. Accordingly, some companies in the USA (such as Kmart and Nature's Recipe) have contracted with AT&T to use geo-fencing location technology to inform consumers who are 2.5 kilometers away from a store. This study also stated that such location-based advertising also generates higher premium of cost per thousand impressions (CPM) for advertisers.

Other studies focused on the various factors influencing the impact of location-based advertising. These factors involved the context advertisements are received in leisure or work situation, and at a private or public space (Banerjee, et al., 2008; Benisch, 2010).Some disclosed that personal characteristics such as gender and attitude towards LBA were also relevant. Some claimed that subjective norms, perceived control and confidence (self-efficacy) in use play critical positive roles (Bruner & Kumar, 2007).

Generally speaking there are four techniques which help advertisers to target consumers accurately (Verve, 2012).The targeting function of LBA is the most frequently listed advantage for advertisers. It is more efficient for advertisers to arrange media so that resources will not be wasted in places where the advertisees have little or even no interests.

1. Geo-Aware

The geo-aware technique uses real-time location information to send specific and dynamic messages to users on the basis of their distance from the nearest business/service location or vicinity to a certain location.

2. Geo-Fencing

Advertisers take advantage of this technique to target users according to a set distance from a place of interest including a shop or a site where known users frequently visit. Users out of the geo-fence are not served with advertisements because they are not accounted inside the location parameters.

3. Location-Based Consumer Data Targeting

Advertisers target potential consumers by utilizing third party information like demographics and transactional patterns and then combine them with highly accurate location information to decide consumer clusters.

4. Place-Based Targeting

Place-based targeting is among the most advanced LBA techniques. Advertisers target consumers in accordance with consumers' fractions which are generated according to their location and time. For example, Verve[1] has been working with a third party information provider, and created a new platform which separated the entire US into small grids and connected multiple information points with every grid. So when advertisers want to target a certain type of consumer, the platform will suggest targeting the grids which have the highest scores for that type.

DEVELOPMENT OF LBS IN CHINA

The first modern commercial LBS was initiated by DoCoMo Company in Japan in 2001. Since then, LBS has been widely spread and gained magnificent popularity around the world. Especially the release of Apple's iPhone 3 and Google's Android system have enabled developers to introduce LBSs to millions of consumers. According to the 2008 fourth-quarter report from Nielsen Mobile, LBSs accounted for 58% of the total download applications revenue for mobile phones in North America[2]. A research report by Jungwon Min (2008) pointed out that the key factors to the LBS success were compelling applications, handset availability, Internet-friendly, good business model and user awareness and promotion.

The development of LBS in China followed a similar path of Internet and social media expansion. Chinese companies just adopted the successful business models of American companies and transformed them into the Chinese market. After LBS corporations such as Foursquare, Yelp and Facebook Check-in became popular in the United States; Chinese counterparts then imitated the business models and successfully adopted them in the Chinese market. Mostly these LBS companies

Figure 1. The growth of LBS users in China
Source: 2012Q2 China LBS Market Report by iiMedia Research URL: http://www.iimedia.cn/32655.html

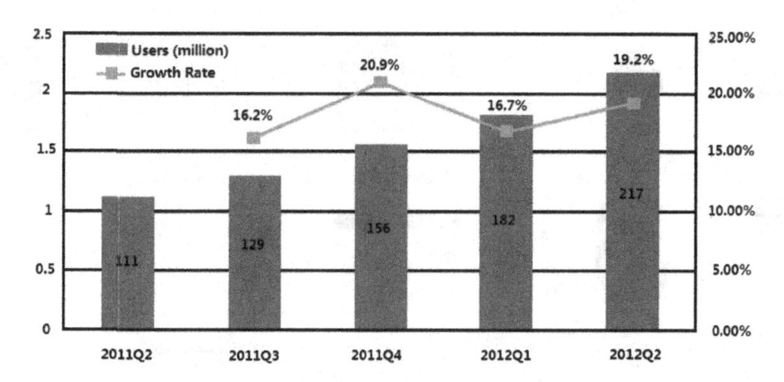

copied the business model of Foursquare, but they also introduced some innovative attributes to fit into the Chinese market. For example, Youwan was dedicated to assist users to track, find and share the information of travel and entertainment. Ququxiaoshi.com enabled users to form different communities so that members could share ideas and organize events with nearby users. However, after the initial interest in the new applications, LBS websites/applications are losing their users because users are inclined to get bored by pure check-ins. According to a survey conducted by iiMedia, seen in Figure 1, for the second quarter of 2012, LBS users in China reached 217 million; but only 75.95 million of them were active users.[3]

According to the report, 33% of the users engaged in map service and orientation, 29.2% of them were for social needs and 9.1% were merely trend followers. Discount bargain hunters stood for 14.5% and the rest 14.2% had other purposes, seen in Figure 2.

The development history of LBS in China can be divided into three stages: the First Generation (2001-2007), the Second Generation (2008-2010) and the Third Generation (2011-present).

Figure 2. Purposes of using LBS
Source: 2012Q2 China LBS Market Report by iiMedia Research. URL: http://www.iimedia.cn/32655.html

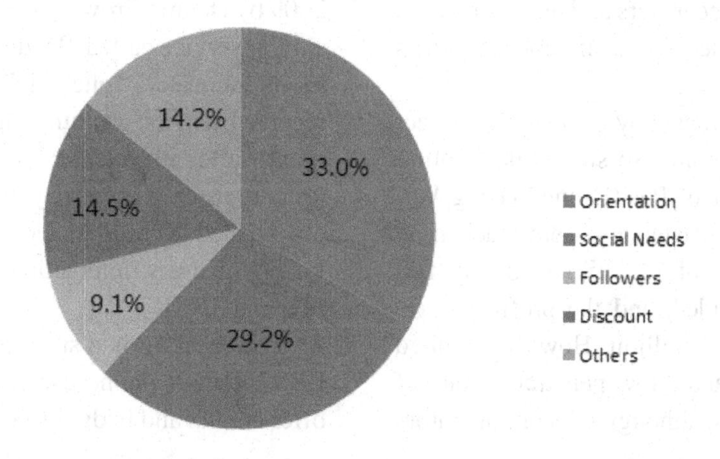

Figure 3. 2nd generation LBS in China, but only few are still active

First Generation: Telecom Operators Took the Lead (2001—2007)

In November 2002, China's biggest telecom operator China Mobile launched the first location-based service. This service contained three functions: "where are you (to locate your friends)", "where am I (to find user's location)" and "search neighborhood (to inquire about the information of the place where users are in vicinity)". China Mobile claimed that the personalized location service were able to protect users' privacy, the positioning function could be manually switched on and off by the users. However due to the uneven density of base stations, the accuracy error could be as large as a few hundred meters and the error would be even larger if the user is in suburban areas (iiMedia, 2011).

In July 2003, attracted by the bright prospect of LBS, China Unicom also started its location service called "Star of Positioning" (Ding Wei Zhi Xing). One year later, its users reached an astonishing number of 1.2 million, nearly 100 services were included; and the profit was reported to be CNY15 million. However, limited by the bandwidth and a low penetration rate of GPS mobile phones, although telecom operators were enthusiastic, the market demand for LBS remained low (Sina, 2003).

Second Generation: Explosive Growth (2008—2010)

Owing to the popularization of GPS mobile phones, users began to perceive the usefulness of LBS. Especially with the advent of 3G and smart phones, LBS-based social networking services, e-commerce, games and other value-added services emerged in an endless stream. One of the most distinctive catalysts for the explosive growth in the Chinese market was the smashing success of Foursquare. Foursquare was created in March 2009 by Dennis Crowley who previously set up a similar service called "Dodgeball"[4]. Empowered by its innovative "check-in" model, Foursquare received its first venture capital investment six month after its release; averaging 15,000 new users registered every day. By April 2012 its users reached 20 million. Foursquare successfully linked the users' interaction with their real location and SNS.

The check-in is a specific feature offered by LBS to smart phone users. LBS providers will offer credits and badges to reward the user when

Figure 4. Digu's partners

he or she checks in to a specific place by a smart phone application, which will use the phone's GPS or base station signal to posit the current location. With check-ins, users could get gifts or discounts through the way of credits or badges. Meanwhile, the LBS providers can carry out target marketing for companies.

Stimulated by Foursquare's success, Chinese companies started to imitate the business models from the US and creatively adapted them into the Chinese market. For example, Digu was one of the successful companies. Founded in 2009, Digu was just a Twitter-like microblog website at the beginning. However, it served as a platform for various third-party applications and mobile clients. This made Digu a popular website. However since 2010, big players of Sina, Sohu, Tencent, Netease and other portal sites started to launch their own microblog websites. These portal giants took advantage of their huge number of users and quickly carved up the market. Under such pressure, Digu switched itself from microblog to a website in July 2010 that specifically "enables users to get and share more happiness and benefits by check-ins" (Tencent, 2010). This indicated that Digu turned into a LBS website. Its transition could be considered a great success, according to 2011Q3 report from EnfoDesk[5], Digu occupied the largest market share of 20.8%. The number of its users was more than 2 million. It partnered with many famous international brands such as MacDonald's, MoTo, Nike, Pepsi, Lenovo, FAB, etc. (Baike, 2013). One of the illustrative examples of Digu's operational model is with Starbucks. Digu will offer a Starbucks badge to the user if he or she checks in a Starbucks' shop and shares its promotion activity on Sina blog or Renren. With this badge, the user will get a free cup of coffee when spending more than CNY50 in Starbucks. And if the user spent more than CNY50 for seven times, Starbucks will award the user a coffee cup. Furthermore if the user checks in Starbucks shop for five times during a certain period, he or she can enjoy a 20% discount (Digu, 2010). Basically the business model of all the LBS companies in this period was same as that of Foursquare.

Third Generation: Emergence of New Lbs (2011—Now)

Since 2011, many LBS websites and applications started losing their users because of a decline in users' viscosity. During this period, many small LBS providers ceased to exist or turned themselves to other business domains. In October 2011, even

Figure 5. Jiepang

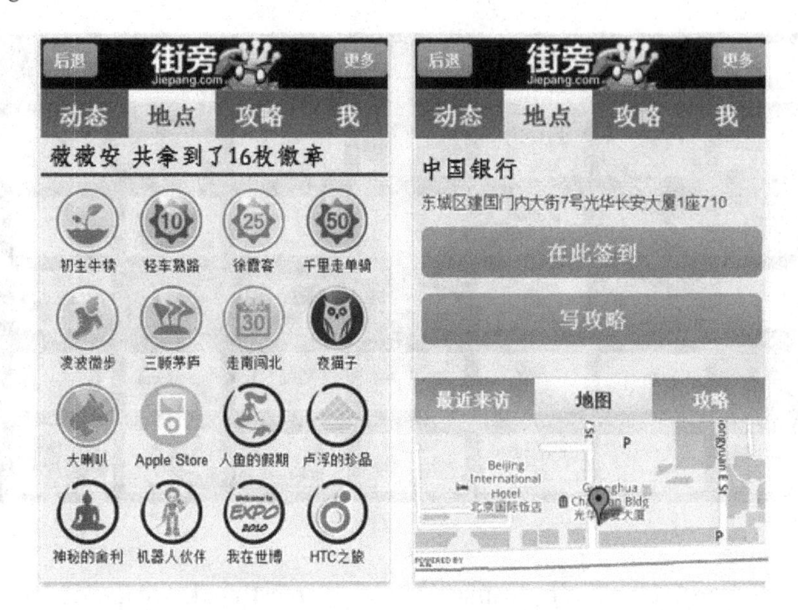

Digu, the most successful LBS company in China, cut 70% of its staff (Southern Daily, 2011).

Another representative example is Jiepang. Jiepang was launched in 2010 and was the third popular LBS in China. After a high-speed development in the early stage, Jiepang did win the favor from business partners such as Nike, Louis Vuitton, Burberry and etc. However some business insiders pointed out that even in its flourishing stage, most of cooperation between Jiepang and brand partners were free; the price to advertise on Jiepang was only between CNY50,000-100,000 (Cyzone, 2012). Moreover, most companies were quite reserved about doing business with independent LBS providers. Normally they preferred to cooperate with multiple companies simultaneously so that more consumers could be covered.

Meanwhile the value of virtual check-in as a sustaining business model has been questioned. Offering badges alone could not keep users' viscosity; and once users linked check-ins with microblog, his or her location privacy would be in danger. However, the most crushing blow to independent LBS companies came from super players such as Sina, QQ and other big players, who already had a large number of users and possessed multiple services other than LBS. These players made Jiepang, which was a pure check-in LBS company, less useful.

Such phenomenon can be attributed to several reasons:

1. The independent LBS players have been threatened by giant corporations. Giant corporations, especially Internet social media corporations have already had their own extensive user networks. It is much easier and more cost-effective for them to have instant user groups than independent LBS companies who have to establish users' network from the very beginning.
2. The single "check-in" model cannot keep users' interest. In China, the LBS business model predominantly relies on the "check-in" type, which could not attain users' interest for a long time. Even Foursquare confessed that check-in is not the best and mature business model. (AllthingsD, 2013)

It is reported that some industry experts considered that the future of LBS business lies in new business models. Since check-ins cannot offer enough user viscidity, more combinations should be tried to create new attractions, such as LBS+SNS, LBS+photo sharing, LBS+groupon, etc. All of these innovations are connected to the emergence of the 3rd generation LBS. (ifeng, 2010)

One of the examples is Lashou, which is considered a shopping app. It created the G+F model (Groupon+Foursquare) which enables users to search the nearby business, become the fan of it, and check in both online and offline. After becoming a VIP of a business, the user could enjoy discounts and other benefits. Furthermore, the badges and credits won by check-ins could also be used to exchange for the shopping quota. By doing so, the users are driven to check in and win the credits continuously. Lashou integrates both online and offline resources, pushes users with discounts and group purchase of entertainment, dining, cosmetics and other products. Except the group purchase, Lashou also upgrades its shopping experience by paying attention to the product quality, credibility and after sales service. In order to enhance consumers' loyalty, Lashou launched several attractive services such as a call center, experience store and etc. By September 2011, its registered users reached 16.8 million; among them 3.3 million were active users. (163, 2011)

Another novel model is Game+LBS. In 2010, Wiselinc launched its smart phone application called "16 fun". This is Monopoly-style[6] which allows users to get credits and upgrades by checking in. With enough credits and levels, the user could "buy" the place with prospects and charge other users who passed by. Through periodic calculation and statistics, the value of every place fluctuates according to the users' activity, which brings the users a feeling of connecting with the reality. The users could win gifts by check-ins and buy tools from virtual shops in the game. The users could also email and invite friends to play the game together.

Another example, which is also considered as the most successful 3rd generation LBS, is Momo. Momo, a location-sharing SNS app was inaugurated on August 3, 2011. Different from other LBS+SNS applications, it allows users to see others in the vicinity as well as provides precise proximity to others by GPS. Less than one month after its release, Momo already occupied the No.3 place among the free social networking apps in the App Store. On its 1st anniversary, the number of its users exceeded 10 million; and now it has more than 50 million users worldwide (Baike, 2013).Apart from the precise positioning function, its success could be attributed to several factors. First the user does not have to apply to be a friend of another user (which means apply first and chat later); the user could start the conversation immediately with anyone. By doing so, the frustration of being rejected, especially for male users, decreases.[7] Second it integrates with SinaWeibo and Renren, so users can browse others' timeline or profile within Momo to obtain more information about people in mutual proximity. The users could also switch themselves off as "invisible" to avoid annoying disturbance. Furthermore one product placement advertising of Momo, which described Momo as a hook-up tool, also attracted people's attention.[8]

CASE STUDY: INDEPENDENT LBS PROVIDER "SBK"

A case study of a LBS app "SBK"[9] has been conducted since summer 2013. The information provided in this section is based on a series of interviews with managers of the company. A full-scale user analysis is still under design and will not be provided in this study.

The background of the company: The company is a commercial metropolitan weekly (DCCB), a subsidiary of a provincial newspaper group in the capital city of a Southwestern province. The weekly, established in 1998, aims at middle income urban readers, providing in-depth national and local news coverage, transportation information, fashion, entertainment and consumer lifestyle. The weekly carries a mobile phone version and is currently running a LBS app "SBK" as one of its mobile service. The SBK is the first LBS ever engaged by the company.

The interviewees were at the management level. The person who provided a major part of the information was the Deputy Editor of DCCB.

Background of the LBS app, SBK: The application, SBK, has a literary meaning: customers at your side. It was originally developed by university students of IT technology. DCCB discovered it and contacted the team to request a customized design. After integrating with the information emission device, the company then purchased the app at CNY30, 000 and obtained a national patent. This app, focusing on a particular hot tourist area, uses three communication base stations to position within a radius of three kilometers all the users of mobile phones QQ, and WeChat. It then provides services of information search and dissemination to these users within its reach. It has a capacity of information emission of 500 messages per hour. This app is primarily intended for tourists to the province aged between 18-35 years old. The advantage of SBK is that the company has obtained a national patent and all its intellectual property rights. Other companies, who later want to get involved in this kind of service, should also obtain consent from the company.

The operational team has 12 employees: three are in charge of marketing, stationed in the target city permanently. Two for technical support, one in charge of advertisement production, one editor, and two administrative staff. The group also has three partners in the target city.

LBS App Development

There are a large number of student teams in the Beijing Post and Telecommunication University and Chongqing Post and Telecommunication University who work on smartphone app development. They can be found in many IT forums such as the App Community (http://BBS.appmedia.cn), Apple Forum (http://BBS.app111.com), 58 (http://www.58.com), Zhubajie Net (http://www.zhubajie.com). Anyone who needs an app development can post a message in these forums. In many cases, just a single contract message will attract a large number of replies from many teams.

The Target and Push Advertising

SBK represents a type of target and push advertising. It does not require a check-in. SBK will first locate accommodations of tourists, and then differentiate them into various levels of consumer groups based on the grades of the accommodations. If customers stay at the lower end inns and budget hotels, they would be sent advertising of cheap bars, or discounts and promotional activities. On the other hand, if customers are located at expensive hotels, they would be sent advertising of high-end products like air travel packages.

SBK was launched in September 2012. The company recruited a special team for marketing. They targeted business entities within the location covered by the LBS who look for tourists such as hotels, bars, restaurants and entertainment places. The promotion efforts were predominantly offline. The marketing team would visit these businesses personally, exhibit the data and PPT presentation, and recommend them to have a try. By July 2013, SBK had 600 information emitters working daily, serving more than 2,000 merchants. Among the customers, about 100 bought directly from the SBK provider the information emitter so that they are able to send out advertising to users by themselves. Others entrusted the company to send out advertising messages on their behalf.

SBK's business model relies on monthly fees charged to businesses who subscribe to the service. There are three tiers of fees, Type A, B, and C. At the time of the case study, the service was still in its initial promotional period with all the prices offered at a 50% discount. Type A customers are charged a monthly fee of CNY8000. The SBK service would cover the areas that are mostly frequented by tourists including the airport, railway station, long distance bus station, major tourist attractions and shopping avenues in the city proper. Type B customers are charged a monthly fee of CNY5000 with the service covering only the city proper. The Type C service is CNY2000 a month, covering a number of particular locations like specific streets or scenic spots. According to the company statistics, the 2000 merchant customers account for 20-30% of the total number of restaurants, bars, entertainment places and tourist accommodations in the city. More than 400 companies subscribed to the service for more than three months.

The company received feedback from merchants every week on the effect of advertisements. For example, for Type A customers, SBK sent 7,000 messages per day (1,000 per hour, 7 hours per day). According to the feedback, the ratio between the number of messages and users' response was 1.5%.

Advertisements were designed and produced as multimedia messages. They were made to fit the display attributes of mobile phones; for instance, the size of a message must be small so that users were able to open easily, read conveniently with only one glance in one page. The ads were designed and produced by the in-house staff.

The Business Model

The check-in model like Foursquare is not popular in China. In China, the usual business model is: discover an app > purchase the app > promote the LBS > attract venture capitals > sell the business to a bigger investor. A company would first investigate an app, then purchase the app and started a promotion campaign for the service. The goal is to attract venture capitals with a larger size of customers. Generally speaking, any app which registers more than 50,000 downloads in the Appstore could be sold to venture capital investors. A smaller investor, after acquiring the business, would continue the promotion, aiming to attract bigger investors. Several known LBS businesses like Momo followed this model.

DCCB's plan for SBK follows the previously-stated model. Said Mr. Li, the Deputy Chief Editor:

Our promotional period for SBK is set to be three years. Our primary goal is to achieve as many customers as possible. With a large number of customers, we can obtain more venture capital. We have to win over more customers so that they can have a better understanding of our service and occupy as much as possible a larger share of the market before SBK is being eliminated by similar services with the next generation technology. Moreover, we need to take advantage of our customer resources, quickly develop the next generation of LBSs and adapt them into the market. (Li, 2013)

The company has a three-step plan for the development of LBS. First it plans to continue expanding SBK so that its service could cover several major tourist cities and even the whole province. Second, it plans to develop and increase more features and functions in SBK, for example, to increase SBK's ability to provide more tourist and business information. Third, based on the model of SBK, it will develop new LBS applications so that it could link to the WeChat platform. SBK is not yet a public account in WeChat. The company plans to develop not only a public account of SBK, but also as a third-party app plugged in WeChat.

In addition to the above model, another business model is called "information marketing". It means with a LBS app similar to SBK, a company would perform mobile users' data collection and classification, and sell them directly to merchants. It is up to the merchants' disposition to use these data. In this case, all the users' private data will be fully disclosed. In China, such practice is relatively common although it is operated within a gray area of legality, according to Mr. Li.

Mr. Li concluded, approaches that are applicable domestically do not work in foreign countries, such as the approach of "information marketing". However, approaches that are popular in other countries may not work in China, such as Foursquare's "check-in" and paid apps.

Problems and Challenges

The biggest challenge, according to Mr. Li, is that their products are not accepted by users and businesses. Although the company offers a wide range of free experience activities, promotional campaigns are still very difficult. Most people view the functions of mobile phones only at the level of communication and entertainment, and have little understanding of their commercial functions.

They also believed that their platform is not big enough.

If SBK could be promoted in the national platforms like Sohu, Sina, the outcome would be much better than it is now. However, in that case, the costs would be very high. Our products have no problem in the technical aspect. The major obstacle is that it is run on too small a platform. Additionally, although the province is known for tourism, the market demand for our products is not big enough. If it is in Beijing, it will have a much larger market than it is now. (Li, 2013)

There are problems at the macro level as well. China's digital map was produced and published by the state government. The government has unified standards and specifications for digital maps. However, the government does not have similar standards for mobile phone maps. At the same time, the government does not permit non-governmental surveying and mapping, and the publication of maps with scale below 1:3200. Therefore, LBSs in China cannot freely publish precise location data as LBSs on Google maps in foreign countries do. In China, LBS positioning is based largely on telecommunication base stations. Limited by the distribution of base stations, positioning error value is very high, from dozens of meters to hundreds of meters, and in some areas even thousands of meters.

In Mr. Li's opinion, the LBS market in China is not very optimistic because: first, consumers demand for LBS is low. Second, technology is not as good as Google maps. Third, the insufficient bandwidth makes the connection speed not fast enough, it is difficult to meet the requirements of LBS. Another factor that obstructs the LBS development in China is that Chinese mobile phone users were very much repulsed by the spam bombardment in previous years. They are very suspicious of LBS right now. Mr. Li attributed the lesser development of LBS in China to four reasons: 1) there are many Internet policy barriers; 2) the bandwidth is insufficient, 3) there is a lack of mature business models, and 4) laws in related areas are inadequate.

DISCUSSIONS AND CONCLUSION

The case study, although with an analysis of smartphone users still being undertaken, is already explicit of several important points:

1. The LBS is distinguished in its intrinsic quality of "networked locality", which provides affordance for the dissemination and communication of location-based and location-oriented information. It enables the propagation of local information, specifically commercial information of local businesses to people who are in physical proximity. In particular, it functions to position people who move into the area. Before the LBS, location-based merchants were unable to acquire information about such group of

people. They did not know who they were, where they were, when they would move into this area, where they would move about and what their consumption patterns were. Location-based applications provide technical affordance for a precise positioning of such moving population and timely and target distribution of commercial information to these people on the move.

2. The LBS brings forth a new form of advertising for business entities which engage in location-based services and cater particularly to a mobile population such as tourists. Marketers who are most interested in LBA are in the tourism industry, such as restaurants, bars, hotels, scenic attractions, taxi services and so on. They rely on "real" people, not "virtual clients" to physically come into their premises to consume. Thus they have a demand for positioning and advertising of potential customers who migrate into the vicinity. They are major customers for commercial LBSs like SBK.
3. Push advertising still dominates in our case study. The advertising is not performed via user-initiated check-ins. Instead it is similar to the mass dissemination of SMS to mobile phones. The location-based application takes advantage of positioning technology and automatically locates smartphone users who move into the proximity. It then uses information emitters to propagate commercial messages directly to these users. However, as expected, SBK users' locative mobile mediated interactions are very limited; users accepted them passively and responded with a very low rate. In our case, the response rate is only 1.5%. Such phenomenon resulted from the deficient understanding of mobile phones' business functions and the spam bombardment.

   The location-based application in our study can automatically perform some kind of classification of the users, specifically their consumption levels for target advertising. The app categorizes the users' spending capacity based on the accommodations they stay in and locations they move about. Accordingly the marketers decide on what types of advertisements to be disseminated. This represents a clear example of target advertising with the new media technology.
4. There is evidence that the technological affordance of "networkedness" in the LBS has not been tapped in our case study. As we elaborated previously that the locative media is distinctive in its nature of being "networked". Every smartphone user is linked to his/her own network of social and virtual friends. However, in the case of SBK, the marketers did not seem to pay too much attention to the characteristic of "networkedness" in their marketing strategy. They were fully dependent on mass messaging to phone users, rather than taking advantage of the users' social connectivity for advertising propagation.

This is in contrast with the strategy of some other LBSs. In another case study of ours, which is on a taxi LBS, we have discovered an interesting pattern of "networked" advertising. In this case, a taxi driver had a wireless router in his car. He would encourage passengers to download the taxi app in his car and then be awarded CNY10 when a passenger starts using the app for taxi service for the first time. According to him, he received around CNY300 per month through this business alone. In this case, individual taxi drivers play a marketing role within the networked circle of his clientele.

To conclude, this study analyzes one type of LBA in China, especially from the marketers' perspective. However, we would be cautious to claim its representativeness. The limitations of the research are obvious. First, a survey of smartphone users in the SBK case who are target of LBA is necessarily important for a greater understanding of user patterns of LBS and effect of push advertising. We are currently working on the second phase of the investigation. Second, a comprehensive study is required on an inquiry of a large number of LBA cases. There are a multitude of LBS applications offering a variety of functions and services, this essay provides an examination of only one of them. For further research, more cases in various types of LBS should be included. Finally, this case study applies only a qualitative method of interviews. Our future study plans to employ a method of app analytics to collect metric data for quantitatively measuring interactivity between phone users and apps so that it is possible to obtain a deeper understanding of how users perceive and react to LBA. In other words, a quantitative app analysis is necessary to have a full evaluation of how LBA influences users and how users evaluate, respond and interact with commercial information they receive on their smartphones.

REFERENCES

Allthings, D. (2013). *CEO Dennis Crowley on Foursquare's biggest mistake*. Retrieved on May 24, 2013 from http://allthingsd.com/20130311/ceo-dennis-crowley-on-foursquares-biggest-mistake/

Baike. (2013). *Digu*. Retrieved on June 1, 2013 from http://baike.baidu.com/link?url=pW9Mf-Oj0SMJlRkwg76CBC8jMxI-yDVAMrK-i9gxZfb-wJJjHGdp9RbTDbuEy0Ew-2lCaFxkQJOqTCyo5KluctGQ0l2cj8YzicuyUnL-lO8Rq-SopP5pdweKBdtD_U1wIdr

Baike. (2013). *Momo*. Retrieved on June 1, 2013 from http://baike.baidu.com/link?url=j7M3z1E-i9fUKrBmSBwkmgDnWQdlhRvrnSb74U-dy70t4QdQG1JIA19IiHfGorPx6

Banerjee, S. A. (2008). Does location based advertising work. *International Journal of Mobile Marketing*, *3*(2), 68–75.

Benisch, M. K. (2010). Capturing location-privacy preferences: Quantifying accuracy and user-burden tradeoffs. *Journal of Personal and Ubiquitous Computing*, *15*(7), 679–694. doi:10.1007/s00779-010-0346-0

Bruner, G. A. (2007). Attitude towards location based advertising. *Journal of Interactive Advertising*, *7*(2), 3–15. doi:10.1080/15252019.2007.10722127

Cyzone. (2012). *Dilemma of Jiepang's check-in*. Retrieved on May 18, 2013 from http://www.cyzone.cn/a/20120425/226227.html

(2011). *Data of Lashu.net*. Retrieved on June 14, 2013 from http://tech.163.com/11/1029/08/7HH5IFRJ000915BF.html

de Souza e Silva, A., & Frith, J. (2010). Locational privacy in public spaces: Media discourses on location-aware mobile technologies. *Communication, Culture & Critique*, *3*(4), 503–525. doi:10.1111/j.1753-9137.2010.01083.x

de Souza e Silva, A., & Frith, J. (2010). Locative mobile social networks: Mapping communication and location in urban spaces. *Mobilities*, *5*(4), 485–505. doi:10.1080/17450101.2010.510332

de Souza e Silva. (2013). Location-aware mobile technologies: Historical, social and spatial approaches. *Mobile Media & Communication*, *1*(1), 116–121. doi:10.1177/2050157912459492

Dibdin, P. (2001). *Where are Mobile Location Based Services*. Retrieved on April 3, 2013 from http://mms.ecs.soton.ac.uk/mms2002/papers/4.pdf?origin=publication_detail

Digu. (2010). *Digu and Starbucks*. Retrieved on April 3, 2013 from http://wenku.baidu.com/view/1f943749f7ec4afe04a1df53.html

Foursquare. (2013). In *Wikipedia*. Retrieved on April 5, 2013 from http://en.wikipedia.org/wiki/Foursquare

Gordon, E., & de Souza e Silva, A. (2011). *Net locality: Why location matters in a networked world.* Boston, MA: Blackwell Publishers.

ifeng. (2010). *The enlightenment from Foursquare*. Retrieved on May 6, 2013 from http://tech.ifeng.com/magazine/local/detail_2010_12/21/3615387_0.shtml

iiMedia. (2011). *What lies behind LBS.* Retrieved on June 20, 2013 from http://www.iimedia.cn/16056_2.html

Jungwon, M., Byung, K., & Shu, W. (2008). *Location Based Services for Mobiles: Technologies and Standards.* Retrieved on May 12, 2013 from http://blue-penguin.org/cache/location-based-services-for-mobiles.pdf

Leppäniemi, M. S. (2006). A review of mobile marketing research. *International Journal of Mobile Marketing*, *1*(1), 30–40.

Licoppe, C., & Inada, Y. (2006). Emergent uses of a location aware multiplayer game: The interactional consequences of mediated encounters. *Mobilities*, *1*(1), 39–61. doi:10.1080/17450100500489221

Okazaki, S. (2004). How do Japanese Consumers Perceive Wireless Ads? A Multivariate Analysis. *International Journal of Advertising*, *23*(4), 429–454.

Paavilainen & Jouni. (2002). *Mobile Business Strategies*. London: IT Press.

Rainie, L., & Wellman, B. (2012). *Networked: The new social operating system*. Cambridge, MA: MIT Press.

Ramaprasad, U., & Robert, H. (2003). Location-Based Services: Models for Strategy Development in M-Commerce. In Proceedings of Management of Engineering and Technology (pp. 416-424). IEEE.

Ramaprasad, U., & Robert, H. (2007). Perceived Effectiveness of Push vs. Pull Mobile Location Based Advertising. *Journal of Interactive Advertising*, *7*(2), 28–40. doi:10.1080/15252019.2007.10722129

Rao. (2011). *Discovering the distance to discount ratio*. Retrieved on July 8, 2013 from http://techcrunch.com/2011/02/04/distance-discount-ratio/

Rheingold, H. (2002). *Smart mobs*. New York: Basic Books.

Sina. (2003). *China Unicom to start new business*. Retrieved on July 1, 2013 from http://tech.sina.com.cn/it/t/2003-08-14/1420221020.shtml

Southern Daily. (2011). *Digu to cut off its 70% staff.* Retrieved on July 2, 2013 from http://epaper.nfdaily.cn/html/2011-10/19/content_7016439.htm

Tang, K., Lin, J., Hong, J., Siewiorek, D., & Sadeh, N. (2010). Rethinking location sharing: Exploring the implications of social-driven vs purpose-driven location sharing. [Copenhagen, Denmark: ACM Press.]. *Proceedings of UbiComp, 2010*, 85–94. doi:10.1145/1864349.1864363

Tencent. (2010). *Digu to transform from Weibo to LBS.* Retrieved on May 5, 2013 from http://tech.qq.com/a/20101104/000171.htm

Teri, R. (2000). *Wireless Marketing is about Location, Location, Location.* Retrieved on May 14, 2013 from http://www.dimensiondata.com/Global/Downloadable%20Documents/Location%20Location%20Location%20-%20Using%20Wireless%20to%20Deliver%20Process%20Improvement%20and%20Innovation%20Opinion%20Piece.pdf

Verve. (2012). *Location Powered Mobile Advertising Report 2012 Annual Review.* Retrieved on June 15, 2013 from http://www.vervemobile.com/pdfs/LIR/LIR_web.pdf

Wellman & Rainie. (2013). If Romeo and Juliet Had Mobile Phones. *Mobile Media & Communication*, *1*(1), 166–171. doi:10.1177/2050157912459505

KEY TERMS AND DEFINITIONS

App: An app is a piece of software programmed to fulfill a certain purpose on computer, smartphone, PDA or other electronic devices.

Check-In: The check-in is a specific feature offered by LBS which enables smart phone users to report their location and arrival time.

Locality: The quality or condition of having a location in time and space.

Location-Aware: It refers to the mobile app which can determine the physical location of a device.

Location-Based Advertising: It is a service or mobile app that offers advertisements to a mobile user depending on the user's physical position.

Location-Based Service: It is a type of service that makes use of geographical position of a mobile user.

Mobile Advertising: Mobile advertising is a form of advertising that orients toward the mobile phone users.

Networked: Connected with Internet, Wi-Fi or phone network.

Privacy: The condition of secrecy, especially refers to the spatial, time and other individual information of a mobile phone user in this chapter.

Smartphone: Smartphone is a mobile phone run on certain operating system such as Android and iOS. It possesses the functions of feature phone and more advanced computing capability.

ENDNOTES

1. Verve is the largest location-based mobile advertising business in the US.
2. Source: http://www.networksinmotion.com/newsroom/02_11_2008_Nielsen_nim_Q4.php
3. Source: http://www.iimedia.cn/32655.html
4. Dodgeball was the first LBS in the US; it was based on SMS which enabled users to send their location and then receive nearby friends and interesting information. Dodgeball was taken over by Google in 2005, discontinued in 2009, and then replaced by Google Latitude.
5. Established in 2000, EnfoDesk engages in providing data, information, and advice to the senior management around the world in the industries of the Internet and information technology, telecom operators, investment organizations, government agencies, etc.
6. Monopoly is a game that enables players with buying, trading and developing properties such as hotels and charge rent from other players.
7. This service thus is popular with male users as Chinese female users tend to be more conservative in being added into male users' friend list. If a male user wants to add a female to his friend list, the possibility of being rejected is relatively high. However, this service may also pose a problem of privacy intrusion.
8. See this advertisement at http://www.56.com/u89/v_OTg5MDgxMTg.html
9. The company requested to be anonymous for privacy protection. Therefore all names and brands used in this case are acronyms or pseudo names.

Chapter 14
"Checking Into" Outdoor Lifestyle?
Mobile Location-Based Games as a Site of Productive Play in Marketing Campaigns

Elaine Jing Zhao
University of New South Wales, Australia

ABSTRACT

The increasing adoption of mobile media in place-making and space-constructing practices as well as participatory culture open up new opportunities in marketing. This chapter approaches mobile location-based games as sites of brand experience, productive play, and value co-creation. It examines a mobile location-based game launched by The North Face (TNF) in China to increase its consumer base by shaping people's lifestyle aspirations. It offers insights into the use of mobile to engage more participants in the emerging market; the role of mobile as the central tool, platform, and the interface of hybrid spaces (de Souza e Silva, 2006) in the integrated campaign; and the value co-creation process by consumers for the brand. It also points to unintended consequences of the campaign, evidenced in consumers' deconstruction and reconstruction of the marketer-designed game process and their distortion of space. More generally, it reveals the importance of understanding the opportunities and challenges in addressing creative consumers in utilising mobile location-based game for marketing purposes.

INTRODUCTION

The rapid development of mobile and Internet technologies means that many people now live increasingly hybrid lives where the physical and the digital, the real and the virtual, interact. While earlier studies suggest that mobile phones withdraw users from physical space (Gergen, 2002; Puro, 2002), the need to transcend the online-offline dichotomy has become widely acknowledged now. For example, de Souza e Silva (2006) argues that a hybrid space arises when mobile technologies are used as social devices, resulting in the merging of borders between physical and digital spaces. According to de Souza e Silva, mobile phones strengthen users' connections to physical space by enfolding remote contexts inside the present context, and promoting sociability and

DOI: 10.4018/978-1-4666-6166-0.ch014

communication in urban spaces. Moreover, a new mobility is arising from the mixing of physical and virtual mobility and breeds new forms of places as a result of the relationship between informational and other territories that constitute them (Lemos, 2010). In discussing what they call 'net locality', Gordon and de Souza e Silva (2011) point out that location awareness as a social agent creates a geographical context to networked data and facilitates interactions among physical and virtual communities.

Mobile location-based services such as Foursquare, launched in March 2009 in the U.S. are one example of bridging online and offline spaces. As users share their locations with friends by 'checking in' at various physical venues via a smart phone application or by text message, they are awarded points and can collect virtual badges as cultural capital. For example, users who check in most times at a certain venue will be crowned 'Mayor' until someone supersedes them. A variety of virtual badges are awarded to people who check in at places more frequently, with more friends, or at a particular type of place. The rapid growth of Foursquare has led Facebook to enter the market with its own service, dubbed 'Places', which allows users signal where they are to their friends on the network. Location sharing has thus become embedded in major social networks. In addition to the increasing adoption of virtual items as a gaming element in location-based services, there are a growing number of location-based mobile games and hybrid reality games. As de Souza e Silva and Hjorth (2009) argue, these games erode the notion of a magic circle or dedicated game-space. They open up new possibilities of identity building in place-making and space-constructing process. However, privacy and security remain contentious issues in these contexts (Dourish& Anderson, 2006; Wilken& Sinclair, 2009).

The culture of 'checking-in' and location-sharing via mobile phones has also become increasingly popular in China, the world's biggest mobile market. According to statistics from China Internet Network Information Center (CNNIC) (2013), mobile internet users reached 464 million, accounting for approximately 78.5% of all internet users by the end of June 2013. Local service providers have followed the Foursquare model and launched similar services since 2009, such as Jiepang, Kaikai and Qieke among others. These sites also use virtual badges to encourage users to use the services more. Users' enthusiasms toward virtual items are evident in China, where transactions of virtual items including avatars, garments, accessories, and gaming equipment have created a 'mini-economy' with its own culture (Castronova, 2005; Kshetri, 2009; Nystedt, 2005). With the increasing penetration of mobile internet, the real economy around virtual items in China is now extending to mobile platforms. According to statistics released by Analysys International (2011), by the end of June, 2011, the number of accumulate accounts of China location-based services has reached 105 million. The post-80s generation is the main user group of mobile locative media games, where they create new forms of ambient, emplaced, social visualities while maintaining older social ties (Hjorth & Arnold, 2013).

Given the rapid adoption of mobile and locative media, brands have also begun to tap into mobile marketing with a geographic turn. Brands such as Startbucks, PepsiCo and Mcdonald's have used the check-in service with a reward mechanism to lure customers to check in at the bricks-and-mortar stores more often and participate in events. Compared to earlier location-based marketing propositions based on monitoring, encouraging people to check in via mobile is a fundamentally different approach. A typical example of the earlier propositions is sending promotional offers via Bluetooth to people within a certain radius or proximity (Leek & Christodoulides, 2009). This kind of proximity marketing adopts a broadcasting approach in the hope of driving foot traffic to stores and realizing impulse purchases. However, that caused marketers' concerns over privacy intrusion (Okazaki & Taylor, 2008) and the risks of overuse and spam (Wilken & Sinclair, 2009). Compared to such sales-driven efforts in proximity marketing, location-based marketing based on 'self-positioning' (Benford et al., 2004) via the mobile interface is a step forward. However, the commercialisation of such services and the goal-oriented nature in collecting locations have raised concerns about dilution of the playful qualities of the platform (Gazzard, 2011).

Along with the growing use of mobile and locative media among consumers, researchers (Firat & Venkatesh, 1995; Prahalad & Ramaswamy, 2003; 2004; 2005; Sheth & Uslay, 2007; Venkatesh, 1999) have increasingly acknowledged consumers as creative agents participating in the co-production of value. User-generated content is an important part of value co-creation. In the case of mobile communications, it allows an efficient integration of the individual customer as a co-creator in content production owing to the inherently personal nature and portability of the mobile phone (Ito, Okabe, & Matsuda, 2005; von Hippel & Katz, 2002). Therefore user-generated content plays an increasingly important role in the consumer's relationship to the brand (Burmann & Arnould, 2009). While opportunities exist, user-generated content also brings new challenges to brand owners, who have less control over the process (Christodoulides & Jevons, 2011).

While previous studies on mobile location-based games have contributed to our understanding of the mobile interface in gaming (de Souza e Silva, 2006), mobile gaming's role in the reconstruction of urban space and new ways of experiencing space (de Souza e Silva & Hjorth, 2009; de Souza e Silva, A. & Sutko, 2008; de Souza e Silva, A. & Sutko 2011; Evans, 2011; Gordon & de Souza e Silva, 2011), there is a paucity of literature on mobile location-based games for advertising or marketing purposes. Meanwhile, in the existing advertising or marketing literature on mobile location-based services, the focus is more on users' willingness or intention to use such services (e.g., Meuli & Richard, 2013; Zhang & Mao, 2012) and limited attention has been paid to the situated brand experiences.

This chapter aims to fill the literature gap and approaches mobile location-based games as spaces of brand experience and value co-creation. It examines a mobile location-based game launched by The North Face (TNF) in China, which utilised the mobile check-in culture and the virtual items to interact with its target consumers. Through this case study, the chapter aims to reveal how a multinational corporation utilises mobile media to adapt to the emerging market in China; how consumers co-create value for the brand as they build collective identity and memory by interacting around shared values and codes of meaning embodied in the mobility; and how consumers deviate from the marketers-designed online-offline dynamics and construct their own playground online which paradoxically aids the marketing effort with a viral effect. By doing so, it reveals the importance of understanding the opportunities and challenges in addressing creative consumers in utilising mobile location-based game for marketing purposes.

The following section starts with an introduction of the background on TNF's challenges in China's emerging market. Then it presents a brief description of TNF's campaign objectives, design and the logic behind utilising the mobile locative game. Next, the article analyses users' productive play and value co-creation in the game, where they used the mobile as an interface of a hybrid space (de Souza e Silva, 2006), the central tool and platform, and served the interests of the corporate. It also examines the cross-media dynamics constructed by the marketers in online and offline spaces to enhance the effectiveness of the campaign. Following that it discusses the playful activities performed by online users as a deviation from the marketer's intention before presenting the conclusions of this study.

WHEN TNF MEETS CHINA: CHALLENGES IN AN EMERGING MARKET

Founded in the U.S. in 1968, TNF has become well-known in the west as an explorer's brand for outdoor and endurance sports over more than 40 years. Never Stop Exploring is the line behind the brand. As TNF entered the Chinese market, it faced a landscape where the outdoor industry only started to emerge in the late 1990s. Despite the short history, the outdoor products market has developed rapidly in China. According to Himfr.com (2008), one of China's leading B2B search platforms, China's outdoor clothing and equipment sales reached 260 million Yuan (around 39 million USD), the figure was a mere 10 million Yuan (around 1.5 million USD) in 2003. Outdoor professional dealers grew from 400 in 2003 to 2,125 in 2007.

However, the primary dynamics driving growth in China – rapidly rising incomes and conspicuous consumption – are different from the forces that drove the U.S. industry in the last century (Outdoor Industry Association, 2010). Common leisure activities in China include internet surfing, Karaoke, dining out, and going to night clubs. Card and board games such as Sanguosha (Killers of the Three Kingdoms) continue to be popular entertainment among people; these are cheaper options, particularly among young people (Zhou & Peng, 2010). A smaller percentage of people participate in 'regular sports,' let alone the more challenging 'outdoor sports.' Executives in the outdoor industry also notice they are hard pressed to find core outdoor enthusiasts behind the thousands of shops in China (Outdoor Industry Association, 2010). Therefore outdoor products market remains an emerging market in China.

Against this background, China is a great challenge for the company. Therefore the brand has adjusted its global positioning from 'Never Stop Exploring' to 'Start Exploring' in the China market.The corporate regarded the strategy of educating the market and encouraging outdoor participation especially important internationally as the brand expands into markets with a less defined and developed outdoor culture such as China (VF Corporation, 2010, p. 17). This could be seen as an approach of a multinational brand to adapt to the local market and make itself relevant to a broader audience.

THE VIRTUAL RED FLAG CAMPAIGN: A LOW BARRIER TO OUTDOOR

The virtual red flag campaign was launched over 18 days from 15 October to 1 Nov 2009 in China. It aimed to encourage urban dwellers, mostly the Internet explorers, to get outdoors and take the first step to 'Start Exploring'. By doing so, it hoped to connect with a larger population of potential consumers than those hardcore outdoor enthusiasts such as endurance athletes, mountain climbers and extreme skiers among others. The campaign targeted Chinese urban consumers, with a key target audience between 18 and 24 years old; inspire even non-outdoor enthusiasts to step out and venture into the unknown, and turn them into brand advocates.

The campaign evolved around a location-based game to get participants to use their mobile phones to plant virtual 'Red Flags' with short messages to claim a virtual piece of China over 18 days. The iconic action of flag planting is very established in the world of outdoor adventure, where an explorer would lay claim to a summit by planting a flag, and he would continue to do so at every other summit he manages to conquer. In a race to see who would be the conqueror, a virtual map of China was gradually covered in a sea of red flags. During the campaign, the mobile acted as the central tool and platform for individual players to participate in the game, and also facilitated a communal experience for players in both the competition and collaborative mapping.

While outdoor and endurance sports is still a distant concept for average people living in urban China, the mobile location-based game offered a lower barrier to participation. One of the frequently used marketing strategies for TNF is organizing sports challenges such as skiing, trail running, and sponsoring expeditions to some of the most far-flung, still largely untouched corners of the globe. By involving people including professional athletes in outdoor sports, the brand enhances its image among outdoor sports fans. The Red flag campaign however adopted a quite different approach. It was much easier for average people to participate: it did not require professional skills or equipment. Anyone with a mobile phone could take part in the campaign to plant the digital flags.

Participants simply needed to send a message to a short messenger service (SMS) code stating their current locations, their names, and the cities they are in. This would enable them to plant a virtual flag on a live dynamic map of China. Upon sending the message, people would get a reply message from the brand, which included a username and a password to enable them to log on to the website www.thenorthface.com.cn/redflag or mobile site m.thenorthface.com.cn to check out the dynamic map and the flag-planting ranking list. They could also upload photos and share their discoveries or experiences with comments. Each location could only be claimed by one person, and the location claimed was then named by the conqueror.

The portability of the mobile played an important role in the game. As people checked in at different locations, the power of 'here and now' of the mobile media gave people a taste of the great feeling of exploring a new place and claiming credit for being there first. The virtual red flags represented people's footprint in the real world, and meanwhile reflected their claim of the piece of land in the virtual world. The line between the real and virtual blurred as people replicated their presence via mobile, which acted as a tool of location sharing and the interface of hybrid space (de Souza e Silva, 2006). As the interface of the hybrid space, the mobile not only connected the game experience with the ordinary life by bringing the game outside of the traditional magic circle, but also encouraged people's mobility in the urban space (de Souza e Silva & Sutko, 2008). In this case, people's physical mobility serves the marketer's need to build its brand image associated with outdoor activities.

While outdoor and endurance sports is still a distant concept for average people living in urban China, the brand managed to promote the brand awareness and spread the brand ethos among ordinary people by lowering the barrier to participation. In total, 651,540 flags were planted over 18-day campaign period, beyond the expectation of 500,000 flags. The top ten flag planters planted an average of over 6000 flags. For the brand owner, such an approach allowed it to address the previously under-tapped emerging market segments and potentially increase its consumer base.

While the widely adopted SMS was simple and easy for people to participate, the game did not integrate it with existing location-based mobile social networks. This could be attributed to the marketers' concern over the risk in privacy issues. While Renren.com, a major social networking site in China, published the call for participation of the game, users could not use their Renren account to participate in the game. This granted users anonymity in the campaign, which could dispel their concern about privacy and encourage more participation. As privacy is recognized as one of the thorny issues of mobile location-based services and mobile marketing (Evans, 2003; Dourish & Anderson, 2006), the anonymity embedded in the user name allocated to participants via SMS circumvented the issue.

In this campaign, people initiated the interaction with brands, and thus had more agency than being targeted via Bluetooth without the concerns over receiving interruptive sales messages on their mobile, if they had their Bluetooth turned on at all. Compared to the sales-driven efforts in proximity marketing, this game enabled a consumer experience in hybrid space through 'self-positioning' (Benford et al., 2004) via the mobile interface. By allowing ordinary people to participate in the campaign with mobile phones while getting active outdoors, the campaign did not treat potential consumers as passive audience. Instead, the brand built a closer connection with participants as they moved around in the urban space and made their marks on the representational map.

Productive Play and Value Co-Creation

By leveraging the convergence of mobile devices and locative media, the game went beyond the mobility of the device, and encouraged the mobility of people in urban space. According to de Certeau (1984, p. 117), 'space is composed of intersections of mobile elements. It is in a sense actuated by the ensemble of movements deployed within it. [...] In short, space is a practiced place'. Through the act of walking in the city, the experience of the city shifts from an act of consumption to an act of production (de Certeau, 1984). Therefore encouraging participants' mobility in urban exploration facilitates the act of production, which can be understood in several ways in this case.

First, people's movement in the real urban space and their use of mobile as an interface of hybrid spaces allows new ways of interactions among people. This reflects the shift in focus of mobility from device to person: their physical capabilities and technological means by which they communicate with others (Lane, Thelwall, Angus, Peckett, & West, 2006, p. 23). In the Red Flag Campaign, participants interacted with each other both as competitors in the race and as co-creators in collaborative mapping.

Owing to the portability and internet capability of the mobile phone, participants could keep up to date with the race by checking the virtual interactive map. The functions of searching by city and zooming in and out enabled participants to gain knowledge of their local surroundings and beat others in the race. Location became one of the filters to help participants to acquire the most relevant information then and there. The search function by username allowed them to generate a map of their own or other participants' footprints, and have an idea of the remaining territory in hybrid spaces. While traditional maps represent reality, dynamic locative maps work *in connection to* reality (de Souza e Silva & Sutko, 2011, p. 31). The maps were constantly updated in response to the participants' actions in a dynamic, participatory cartography. The update in the map in turn influenced other participants' actions in the race. As Kabisch (2010, p. 51) pointed out, 'To emphasise objects and coordinates is to be concerned with nouns - with entities that are acted upon and known about. While this focus may elucidate the known, perhaps a focus on action - on verbs -might bring into view the act of knowing.' It was through walking in the city that participants choose to view content relevant to them, either geographically or culturally, thus 'actualize' the information through the interface of the locative medium (de Souza e Silva & Sutko, 2011, p. 35).

Second, the meaning-making practices of checking in, naming, commenting in the process also gave meaning to the places and urban space, which became associated with the brand endorsing the outdoor lifestyle. As these data and content from individual participants were aggregated on the map, the dynamic user-generated map became a temporal and spatial visualization of participants' data-generating activities. Therefore, as participants responded to the brand's call for the outdoor lifestyle, people's engagement in real-time cartography along with content and data generated was itself an annotation to the value-creating process for the brand. Here, the boundaries between commerce, content and information are being redrawn, where user-generated content and data served the commercial interests (van Dijck, 2009).

Third, the shared values and codes of meaning embodied by the flag-planting led to the 'emergence of collective identity' (Castells, Fernandez-Ardevol, Qiu, & Sey, 2004, p. 144) among the participants. While geographically dispersed, participants experienced visual co-presence (Ito & Okabe, 2005) on the virtual dynamic map in the mobile mediated game. Moreover, they experienced 'shared encounter', 'where a sense of performative co-presence is experienced and which is characterised by a mutual recognition of spatial or social proximity' (Willis, Roussos, Chorianopoulos, &Struppek, 2010, p. 4). The iconic action of digitally planting the flags enhanced the connections among the participants, who shared an understanding of the cultural meanings of the action. In this case, such a shared cultural understanding was utilised by the brand for the marketing purpose, which received cooperation from participants.

'The shared encounter' established connections among participants and constituted a memory building process. As people planted virtual flags when moving around, urban spaces were reconfigured (de Souza e Silva & Sutko, 2008) and those locations gained a new layer of meaning for participants, especially for the conquerors. The user-generated map became a collective narrative and memory of co-creative experience. According to Castells (2000), space is the expression of society. In this game the intimate strangers in hybrid spaces demonstrated a collective aspiration brought out by the brand for an outdoor culture and explorative spirit. As participants interacted with each other via mobile, it also brought them a new sense of places (Lemos, 2010). This new sense of places goes beyond the physical and geological levels, and brings about emotional connections and cultural associations in hybrid spaces. All this generated value for the brand through collective identities and memories associated with the meanings of the brand.

As Lindtner and Douris (2011) propose, it is important to understand the 'productive play' (Malaby, 2007; Pearce, 2006; Yee, 2006) that does not end with material creation but in relation to political, economic and social developments. In this case, the mobility of participants and pedestrians, the real-time data generated by the communities, and the interactions in hybrid spaces together demonstrates how the Chinese urban dwellers respond to a multinational brand's call for the outdoor lifestyle to service its aim of expanding its local market. In this process participants become co-promoters of the outdoor lifestyle and the brand itself.

By tying the mobile media with the local experience, and facilitating user-generated content, the experience of urban space consumption doubled up as a process of producing urban space narrative, building collective memory, and constructing collective identity.Participants in this campaign produced value for the brand, as their place-making, space-constructing, meaning-making and identity-building practices evolve around the brand ethos of outdoor lifestyle and exploration. This then served the campaign's purpose of educating the local market and increasing its local consumer base.

MARKETER-ORCHESTRATED ONLINE-OFFLINE DYNAMICS

It has been argued that mobile devices when integrated into campaigns rather than as stand-alone channels can achieve better effects in developing relationships between brands and consumers (IAB, 2008; Steinbock, 2005). In this campaign, marketers have utilised other media to construct an online-offline dynamics in addition to the mobile-mediated game in hybrid spaces.

While participants engaged themselves in the game in hybrid spaces via mobile screens, outdoor screens in Beijing and Shanghai, two mega-cities in China, also facilitated connections between physical and virtual spaces. In addition to following the game online with the assigned user name and password, people could keep track of the game via outdoor electronic boards, where live counters displayed the total number of flags planted across China, by individuals and by location. The synchronized counters in online and physical spaces enabled not only participants but also other pedestrians to follow the race. By attracting pedestrians' attention, outdoor screens also encouraged them to join in the game.

While the outdoor screen drove people to the mobile-mediated game, mobile drove people to outdoor events as well. Following two weeks' of red flag planting in hybrid spaces, the on-ground events in Beijing and Shanghai brought people together to experience the fun of outdoor sports. Those who previously participated in the flag-planting race received the short messages alerts, which drew them to the events. The 7-meter high climbing wall and a waving red flag called upon people to try their hands. Mobile started a journey of exploration for people, and continued to drive them to experience together the fun of outdoor sports in the events.

There were also photo areas at the events, where the background settings were all locations for outdoor exploration, such as snow mountains, forests and deserts. People could take pictures with various background settings as if they were exploring different parts of the world, which could be as far as Antarctica. Despite the non-authentic nature of the experience, the opportunity to take photos with such background attracted many people. After the TNF staff members took photos of them, people were able to download it on the campaign website or WAP site. People also captured moments of simulated experience of exploration with their own mobile phones, and shared photos with friends.

It can be seen that while the mobile played an important role in virtual flag planting; other media such as outdoor screens were used to drive people to the mobile platform. Meanwhile, the mobile continued to play an important role in driving participants to outdoor events, creating shared memories, and igniting their passion for outdoor sports and natural world exploration. Such an online-offline dynamics added to the interactions among participants and pedestrians in hybrid spaces in the mobile mediated game. The integration of the mobile with othermedia brings into play each medium's unique features and enables complementarity (Nysveen, Pedersen, Thorbjornsen, & Berthon, 2005; Wang, 2007).

Further, while the mobile offered a low barrier to participation, participants were able to engage with the existing outdoor culture online which was populated with hardcore outdoor fans and travel lovers. Despite that the outdoor culture remains a niche in China; the early adopters have already established their presence online. Participants could opt in for an outdoor photo sharing competition on www.lvren.cn, a leading community for outdoor sports and travel lovers in China. Community members share their outdoor experiences, offer travel tips, and exchange review of outdoor gear and equipment on the site. The online community also extends offline, when members organise outdoor trips themselves. In association with the TNF campaign, game participants and users of lvren were encouraged to upload and share their photos of outdoor activities, and those whose photos were rated most popular by users would win the competition. People responded well to the photo-sharing activity, and the top winner garnered more than 130,000 votes (Lvren.cn, 2009). This promoted awareness of the mobile locative game and helped drive participation.

The connection with online outdoor culture built a bridge between hardcore outdoor fans and those who have yet to adopt the outdoor lifestyle. This is important in the sense that such an exposure to the outdoor culture may lead ordinary people to be exposed to and get interested in the outdoor lifestyle. While the mobile location-based games lasted for a limited period, the connection and engagement with the outdoor culture online and offline can extend longer. Those hardcore outdoor fans frequenting the online community thus became 'co-promoters', which is an important part of value co-creation (Sheth & Uslay, 2007).

UNINTENDED CONSEQUENCES

While the mobile media provided a low barrier to participation, there were different degrees of participation. Participants could plant virtual flags when they went to a park nearby, which was of course quite different from outdoor trips or sports. Moreover, they could even check into places located within an area in which they may not physically be. In other words, they could cheat the system. This, according to Gazzard (2011), leads to the distortion of space owing to the need to check in and claim virtual symbols. It then dilutes the playfulness in the game and turns the promoted idea of *exploring spaces* into *collecting places*. In discussing the distortion of space in the use of Foursqure, Gazzard pointed to the concept of 'stalking'. As Sinclair noted, '[t]he concept of ''strolling'', aimless urban wandering, the flâneur, had been superseded. We had moved into the age of the stalker, journeys made with intent ... This was walking with a thesis' (Sinclair, 1997: 75). In this campaign, while it was possible to navigate the virtual map and zoom in or out to explore the surrounding spaces, the goal oriented nature of collecting places might supersede the exploring and discovering aspects of the journey, which the brand claimed as its ethos.

While marketers used urban spaces as the game board and aimed to facilitate the brand experience in hybrid spaces, people constructed their own playground online. This was evidenced in various user-generated maps on the internet, with red flags being arranged to form different sorts of images on the maps. These images ranged from a map of China with the Superman in it to one with an image of celebrations to mark the 60th anniversary of the founding of PRC, which was around the time of the campaign. By using image editing software such as Photoshop, internet users reconfigured the game space. Unlike participants of the mobile mediated game, these internet users played with the map, generating various versions at their own will and *independent of* their mobility in the urban space. Interactions among internet users around these user-created images were, like the distortion of space discussed above, unintended consequences for the marketers. By extending the mobile-mediated game and constructing their own playground online, these users turned the campaign on its head. While they injected more playfulness and creativity in their own game, sitting in front of the computer playing with images is a far cry from the brand's call for outdoor exploration. Instead, the campaign invited online nationalism embodied in the maps generated by users with the assistance of image processing software. The commercial nature of the campaign was thus mixed with the nationalist discourse online, demonstrating the productive play going beyond the material creation and inviting political discourse (Lindtner & Douris, 2011).

Therefore the extent to which participants live the brand ethos of outdoor exploration in the campaign remains a question. Moreover, digital media and software tools allowed consumers to deviate from the campaign process designed by marketers and construct their own playground. The mobile location-based game, originally utilised by the multinational brand to expand its local market, turned to some extent into a playful experience going beyond the boundaries set by marketers. Such boundary-crossing practice results in the reconfiguration of the game space and the productions of maps as cultural artefacts. In spite of that, such deviation may paradoxically promote the awareness of the campaign among a wider audience owing to the viral effect online. This however may not be the case in other situations, where brand image could potentially be altered or subverted by users (Christodoulides & Jevons, 2011). This is why user creativity could be a risk to marketers, who now have less control over the communications process.

CONCLUSION

With the increasing adoption of mobile and locative media in the context of participatory culture, mobile location-based games have the potential to become a novel marketing vehicle. This article draws on the concepts of hybrid spaces, value co-creation, and productive play in analysing the mobile location-based game in the international marketing context. The empirical evidence in the TNF case illustrates a multinational brand's efforts to expand its local consumer base in China by shaping their lifestyle aspirations. By utilising the mobile media as a vehicle for ordinary people to participate in the campaign easily, the brand departed from its traditional marketing approach. It made a step forward towards changing local people's perception of outdoor lifestyle being often associated with hardcore outdoor fans, professional athletes and extreme sports, as in the brand's approach to more mature markets.

As marketers used urban spaces as the game board, the mobile acted as an interface in hybrid spaces which were connected, mobile and social (de Souza e Silva, 2006). Mobile as the central tool and a platform facilitated competition in conquering the virtual space and collaborative mapping among participants. The game experience is both physical and virtual, both material and symbolic. The mobile locative game not only facilitates a place-making and space-constructing experience, but also constructs a collective identity building process. The case reveals the importance of understanding the mobility of people, their interactions with each other and with virtual dynamic maps, and the consumption and generation of spatio-temporal data as a productive play and value co-creation for brands. The case also shows the cross-media dynamics designed by marketers drove participation and facilitated a continuous experience online and offline during the game period.

While the case reveals the positive impacts of value co-creation and productive play for the brand owner afforded by the mobile media, it also points to the unintended consequences of the campaign. This is evidenced by consumers' deconstruction and reconstruction of the marketers-designed game process and their distortion of space. More generally, it reveals the importance of understanding the role of gamers as re-producers of play, and the opportunities and challenges in addressing creative consumers in utilising mobile location-based game for marketing purposes. With a new breed of consumers and their increasing adoption of the Smartphone and mobile locative media, they offer an important research venue to understand the evolving relationship between people, space, and technology as well as their implications for marketing research and practice.

REFERENCES

Analysys International. (2011). *China LBS Accumulate Account Exceeded 10 Million in Q2, 2011*. Retrieved July 26, 2012, from http://english.analysys.com.cn/article.php?aid=112335

Benford, S., Seager, W., Flintham, M., Anastasi, R., Rowland, D., Humble, J., et al. (2004). The error of our ways: The experience of self-reported position in a location-based game. In N. Davies, E. D. Mynatt, & I. Siio (Eds.), *Proceedings of the UbiComp 2004* (pp. 70-87). Nottingham, UK: Springer.

Burmann, C., & Arnould, U. (2009). *User-generated branding, state of the art research*. Münster: Lit.

Castells, M. (2000). *The rise of the network society*. Oxford, UK: Blackwell.

Castells, M., Fernandez-Ardevol, M., Qiu, J., & Sey, A. (2004). *The mobile communication society: A cross-cultural analysis of available evidence on the social uses of wireless communication technology*. Annenberg: Annenberg Research Network on International Communication.

Castronova, E. (2005). *Synthetic worlds: The business and culture of online games*. Chicago, IL: University of Chicago Press.

Christodoulides, G., & Jevons, C. (2011). The voice of the consumer speaks forcefully in brand identity. *Journal of Advertising Research, 51*, 101–108. doi:10.2501/JAR-51-1-101-111

CNNIC. (2013). *The 32nd statistical report on internet development in China*. Retrieved December 5, 2013, from http://www1.cnnic.cn/IDR/ReportDownloads/201310/P020131029430558704972.pdf

Corporation, V. F. (2010). *VF Corporation Annual Report 2009*. Retrieved March 2, 2010, from http://reporting.vfc.com/2009/pdfs/vfc_ar09.pdf

de Souza e Silva, A. (2006). From cyber to hybrid: Mobile technologies as interfaces of hybrid spaces. *Space and Culture*, *9*(3), 261–278. doi:10.1177/1206331206289022

de Souza e Silva, A., & Sutko, D. M. (2008). Playing life and living play: How hybridreality games reframe space, play, and the ordinary. *Critical Studies in Media Communication*, *25*(5), 447–465. doi:10.1080/15295030802468081

de Souza e Silva, A., & Hjorth, L. (2009). Playful urban spaces: A historical approachto mobile games. *Simulation & Gaming*, *40*(5), 602–625. doi:10.1177/1046878109333723

de Souza e Silva, A., & Sutko, D. M. (2011). Theorizing locative technologies through philosophies of the virtual. *Communication Theory*, *21*(1), 23–42. doi:10.1111/j.1468-2885.2010.01374.x

deCerteau, M. (1984). *The Practice of Everyday Life*. Berkeley, CA: University of California Press.

Dourish, P., & Anderson, K. (2006). Collective information practice: Exploring privacy and security as social and cultural phenomena. *Human-Computer Interaction*, *21*(3), 319–342. doi:10.1207/s15327051hci2103_2

Evans, L. (2011). Location-based services: Transformation of the experience of space. *Journal of Location Based Services*, *5*(3-4), 242–260. doi:10.1080/17489725.2011.637968

Evans, M. (2003). The relational oxymoron and personalisation pragmatism. *Journal of Consumer Marketing*, *20*(7), 665–685. doi:10.1108/07363760310506193

Firat, A. F., & Venkatesh, A. (1995). Liberatory postmodernism and the reenchantment with consumption. *The Journal of Consumer Research*, *22*(December), 239–267. doi:10.1086/209448

Gazzard, A. (2011). Location, location, location: Collecting space and place in mobile media. *Convergence*, *17*(4), 405–417. doi:10.1177/1354856511414344

Gergen, K. (2002). The challenge of absent presence. In J. Katz, & M. Aakhus (Eds.), *Perpetual contact: Mobile communication, private talk, public performance* (pp. 227–241). Cambridge, UK: Cambridge University Press.

Gordon, E., & de Souza e Silva, A. (2011). *Net locality: Why location matters in a networked world.* Boston, MA: Blackwell Publishers.

Himfr. (2008). *Himfr.com reports that China's outdoor sports brandshave enormous market opportunities*. Retrieved November 26, 2009, from http://www.reuters.com/article/2008/11/17/idUS146919+17-Nov-2008+PRN20081117

Hjorth, L., & Arnold, M. (2013). The place of intimate visualities: Ba ling hou, LBS and camera phones (Shanghai). In *Online@AsiaPacific: Mobile, social and locative media in the Asia-Pacific*. London: Routledge.

IAB. (2008). *IAB platform status report: A mobile advertising overview*. Retrieved October 29, 2008, from http://www.iab.net/media/file/moble_platform_status_report.pdf

Ito, M., & Okabe, D. (2005). *Intimate visual co-presence*. Paper presented at theUbicamp 2005. Retrieved January 3, 2010, from http://www.itofisher.com/mito/archives/ito.ubicomp05.pdf

Kabisch, E. (2010). Mobile after-media: Trajectories and points of departure. *Digital Creativity*, *21*(1), 46–54. doi:10.1080/14626261003654996

Kshetri, N. (2009). The evolution of the Chinese online gaming industry. *Journal of Technology Management in China*, *4*(2), 158–179. doi:10.1108/17468770910965019

Lane, G., Thelwall, S., Angus, A., Peckett, V., & West, N. (2006). *Urban tapestries: Public authoring, place & mobility*. Retrieved March 26, 2009, from http://socialtapestries.org/outcomes/reports/UT_Report_2006.pdf

Leek, S., & Christodoulides, G. (2009). Next-generation mobile marketing: How young consumers react to Bluetooth-enabled advertising. *Journal of Advertising Research*, *49*(1), 44–53. doi:10.2501/S0021849909090059

Lemos, A. (2010). Post-mass media functions, locative media, and informational territories: New ways of thinking about territory, place, and mobility incontemporary society. *Space and Culture*, *13*(4), 403–420. doi:10.1177/1206331210374144

Lindtner, S., & Douris, P. (2011). The promise of play: A new approach to productive play. *Games and Culture*, *6*(5), 453–478. doi:10.1177/1555412011402678

Lvren.cn. (2009). *Photolist of The North Face Campaign*. Retrieved November 10,2010, from http://active.lvren.cn/chuqizhisheng-0909/photolist.php

Malaby, T. M. (2007). Beyond play: A new approach to game. *Games and Culture*, *2*(2), 95–113. doi:10.1177/1555412007299434

Meuli, P. G., & Richard, J. E. (2013). Exploring and modellling digital natives' intention to use permission-based location-aware mobile advertising. *Journal of Marketing Management*, *29*(5/6), 698–719.

Nystedt, D. (2005). *Online gaming growing fast in China, study says*. Retrieved August 6, 2010, from http://www.macworld.com/article/44065/2005/04/chinagaming.html

Nysveen, H., Pedersen, P. E., Thorbjornsen, H., & Berthon, P. (2005). Mobilizing the brand - The effects of mobile services on brand relationships and main channel use. *Journal of Service Research*, *7*(3), 257–276. doi:10.1177/1094670504271151

Okazaki, S., & Taylor, C. R. (2008). What is SMS advertising and why do multinationals adopt it? Answers from an empirical study in European markets. *Journal of Business Research*, *61*(1), 4–12. doi:10.1016/j.jbusres.2006.05.003

Outdoor Industry Association. (2010). *China's outdoor market spawns intense competition*. Retrieved December 16, 2010, from http://www.outdoorindustry.org/news.ceo.php?newsId=13116&newsletterId=158&action=display

Pearce, C. (2006). Productive play: Game culture from the bottom up. *Games and Culture*, *1*(1), 17–24. doi:10.1177/1555412005281418

Prahalad, C. K., & Ramaswamy, V. (2003). The new frontier of experience innovation. *MIT Sloan Management Review*, *44*(4), 12–18.

Prahalad, C. K., & Ramaswamy, V. (2004). Co-creation experiences: The next practice in value creation. *Journal of Interactive Marketing*, *18*(3), 5–14. doi:10.1002/dir.20015

Prahalad, C. K., & Ramaswamy, V. (2005). *The future of competition: Co-creating unique value with customers*. Boston: Harvard Business School Press.

Puro, J. P. (2002). Finland: A mobile culture. In J. Katz, & M. Aakhus (Eds.), *Perpetual contact: Mobile communication, private talk, public performance* (pp. 19–29). Cambridge, UK: Cambridge University Press.

Sheth, J. N., & Uslay, C. (2007). Implications of the revised definition of marketing: From exchange to value creation. *Journal of Public Policy & Marketing*, *26*(2), 302–307. doi:10.1509/jppm.26.2.302

Sinclair, I. (1997). *Lights out for the territory*. London: Penguin Books.

Steinbock, D. (2005). *The mobile revolution: The making of mobile services worldwide*. London: Kogan Page.

van Dijck, J. (2009). Users like you? Theorizing agency in user-generated content. *Media Culture & Society*, *31*(1), 41–58. doi:10.1177/0163443708098245

Venkatesh, A. (1999). Postmodern perspectives for macromarketing: An inquiry into the global information and sign economy. *Journal of Macromarketing*, *19*(2), 153–169. doi:10.1177/0276146799192006

von Hippel, E., & Katz, R. (2002). Shifting innovation to users via toolkits. *Management Science*, *48*(7), 821–833. doi:10.1287/mnsc.48.7.821.2817

Wang, A. (2007). Branding over mobile and internet advertising: The cross-media effect. *International Journal of Mobile Marketing*, *2*(1), 34–42.

Wilken, R., & Sinclair, J. (2009). 'Waiting for the kiss of life': Mobile media andadvertising. *Convergence: The International Journal of Research into Newmedia Technologies*, *15*(4), 427–446. doi:10.1177/1354856509342343

Willis, K. S., Roussos, G., Chorianopoulos, K., & Struppek, M. (Eds.). (2010). *Sharedencounters*. London: Springer.

Yee, N. (2006). The labor of fun: How video games blur the boundaries of work and play. *Games and Culture*, *1*(1), 68–71. doi:10.1177/1555412005281819

Zhang, J., & Mao, E. (2012). What's around me? Applying the theory of consumption values to understanding the use of location-based services (LBS) on smart phones. *International Journal of E-Business Research*, *8*(3), 33–49. doi:10.4018/jebr.2012070103

Zhou, W. T., & Peng, Y. N. (2010). Breaking free from virtual reality and Web of boredom. *China Daily*. Retrieved March 2, 2011, from http://www.chinadaily.com.cn/china/2010-10/12/content_11396450.htm

KEY TERMS AND DEFINITIONS

Co-Creation: A form of collaborative creativity, which is usually initiated by organizations to create with, rather than simply for consumers in order to build relationships with them.

Collaborative Mapping: The aggregation of user-generated geospatial data into online maps.

Hybrid spaces: The spaces where virtual communities migrate to physical spaces through mobile technologies.

Locative Media: Digital media that are linked to particular geographical locations through location sensing technologies.

Mobile Location-Based Game: A type of game which evolves based on a player's location in the real world and where mobile devices act as interfaces between the physical and the virtual spaces.

Participatory Cartography: Map production undertaken by communities mainly for their use or entertainment.

Participatory Culture: A culture with relatively low barriers to artistic expression and civic engagement, strong support for creating and sharing creations, some type of informal mentorship, a belief among members that contributions matter, and a sense of social connection (See more at http://www.newmedialiteracies.org/wp-content/uploads/pdfs/NMLWhitePaper.pdf).

Productive Play: Play experiences where creative production for its own sake is an active and integral part.

Section 6
Mobile Uses and Usability

Mobile has been widely used in increasingly diversified areas, such as learning, identity-building, gender rituals, solution to intimate conflicts, audience engagement, civil movement, parenting, location-based marketing, and location-based advertising. More uses of mobile include gaming, healthcare, and journalism, three areas examined in this section. Behind different mobile uses lies the same need to enhance mobile usability. In this section, location-based mobile gaming, mobile healthcare, mobile news, and mobile usability are the selected topics for investigation. Mobile gaming is one of the most popular mobile activities. With the introduction of mobile location-based technologies, mobile gaming has substantially changed the way mobile gamers interact with other people and objects in a gaming environment. After critically reviewing earlier studies, the author of Chapter 15 investigates how mobile location-based games reconfigure people's relationships with other people and objects in their environment, and presents a model to understand this reconfiguration. Mobile has been actively used in news reporting and writing by both professional and amateur journalists. Equally, mobile has become a popular device for users to seek and follow news on a regular basis. When and under what circumstances, however, do mobile users use mobile as a device for news? The answer can be found in Chapter 16. As a fast-growing market, mobile health has become one of the big businesses. Looking into the use of mobile for healthcare in the context of global wirelessly connected health sector, the author of Chapter 17 offers timely and valuable observations of and insights on the exponentially expanded market as well as recommendations from a socio-economic and political standpoint for responsible and effective future industry growth. Mobile usability is a worldwide concentration of efforts to enhance mobile use for all purposes. To facilitate efforts to enhance mobile usability, the authors of Chapter 18 explain various usability testing methods, contextual complexity, audio interfaces, eye and hands-free interactions, augmented reality, and recommendation systems. The chapter also examines methodological challenges in mobile usability as well as the social and economic implications of mobile usability.

Chapter 15
Socio-Spatial Relations in Mobile Gaming:
Reconfiguration and Contestation

Paul Martin
University of Nottingham, Ningbo, China

ABSTRACT

This chapter explores the opportunities of mobile games to critique and constitute the networks of which they are a part, attending particularly to location-based games. It discusses how these kinds of mobile games reconfigure people's relationships with other people and objects in their environment. In order to understand this reconfiguration, a model is put forward that clarifies the various ways in which people and objects are presented to the mobile game player. Using this model, examples are discussed of games that make interactions available that are disruptive of a social or political order, arguing that this disruption may be drafted into socio-political critique. Other examples demonstrate how mobile games bring everyday life within a capitalist logic, monetizing leisure and the mundane. This suggests that mobile gaming as a technology, practice, or product is neither fundamentally emancipatory nor fundamentally regressive but rather can be employed in various ways.

INTRODUCTION

In their 2005 white paper, the International Game Developers Association (IGDA) defined mobile games as those that 'are delivered via wireless networks to devices whose primary function is a mobile phone' (Wisniewski et al., 2005, p.4). Since the release of the iPad in 2010, the term has tended to include tablet devices too. In most game review websites, such as IGN, Eurogamer and Edge, as well as game developer website Gamasutra, the term essentially refers to games for operating systems such as the iOS and Android, whether they are played on tablets or smartphones.

This definition distinguishes the mobile from other kinds of digital game platforms; consoles, such as Microsoft's Xbox, that are linked up to TV screens; dedicated portable, or handheld, gaming devices such as Sony's Vita; and multifunctional PCs. There are two distinctions here: the player can access mobile games in a range of places and times and mobile game platforms – mobile

DOI: 10.4018/978-1-4666-6166-0.ch015

phones – are not primarily designed or purchased for gaming.

The first of these distinctions seems most important. The fact that players can play in different locations, move between locations while playing, and that games can register these movements and incorporate them into gameplay is of particular relevance to the way the games discussed here reconfigure, contest and reinforce socio-spatial relations. Not all mobile games do register player movement through the world. Many popular mobile games take advantage of the mobile phone's ubiquity without taking advantage of its ability to locate and track player movements. This chapter will focus on games that do register player mobility as it is in these mobile games that some of the most interesting socio-spatial reconfigurations occur.

The second distinction is also relevant. The fact that mobile phones are multi-functional means that far more people own mobiles than portable gaming devices such as the Vita. This situates mobile games in a very different network of possible players. Certain game forms that could not work on portables can work on mobiles, for example games such as *Ingress* (Niantic Labs, 2012), discussed later in the chapter, which require a critical mass of players to function. Games that function by disrupting players' everyday routines, such as *Mogi: Item Hunt* (Castelli, 2003), require a platform that is always with the player. People who own mobile phones usually carry them around with them (Mainwaring, Anderson, & Chang, 2005, p.278) but, since portable gaming devices are usually slightly bulkier than mobiles and only serve one purpose, people are less likely to routinely carry portables about (Tassi, 2012; Totilo, 2012).

While some of the following argument could be made in relation to portable gaming devices, there are substantial differences in the socio-spatial relations that the two platforms make possible.

LITERATURE REVIEW

Socio-Spatial Relations and Political Critique

Over the last 30 years there has been a sustained academic interest in the importance of space in shaping social and political relations and the ways in which the spatial is shaped by political, ideological and economic interests (for a summary, see Warf & Arias, 2008). Two of the most influential texts in this spatial turn are Henri Lefebvre's *The Production of Space* (1991), and Michel de Certeau's *The Practice of Everyday Life* (1984). The former argues that space is socially produced and is integral to the reproduction of social relations. The latter provides a blueprint for thinking about how the use of space by ordinary people could challenge, resist and perhaps undermine the spatial regimes that are imposed by the powerful and that serve to maintain existing structures of power. This shaping of the character of space from above and below is sometimes termed *spatialization*.

This central idea – that space is produced by and productive of socio-political relations and that people's practices in space are capable of contesting these relations – has been central to much work in cultural studies, particularly as it relates to urban practices such as skateboarding (Borden, 2001) and graffiti (Bowen, 2013). The mobile phone – like the skateboard or the spray-paint can – is a potential tool through which certain spatial practices can be enacted that realise or challenge existing configurations that are simultaneously spatial, social and political. This simultaneity means that these aspects – the spatial, the social and the political – are mutually constitutive. Take the example of a city train during rush hour. The spatial refers to the relationship between people and objects in space: the number and arrangement of seats, the placement of doors, the number of passengers. But the spatial is also socially constituted, determined in part by the social conventions by which people abide in this social situation: the distance observed between passengers, the unwritten rules governing eye-contact or conversation. The spatial is also, however, constitutive of the social. A larger train carriage, fewer seats, more or wider doors, would all contribute to a different space and make available to passengers different repertoires of social behaviour. These different spatial configurations and the practices that they make available help to constitute the character of the space, in terms of privacy, playfulness, comfort, suitability for commerce, suitability for work, and so on. Both the spatial and the social are underpinned by specific political realities. Spatial configurations on trains are in part determined by practical political realities such as changes to government spending on transport. Socio-economic differences between passengers are reflected in the train company's decision to have differently priced carriages or seats, and the meaning of these differences is rhetorically established through forms of separation and naming practices: first class, business class, and luxury all carry different ideological weight.

Socio-Spatial Relations and Mobile Phones

While skateboarding and graffiti-spraying are urban spatial practices frequently seen as politically resistant, behaviour repertoires associated with train journeys are perhaps less so. Nonetheless, as researchers have shown, attention to these repertoires demonstrates that they do have cultural significance (for example, O'Dell, 2009; Symes, 2007). Similarly, mobile phone use is, on the face of it, a far from subversive practice. In its early years the mobile phone was largely understood as 'an elitist device mainly used for business by middle and upper class males' (Lacohee & Wakeford, 2003, p.205). Clearly, the mobile has since then become a more ubiquitous technology, used by a wide range of people in a range of different ways, including facilitating the organisation of disruptive actions, from flashmob (Evans-Cowley, 2010) to insurgency (Adey, 2010), to revolution (Rheingold, 2003). All of these uses – from the mundane to the revolutionary – are shaping ways of being in space, with some conforming to, some contesting prevailing socio-political relations.

As a tool involved in spatialization mobiles have three defining characteristics that are important in understanding the social and political impacts of mobile gaming. First, most are locatable by an off-site network of computers. This locatability is often achieved through the global positioning system (GPS), which can achieve in the right circumstances accuracy of ±3 metres (Trinklein & Parker, 2013). Mobiles equipped with Bluetooth, WiFi, or similar wireless technologies can also locate other similarly equipped and enabled devices within a certain distance. These technologies allow for a greater accuracy than GPS and can be used indoors, but can only be used over a much smaller range (Wang, Yang, Zhao, Liu, & Cuthbert, 2013). Second, mobiles provide a semi-private screen and speaker system. On most handsets sound can be muted or channelled through headphones, but it can also be played through the speakers, and which of these options a player chooses impacts the space in which the game is being played. Visual information is necessary in most mobile games and it is provided through screens that are visible to the player and potentially to others in a small area around the player. Again, depending on how the player uses the game this feature potentially affects the space of the people in this radius. Third, mobiles make people and objects not locally present available for certain kinds of interaction – talking, texting, and a host of game-related actions. These people and objects, which may be real or fictional, present themselves to the player as information, with the player's imagination fleshing them out and giving them life.

Socio-Spatial Relations and Digital Games

These spatializing characteristics of mobile phones are combined with spatializing characteristics that are inherent to games. As has been argued by several game scholars, spatiality is integral to game form (for example, Aarseth, 2000; Nitsche, 2008). Many games involve spatial manipulations (for example, puzzle games such as *Tetris* (Pajitnov, 1984)) or spatial exploration (for example, open world games such as the *Grand Theft Auto* series (DMA Design/Rockstar North, 1997-2013)). We might go so far as to say that the capacity for games to create spaces in which players can perform in various ways leads in all games to a reconstitution of spatial relations that is always potentially political.

Perhaps the most famous take on spatiality in games is the 'magic circle'. This concept, introduced by Johann Huizinga (1955) and developed by several digital games scholars over the last ten years (for example, Salen & Zimmerman, 2004; Klabbers, 2009), attempts to describe the unstable separation of a conceptual *game space* from *non-game space*. While the term has been criticised from a number of directions (Consalvo, 2009; Copier, 2009; Richardson, 2010; Calleja, 2012), it speaks to the basic principle that relations between people and objects that constitute a gaming situation are different than relations between the same people and objects when there is no game going on.

Mobile games potentially make use of the ability that mobile devices have to reconfigure socio-spatial relations *and* the ability that digital games have to reconfigure socio-spatial relations. They are spatializing in these two sets of ways.

Mobile Games and Spatial Metaphors

Mobile games shape the relationship of the player to the space in which the game is being played. That is, mobile games shape the relationship of the player to other people and to other objects. It is not just the player's relationship to other people and objects that changes, but also other people's and objects' relationship to the player and to each other. Furthermore, and on a larger scale, widespread use of technologies have an aggregate effect on the sum total of relationships between phones, people, trains, networks, ideas, couches, markets and so on, in which mobile games nestle. We might say this changes the general character of space, but what is at stake here is the way it contributes to the dynamic shifting of the network of relationships. It shapes how people and objects relate in society at large. This, of course, is a political question. Mobile gaming plays a role within this network of relationships as a practice that responds to and is productive of social and political realities.

Space will be important here, and it is therefore important to be clear on how the term is being used. Rather than talk about 'hybrid spaces' or 'virtual spaces' this argument will proceed by thinking about how mobiles make possible, encourage and foreclose certain kinds of interaction with others. By 'others' I mean here both other people and other objects. This is an attempt to clear out the use of spatial metaphors from talk about digital mediation in order to speak more clearly about space as the physical arrangement of material objects and people and to speak more clearly about the social and political consequences of these arrangements. It is of course possible, and it may sometimes be fruitful, to use spatial metaphors to discuss technological mediation. But slippage can occur. It is not uncommon to talk of 'virtual space', 'cyberspace', 'thought space' or 'fictional space' as though these were ontologically comparable with and even somehow independent of 'physical space'. For conceptual clarity, these are much better thought of as overlapping systems of interactions or *interactional ecologies*.

Adriana de Souza e Silva suggests mobile games provide 'new types of sociability' and 'new perceptions of physical spaces' (Silva, 2006, pp.19-20). This is certainly true. Mobile games do cause changes in how we perceive the arrangement of people and objects and our relationship to this arrangement. The same has been true of many previous technologies, especially those to do with communication and the media (Parikka & Suominen, 2006). This can be the radical reconfiguration of space seen in location-based games as well as the more mundane reconfigurations that take place playing *Tetris* while waiting for the bus. They also put us in touch with friends and strangers in new ways. For example, one market research report suggests that mobile games were particularly good at maintaining low level friendships (PricewaterhouseCoopers, 2012). As game designer and theorist Jane McGonigal observes, a quick game of mobile Scrabble allows people to stay in touch even if they have no news to share (McGonigal, 2011).

In an article on mobile gaming in education de Souza e Silva and Girlie Delacruz frequently describe this reconfiguration of space by contrasting 'physical' and 'digital' space, suggesting that hybrid reality games act as a 'bridge' between the two (Silva & Delacruz, 2006, p.232). While I agree that mobiles provide new types of sociability and perceptions of space, this distinction of 'virtual' and 'physical' is, I would argue, not the most conceptually useful metaphor available. All of the stuff of the virtual is based in physical space. The signals, wires, servers, antennae, ,motherboards are all perfectly physical. There is no need to 'bridge physical and digital spaces' (p. 232) because they were never and could never be apart. What is happening, rather, is that information is being transmitted between more or less distant places in such a way that it makes people and objects in those places available for certain kinds of interaction. That is, it reconfigures the network of people and objects in which the player exists, acts, and finds meaning.

The article goes on to mention several other spaces: 'student's prior knowledge' (Silva & Delacruz, 2006, p. 233), 'real space and the game space' (p. 236) and 'the player's imagination' (p. 237). Clearly, at least some of these are meant in a metaphorical sense, but their use leads to analytical problems, for example in the claim that mobile phone use generally causes people to 'withdraw from physical space' but that mobile games 'strengthen the users' connection to both physical and digital spaces' (p. 237). The use of space here is unhelpful because it mystifies the functioning of mobile phones and mobile gaming. Mobile phone users do not withdraw from physical space. They remain in physical space, becoming absorbed in a particular set of tasks that the phone makes available. That these tasks distract users' attention from other aspects of the immediate environment is no different from the effect of crossword puzzles or novels. Particular kinds of tasks change the character of space for the person engaged in that task. This cannot be measured in terms of strength of connection to some pre-existing 'physical' or 'digital' generic space. To understand how this works the specific character of the space as shaped by these practices must be described. Talk of 'physical' and 'digital' space serves to mask the heterogeneity and dynamism of the network of relationships as it is produced and reproduced through the practices people engage in.

Rather than understand *presence* as the feeling that a player experiences of being in some putative hybrid or virtual space I want to understand it as the various ways in which people and objects *present themselves to* a person playing a mobile game. Here, the term 'space' is reserved for the configuration of people and objects relative to each other, a configuration that presents itself to different people in different ways based on their different interests and capacities. These technologically mediated interests and capacities encourage and foreclose different projects and enterprises – different sets of practices. The player's environment is constituted by all of the people and objects that are presented to the player for potential interaction, whether they are presented through the game interface or not.

This is more in line with Parikka and Suominen's (2006) use of Sadie Plant's (2001) term 'bi-psyche' to discuss the way in which mobile game players split their attention between game tasks and non-game tasks. Players are not in two different spaces, but rather are attending to two different sets of tasks. It is also informed by Ingrid Richardson's (2010) work on mobile gaming. Seeing the magic circle metaphor as 'overly discrete, deterministic and artificial' (2010, p. 439), Richardson borrows the trope of the network from Copier (2009) as an alternative to capture the relational nature of mobile gaming. This is because the network, unlike space, is a 'conduit metaphor that prioritises movement, connection, and exchange' (Richardson, 2010, p. 445). The network trope allows us to get away from the notion of a layering of separate but bridged *spaces*, which fails to adequately account for how these spaces are interconnected. Rather, it focuses on the 'possibilities for action' (p. 433) that different networked configurations of objects and people make available.

A MODEL FOR UNDERSTANDING MOBILE GAME SOCIO-SPATIAL RELATIONS

In drawing out the interactional ecology that is presented to the mobile game player the following questions might serve as a helpful starting point:

1. Can the player interact with objects or people in the interactional ecology without using the game interface? Interaction is understood broadly and may or may not be related to the game being played. The important thing is not whether the interaction takes place but whether it is possible. Interaction includes seeing, touching and otherwise perceiving them and communicating with them.
2. Can the player interact with these objects and people through the game interface? Interaction here might include seeing them, getting information about their position or various aspects of their game state, communicating with them, and changing their game state. The important thing is whether taking away the mobile technology would make the interaction impossible.

Based on these two questions, it is possible to illustrate different kinds of presence in a simple Venn diagram consisting of two sets, A being all of those people and objects that are present for mediated interaction, and B, all those present for unmediated interaction.

This diagram gives three possible groups. Group 1 consists of people and objects that are present to the player only through the game interface. This could be game objects or other players connected over the Internet. Group 2 consists of people and objects not present to the player through the game interface, but physically close enough to the player to be present for unmediated interaction. An example of this might be the person seated beside me on the bus as I play a game on my mobile, whether she occasionally glances at the screen, becomes annoyed at the noise coming through my earphones, or is completely oblivious to my playing. Group 3 consists of those that are present both through the interface and without it. For example, a mobile game interface may present a particular urban landmark, say a museum, as the headquarters of an opponent. The museum is obviously still present to me in a capacity that is unmediated by the mobile.

This model can describe the presence of people and objects as they relate to any game, whether it is mobile or not, single player or multiplayer, or belonging to any genre. All games will have members of groups 1 and 2, though not all games will have members in group 3. However, many mobile games, especially but not exclusively those with an experimental or artistic sensibility, draw their aesthetic power from populating this intersection.

This diagram can be used to discuss the relation of players to other people and to things in their interactional ecology. The following sections will explain how mobile games reconfigure socio-spatial relations in each of these groups.

Group 1: Mediated Interaction Only

There are several different possible relationships between the player and the members of this first group. They may be family or friends known to the player independently of the game. The player may have made some acquaintances in this group in the game, and may never have met them face-to-face. There are members of this group whom the player can interact with, but does not know anything about beyond the fact that they are players in the same multiplayer game. There are also game objects populating this group: objects that are only accessible through the game interface. Included in this last category are non-player characters (NPCs) and player characters, or avatars. This is a special group of objects that present themselves as tools, as enemies, as information sources, and as means of establishing the atmosphere of the fictional game world.

The flexibility of mobile technology allows gaming to be brought into routines and spaces previously untouched by digital games. Mobile gaming is flexible, but this does not mean that mobile games are only played on the go. More research is needed on where people play mobile games, and particularly on why people play mobile games where they do, but existing research suggests that there may be cultural differences. There is some evidence from Japan (Joffe, 2005) and the US (Graft, 2011; MoPub, 2012) that mobile games are primarily played not on the move but at home. A Chinese study reported by Liu and Li, however, cites the commute as the main cite of mobile gameplay in China (iResearch Consulting Group, 2010, cited in Liu & Li, 2011, p.893). Most of this research is industry-based and difficult to verify, but it does serve as a reminder that the flexibility of mobile gaming does not only reconfigure players' relationship to public space but also to domestic space.

Having said this, it is in the realm of public space that this reconfiguration is most marked – or at least that it is most open to scrutiny, and the examples discussed here relate to games that are played in public. This can be understood in terms of the various technologies and performances that people employ to *domesticate* corners of public space. Tom O'Dell, for example, discusses the 'micro-processes of home-making' (p.92) that take place on the train during the commute, including the claiming of a seat and the arrangement of personal effects. Joachim Hoflich, translating Matthias Heine (2006), calls this the 'living-roomisation' of public space (Hoflich, 2010, p.7), and sees the mobile phone as central to the process.

Mobile gaming can present objects and people associated with the home to the player in a public setting. By making friends and acquaintances present for interaction, it establishes a repertoire of behaviours – joking, teasing, celebrating and so on – close to those practiced at home or in some recreational venue. The acquaintances made present by the mobile may in this way serve a significant role in domesticating the public space for the player. Even where there is no pre-existing relationship between two players in a multi-player game, the very act of playing together can be enough to forge social bonds that are felt between fellow players. A mobile game is more than mobile technology, it is mobile technology plus play.

Group 1 Example: *You Get Me*

Blast Theory are a UK art collective who have created several games using mobile technology that examine the relationship between the social, the political and the spatial. *You Get Me* (Blast Theory, 2008) is a game that sits at this crossroads. The game explicitly links space with personal and political narratives, with groups of London teenagers working with Blast Theory to create maps based on important personal experiences. The game was played in two locations: patrons of the Deloitte Ignite arts festival sitting at PCs in the Covent Garden Opera House and, five miles away, a group of eight teenagers gathered with walkie-talkies and mobile phones at Mile End Park. The game was about urban encounters, but encounters in which the social divisions that are a characteristic of the urban are foregrounded. The Opera House is associated with – in spite of its various outreach programmes – privilege. Mile End Park is located in Tower Hamlets, one of central London's most deprived boroughs. Mobile technology is the bridge between people in these socially and economically disparate spaces. The game sets up a brief mediated relationship between patron and teenager in which the presumably adult patron controls an avatar on screen to find the teenager in the park, then initiates a conversation by text and mobile. The relationship is transient, but clearly marked in terms of differentiations of power. The patron in the Opera House is seeing unseen, and gets to choose which teenager to chase, foregrounding questions of surveillance and authority. But the teenager sets the terms of the conversation, and can decide to rebuff the patron's advances, suggesting a kind of tactical resistance similar to that discussed by de Certeau (1984).

While the game is initially about bringing people together, it sustains its interest through an edge of conflict. The players do not come together, the relationship remaining in group 1. They have a brief conversation, but the teenager remains resolutely in Mile End and the patron in Covent Garden. The game does not seek to rectify social inequality, but rather to make this inequality performable or playable, and so to bring it into public consciousness. The final inability to move into group 3 highlights the intractability of socioeconomic division even as a genuine feeling of identity may be achieved. The spatial separation of different classes was observed by Frederick Engels in his description of the layout of 19th century Manchester. Here, Engels claims that the layout of Manchester sharply distinguishes the working class and middle class areas, such that 'a person may live in it for years, and go in and out daily without coming into contact with a working-people's quarter or even with workers' (Engels, 2001[1845], 107-8). Much has changed in English cities since the time of Engels, but *You Get Me* is a reminder of how spatial and socio-economic divisions inter-relate in the city.

Group 2: Local Only

The public spaces in which mobile games are frequently played – waiting in queues and travelling on public transport – are often seen as representing 'dead time' (Symes, 2007). This is time that is both unproductive and devoid of possibilities for pleasure. The people and objects that populate these space/times are in some way not present for interaction. It is a constant in the literature that mobile games are seen as a kind of crutch to get people through these dull spells (Liu & Li, 2011, p.893; PricewaterhouseCoopers, 2012, p.6; Goh, Lee, & Low, 2012, p.800). The technological limitations of mobile games – the relatively poor graphics and simplistic gameplay – would not be tolerated in certain use contexts where there are competing possibilities for entertainment, but are tolerated when no such alternative exists, for example on a long commute (Liu & Li, 2011, p896). The kinds of gaming experiences mobiles provide are seen here as shallow or *casual* – a good way to pass a few minutes waiting for the bus, but instantly forgettable. That mobile games are less demanding of players' time is supported by studies looking at how people play mobile games. One report suggests that more than half US mobile gamers play in bursts of less than 15 minutes, and 87% play for bursts of less than 30 minutes (PricewaterhouseCoopers, 2012, p.17).

These public situations in which players use mobile games to kill time reconfigure the interactional ecology of which the player is a part by involving other locally present people in the gaming situation, though not as fellow players. People in this second group relate to the mobile game player in two potential ways: as strangers to be kept socially distant or as onlookers to be willingly or unwillingly drafted into a game audience. That is, mobile games can intensify the other's absence or make the other present.

The domestication of public space, as well as making domestic objects and repertoires present in the public space, can also be seen in an attempt to socially separate oneself from nearby strangers. Erving Goffman (1972) documented many of the behaviours through which city-dwellers handle the social proximity with strangers that characterises city life. He coined the term 'civil inattention' for patterns of behaviour where a person will acknowledge the presence of a stranger while closing off the possibility of significant social contact. Goffman's work has been influential in the mobilities literature, with the mobile phone seen as a tool drafted into some of the tactics of urban sociality that Goffman describes (for example, Jensen, 2006; Wilken, 2010). Symes, for example, describes mobiles as '[s]elf-containment devices' (p. 451) potentially used to symbolically reject social contact with others in a public space.

Playing a game involves focusing on the tasks that the game presents, and this often leads to a diminishing of attention to one's surroundings. Note that this is not a 'withdrawal' from physical space but is rather a reconfiguration of the network that takes place in physical space based on the particular kinds of tasks that the mobile game is demanding.

Many mobile games, rather than focussing the player's attention away from the immediate surroundings, revitalise the relationship between the player and others, making this second group of people audience members of the mobile game performance. This audience-performer relationship may take a number of different forms, from peeking over the shoulder of a fellow commuter playing a familiar game, to seeing players chase each other around city streets 'shooting' each other with their mobile phones in a game of *Botfighters* (It's Alive!, 2001) or *CitiTag* (Centre for New Media, 2004). Onlookers may be appreciative, finding the mobile game performance an amusing distraction, or be annoyed by the sounds of a fellow commuter playing without headphones or the disruption caused to a routine journey by a location-based urban game.

That mobile games can have these two seemingly contradictory functions with respect to this group – to signal removal from social interaction and to instigate a social encounter – speaks to what Michael Arnold (2003) calls the 'Janus-faced' nature of mobile phones, which provide opportunities for socializing and for separation. But mobile games do not only represent opportunities for socialization and separation, they also illuminate the tension between these that is characteristic of contemporary urban life. Mobile games represent a kind of elevated performance of routine behaviours with mobile technology that brings these taken-for-granted everyday behaviours into critical focus. Not all mobile games do this, of course, and we are more likely to find this sort of 'critical play' (Flanagan, 2009) in games that brand themselves as art games or experimental games. Rather than being a means of killing time and distracting the player's attention from dead time and a routine environment, these games are often planned events that attempt to revitalise this environment.

Group 2 Example: *PacManhattan*

PacManhattan (Tisch School of the Arts, 2004) is one such game. It is specifically designed to create interactions between the player and members of this second group, in ways that cast a critical eye on the meaning of space and sociality in the city.

In *Pac-Manhattan*, a game that took place in New York in 2004, five players in Washington Park – one representing Pac-Man and the other four representing the ghosts who are chasing him – were each teamed up with a 'general' at a computer, to whom they reported their position at every intersection. The general updated the game map according to this information and advised the player on where to go to avoid the ghosts or to hunt down Pac-Man. The players dressed up in costumes reflecting the character they represented, and this put the players in a particular kind of playful relationship to the pedestrians who were not part of the game. Here, space was not being domesticated for the players, but was being made playful for the people in group 2.

In *The Ludic City* (2007) Quentin Stevens discusses how a playful city can act as critique of elite power and can foster a grassroots resistance to such power. Based very much in de Certeau's conception of 'tactics', play is here seen as a bottom-up disruption of the everyday routines of the city. These established routines are seen to privilege commerce and business and are inscribed in spatial structures that are imposed from the top-down. The creation of situations in which a space that has a particular character is made playful has a long history, including the medieval European carnival and the counter-cultural *happenings* of 1960s US hippy culture. It has always had political implications. Mikhail Bakhtin (1941 [1965]) saw carnival as a challenge, but ultimately a reconstitution, of the prevailing Christian social order. The playfulness of 1960s happenings was an intentional and direct challenge to a seemingly humourless, bourgeois space whose function was exhausted by economic productivity. Play is potentially subversive of this kind of commercial functionalism because it is, as anthropologist Roger Caillois has it, 'an occasion of pure waste' (2001 [1961], p. 5). *Pac-Manhattan*, so this line of thinking goes, presents members of this second group – the onlookers in Washington Park – with a radical alternative to conventional ideas of how space should be used. It is one of the means by which people engage in local, bottom-up practices that oppose top-down socio-spatial configurations.

Group 3: Local and Mediated

This intersection is where a lot of the most interesting location-based mobile gaming occurs. People and objects in this group are presented to the player through the game interface but are also present without it. Usually there are differences between the information available through the game interface and the information available without it, and players must combine these two sources of information to play effectively. In *Pac-Manhattan*, for example, the mobile is used to give the approximate location of the opponent, but players must then physically tag each other.

In *Pac-Manhattan*, and in other chasing-type games like *CitiTag* and *Botfighters*, players know who is playing. This is not always the case, and guessing who is in group 1 and who is in group 3 – and acting on that – is part of the risk of a number of mobile games. This risk brings up an ethical question about contact between players who are present to each other in both a mediated and unmediated sense. Players may be comfortable with being present in one sense but not the other, or may be comfortable with certain forms of mediation but not others. This is an issue that is discussed by Christian Licoppe and Yoriko Inada (2006) with respect to the Tokyo-based location aware mobile game *Mogi: Item Hunt* (Castelli, 2003). In this game players can work together to find virtual items hidden across the city. Often these collaborators chat over the game interface without actually meeting, putting them in group 1 of the diagram in figure 1, but opportunities for unmediated co-presence arise naturally since players will from time to time notice that another player is close by. These meetings may also be engineered by one or other player, with or without the other player's approval, leading to ethical issues around privacy and cyber-stalking. Licoppe and Inada document interchanges between players of this game where there is a clear and ethically-charged frisson involved in the potential of moving

Figure 1. Venn diagram illustrating how different groups of people and objects present themselves to the mobile game player

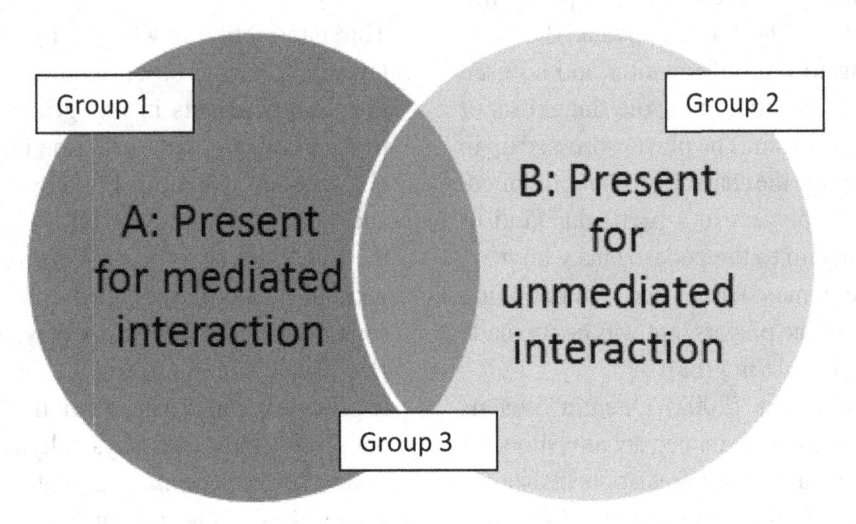

between mediated and non-mediated forms of interaction (2006, pp.53,55).

Looking at this same intersection and thinking about the objects that are present for both mediated and unmediated interaction leads to a consideration of what are sometimes called *augmented reality* games. In these kinds of games the phone registers objects and locations in its vicinity and provides extra information about them, which may be true, false or fictional but always, of course, selective. There is a clear rhetoric in the term 'augmented reality', where mediation is seen as simple addition to what is already there. What is ignored in the term is the extent to which the selectivity of what is added constructs a particular kind of reality and the extent to which the inclusion of extra information removes certain other forms of information either from the user's consciousness or from the environment itself.

By altering the information that is available with respect to a player's immediate environment, a game changes the character of that environment. As mentioned already, a game will often make a space playful, that is, it will bring out the space's potentials for playful interaction. For McGonigal this kind of gaming is transformative of urban experience, in that it can change:

> everyday objects and places into interactive platforms and also activate players by making them more responsive to potential calls to interaction. This is because the act of exposing previously unperceived affordances creates a more meaningful relationship between the actor and the object or space in the world (McGonigal, 2011, p. 236).

There is a dangerous elision here of the role of the designer in creating particular kinds of affordances. McGonigal's thesis has a blind spot that is caused by a kind of naturalism. This naturalism would have it that the game designer is not engaged in the political act of shaping spaces – in spatialization – but rather in revealing the natural potentials hidden in space. This ignores the work being done to bring certain potentials to the surface and submerge others. This is not to say that mobile games cannot be used to enrich people's lives, but that this enrichment is not through the enactment of some dormant and natural set of socio-political relations that are waiting to be revealed.

One of the repeated assertions about location-based games is that they make invisible aspects of the city visible. Richardson cites one player of the

Tokyo-based game *Mogi* who suggested the game allowed him to discover previously unfamiliar parts of the city (Hall, 2004, cited in Richardson, 2010, p. 435). For this player, a renewed vision of Tokyo was a positive thing, but it is important to avoid seeing this as a situation of pure gain. Jordan Frith quotes John Montgomery's definition of legibility, in turn developed from Kevin Lynch's influential *Image of the City* (1960), as: 'the degree to which the different elements of the city (defined as paths, edges, districts, nodes, and landmarks) are organized into a coherent and recognizable pattern' (Montgomery, 1998, p.100, cited in Frith, 2013, p.250). While breaking the city up into these different elements is certainly one way of thinking about the city, it is only one way. As has been noted elsewhere, urban planning theory and representations of the city do not reveal the city as it is but discursively create the city in ways that enshrine particular socio-political relations (for example, Mele, 2000). By looking at the city from the perspective of the urban planner, the legible city becomes the city as it is read by the planner's 'ideal reader' (to adapt Culler, 2001) who is, in fact, an *ideal citizen*. Lynch's city is an ordered city fit for governance and commerce. This is not to say that this image of the city is not useful, but that it has a particular discursive provenance and function. Frith uses the concept of 'legibility' alongside Dourish and Bell's assertion that mobile technology can 'make the invisible visible' (Dourish & Bell, 2011, p.195). But this is not a simple move from invisible to visible or from illegible to legible. Rather it is the discursive formation of reality.

Mobile games, to the extent that they allow for radical spatializations of the city, play with different visions of the city. At their most politically radical they do not reveal the true nature of the city but rather contest the naturalness of any one of these visions. All location-based urban games spatialize the city as playground, but this can mean very different things depending on the game.

Group 3 Example: *Ingress*

For Larissa Hjorth, GPS in mobile technology 'has allowed for multiple cartographies of a sense of space in which the geographic and physical is overlaid with the electronic, the emotional and the social' (2011, p.357). Of course, the emotional and the social have always been involved with the geographic and physical. What is new is the way in which the electronic is folded into this network of relationships. For Hjorth, urban mobile gaming 'seeks to challenge everyday conventions and routines that shape the cityscape' (p.358). While games like *Pac-Manhattan* and *You Get Me* certainly have this emancipatory and resistant project in mind, mobile games may also spatialize in such a way as to conform to and more deeply root down everyday conventions, routines and networks of relationships.

Frith (2013) discusses this possibility with respect to the geotagging game *Foursquare* (Crowley & Selvadurai, 2009). Geotagging is the process of adding geographical data to digital media such as digital photographs or video. Within a game context, players might upload data about locations they visit to a server run by the company hosting the game. Some of this information is then available to other players, and some of it is used by the company or sold to third parties to better understand the behaviour of potential consumers, or to create or enrich digital products such as maps, directories and search engines.

Hjorth compares mobile games to Guy Debord's *dérive*, a tactic for challenging 'the increasing commercialization that Debord viewed within everyday Parisian life' (2011, p.359). While many mobile games may do this, *Foursquare* is certainly not one of them, as it very clearly structures practices in terms of consumption. As Frith quips, '[f]or now at least, geotagging works in a similar fashion to public graffiti – if spray paint required a monthly data plan' (2012, p.145). For Marc Tuters (2012) this kind of game challenges the recent academic tendency to celebrate location-based games as contemporary *dérives*. Perhaps this is simply one more element of the 'Janus-faced' character of the mobile, making location-based games potentially both resistant and conservative. Tuters' argument goes further than this, though. He maintains that because the nature of capitalism has changed since the days of Guy Debord, the kind of practice that could have been understood as resistant has also changed. The kinds of spatial practices that mobile gamers engage in match the situationists' *dérive* and contest the logic of production, but, in a post-Fordist economy, this logic of production is redundant anyway. Caillois' idea of play being opposed to productivity loses its critical force and the gaming practices are rendered politically hollow.

If *Foursquare* rewards consumption by turning it into play, *Ingress* puts players to productive use. *Ingress* is a large scale mobile game commissioned by Google and at the time of writing (October 2013) still in the beta testing phase. It is a sci-fi adventure set in an alternative reality in which two factions of players – the Enlightened and the Resistance – battle to control locations, called 'Portals', around the world. Players gain control of these Portals by checking into selected locations, just as in *Foursquare*. The central server checks the phone's location using GPS and awards the portal to the player's faction. Players in the opposing faction can win it back in a similar way.

Ingress is a game, and judging by the many positive reviews on the Google Play Store where it is available for free download, it is a game enjoyed by many people. It is also a means of gathering potentially valuable information, both about people's behaviour and about things in the world, information that can be monetized in a number of ways (Hodson, 2012). Information about people's behaviour is valuable to advertisers and marketers. Information about things in the world is important in developing mobile search applications and augmented reality products such as Google's *Project Glass*. *Ingress* is a game, but it is also a means of recruiting large numbers of people in diverse locations to assist in Google's corporate objectives.

What is wrong with this? While players of *Ingress* are providing labour to Google without financial remuneration, they are getting something out of the deal. The real question here is how this kind of relationship between a company and its players who are also workers might lead to exploitation of player-workers (Fuchs, 2012; Hong & Chen, 2013; Kücklich, 2005). To what extent are players aware of the work that they are doing and its implications in terms of ownership and privacy? How does this relate to such labour relations concepts as organisation (Kücklich, 2005)? In other words, is the company-consumer relationship – and the legislation and regulations governing this – adequate to ensure fairness in the company-prosumer relationship?

Ingress is also an advertisement in itself. Or rather, it is a demonstration. The conspiracy theory backstory is inspired in part by the convoluted and much debated plot of the TV show *Lost* (Lieber, Abrams, & Lindelof, 2004-2010). While the intricate and baffling plot in *Lost* encouraged a great deal of debate across different websites, the goal of *Ingress* is, in part, to contain this kind of buzz within Google applications. Of course, there is nothing stopping players discussing plot points and sharing clues using different Internet products, but the game is set up for players to conduct collaborative research through products like Google+ and Google Drive. The relationship between alternate reality games, convergence culture and marketing has been discussed elsewhere (Ornebring, 2007) but what is new about *Ingress* is that audience behaviour is converging within a suite of software applications offered by a single corporation. Henry Jenkins takes a positive view of the educational benefits of this kind of alternate reality game. These games, 'teach participants how to navigate complex information environments and how to pool their knowledge and work together in teams to solve problems' (Jenkins, 2006, p. 126-7). *Ingress* teaches participants that the way to do this is with Google software. The game's slogan is 'The world around you is not what it seems', and the suggestion is that Google is the means to reveal the truth behind everyday things.

CONCLUSION

Mobile gaming provides a platform for socio-political performance and critique because it shapes the network of relationships – relationships with people and objects both designed by game companies and artists, involved in the game, and separate from it – within which a player finds meaning.

The model put forward here is intended to clarify how mobile games can shape these relationships. Some people and objects in a player's interactional ecology will be available for unmediated interaction, some for mediated interaction, and some for both unmediated and mediated interaction. Each of these sorts of relationships lends itself to different political interventions, whether these critique, revitalise or reinforce existing socio-spatial relations.

As with all forms of popular culture and all forms of technology, this kind of mobile gaming is neither fundamentally emancipatory nor fundamentally regressive. It is put to use by designers with particular political, economic and aesthetic agendas, used by players with their own agendas, and shaped by the dynamic network of relationships of which it is a part.

REFERENCES

Aarseth, E. (2000). Allegories of Space: The Question of Spatiality in Computer Games. In R. Koskimaa (Ed.), *Cybertext Yearbook*. Jyvaskyla: University of Jyvaskyla.

Adey, P. (2010). Vertical Security in the Mega-city: Legibility, Mobility and Aerial Politics. *Theory, Culture & Society*, *27*(6), 51–67. doi:10.1177/0263276410380943

App. Annie, & IDC. (2013). *The Future of Mobile & Portable Gaming*. App. Annie IDC.

Arnold, M. (2003). On the phenomenology of technology: The 'Janus-faces' of mobile phones. *Information and Organization*, *13*(4), 231–256. doi:10.1016/S1471-7727(03)00013-7

Bakhtin, M. (1941). *Rabelais and His World* (H. Iswolsky, Trans.). Bloomington, IN: Indiana University Press.

Blast Theory. (2008). *You Get Me*. London: Deloitte Ignite.

Borden, I. (2001). *Skateboarding, Space and the City: Architecture and the body*. Oxford, UK: Berg.

Bowen, T. (2013). Graffiti as Spatializing Practice and Performance. *Rhizomes, 25*.

Caillois, R. (2001). *Man, Play and Games* (M. Barash, Trans.). Urbana, IL: University of Illinois Press.

Calleja, G. (2012). Erasing the Magic Circle. In J. R. Sageng, H. Fossheim, & T. M. Larsen (Eds.), *The Philosophy of Computer Games*. London: Springer Press. doi:10.1007/978-94-007-4249-9_6

Castelli, M. (2003). *Mogi: Item Hunt*. Tokyo, Japan: Newt Games.

Centre for New Media. (2004). *CitiTag*. Centre for New Media.

Consalvo, M. (2009). There is No Magic Circle. *Games and Culture*, *4*(4), 408–417. doi:10.1177/1555412009343575

Copier, M. (2009). Challenging the magic circle: How online role-playing games are negotiated by everyday life. In M. v. Boomen, S. Lammes, A.-S. Lehmann, J. Raessens, & M. T. Schafer (Eds.), *Digital Material: Tracing New Media in Everyday Life and Technology*. Amsterdam: Amsterdam University Press.

Crowley, D., & Selvadurai, N. (2009). *Foursquare*. New York: Foursquare.

Culler, J. (2001). *The Pursuit of Signs: Semiotics, Literature, Deconstruction*. New York: Routledge.

de Certeau, M. (1984). *The Practice of Everyday Life* (S. Rendall, Trans.). Berkeley, CA: University of California Press.

de Souza e Silva, A., & Delacruz, G. C. (2006). Hybrid reality games reframed: Potential uses in educational contexts. *Games and Culture*, *1*(3), 231–251. doi:10.1177/1555412006290443

de Souza e Silva. A. (2006). Interfaces of Hybrid Spaces. In A. Kavoori & N. Arceneaux (Eds.), The Cell Phone Reader: Essays in Social Transformation. New York, NY: Peter Lang.

Dourish, P., & Bell, G. (2011). *Divining a Digital Future*. Cambridge, MA: MIT Press. doi:10.7551/mitpress/9780262015554.001.0001

Engels, F. (2001). *The Condition of the Working Class in England*. London: ElecBook.

Evans-Cowley, J. (2010). Planning in the Real-Time City: The Future of Mobile Technology. *Journal of Planning Literature*, *25*(2), 136–149. doi:10.1177/0885412210394100

Flanagan, M. (2009). *Critical Play: Radical game design*. Cambridge, MA: MIT Press.

Frith, J. (2012). Splintered Space: Hybrid Spaces and Differential Mobility. *Mobilities*, *7*(1), 131–179. doi:10.1080/17450101.2012.631815

Frith, J. (2013). Turning Life into a Game: Foursquare, Gamification, and Personal Mobility. *Mobile Media & Communication*, *1*(2), 248–262. doi:10.1177/2050157912474811

Fuchs, C. (2012). The Political Economy of Privacy on Facebook. *Television & New Media*, *13*(2), 139–159. doi:10.1177/1527476411415699

Goffman, E. (1972). *Relations in public*. Transaction Books.

Goh, D. H.-L., Lee, C. S., & Low, G. (2012). I played games as there was nothing else to do: Understanding motivations for using mobile content sharing games. *Online Information Review*, *36*(6), 784–806. doi:10.1108/14684521211287891

Graft, K. (2011). *DICE 2011: EA's Boatman Busts Five Mobile Gaming Myths*. Retrieved 27th June, 2013, from http://www.gamasutra.com/view/news/32977/DICE_2011_EAs_Boatman_Busts_Five_Mobile_Gaming_Myths.php

Hjorth, L. (2011). Mobile@game Cultures: The Place of Urban Mobile Gaming. *Convergence: The International Journal of Research into New Media Technologies*, *17*(4), 357–371. doi:10.1177/1354856511414342

Hodson, H. (2012). Why Google's *Ingress* game is a data gold mine. *New Scientist*, 2893.

Hoflich, J. R. (2010). Mobile media and the change of everyday life: A short introduction. In J. R. Hoflich, G. F. Kircher, C. Linke, & I. Schlote (Eds.), *Mobile Media and the Change of Everyday Life*. Frankfurt: Peter Lang.

Hong, R., & Chen, V. H.-H. (2013). Becoming and ideal co-creator: Web materiality and intensive laboring practices in game modding. *New Media & Society*, 1–16.

Huizinga, J. (1955). *Homo Ludens*. Boston: Beacon Press.

iResearch Consulting Group. (2010). *2009年中国手机游戏用户行为?研报告*. Author.

It's Alive!. (2001). *BotFighters*. It's Alive!.

Jenkins, H. (2006). *Convergence Culture: Where Old and New Media Collide*. New York: New York University Press.

Jensen, O. B. (2006). 'Facework', Flow and the City: Simmel, Goffman, and Mobility in the Contemporary City. *Mobilities*, *1*(2), 143–165. doi:10.1080/17450100600726506

Joffe, B. (2005). Mogi: Location and presence in a Pervasive Community Game. In *Proceedings of Ubicomp Workshop on Ubiquitous Gaming and Entertainment*. Academic Press.

Klabbers, J. H. (2009). *The magic circle: Principles of gaming & simulation*. Rotterdam, The Netherlands: Sense Publishers.

Kücklich, J. (2005). Precarious Playbour: Modders and the Digital Games Industry. *The Fibreculture Journal, 5*.

Lacohee, H., & Wakeford, N. (2003). A Social History of the Mobile Telephone with a View of its Future. *BT Technology Journal*, *21*(3), 203–211. doi:10.1023/A:1025187821567

Lefebvre, H. (1991). *The Production of Space* (D. Nicholson-Smith, Trans.). Oxford, UK: Blackwell.

Licoppe, C., & Inada, Y. (2006). Emergent uses of a multiplayer location-aware mobile game: The interactional consequences of mediated encounters. *Mobilities*, *1*(1), 39–61. doi:10.1080/17450100500489221

Liu, Y., & Li, H. (2011). Exploring the impact of use context on mobile hedonic services adoption: An empirical study on mobile gaming in China. *Computers in Human Behavior*, *27*, 890–898. doi:10.1016/j.chb.2010.11.014

Lynch, K. (1960). *The Image of the City*. Cambridge, MA: MIT Press.

Mainwaring, S. D., Anderson, K., & Chang, M. F. (2005). *Living for the Global City: Mobile Kits, Urban Interfaces, and Ubicomp*. Paper presented at UbiComp. New York, NY.

McGonigal, J. (2011). *Reality is Broken: Why games make us better and how they can change the world*. London: Penguin.

Mele, C. (2000). The Materiality of Urban Discourse: Rational Planning in the Restructuring of the Early Twentieth-Century Ghetto. *Urban Affairs Review*, *35*(5), 628–648. doi:10.1177/10780870022184570

Montgomery, J. (1998). Making a City: Urbanity, Vitality, and Urban Design. *Journal of Urban Design*, *3*(1), 93–116. doi:10.1080/13574809808724418

MoPub. (2012). *Mobile Games Played Most Often at Home*. San Francisco, CA: MoPub Miniclip.

Niantic Labs. (2012). *Ingress*. Google.

Nitsche, M. (2008). *Video Game Spaces: Image, Play, and Structure in 3D Worlds*. Cambridge, MA: MIT Press. doi:10.7551/mitpress/9780262141017.001.0001

O'Dell, T. (2009). My Soul for a Seat: Commuting and the Routines of Mobility. In R. Wilk, & E. Shove (Eds.), *Time, Consumption and Everyday Life*. Oxford, UK: Berg.

Ornebring, H. (2007). Alternate reality gaming and convergence culture: The case of Alias. *International Journal of Cultural Studies*, *10*(4), 445–462. doi:10.1177/1367877907083079

Pajitnov, A. (1984). *Tetris*. Academic Press.

Parikka, J., & Suominen, J. (2006). Victorian Snakes? Towards a Cultural History of Mobile Games and the Experience of Movement. *Game Studies, 6*(1).

Plant, S. (2001). *On the mobile: The effects of mobile telephones on social and individual life*. Motorola.

PricewaterhouseCoopers. (2012). *The evolution of video gaming and content consumption*. Pricewaterhouse Coopers.

Rheingold, H. (2003). *Smart Mobs: The next social revolution*. Cambridge, MA: Basic Books.

Richardson, I. (2010). Ludic Mobilities: The Corporealities of Mobile Gaming. *Mobilities*, *5*(4), 431–447. doi:10.1080/17450101.2010.510329

Salen, K., & Zimmerman, E. (2004). *Rules of Play: Game Design Fundamentals*. Cambridge, MA: MIT Press.

Stevens, Q. (2007). *The Ludic City*. London: Routledge.

Symes, C. (2007). Coaching and Training: An Ethnography of Student Commuting on Sydney's Suburban Trains. *Mobilities*, *2*(3), 443–461. doi:10.1080/17450100701597434

Tassi, P. (2012). Is the Playstation Vita a Dinosaur Already? *Forbes*. Retrieved 27th July, 2013, from http://www.forbes.com/sites/insertcoin/2012/02/17/is-the-playstation-vita-a-dinosaur-already/

Tisch School of the Arts. (2004). *Pac-Manhattan*. New York: Tisch School of the Arts.

Totilo, S. (2012). PlayStation Vita Review. *Techspot*. Retrieved 27th June, 2013, from http://www.techspot.com/review/500-playstation-vita/

Trinklein, E., & Parker, G. (2013). *Combining multiple GPS receivers to enhance relative distance measurements*. Paper presented at the Sensors Applications Symposium. Galveston, TX.

Tuters, M. (2012). From mannerist situationism to situated media. *Convergence: The International Journal of Research into New Media Technologies*, *18*(3), 267–282. doi:10.1177/1354856512441149

Wang, Y., Yang, X., Zhao, Y., Liu, Y., & Cuthbert, L. (2013). *Bluetooth Positioning using RSSI and Triangulation Methods*. Paper presented at the Consumer Communicaitons and Networking Conference. Las Vegas, NV.

Warf, B., & Arias, S. (2008). Spatial Turn, The: Interdisciplinary Perspectives. In *Routledge Studies in Human Geography*. New York: Taylor & Francis.

Wilken, R. (2010). A Community of Strangers? Mobile Media, Art, Tactility and Urban Encounters with the Other. *Mobilities*, *5*(4), 449–468. doi:10.1080/17450101.2010.510330

Wisniewski, D., Morton, D., Robbins, B., Welch, J., DeBenedictis, S., Dunin, E., et al. (2005). *2005 Mobile Games White Paper*. International Game Developers Association.

KEY TERMS AND DEFINITIONS

Augmented Reality Gaming: In these kinds of games the phone registers objects and locations in its vicinity and provides extra information about them, which are used to create opportunities for gameplay.

Critical Play: A term coined by Mary Flanagan (2009), it is used here to think about the ways in which certain games cast a critical eye on conventional uses of space and conventional socio-spatial relations.

Interactional Ecologies: Overlapping systems of interactions that constitute socio-spatial relations.

Mobile Gaming: While wider definitions are possible, for the purposes of this chapter mobile gaming is restricted to digital games played on or with mobile phones and tablets.

Presence: The term is understood differently in this chapter than in the VR and presence literature. Here, it is understood to mean the various ways in which people and objects present themselves to a person playing a mobile game.

Socio-Spatial Relations: Social relations as they are defined by or mediated through spatial relations.

Spatialization: The shaping of the character of space and socio-spatial relations through structures of power and processes of resistance.

Chapter 16
Technology Facility and News Affinity:
Predictors of Using Mobile Phones as a News Device

Xigen Li
City University of Hong Kong, China

ABSTRACT

Based on a theoretical framework drawn from the diffusion of innovation theory, the expectancy-value model, and the technology-acceptance model, this chapter presents an empirical study of technological and informational factors as predictors of the use of second-generation mobile phones as a news device. The study differentiates the initial adoption of a mobile phone as a technology innovation from second-level adoption, which refers to the acceptance of a distinctive new function of a technology device serving a communication purpose different from that for which the device was originally designed. The study found that technology facility and innovativeness were significant predictors of mobile phone use as a news device; further, it partially confirmed the model of predictors of mobile phone use for news access. However, the two informational factors—perceived value of information and news affinity—were found to have no direct effect on mobile phone use as a news device. The study departs from the traditional approach of adoption research and offers a novel perspective on examining the adoption of new media with multiple evolving functions.

INTRODUCTION

The mobile phone, a technology innovation originally designed as a portable device for interpersonal communication, has evolved in its functions as a multimedia device (e.g., camera), a media player, and a media channel through which to access news information. Technological advances have expanded the mobile phone's capacity to enable communication for various purposes (Goggin & Hjorth, 2009). The use of mobile phones as a news device has been possible since the mid 2000s, when the second-generation of mobile phones was configured for digital data transmission and mobile Internet. The third generation mobile phones, such as Apple's iPhone

DOI: 10.4018/978-1-4666-6166-0.ch016

and the phones on Android system have expanded transmission capabilities making it easier for a mobile phone user to browse the Internet and access news information, including photos and videos. More recently, mobile phones have become a functional device to access news information due to the growing information channels available through mobile phone applications. While traditional media, such as newspapers and television, and the Internet still serve as mainstream media to deliver news information to a large number of people, the advancement of media technology and mobile phone applications has caused the mobile phone to gradually emerge as a multimedia device that enables users to browse information from news websites and get news information through various applications and social media. As a result, more people have started to access news information through mobile phones. For those people who need quick access to news information and who tend to read or watch headline news, the second-generation mobile phones worked as an efficient media channel. With an increasing amount of information accessible through mobile phones combined with the flexibility of access to news through them, the mobile phone has become a viable alternative to traditional media and the Internet as a media channel through which to obtain news information.

The changes brought about by the use of mobile phones as a news device could have a significant impact in the long term on how news information is distributed and on how people access and react to news information (Bivens, 2008). As an effective channel of delivering news information, mobile phones have already brought about significant changes to both news consumers and society. For news consumers, because mobile phones are always at hand, news information is accessible immediately through various platforms and applications as soon as it is distributed. To get the latest news, mobile phone users can open a news application on the phone any time. Using mobile phones to access news has become a daily routine for many news consumers. When mobile phones are used by a large number of people as devices to access news, the ways that news information is produced and delivered and the extent to which media content influences its audience could change significantly due to the pattern of news consumption and media exposure.

At the society level, the use of mobile phones as a news device will result in quicker responses to social events and changes. For example, mobile phones could serve as a more effective channel in the situation of a disaster or a crisis than most of the regular information channels. Before the Internet became widely available, people used to rely on traditional media for information during a disaster (Neal, 1998; Piotrowski & Armstrong, 1998). More recently, they have turned to the Internet for information. Now people can turn to mobile phones to get news information and explanations and interpretations of the event. Mobile phones could become the most widely used and depended on media when people need to obtain information urgently. Immediate access to news information, especially information about a disastrous event or a crisis, enables people to react to the situation in a quicker and more reasoned way, resulting in a significant reduction in the impact of the crisis. The information distributed through mobile phones to the public in a crisis or an important social event will have significant positive effects on public interest and social order.

As more people start to use mobile phones as news devices, many important questions regarding the use of mobile phones as a multimedia communication device remain unanswered. Few studies have explored the use of mobile phones as a multimedia device (Ling, 2004; Wei & Lo, 2006), and even fewer have examined the use of mobile phones as a news device (Humphreys, Von Pape, & Karnowski, 2013; Westlund, 2008). The use of mobile phones as a news device is based on an adopted innovation, but most of the adoption studies have examined such multimedia media adoption as a one-time event. Few studies have explored the factors that influence the adoption of a distinctive new function of an adopted technology device beyond its primary function. This study examines the adoption and use of mobile phone for news access in the late 2000s when second-generation mobile phones became functional devices for people to access news. It explores the factors that influence the adoption of a distinctive multimedia communication function of mobile phones, which were originally designed for interpersonal communication.

NEWS CONSUMPTION AND MEDIA TECHNOLOGY

News informs people about the latest developments of current events and issues and provides useful information to people in an ever-changing society. Regular consumption of news affects people's social and everyday life (Allan, 1999). News was primarily disseminated through traditional media, such as newspapers and television, before the emergence of the Internet in the 1990s. In the case of traditional media, news is usually accessed at home and/or in the after-work hours. Because the availability and accessibility of media channels at a given time are the main reasons for media selection, the patterns of news consumption change with the development of media technology. As media technology advances, consumers have more options and the media channels on which they rely for news consumption gradually shift to those that are superior in functionality and content compared to traditional media. The Internet, a new medium that works as a platform to deliver news information like the traditional media, has changed profoundly the ways in which news content is consumed and perceived (McLuhan, 1994). Wilzig and Avigdor (2004) examined the evolution of the Internet, with special emphasis on its impact on the traditional media and the consequences of new media growth. They highlighted a six-stage natural life cycle model of new media evolution, comprising birth (technical invention); penetration; growth; maturity; self-defense; and adaptation, convergence or obsolescence. In the early years of the Internet, the World Wide Web was used by individual consumers mainly for entertainment and was considered to be only complimentary to news from traditional media (Althaus & Tewksbury, 2000; Vyas, Singh, & Shilpi, 2007). For example, Ahlers (2006) noted that only 12% of news consumers moved directly from traditional news media to electronic news media. Furthermore, another 22% of U.S. adults reported that they used the Web as a complement to rather than a substitute for traditional news media (Ahlers, 2006; Dutta-Bergman, 2004). Thus, in the early 2000s, the Internet as a new medium for news was still in the stages of penetration and growth.

However, new media such as the Internet are significantly different from the traditional media in the scope of content they provide and in their means of information delivery. The amount of news information available on the Web far exceeds and is far more accessible than that provided by traditional media. Further, compared to traditional news media, new media such as the Internet offer a higher level of selectivity and therefore much more value in the information obtained and the extent to which one's information needs are fulfilled. In the late 1990s, a variety of media channels entered the scene previously dominated by traditional media to provide news information to consumers, beginning with the various news platforms on the Internet, such as websites of traditional media and news portals. The interactive features of these channels, combined with the ability to offer up-to-date information, make newspaper websites an attractive source for news (Vyas, et al., 2007). Later all types of social media came to serve as sources of news information (Hermida, 2010). As a result, studies have found a diminishing role of newspapers (Schrøder & Steeg Larsen, 2010). New media as platforms to deliver news demonstrated the potential to replace traditional media as the main channels for news (Kayany & Yelsma, 2000).

In the late 1990s, media convergence – the integration of media content from multiple media platforms to provide information in a more effective manner than one single media platform and serve audiences of different types – emerged as a visible trend; it has been an increasingly common practice in the news industry ever since (Grant & Wilkinson, 2009; Jensen, 2010). The use of mobile phones for news access is part of this media convergence, providing a new means of news consumption made possible by the advent of the Internet as a platform to deliver news information. As a result of the development of mobile technology, mobile phones have come to possess all the advantages of the Internet in terms of the scope of information provided and the methods of information delivery. In addition, mobile phones are more accessible than personal computers in obtaining news from various sources. Ecological approaches to mediated communication suggest that changes in communication technology would alter the human experience of media use in various aspects (Gibson & Pick, 2000). The growth of mobile phone use for news access reflects such changes in the human experience of media use due to technology advances.

The use of mobile phones for news access offers audiences a new experience in obtaining news and affords additional advantages for people to access news about current events in the world. Frequent usage of mobile news services has been associated with specific lifestyles, including the lifestyles of people who often engage in activities outside their home and business and/or whose jobs require regular travel (Westlund, 2008). Mobile phone use for news access has also gradually gained popularity as the functionality of the mobile phone as a news device has improved. In a study conducted in 2006, Westlund (2008) found that about 20% of mobile phone users expressed uncertainty regarding the issue of whether a mobile device was a good way to stay up to date about the news. In the late 2000s, mobile phones were more likely to be used in situations where no other news media are accessible and could be considered supplementary to other news media for certain groups of people. These people actively sought information to satisfy their needs, orientations, and motivations; they were always connected to the Internet and had a special need for news wherever they went (Westlund, 2008). Scholars also found that mobile channels occupied a new niche: access in the interstices, or the gaps that media users need to fill between scheduled activities in their daily routines (Dimmick, Feaster, & Hoplamazian, 2011). A more recent study found that the perceived relative advantage (especially content), utility, and ease of use of mobile news were positively related to mobile news adoption; further, young adults' news consumption patterns and preferences, as well as media usage, were all found to play a role in the adoption of mobile news (Chan-Olmsted, Rim, & Zerba, 2013).

With more people using mobile phones to access news, mobile phone use for news has entered the stages of penetration and growth. For some of those users, it is expected that they may restore their previous media habits with the new medium once they have become accustomed to the technology (LaRose & Atkin, 1988). As the use of mobile phones for news access continues to grow, it is imperative to explore to what degree technological and informational factors influence the adoption and use of mobile phones as a news device. By examining the predictors of mobile phone use for news access through an adopted technology device, this chapter will contribute to the understanding of the adoption and use of a technology device for communication purposes other than that for which the device was originally designed.

THEORETICAL FRAMEWORK ON ADOPTION OF INNOVATIONS

Adoptions of technology innovations for communication needs have been studied from a variety of perspectives. The theoretical framework that guides this study comes from the diffusion of innovation theory, the technology acceptance model, and the expectancy-value model. We will review the three theoretical perspectives and discuss how they inform our examination of the effect of technological and informational factors on the use of mobile phones as a news device.

Diffusion of Innovation

The diffusion of innovation theory explains why people adopt technical innovations and what factors play a role in the adoption process (Rogers, 2003). According to Rogers (1986), the decision to adopt or reject an innovation is subject to a wide variety of factors that can be categorized in four groups: (1) adopter-related personality traits; (2) socioeconomic influences; (3) interpersonal channels and mass media use; and (4) perceived attributes of the innovation. The perceived attributes of the innovation are particularly influential in the adoption decision, explaining between 49 and 87 percent of the variance in the rate of adoption (Rogers, 2003). Five attributes relate to people's decision to adopt technology innovations: relative advantage, compatibility, complexity, trialability, and observability. Innovations will be adopted more quickly by individuals when they are perceived to have less complexity and more of the other four attributes (Rogers, 2003). Whereas some studies have confirmed that these five perceived attributes are significant predictors of the adoption of innovations (Kaminer, 1997; Lai & Guynes, 1994), other studies have showed that the effects of perceived attributes on the adoption of innovations are inconsistent. Tornatzky and Klein (1982) found that compatibility and relative advantage were usually, but not always consistently, related to the rate of adoption in a positive direction, and complexity was negatively related to the rate of adoption. A study on mobile phone adoption found a significant impact of compatibility and observability on the likelihood of adoption (Leung & Wei, 2000). Wei (2001) extended the earlier study of cell phone adoption longitudinally and found that only observability continued to have a significant impact on the likelihood of adoption. The perceived attributes of the innovation were found to have an indirect influence on the potential adoption of mobile phones by current non-adopters (Vishwanath & Golohaber, 2003).

Studies of diffusion of innovations also explored the effect of characteristics of the adopters of the innovations and other factors (Rogers, 2003). The relative influences of adopter personality traits, socioeconomic influences, and media exposure have been identified in diffusion research on a wide range of technologies, including videotex (Ettema, 1984, 1989), the Internet (Atkin & Jeffres, 1998), high-definition television (HDTV) (Dupagne, 1999), mobile phones (Leung & Wei, 2000), and digital cable (Kang, 2002). Personal attributes such as innovativeness and media usage were found to be significant predictors of online service (Lin, 2001) and Webcasting adoption (Lin, 2004). In light of the past research on the adoption of innovations, it seems clear that a user's perceptions about each of these characteristics of an innovation will affect the extent to which he or she adopts that innovation. However, Lin (1998) points out that Rogers' five attributes were constructed from a broad perspective, encompassing traditional agrarian innovations; thus, some of the attributes may be not applicable to newer technologies, such as personal computers. Lin's rationale for dropping three attributes (compatibility, trialability, and observability) was that, because personal computers were widely available, easily trialable and highly observable at that time, the discriminative power of the three attributes was significantly diminished and therefore negligible.

The same argument can be made for the use of mobile phones for news access. People who begin to use their mobile phones to access news have likely already used them for some time for their original purpose of making calls. Thus, it is reasonable to assume that compatibility, trialability, and observability are almost negligible in using mobile phones for news access.

Expectancy-Value Model

As part of the uses and gratifications theory, the expectancy-value model suggests that audiences are actively looking for information from media to fulfill their needs (Levy, 1978, 1987; Lin, 1993a; Rubin, 1983; Swanson, 1979). Palmgreen and Rayburn (1985) reformulated the uses and gratifications theory by specifying an increment of valued satisfactions obtained relative to an original expectation, thereby developing an expectancy-value model of media gratifications sought and obtained. The expectancy-value model proposes that media use can be explained by comparing the perceived benefits offered by the medium and the gratification derived from its use. The model has been widely tested in studying the adoption and use of both traditional media and new media (Babrow & Swanson, 1988; Jeffres & Atkin, 1996; LaRose & Atkin, 1992; Lin, 1993b; Philip Palmgreen & Rayburn, 1982; P. Palmgreen & Rayburn, 1985; Perse & Ferguson, 1994; Rayburn & Palmgreen, 1984; Rubin, 1983). In terms of new media adoption and use, the greater are the relative values or advantages provided by an innovation or technology, the greater will be the incentive to adopt it (Lehmann & Ostlund, 1974). Expectancy value orientations affect both intentions to use media and the frequency of use, mainly through their influence on attitudes (Babrow & Swanson, 1988). In their study on using a pager to seek news, Leung and Wei (1999) found that the expected value, especially the information-seeking benefit, significantly predicted the level of exposure to news via the pager. Perse and Ferguson's (1994) study found that the use of technologies such as cable and VCRs affects benefits perceived from television viewing and hence to increase the gratification received from watching television. The studies on new media adoption and use based on the expectancy-value model demonstrate that the perceived benefit of using a particular media channel positively predicts its adoption and use.

Technology Acceptance Model

The Technology Acceptance Model (TAM) shares some ideas similar to that of the expectancy-value model. While both include perception of benefits as a predictor of the use of new technologies, the TAM has been found to be simpler, easier to use, and more powerful in determining user acceptance of computer technology (Igbaria, Guimaraes, & Davis, 1995). Diffusion of innovations and TAM are similar in their emphasis of the influence of beliefs and external variables on the decision to adopt a technology. However, the TAM differs from Diffusion of innovation in its inclusion of individual attitudes towards the decision. According to the TAM, the decision to accept a technology is determined primarily by the user's attitudes towards the decision, which in turn mediate the impacts of beliefs and external variables on acceptance behavior (Davis, 1989).

Perceived usefulness and ease of use are two key constructs in the TAM (Davis, 1993; Davis, Bagozzi, & Warshaw, 1989; Karahanna & Straub, 1999). The model hypothesizes that perceived usefulness will correlate to the adoption and use of a given information technology. Numerous studies testing this hypothesis have shown that perceived usefulness has a strong influence, both directly and indirectly, on the use of computers and other technology devices (Davis, 1989; Dishaw & Strong, 1999; Karahanna & Straub, 1999). Perceived usefulness and perceived ease of use were also found to be important factors in producing a positive attitude toward online banking (Chau

& Lai, 2003) and to have an impact on online purchasing (McCloskey, 2003).

These theories and models offer ample insights into the factors that influence the adoption of technology innovations. However, as the technology advances and more innovations with multiple functions emerge, these theories and models appear to be inadequate in explaining the adoption of innovations with evolving distinctive functions, because none of them differentiates the initial adoption of a technology innovation from the subsequent adoption of its evolving functions. Diffusion of innovation theory and the TAM mainly focus on the adoption of a specific technology or a technology-based service. However, the use of mobile phones as a news device is not simply an adoption of a new technology innovation; rather, it is an adoption based on a previously adopted technology device. We refer to the adoption of a distinctive multimedia function based on an adopted technology device not originally designed for the new function as a second-level adoption. The above-discussed theories do not deal with such second-level adoptions. The expectancy-value model explores the effect of perceived benefits of media content on media use, but the model explains media use without considering the facilitation of technological factors in the process, whether from the technology device or from users.

Unlike innovations in the 1980s and the 1990s, most technology innovations today have multiple evolving functions, and adoptions of distinctive new functions take place at different times. Previous research has mostly treated the adoption of innovations as a one-time action; while the process of adoption could take a long time and the adoption curve explains the progress along the way, these both relate to the adoption of one innovation. Such a perspective in understanding adoption was suitable at a time when most media were developed with only one function and no further adoption occurred after the initial adoption. However, in the modern digital age, when most media have the capacity for further development to include distinctive new functions, the adoption process is no longer a one-time activity. The scope of media functions is the key concept that distinguishes the initial and subsequent adoptions based on the same technology device. When a distinctive function switch occurs, such as the use of mobile phones as a multimedia communication device for accessing news, the adoption of the distinctive new function has a specific application context, and the influencing factors may change from those involved in the initial adoption. Users are now constantly facing evolving distinctive functions associated with a medium they already own, and they must make decisions regarding whether to adopt the new functions and/or services based on the previously adopted medium. Thus, instead of examining media adoption as a one-time occurrence, studies of new media adoption need to explore subsequent adoptions based on the same technology device in order to better understand the adoption process in the digital age, especially when such a switch changes the scope of the device's function from interpersonal communication to multimedia communication, such as in the case of mobile phone use as a news device. Therefore, a new theoretical perspective is needed when examining technology innovations with evolving and distinctive new functions. Following this reasoning, we argue that the perspective of second-level adoption will expand our vision when examining the adoption and use of a distinctive function serving a communication purpose other than that for which the associated technological device was originally designed.

Second-Level Adoption

The concept of second-level adoption is advanced to differentiate the acceptance of a distinctive communication function serving a purpose other than that for which a technology device was originally designed from the initial adoption of that device. Adoption of new media usually takes place in one of two ways. The first is adopting a new medium as an individual innovation, such as in the case of television, cable or the VCR. The adoption happens when a new medium emerges and is considered to have advantages over the media currently in use. Such adoption is often a one-time occurrence, not evolved from a technology device already adopted, and independent from earlier adoptions of other devices. The second form of adoption is that of adopting additional media functions or services based on an adopted media device, such as adopting email, photo-editing, web browsing, or multimedia player functions on an adopted computer. The additional adopted functions of a computer are still within the computer's capacity for processing digital information.

Second-level adoption is different from the types of new media adoption discussed above in that it involves a distinctive function switch based on a previously adopted technology device. The initial adoption involves any technology innovation that has a primary function, such as the function of television for watching broadcast programs or the function of an MP3 player for playing digital music. As the technology advances, more functions could be developed and added to the initial technology innovation. If the additional functions are within the scope of the initial function, adoptions of the additional functions are still considered part of the initial adoption. Second-level adoption occurs only when a distinctive new function emerges for the adopted device, resulting in a use of the device for a purpose different from that for which it was originally designed. A mobile phone is one such technology device that contains additional functions distinctive from the initial function. The primary function of mobile phones is making calls, but they are now being used as a news device to access all types of information. Making calls and functioning as a news device are two distinctive functions based on the same technology innovation. The adoption of a mobile phone as an interpersonal communication device (i.e., making calls) does not necessarily lead to the use of the same mobile phone as a news device (Westlund, 2008). The use of mobile phones as a news device is a multimedia function that occurs after the initial adoption and drastically differs from the function for which the device was originally designed; thus, it can be considered a second-level adoption.

Second-level adoption is also different from reinvention. Reinvention is the degree to which an innovation is changed by the adopter in the process of adoption and implementation after its original development (Rice & Rogers, 1980). By reinventing, the adopter takes the initiative to change the innovation. Usually, such change is constrained to the range allowed by the original innovation and is limited in scope due to the capacity and resources the adopter possesses. Second-level adoption opens a new means of using the adopted technology devices for communication purposes other than those for which they were originally designed.

The idea of differentiating the stage of adoption has important implications for the study of new media. As technology advances, new functions continuously emerge and are added to the technology devices already in use. The adoption of such additional functions may be either initial or second-level adoption, which differ in the scope of media functions and the application context of technology devices. Past research examined the adoption of innovation from different perspectives, but few studies have examined adoptions from the perspective of changing the scope of media functions and the application context of initial and second-level adoptions. The scope of media functions, which refers to the degree to which media fulfill distinctive goals, defines the level of adoption. The application context, which refers to the setting in which a technology innovation is adopted and used, defines key factors that play a role in the process of adoption. Although studies have examined technology adoptions under different settings and found that the application context has an effect on the adoption of innovations consistently (Calantone, Griffith,

& Yalcinkaya, 2006; Nystrom, Ramamurthy, & Wilson, 2002), the differences in media functions and the application context between the initial and second-level adoptions have never been examined. Second-level adoptions are usually treated as initial adoptions, such as in the studies of online shopping (Soopramanien & Robertson, 2007), Webcasting (Lin, 2004), and mobile news service (Westlund, 2008). In the age of multi-functional media, the media that are adopted are most likely to be those with evolving distinctive functions. Examining the distinctive new functions of an adopted medium and the subsequent adoptions as second-level adoptions will bring a more illustrative perspective to the understanding of new media with evolving functions and the factors that influence the adoption of multi-functional media in the digital age.

Predictors of Second-Level Adoption

Second-level adoption evolves from the initial adoption of an innovation. The media device that offers new functions and services is already in use, and the distinctive function switch serves a totally different need. The factors that lead to initial adoption may not be crucial for second-level adoption. It is therefore necessary to identify and explore the predictors of second-level adoption.

Rogers (2003) noted that five technology attributes contribute to the adoption of an innovation. In this study, two of these five attributes – device advantage and functionality – are considered applicable to the new application context. In addition, information accessibility, a third factor specifically applied to the distinctive new function based on the medium that has been adopted, is introduced. Together, these three factors form a new construct, technology facility. When the adoption relates to the distinctive function or service based on an adopted innovation, the general personal traits and other general factors that played a role in the initial adoption may be of decreased importance in the process, while the factors that are associated with the distinctive new media functions will play a more important role. In the case of the use of mobile phones as a news device, news affinity is considered a factor that plays an important role in the new application context of the adopting a distinctive new function. The exploration of the function-related factors is necessary when examining the determinants of the adoption of distinctive functions or services based on an adopted innovation. A mobile phone user may not automatically adopt a distinctive function or service developed based on the adopted mobile phone; instead, certain personal traits of mobile phone users could be influencing factors in choosing to adopt the distinctive function. The specific factors linked to the new functions, such as news affinity, need to be considered when examining the potential adopters of a distinctive function or service based on an adopted innovation.

Based on the conceptual work of second-level adoption and the discussion of the changing context of second-level adoption, this study tries to identify and test the influencing factors in second-level adoption. Three such factors are identified and elucidated as follows.

Technology Facility

As discussed earlier, when examining computer adoption, Lin (1998) pointed out that the discriminative power of three of the five attributes suggested by Rogers (2003) – compatibility, trialability and observability – was significantly diminished and therefore negligible due to technological advances. Technically, a distinctive new function of an adopted innovation usually does not need to be tried before it can be used properly; if a trial period exists, it is likely to be so short as to be considered negligible. The attributes that remain applicable are relative advantage and complexity. The construct of technology facility was conceived as a leading factor in lieu of the technology attributes in the diffusion of innovation theory. Technology facility refers to the level of capacity of a technology device that enables a distinctive new function beyond its primary function. In the case of mobile phone use as a news device, this is the capacity of a mobile phone that allows users to access news information. This factor comprises three aspects: device advantage, functionality and information accessibility. Device advantage and functionality are similar to relative advantage and complexity but contain the components associated with the application context of second-level adoption. Information accessibility, an aspect applied specifically to mobile phone use as a news device, is often the most important factor in the use of information sources and media as well as in decision-making (Tybout, Sternthal, Malaviya, Bakamitsos, & Park, 2005).

Innovativeness

Innovativeness, a commonly used concept in the literature of adoption, refers to the degree to which one actively seeks and tries new technology innovations. Innovativeness involves risk-taking, creativity, and a desire for change. Innovators have an obsession with "venturesomeness" and risk, which leads to a proclivity to adopt (Rogers, 2003); as a result, early adopters have been found to possess a higher degree of personal innovativeness compared to other users. A valid measure of innovativeness involves the intelligent, creative, and selective use of communication for solving problems (Midgley & Dowling, 1978) and willingness to change (Hurt, Joseph, & Cook, 1977). These characteristics play a significant role in the adoption of innovations. The adoption of evolving distinctive new functions of an adopted technology device, although not as technologically challenging as those in the initially adopted technology innovation, as suggested by the diffusion of innovation theory, is still considered to be influenced by innovativeness.

News Affinity

Media use is often considered to be a predictor of adoption decisions. A study have found that media use is related to the adoption of news-related devices and new media (Li, 2004). Use of a new medium for news information is associated with news access from other media (Perse & Dunn, 1998). Media use, usually measured by the time spent with media, has often been found to be a weak predictor of adoption. In the case of the use of mobile phones as a news device, the driving forces for using the new medium to access news information are well beyond simple media use. The media dependency theory offers some rationale to identify the influencing factors. This theory predicts that people depend on media information to meet certain needs and achieve certain goals; thus, they depend more on those media that meet more of their needs (Ball-Rokeach & DeFleur, 1976). News affinity, a construct based on the theory of media dependency, is identified for the case of mobile phone use as a news device. News affinity refers to the degree to which one's use of and reliance on news across media constitutes an indispensable need in one's life. Two aspects form this construct: 1) news reliance, the degree to which one feels his/her personal fulfillment in life relies on news media; and 2) news access from the general media. In the case of mobile phone use as a news device, news affinity is expected to be a specific factor that plays a role in predicting the adoption of a distinctive new function regarding news access.

HYPOTHESES AND A MODEL OF SECOND-LEVEL ADOPTION

Based on the insight drawn from the three theoretical perspectives and previous research, the following hypotheses are presented:

H1: Technology facility will positively predict the use of mobile phones as a news device.

Technology facility, similar to perceived attributes of technology innovation in the diffusion of innovation theory, is identified for the application context of mobile phone use as a news device. It is expected to have a significant effect on the adoption of a distinctive new function based on an adopted innovation. With its contextual dimension regarding information accessibility, technology facility is likely to affect the adoption of a technology innovation for news access.

H2: Perceived value of information will positively predict the use of mobile phones as a news device.

The likelihood of using a medium is predicted by the relative values provided by an innovation or technology (Lehmann & Ostlund, 1974). Expected value was found to significantly predict the level of exposure to news via a pager (Leung & Wei, 1999). The perceived value of information accessible through a mobile phone, a benefit associated with the distinctive new function of a technology device beyond the initial adoption, is expected to play an important role in mobile phone use as a news device.

H3: Innovativeness will positively predict the use of mobile phones as a news device.

Innovativeness, which involves risk taking, creativity, and desire for change, is expected to have a noticeable effect on mobile phone use as a news device.

H4: News affinity will positively predict the use of mobile phones as a news device.

Early adopters access media content through more communication channels (Kang, 2002). Users of one medium are likely to use other media, especially newly emerged media (Holbert, 2005). Use of a new medium has been associated with past experience of media use (Perse & Ferguson, 1994). News affinity, a construct identified as a key factor with specific relevance to news access, is expected to have an effect on mobile phone use as a news device.

H5: Innovativeness will positively predict perceived technology facility of a mobile phone as a news device.

A user taking technology initiatives may be assumed to possess a certain level of technology savvy and the motivation to learn about the features of the innovation. It is likely that a user with a higher level of innovativeness would better understand and appreciate the technological features and attributes that facilitate mobile phone use as a news device, which will result in the user's higher perception of the technology facility of a mobile phone as a news device.

In light of the conceptual foundation of second-level adoption, we propose an integrated model of predictors of the use of mobile phones as a news device. The model which contains relevant components of the above-mentioned theoretical perspective and related variables in the specific case of mobile phone use as a news device, highlights the relationship between the key variables as specified in the hypotheses. The model is illustrated as follows in Figure 1.

METHOD

A cross-sectional survey was conducted to test the model of predictors of mobile phone use for news access and the related hypotheses. The study was conducted in November 2007 when the second-generation mobile phones started to deliver news updates and allowed users to go to the Internet to get news information. It was around the time when people started to use mobile phones to access news. The population of interest of this study is college students. In the mid-2000s, 89% of college students were found to own a mobile phone (Student Monitor, 2005b). In addition, *college students* constitute an important sample of *news* users because they are in the midst of an important period of socialization with *news* media. Henke (1985) found that general media access and its perceived importance both increase with year in

Figure 1. A model of adoption of a distinctive new function in the case of mobile phone use as a news device

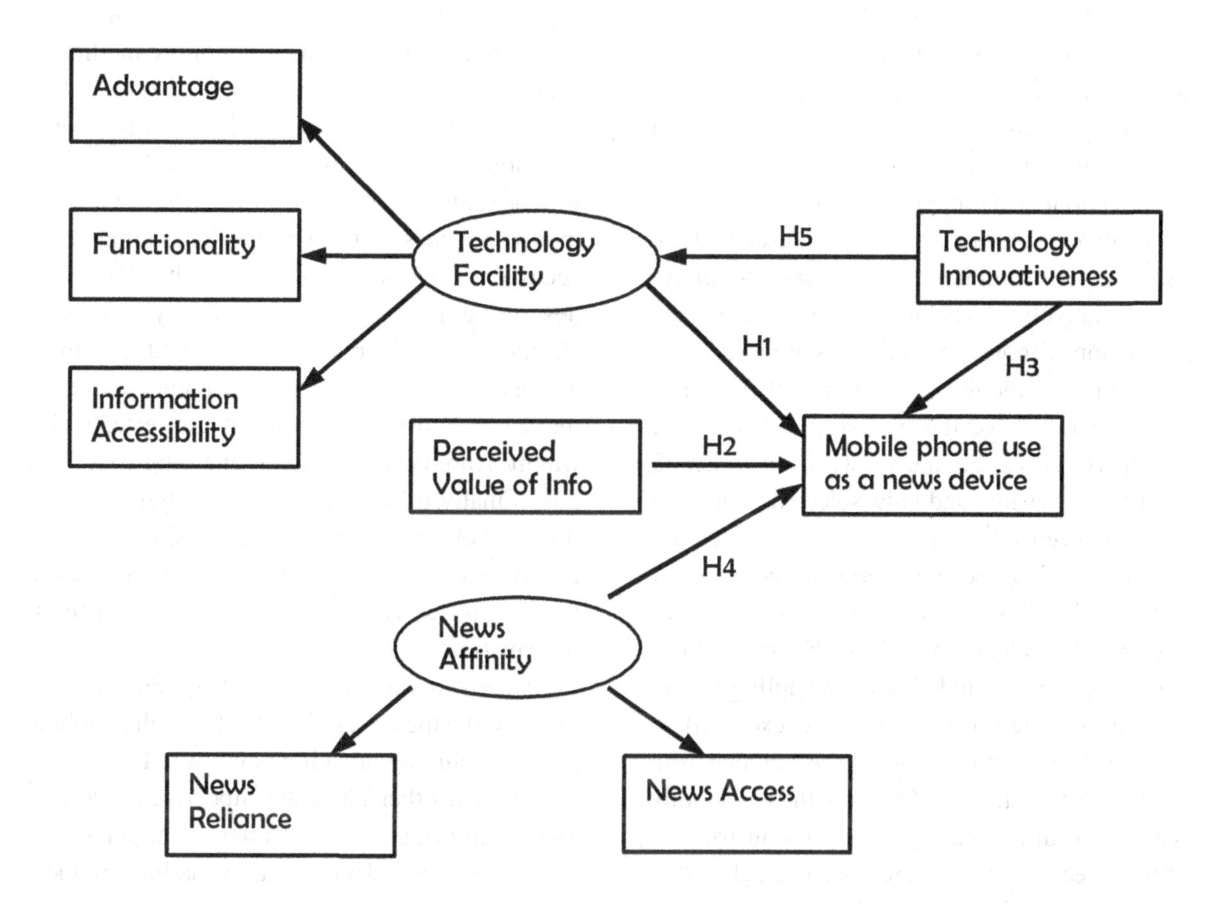

college. While the readership of newspapers was declining among college students in the 1990s (Atkin, 1994), more recently there was some indication that this trend had begun to reverse (Student Monitor, 2005a). College students have easier access to new media than do other groups, and they tend to be among the active users of new media (Vincent & Basil, 1997). Use of new media as information sources increased among college students since the Internet become widely available (Diddi & LaRose, 2006). For these reasons, the use of college students for the study will reveal useful information about a population that tends to access news information through new media.

This study chose the student body of a Southern U.S. university as a subset of the population; it had a total of 31,582 students at the time of the study. Although the use of a student sample is not ideal and will limit the generalizability of the findings, it can nonetheless contribute to theory building for three reasons. First, this study explores relationships between the variables informed by theories. The relationships examined are multivariate in nature instead of univariate. Because the study does not try to estimate the univariate values in the population, a non-probability sample may still serve the purpose of testing multivariate relationships. Second, a prior study show that results regarding multivariate relationships from a student sample were consistent with those from a random sample of the general population (Basil, Brown, & Bocarnea, 2002). Third, if a theory is expected to be valid in the population, it should also apply to a student sample. Thus, the test of the theory through a student sample could also be valuable for understanding the predictors of mobile phone use for news access.

Multistage cluster sampling was used to draw a sample. The sampling method allows for drawing a relatively representative sample from the population without the need of a sampling frame containing all members of the population. The sampling frame for the first stage consisted of all departments at the university, from which 25 departments were randomly selected. The next stage involved selecting one class from each department. Using a schedule book of the semester during which the survey was conducted, all classes with 30-60 students in each of the selected departments were included in the sampling frame. Classes with fewer students were excluded to ensure at least 30 students would be included from each department. One class was then randomly selected from a department, and the instructors of the selected classes were contacted. If a class was not available, another class was randomly selected. A total of 25 classes were selected for the study. The number of students in attendance in each class varied from 20 to 58 on the day the survey was conducted. A two-page questionnaire containing 48 questions was used for the data collection. A total of 928 students out of about 1125 enrolled students in the 25 classes in the selected departments completed questionnaires. Since a multistage cluster sampling method was used to draw a sample, and students in the classes were not contacted beforehand, an accurate response rate could not be calculated.

Measurement of Key Variables

Technology facility refers to the level of capacity of a technology device that enables a distinctive new function beyond its primary function. In this study, this term specifically refers to the level of capacity of a mobile phone that enables users to access news information. Technology facility is a latent variable measured with 13 observable items in three dimensions: device advantage, functionality and information accessibility. Device advantage, defined as the degree to which a mobile phone is seen to be superior to other news accessing devices, was measured with five items describing the ease of use, portability and availability of a mobile phone. Functionality, defined as the degree to which a mobile phone performs the task of delivering news information effectively, was measured with four items regarding effectiveness. Finally, information accessibility, defined as the level of ease in obtaining and using information through a mobile phone, was measured with four items describing ease of access and use of information.

Perceived value of information, information refers to the messages obtained through a mobile phone, including breaking news, weather, sports scores, and other news and updates. Perceived value of information, which refers to the perceived benefits associated with the information obtained through a mobile phone, was measured with four items regarding usefulness, relevance, updatedness and conciseness of information.

Innovativeness, which refers to the degree to which one actively seeks and tries new technology innovations, was measured with six items associated with a user's willingness and activeness in exploring new technology devices. Since the adoption of a distinctive new function is not as technologically challenging as the initial adoption, the measure contained items reflecting the application context such as "I like to use new technology devices," and "I am not intimidated by the new features of technology devices." The user's activeness was measured by the item "I always take initiative to try new features of technology devices."

News affinity refers to the degree to which one uses and relies on news across media as an indispensable need in one's life. News affinity is a latent variable measured with nine observable items in two dimensions: news reliance and news access. News reliance, defined as the degree to which a mobile phone user sees news as indispensable in advancing various aspects of his or her life, was measured with six items regarding how news is involved in and helps with one's life. News access, which refers to the extent to which one uses communication channels other than a mobile phone to obtain news information, was measured with the frequency with which one uses three channels to obtain news: newspapers, television, and the Internet.

A 5-point Likert scale, ranging from strongly agree to strongly disagree, was used to measure the first three variables – technology facility, perceived value of information, and innovativeness – as well as one of the dimensions of news affinity, news reliance. The other dimension of news affinity, news access, was measured with a 5-point verbal frequency scale, ranging from every day to never.

Mobile phone use as a news device, which refers to the frequency of using a mobile phone to obtain news-related information, was measured with two aspects regarding accessing news information through a mobile phone: a) getting news updates on a mobile phone (accessed as text messages delivered by an information service) and b) getting news from the Internet through a mobile phone (accessed through news websites). Getting news updates and getting news from the Internet were both measured with a 5-point verbal frequency scale, ranging from "every day" to "never." A composite index of the use of mobile phones as a news device was created with the means of the two items.

Data Analysis

A factor analysis and a reliability test were conducted to confirm the measurement of the underlying concepts of the two key constructs: technology facility and news affinity. The items measuring the two constructs were respectively unidimensional and reliable (see Table 1). Reliability tests were also run for the other two independent variables, perceived value of information ($\alpha = .92$) and innovativeness ($\alpha = .94$), and the dependent variable, mobile phone use as a news device ($\alpha = .85$), with acceptable results.

Three factors with Eigenvalue above .90 were extracted for Technology Facility with 85.27% variance explained. Two factors with Eigenvalue above 1.0 were extracted for news affinity with 62.53% variance explained.

FINDINGS

Among the 928 respondents, 84.6% used mobile phones quite often on a 5-point verbal frequency scale ranging from "all the time" to "never," and 76.3% used them as a regular phone to make calls. A total of 65% of the respondents could connect to the Internet using their second-generation mobile phones. Regarding use as a news device, 19.5% of the respondents indicated that they got news updates through their mobile phones at least sometimes, and 21.2% connected to the Internet through their mobile phones to get news at least sometimes. These findings reflect the relatively early stage of mobile phone use as a news device. Getting news updates through mobile phones directly and getting news through mobile phones connected to the Internet were correlated ($r = .74$, $p < .01$) but still distinct from each other.

The model of predictors of mobile phone use for news access and the hypotheses were tested through structural equation modeling and path analysis. The goodness of fit of the model of predictors of mobile phone use for news access was

Table 1. Factor analysis of items measuring technology facility and news affinity

Items	Technology Facility			News Affinity	
	Advantage	Function	Access	Reliance	Usage
Ease of use	.79				
Always at hand	.91				
Available any time	.92				
Use immediately	.91				
Use anywhere	.80				
Efficient to get news information		.83			
Satisfies needs for news information		.88			
Always get information I look for		.77			
Keeps me updated		.84			
Easy to find what I need			.80		
Easy to choose what I need			.81		
Easy to store what I need			.81		
Easy to share what I need			.80		
I am a news junkie				.63	
I need to get updated news				.68	
News helps make decisions				.84	
News makes me smarter				.83	
News helps me learn better				.85	
News helps my career				.82	
Read newspapers					.69
Watch TV news					.75
Get news from the Internet					.65
Eigenvalue	8.14	2.05	0.90	4.46	1.16
Variance explained	62.60%	15.78%	6.89%	49.59%	12.94%
Cronbach's alpha	.95	.93	.95	.89	.57

evaluated using a chi-square test, which initially rejected the model ($X_2 = 133.73$, $df = 16$, $p < .01$; $CMINDF = 8.36$). Since the chi-square test is considered overly stringent and requires that the model fit exactly in the population, several alternative tests were conducted, which collectively provided a reasonable estimation of overall model fit (Browne & Cudeck, 1993; Holbert & Stephenson, 2002). The results showed that the model had a Normed Fit Index (*NFI*) of .99, Comparative Fit Index (*CFI*) of .99, and Tucker-Leris Index (*TLI*) of .98, all which were above the cutoff value of .95 (Hu & Bentler, 1999). The consistent results of *NFI*, *CFI* and *TLI* indicated a relatively good fit of the model. The standardized regression coefficients were reported to show the relationships between the variables in the model (Figure 2).

H1, that technology facility will positively predict the use of mobile phones as a news device, was supported. The model showed that technology facility had a moderate effect on mobile phone use as a news device ($\beta = .38$, $p < .01$). A multiple

Table 2. Correlation matrix of the variables examined in the model, and the means and standard deviations of the variables (N = 928)

	Advantage	Functionality	Accessibility	Perceived Value	News Reliance	News Access	Innovative	News Use
Advantage	-							
Functionality	.48**	-						
Accessibility	.62**	.72**	-					
Perceived Value	.64**	.69**	.81**	-				
News Reliance	.02	.17**	.09**	.11**	-			
News Access	.12**	.05	.04	.01	.48**	-		
Innovative	.18**	.27**	.23**	.22**	.31**	.24**	-	
Mobile phone Use for News	.30**	.38**	.35**	.35**	.07	.09**	.18**	-
Mean	17.88	10.94	11.45	11.48	19.90	11.24	21.04	3.08
SD	6.09	4.15	4.43	4.03	5.12	2.39	5.84	1.89

* p <.05; ** p <.01

regression analysis was also conducted with the three aspects of technology facility entered into the equation. The result was statistically significant ($R^2 = .17$, $F = 43.75$, $p < .01$). Device advantage ($\beta = .13$, $p < .01$), device functionality ($\beta = .23$, $p < .01$) and information accessibility ($\beta = .12$, $p < .05$) were all significant predictors of mobile phone use as a news device.

H2, that perceived value of information will positively predict the use of mobile phones as a news device, was not supported. In the computed model, the standardized regression coefficient between perceived value of information and mobile phone use as a news device was not statistically significant ($\beta = -.05$, $p > .05$).

H3, that innovativeness will positively predict the use of mobile phones as a news device, was weakly supported. In the computed model, the link between innovativeness and mobile phone use as a news device was statistically significant ($\beta = .07$, $p < .05$).

H4, that news affinity will positively predict the use of mobile phones as a news device, was not supported. In the computed model, the link between news affinity and mobile phone use as a news device was not statistically significant ($\beta = .01$, $p > .05$).

H5, that innovativeness will positively predict perceived technology facility of a mobile phone as a news device, was supported. The model showed that the standardized regression coefficient between innovativeness and technology facility was statistically significant ($\beta = .25$, $p < .01$).

The revised model of predictors of mobile phone use for news access also revealed that technology facility strongly predicted perceived value of information ($\beta = .88$, $p < .01$) and that perceived value of information positively predicted news affinity ($\beta = .13$, $p < .01$).

DISCUSSION

Based on a theoretical framework drawn from the diffusion of innovation theory, the technology acceptance model and the expectancy-value model, this study examined the predictors of the use of mobile phones as a news device. The study departed from the traditional approach of adoption research and offered a novel perspective

Figure 2. A revised model of adoption of a distinctive new function in the case of mobile phone use as a news device

X_2 = 133.73, df = 16, p < .01; CMINDF = 8.36; NFI = .990; CFI = .991; TLI = .980; RMSEA = .10
Standardized regression coefficients with solid line, p < .05; Standardized regression coefficients with dotted line, p > .05.
** Standardized regression coefficient larger than one, see (Jöreskog, 1999)*

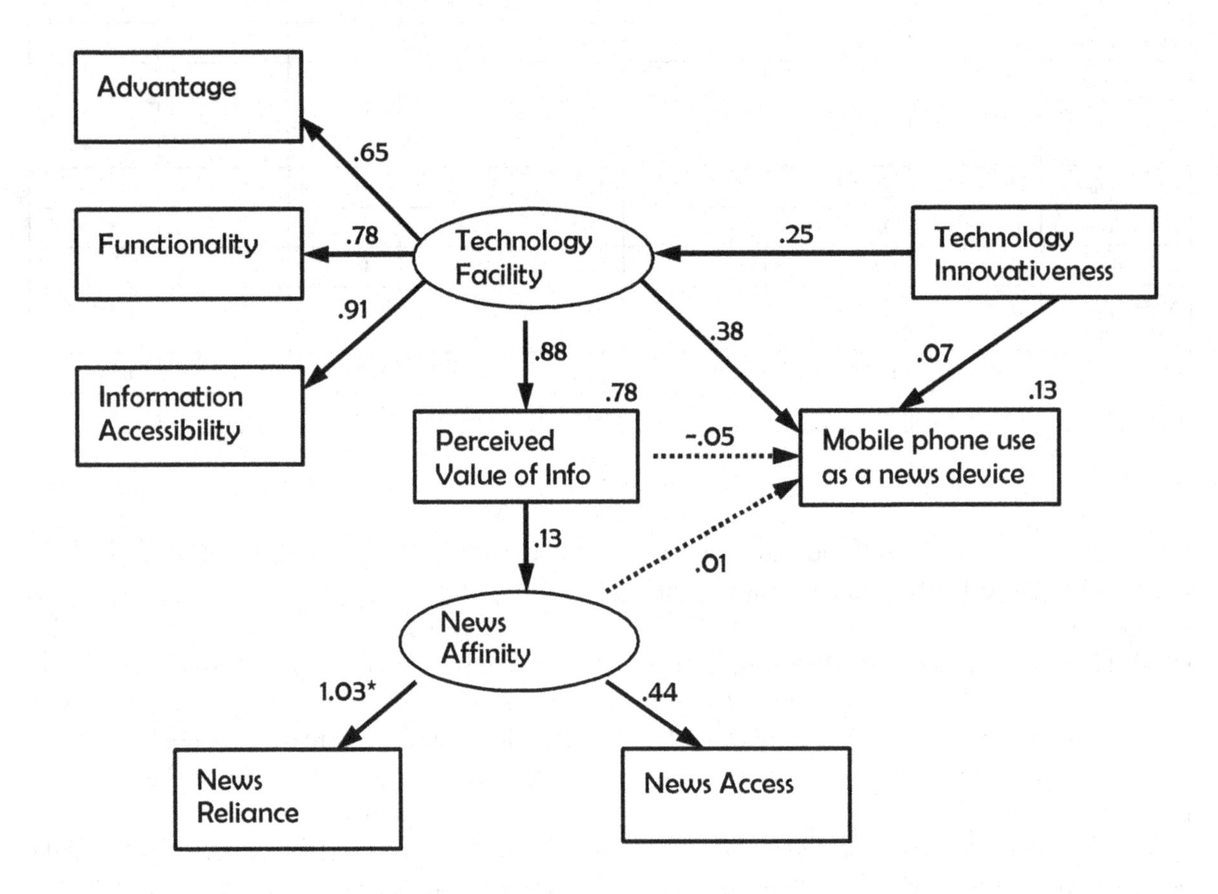

in examining the adoption and use of a distinctive function serving a communication purpose other than that for which a technology device was originally designed.

The findings of the study partially supported the hypotheses with regard to the predictors of the use of mobile phones as a news device. Two constructs – technology facility and perceived value of information – were adapted from the related theories and tested as predictors of mobile phone use as a news device. The findings revealed that technology facility (the construct similar to technology attributes in the diffusion of innovation theory) played an important role in predicting mobile phone use as a news device. Device advantage, one of the major attributes of an innovation and a significant predictor in initial adoption, also had an effect on the adoption of the distinctive new function based on a previously adopted technology device. The measurement of device advantage differed from that of the same variable as a predictor of the initial adoption in its relevance to the application context of the adoption of a distinctive new function. Device advantage measured the superiority of mobile phones in news information access instead of the advantage of mobile phones as the technology innovation to make calls. The effect of functionality and information accessibility on mobile phone use as a news device further explained the importance of the application context of the adoption of the distinctive new function. Both of these aspects were tied closely to the adoption of a distinctive new function beyond the initial adoption of the mobile phone. The effects of the three dimensions of technology facility suggest that application context, which defines technology attributes of new media, made a difference in explaining the effects of the leading factors in subsequent adoptions of a distinctive new function based on a previously adopted technology device.

Perceived value of information was found to have no direct effect on mobile phone use as a news device. This result contradicts previous findings regarding the effect of perceived value on the adoption of innovations. In the adoption of a distinctive new function, the scope of the perceived value of information is defined by the application context. The measure of perceived value was aimed at gauging the usefulness and benefit of the information obtained from mobile phones. A possible reason for the lack of effect could be the nature of mobile phones as a peripheral channel in accessing news information in the relatively earlier stage of penetration. As the majority of phones in use at the time of this study were second-generation phones, a user of such a phone could perceive news information obtained via the phone to be useful but still choose not to use it as a news device because the information available via a second-generation mobile phone is not comparable to that from other media in content and accessibility. As the findings of this study regarding the effect of perceived value of information on the use of mobile phones as a news device are inconsistent with those of previous studies of the expectancy-value model and the technology acceptance model, this area requires further exploration in future studies.

The results further showed that innovativeness was a weak predictor of mobile phone use as a news device. A mobile phone user might be innovative, but did not necessarily take the initiative to get news information through a mobile phone. There are two possible reasons for this weak effect. First, the capacity of second-generation mobile phones for receiving news information from the Internet was still limited at the time of this study. Although the prevalence of Internet availability through mobile phones has grown significantly in the late 2000s, 35% of the respondents still could not connect to the Internet through their mobile phones. At the time of the study, the capacity of second-generation mobile phones to connect to the Internet and the volume of their data transmissions were still limited. In addition, the data service required to connect to and receive large amounts of data from the Internet were still relatively expensive, which might restrict mobile phone users' use of the Internet to obtain news information even though phones equipped with such functions. Second, the amount of information that can be displayed on the screen of a second-generation mobile phone, either through text messages or an Internet browser, was also relatively limited. Unlike the third generation mobile phones, the second-generation mobile phones were designed primarily for receiving simple text messages. This technological limitation might have prevented a number of innovative users from accessing news information through their mobile phones.

News affinity, the informational factor that drives consumers to use mobile phones for news access, was not found to be a significant predictor of the use of mobile phones as a news device. This finding suggests that the limitation of the technology device in its capacity to provide news information compatible to that of traditional media and the Internet may negate the effect of news affinity on the use of new media as a news device. The second-generation mobile phones' utility as a news device was limited in their content richness and information accessibility and the speed with which they could receive a large volume of information; therefore, they were unable to fulfill the essential needs of the news audience. When heavy media users look for information from a new media channel, they usually try to fulfill two needs. First, they seek to access information in a different way, such as going to the Internet to find information that is not available from other media channels. Second, they seek to access information in addition to the information that they obtain from other media. In the late 2000s, the capacity of second-generation mobile phones for news access was still unable to meet either of these needs adequately. Although the information available through a mobile phone was continuously expanding, the second-generation mobile phones used by the respondents in this study were restricted in their capacity to receive news information. A mobile phone could serve as a source for news information and provide news updates on a timely manner. But it was restricted in the scope of news information delivered and was unable to provide news information compatible with that provided by traditional media and other new media such as the Internet. The bottleneck in information accessibility of mobile phones was later broken by the third-generation mobile phones equipped with functional applications, the capacity of transferring large volume of data at low cost, and the efficient operation for information seeking and browsing.

The finding of the study suggests that the effect of news affinity is contingent upon the content richness and service adequacy experienced through mobile phones. After the third-generation mobile phones emerged, a mobile phone's function of delivering news information was enhanced in both content and service quality. Faster connection speeds allow for quick data transmission, and the mobile phone display facilitates website browsing and information access in a variety of formats. Mobile phones have gradually become functional devices for news access. As the information richness and the capacity to receive useful information through mobile phones improve significantly, news affinity is likely to become an influencing factor that leads to more regular use of mobile phones as a news device. In 2012, the 4G mobile phones started to replace the 3G phones. Mobiles phones as a functional alternative, are now more than capable of providing news information that is compatible and complementary to what is offered by the mainstream media.

The findings of this study show that technological factors play a more important role in the adoption and use of mobile phones for news access than do informational factors. This could be due to the technical limitation of the second-generation mobile phones in accessing news information effectively. In the relatively earlier stage of using mobile phones for news access, technological factors constrained users to use mobile phones as a functional channel to obtain news. But as media technologies advance, and mobile phones were equipped with more powerful applications and quicker access to data, technology was no long a bottleneck in the case of mobile phone use for news access. With the third- and fourth-generation mobile phones, informational factors would play a more important role in the adoption and use of mobile phones for news access. Therefore, the stage defined by the media technology may change the weight of importance of technology and informational factors in the adoption of innovations, especially in second-level adoption, in which technology is often more advanced but less challenging to users who already have experience in using the adopted device. With more advanced mobile phone applications and larger-screen mobile phones, more users have already started to use mobile phones as an alternative yet effective channel to access news information.

The findings of this study offer some insights into the applicability of the theories about the adoption of innovations to multimedia devices with evolving functions. Technology facility, as a factor similar to technology attributes in the diffusion of innovation theory, still plays an important role in the adoption and use of mobile phones for news access. The finding regarding technology facility confirms what the diffusion of innovation theory suggests with regard to the effect of technology attributes. However, innovativeness, although also a technology-related factor, was not a strong predictor, contrary to earlier studies of the adoption of innovations. The context of second-level adoption could explain the weak role of innovativeness in the use of mobile phones for news access; because mobile phone use for news access is based on a technology device that has already been adopted, innovativeness is no longer as important as when adopting an innovation with only one function. The finding suggests that the context of second-level adoption weakens the effect of innovativeness on the adoption of distinctive functions of multimedia devices. Based on the expectancy-value model and technology acceptance model, perceived value of information was expected to be a positive predictor of mobile phone use for news access. However, this study did not confirm the effect of perceived value of information. The limited value of the information accessible through second-generation mobile phones could be the main reason for this finding. The third- and fourth-generation mobile phones, with their more powerful functions to access news information, can be expected to improve the predictive power of perceived value of information on the adoption and use of mobile phones for news access.

The results of this study should be treated with caution due to the limitations of the study. First, the study was conducted using a student sample. Although the results of a study with a student sample can be illuminating in understanding multivariate relationships, they require further testing to establish the claim regarding the effects of those factors that specifically apply to the adoption of a distinctive new function beyond the initial adoption of the technology device. In order to demonstrate the significant difference between the initial and subsequent adoption of a distinctive new function and the corresponding influencing factors, future studies could include the factors applicable to both initial adoption of a technology innovation and the second-level adoption. They will reveal to what degree the predictors differ in their effects on the adoption of a distinctive new function evolved with an adopted technology device from the effects on the one-time adoption or the initial adoption of a technology device. In addition, the study was conducted in the relatively early stage of mobile phone use for news access. With the advent of third- and fourth-generation mobile phones, it is necessary to conduct new studies to determine the extent of the role of informational factors such as perceived value of news and news affinity in the process, in comparison with technological factors such as technology facility and innovativeness. When technological factors are no longer a barrier for people to access news through mobile phones, informational factors can be expected to take the lead in making mobile phones mainstream media for news access.

CONCLUSION

This study tested the predictors of the use of mobile phones as a news device based on a theoretical framework of technology adoption. The study attempted to contribute to the understanding of the adoption and use of mobile phones as a news device in several aspects.

First, this study examined the adoption of a distinctive new function of a technology device beyond its initial function. The concept of second-level adoption – the adoption of a distinctive new function for an adopted technology device, with a communication purpose other than that for which the device was originally designed – offers a novel perspective on examining the adoption of innovations. Instead of treating all adoptions of innovations equally, the adoption of a distinctive new function based on a previously adopted innovation needs to be examined from its application context of second-level adoption. This is especially important for the studies of new media with multiple evolving functions.

Second, this study elaborated upon the difference between initial adoption and subsequent adoptions of distinctive new functions in several important aspects; it also identified several technological factors applicable to the adoption of a distinctive new function based on an adopted technology device. The ability to differentiate influential factors in the subsequent adoption of distinctive new functions from those in the initial adoption enables researchers to provide a more insightful understanding of new media adoption.

Third, the findings show that the importance of technological factors and informational factors may vary in different stages of the adoption and use of evolving new functions based on an adopted technology device. In the relatively early stage of mobile phone use for news access, technological factors still played an important role in the adoption and use of mobile phones as a news channel. As the technology advances and the capacity of mobile phones to access news grows, the informational factors would play a more important role in affecting second-level adoption compared to the technological factors.

Mobile phones have entered their fourth generation. There are now few technical problems that will prevent people from accessing news information through mobile phones. People read news from mobile phones regularly in a variety of formats. With the enhanced capacity to deliver information in text, graphic and video formats, mobile phones have become a member of mainstream media. In fact, mobile phones are taking up the media market rapidly. Media organizations develop mobile phone applications to allow users to access news information through mobile phones more efficiently. News information is now delivered to mobile phones on a daily basis, and people access news information easily through mobile phone applications. With easy access to social media on a variety of platforms through mobile phones, the channels through which to obtain news information are much broader than those for accessing news from traditional media or even from the Internet via a personal computer. The media landscape is changing rapidly due to mobile phones working as a functional channel to deliver news information. Such changes are bringing significant impacts on news production and delivery, the structure of the media industry, and the whole media landscape. This study suggests, technological factors played a role in the relatively early stage of mobile phones for news access. Although technology factors is considered less influential in the second-level adoption, easy and functional ways of accessing news information remain as an important aspect for news organizations to attract audience to the news delivered through mobile phones.

As technology advances, the variation in technology advantage gradually diminishes, and the new functions based on an adopted media device require less technology savvy from their users. The role of technological factors decreases in the adoption and use of distinctive functions and services based on adopted innovations. Instead, informational factors could play a more important role in attracting audiences to new media and the various evolving functions of multimedia devices. In the age of media convergence, second-level adoption will be the main trend of new media adoption; in other words, the innovative functions and services most likely to be adopted will be those based on the technology devices that have already been adopted. Informational factors will gradually exceed the technological factors in affecting the adoption and use of distinctive functions of multimedia devices serving a communication purpose other than that for which the devices were originally designed.

REFERENCES

Ahlers, D. (2006). News consumption and the new electronic media. *The Harvard International Journal of Press/Politics*, *11*(1), 29–52. doi:10.1177/1081180X05284317

Allan, S. (1999). *News culture*. Buckingham, UK: Open University Press.

Althaus, S. L., & Tewksbury, D. (2000). Patterns of Internet and traditional news media use in a networked community. *Political Communication*, *17*(1), 21–45. doi:10.1080/105846000198495

Atkin, D. J. (1994). Newspaper readership among college students in the information age: The influence of telecommunication technology. *Journal of the Association for Communication Administration*, *2*, 95–103.

Atkin, D. J., & Jeffres, L. W. (1998). Understanding Internet adoption as telecommunications behavior. *Journal of Broadcasting & Electronic Media*, *42*(4), 475–490. doi:10.1080/08838159809364463

Babrow, A. S., & Swanson, D. L. (1988). Disentangling antecedents of audience exposure levels: Extending expectancy-value analyses of gratifications sought from television news. *Communication Monographs*, *55*(1), 1–21. doi:10.1080/03637758809376155

Ball-Rokeach, S. J., & DeFleur, M. L. (1976). A dependency model of mass-media effects. *Communication Research*, *3*(1), 3–21. doi:10.1177/009365027600300101

Basil, M. D., Brown, W. J., & Bocarnea, M. C. (2002). Differences in univariate values versus multivariate relationships: Findings from a study of Diana, Princess of Wales. *Human Communication Research*, *28*(4), 501–514. doi:10.1111/j.1468-2958.2002.tb00820.x

Bivens, R. K. (2008). The Internet, mobile phones and blogging. *Journalism Practice*, *2*(1), 113–129. doi:10.1080/17512780701768568

Browne, M. W., & Cudeck, R. (1993). Alternative ways of assessing model fit. In K. A. Bollen, & J. S. Long (Eds.), *Testing structural equation models* (pp. 445–455). Newbury Park, CA: Sage.

Calantone, R. J., Griffith, D. A., & Yalcinkaya, G. (2006). An empirical examination of a technology adoption model for the context of China. *Journal of International Marketing*, *14*(4), 1–27. doi:10.1509/jimk.14.4.1

Chan-Olmsted, S., Rim, H., & Zerba, A. (2013). Mobile news adoption among young adults: Examining the roles of perceptions, news consumption, and media usage. *Journalism & Mass Communication Quarterly*, *90*(1), 126–147. doi:10.1177/1077699012468742

Chau, P. Y. K., & Lai, V. S. K. (2003). An empirical investigation of the determinants of user acceptance of Internet banking. *Journal of Organizational Computing and Electronic Commerce*, *13*(2), 123–145. doi:10.1207/S15327744JOCE1302_3

Davis, F. D. (1989). Perceived usefulness, perceived ease of use, and user acceptance of information technology. *Management Information Systems Quarterly*, *13*(3), 318–340. doi:10.2307/249008

Davis, F. D. (1993). User acceptance of information technology: System characteristics, user perceptions and behavioral impacts. *International Journal of Man-Machine Studies*, *38*(3), 475–487. doi:10.1006/imms.1993.1022

Davis, F. D., Bagozzi, R. P., & Warshaw, P. R. (1989). User acceptance of computer technology: A comparison of two theoretical models. *Management Science*, *35*(8), 982–1003. doi:10.1287/mnsc.35.8.982

Diddi, A., & LaRose, R. (2006). Getting hooked on news: Uses and gratifications and the formation of news habits among college students in an Internet environment. *Journal of Broadcasting & Electronic Media*, *50*(2), 193–210. doi:10.1207/s15506878jobem5002_2

Dimmick, J., Feaster, J. C., & Hoplamazian, G. J. (2011). News in the interstices: The niches of mobile media in space and time. *New Media & Society*, *13*(1), 23–39. doi:10.1177/1461444810363452

Dishaw, M. T., & Strong, D. M. (1999). Extending the technology acceptance model with task-technology fit constructs. *Information & Management*, *36*(1), 9–21. doi:10.1016/S0378-7206(98)00101-3

Dupagne, M. (1999). Exploring the characteristics of potential high-definition television adopters. *Journal of Media Economics*, *12*(1), 35–50. doi:10.1207/s15327736me1201_3

Dutta-Bergman, M. J. (2004). Complementarity in consumption of news types across traditional and new media. *Journal of Broadcasting & Electronic Media*, *48*(1), 41–60. doi:10.1207/s15506878jobem4801_3

Ettema, J. S. (1984). Three phases in the creation of information inequities: An empirical assessment of a prototype videotex system. *Journal of Broadcasting*, *28*(4), 383–395. doi:10.1080/08838158409386548

Ettema, J. S. (1989). Interactive electronic text in the United States: Can videotex ever go home again? In J. L. Salvaggio, & J. Bryant (Eds.), *Media use in the information age: Emerging patterns of adoption and consumer use* (pp. 105–123). Hillsdale, NJ: Lawrence Erlbaum Associates.

Gibson, E. J., & Pick, A. D. (2000). *An ecological approach to perceptual learning and development*. Oxford, UK: Oxford University Press.

Goggin, G., & Hjorth, L. (2009). *Mobile technologies: From telecommunications to media*. New York: Routledge.

Grant, A. E., & Wilkinson, J. (2009). *Understanding media convergence: The state of the field*. New York: Oxford University Press.

Henke, L. L. (1985). Perceptions and use of news media by college students. *Journal of Broadcasting & Electronic Media*, *29*(4), 431–436. doi:10.1080/08838158509386598

Hermida, A. (2010). Twittering the news. *Journalism Practice*, *4*(3), 297–308. doi:10.1080/17512781003640703

Holbert, R. L. (2005). Intramedia mediation: The cumulative and complementary effects of news media use. *Political Communication*, *22*(4), 447–461. doi:10.1080/10584600500311378

Holbert, R. L., & Stephenson, M. T. (2002). Structural equation modeling in the communication sciences, 1995-2000. *Human Communication Research*, *28*(4), 531–551. doi: doi:10.1111/j.1468-2958.2002.tb00822.x

Hu, L.-T., & Bentler, P. M. (1999). Cutoff criteria for fit indexes in covariance structure analysis: Conventional criteria versus new alternatives. *Structural Equation Modeling*, *6*(1), 1–55. doi:10.1080/10705519909540118

Humphreys, L., Von Pape, T., & Karnowski, V. (2013). Evolving mobile media: Uses and conceptualizations of the mobile Internet. *Journal of Computer-Mediated Communication*, *18*(4), 491–507. doi:10.1111/jcc4.12019

Hurt, H. T., Joseph, K., & Cook, C. (1977). Scales for the measurement of innovativeness. *Human Communication Research*, *4*(1), 58–65. doi:10.1111/j.1468-2958.1977.tb00597.x

Igbaria, M., Guimaraes, T., & Davis, G. B. (1995). Testing the determinants of microcomputer usage via a structural equation model. *Journal of Management Information Systems*, *11*(4), 87–114.

Jeffres, L., & Atkin, D. (1996). Predicting use of technologies for communication and consumer needs. *Journal of Broadcasting & Electronic Media*, *40*(3), 318–330. doi:10.1080/08838159609364356

Jensen, K. B. (2010). *Media convergence: The three degrees of network, mass, and interpersonal communication*. London: Routledge.

Jöreskog, K. (1999). *Advanced topics on Lisrel by Professor Karl Jöreskog: How large can a standardized coefficient be?* Retrieved from http://www.ssicentral.com/lisrel/techdocs/HowLargeCanaStandardizedCoefficientbe.pdf

Kaminer, N. (1997). Scholars and the use of the Internet. *Library & Information Science Research, 19*(4), 329–345. doi:10.1016/S0740-8188(97)90024-4

Kang, M.-H. (2002). Digital cable: Exploring factors associated with early adoption. *Journal of Media Economics, 15*(3), 193–207. doi:10.1207/S15327736ME1503_4

Karahanna, E., & Straub, D. W. (1999). The psychological origins of perceived usefulness and ease-of-use. *Information & Management, 35*(4), 237–250. doi:10.1016/S0378-7206(98)00096-2

Kayany, J. M., & Yelsma, P. (2000). Displacement effects of online media in the socio-technical contexts of households. *Journal of Broadcasting & Electronic Media, 44*(2), 215–229. doi:10.1207/s15506878jobem4402_4

Lai, V., & Guynes, J. (1994). A model of ISDN (integrated services digital network) adoption in U.S. corporations. *Information & Management, 26*(2), 75–84. doi:10.1016/0378-7206(94)90055-8

LaRose, R., & Atkin, D. (1992). Audiotext and the re-invention of the telephone as a mass medium. *The Journalism Quarterly, 69*(2), 413–421. doi:10.1177/107769909206900215

LaRose, R., & Atkin, D. J. (1988). Satisfaction, demographic, and media environment predictors of cable subscription. *Journal of Broadcasting & Electronic Media, 32*(4), 403–413. doi:10.1080/08838158809386712

Lehmann, D. R., & Ostlund, L. E. (1974). Consumer perceptions of product warranties: An exploratory study. *Advances in Consumer Research. Association for Consumer Research (U. S.), 1*(1), 51–65.

Leung, L., & Wei, R. (1999). Seeking news via the pager: An expectancy-value study. *Journal of Broadcasting & Electronic Media, 43*(3), 299–315. doi:10.1080/08838159909364493

Leung, L., & Wei, R. (2000). More than just talk on the move: Uses and gratifications of the cellular phone. *Journalism & Mass Communication Quarterly, 77*(2), 308–320. doi:10.1177/107769900007700206

Levy, M. R. (1978). The audience experience with television news. *Journalism Monographs (Austin, Tex.), 55*, 1–29.

Levy, M. R. (1987). VCR use and the concept of audience activity. *Communication Quarterly, 35*(3), 267–275. doi:10.1080/01463378709369689

Li, S. S.-C. (2004). Exploring the factors influencing the adoption of interactive cable television services in Taiwan. *Journal of Broadcasting & Electronic Media, 48*(3), 466–483. doi:10.1207/s15506878jobem4803_7

Lin, C. A. (1993a). Adolescent viewing and gratifications in a new media environment. *Mass Communication Review, 20*(1/2), 39–50.

Lin, C. A. (1993b). Modeling the gratification-seeking process of television viewing. *Human Communication Research, 20*(2), 224–244. doi:10.1111/j.1468-2958.1993.tb00322.x

Lin, C. A. (1998). Exploring personal computer adoption dynamics. *Journal of Broadcasting & Electronic Media, 42*(1), 95–112. doi:10.1080/08838159809364436

Lin, C. A. (2001). Audience attributes, media supplementation, and likely online service adoption. *Mass Communication & Society*, *4*(1), 19–38. doi:10.1207/S15327825MCS0401_03

Lin, C. A. (2004). Webcasting adoption: Technology fluidity, user innovativeness, and media substitution. *Journal of Broadcasting & Electronic Media*, *48*(3), 446–465. doi:10.1207/s15506878jobem4803_6

Ling, R. S. (2004). *The mobile connection: The cell phone's impact on society*. San Francisco, CA: Morgan Kaufmann.

McCloskey, D. (2003). Evaluating electronic commerce acceptance with the technology acceptance model. *Journal of Computer Information Systems*, *44*(2), 49–57. doi: doi:10.4018/978-1-59140-066-0.ch101

McLuhan, M. (1994). *Understanding media: The extensions of man*. Cambridge, MA: MIT Press.

Midgley, D. F., & Dowling, G. R. (1978). Innovativeness: The concept and its measurement. *The Journal of Consumer Research*, *4*(4), 229–242. doi:10.1086/208701

Neal, A. G. (1998). *National trauma and collective memory: Major events in the American century*. Armonk, NY: M.E. Sharpe.

Nystrom, P. C., Ramamurthy, K., & Wilson, A. L. (2002). Organizational context, climate and innovativeness: Adoption of imaging technology. *Journal of Engineering and Technology Management*, *19*(3/4), 221–247. doi:10.1016/S0923-4748(02)00019-X

Palmgreen, P., & Rayburn, J. D. (1982). Gratifications sought and media exposure: An expectancy value model. *Communication Research*, *9*(4), 561–580. doi:10.1177/009365082009004004

Palmgreen, P., & Rayburn, J. D. (1985). An expectancy-value approach to media gratifications. In K. E. Rosengren, L. Wenner, & P. Palmgreen (Eds.), *Media gratification research* (pp. 61–72). Beverly Hills, CA: Sage Publications.

Perse, E. M., & Dunn, D. G. (1998). The utility of home computers and media use: Implications of multimedia and connectivity. *Journal of Broadcasting & Electronic Media*, *42*(4), 435–456. doi:10.1080/08838159809364461

Perse, E. M., & Ferguson, D. A. (1994). The impact of the newer television technologies on television satisfaction. *Journalism Quarterly*, *70*(4), 843–853. doi:10.1177/107769909307000410

Piotrowski, C., & Armstrong, T. R. (1998). Mass media preference in disaster: A study of Hurricane Danny. *Social Behavior & Personality: An International Journal*, *26*(4), 341–346. doi:10.2224/sbp.1998.26.4.341

Rayburn, J. D., & Palmgreen, P. (1984). Merging uses and gratifications and expectancy-value theory. *Communication Research*, *11*(4), 537–562. doi:10.1177/009365084011004005

Rice, R. E., & Rogers, E. M. (1980). Reinvention in the innovation process. *Science Communication*, *1*(4), 499–514. doi:10.1177/107554708000100402

Rogers, E. M. (1986). *Communication technology: The new media in society*. New York: Free Press.

Rogers, E. M. (2003). *Diffusion of innovations* (5th ed.). New York: Free Press.

Rubin, A. M. (1983). Television uses and gratifications: The interactions of viewing patterns and motivations. *Journal of Broadcasting*, *27*(1), 37–51. doi:10.1080/08838158309386471

Schrøder, K. C., & Steeg Larsen, B. (2010). The shifting cross-media news landscape. *Journalism Studies*, *11*(4), 524–534. doi:10.1080/14616701003638392

Soopramanien, D. G. R., & Robertson, A. (2007). Adoption and usage of online shopping: An empirical analysis of the characteristics of buyers? browsers? and non-Internet shoppers? *Journal of Retailing and Consumer Services*, *14*(1), 73–82. doi:10.1016/j.jretconser.2006.04.002

Student Monitor. (2005a). *College student readers of national newspapers gain*. Retrieved from http://www.studentmonitor.com/press/05.pdf

Student Monitor. (2005b). *Study finds record number of student cell phone owners*. Retrieved from http://www.studentmonitor.com/press/02.pdf

Swanson, D. L. (1979). Political communication research and the uses and gratifications model: A critique. *Communication Research*, *6*(1), 37–53. doi:10.1177/009365027900600103

Tornatzky, L. G., & Klein, K. J. (1982). Innovation characteristics and innovation adoption–implementation: A meta-analysis of findings. *IEEE Transactions on Engineering Management*, *29*(1), 28–45. doi:10.1109/TEM.1982.6447463

Tybout, A. M., Sternthal, B., Malaviya, P., Bakamitsos, G. A., & Park, S.-B. (2005). Information accessibility as a moderator of judgments: The role of content versus retrieval ease. *The Journal of Consumer Research*, *32*(1), 76–85. doi:10.1086/426617

Vincent, R. C., & Basil, M. D. (1997). College students' news gratifications, media use, and current events knowledge. *Journal of Broadcasting & Electronic Media*, *41*(3), 380–392. doi:10.1080/08838159709364414

Vishwanath, A., & Golohaber, G. M. (2003). An examination of the factors contributing to adoption decisions among late-diffused technology products. *New Media & Society*, *5*(4), 547–572. doi:10.1177/146144480354005

Vyas, R. S., Singh, N. P., & Shilpi, B. (2007). Media displacement effect: Investigating the impact of Internet on newspaper reading habits of consumers. *Vision: The Journal of Business Perspective*, *11*(2), 29–40. doi: doi:10.1177/097226290701100205

Wei, R. (2001). From luxury to utility: A longitudinal analysis of cell phone laggards. *Journalism & Mass Communication Quarterly*, *78*(4), 702–719. doi:10.1177/107769900107800406

Wei, R., & Lo, V.-H. (2006). Staying connected while on the move: Cell phone use and social connectedness. *New Media & Society*, *8*(1), 53–72. doi:10.1177/1461444806059870

Westlund, O. (2008). From mobile phone to mobile device: News consumption on the go. *Canadian Journal of Communication*, *33*(3), 443–463.

Wilzig, L. S., & Avigdor, C. N. (2004). The natural life cycle of new media evolution. *New Media & Society*, *6*(6), 707–730. doi:10.1177/146144804042524

KEY TERMS AND DEFINITIONS

News Affinity: The degree to which one uses and relies on news across media as an indispensable need in one's life. News affinity is a latent variable in two dimensions: news reliance and news access. 1) News Reliance: The degree to which a mobile phone user sees news indispensable in advancing various aspects of life. 2) News Access: The extent to which one uses communication channels other than a mobile phone to obtain news including three channels: newspapers, television, and the Internet.

Diffusion of Innovation: The theory explains why people adopt technical innovations and what factors play a role in the adoption process. According to Rogers (1986), the decision to adopt or reject an innovation is subject to a wide variety of factors, including: (1) adopter-related personality traits; (2) socioeconomic influences; (3) interpersonal channels and mass media use; and (4) perceived attributes of an innovation.

Expectancy-Value Model: As part of the uses and gratifications theory, the model suggests that audiences are actively looking for information from media to fulfill their needs. The model proposes that media use could be explained by comparing perceived benefits offered by the medium and the gratification derived from media use.

Innovativeness: The degree to which one actively seeks and tries new technology innovations. Innovativeness involves risk-taking, creativity, and a desire for change, and demonstrates a user's willingness and initiative in exploring new technology devices.

Perceived Value of Information: The perceived benefits associated with the information obtained through a mobile phone. Information here refers to the messages obtained through a mobile phone including breaking news, weather, sports scores, and other news and updates.

Second-Level Adoption: Acceptance of a distinctive communication function of a technology device serving a communication purpose different from that for which a technology device was originally designed. Mobile phone use as a news device, a multimedia function drastically different from the calling function originally designed for the technology device, and occurs after the initial adoption, makes second-level adoption.

Technology Acceptance Model: The model (TAM) specifies that perceived usefulness and ease of uses are two key constructs in determining user acceptance of computer technology. According to TAM, the decision to accept a technology is determined primarily by attitudes towards the decision, which in turn mediate the impacts of beliefs and external variables on acceptance behavior.

Technology Facility: The level of capacity of a technology device that enables a distinctive new function beyond its primary function. In the case of mobile phone use as a news device, it is the capacity of a mobile phone to enable a user to get news information. It contains three aspects: device advantage, functionality and information accessibility. 1) Device Advantage: The degree to which a mobile phone is seen superior to other news accessing devices. 2) Functionality: The degree to which a mobile phone performs the task of delivering news information effectively. 3) Information Accessibility: The level of easiness in obtaining and using the information through mobile phones.

Chapter 17
Wireless Connected Health: Anytime, Anyone, Anywhere

Florie Brizel
Brizel Media, USA

ABSTRACT

Wireless connected health is the most current, inclusive phrase to describe healthcare that incorporates wireless technologies and/or mobile devices. It represents one of the fastest growing sectors in the global mobile and wireless ecosystem, with extraordinary change occurring daily. According to the World Health Organization, 80 percent of people in greatest medical need live in low- to middle-income countries. Not enough has been written about how they will afford wireless connected health, or how it can bring positive benefits to patients everywhere with non-lethal chronic illnesses. It also remains to be seen whether people outside the healthcare industry, without any special interest in science, technology, medicine, or illness prevention, will adopt new and future behavior-changing connected health technologies. This chapter provides a current overview of the global health crises created by noncommunicable diseases, explains the evolution of the global wireless connected health sector, includes information about BRICS nations, and offers observations, insights, and recommendations from a socio-economic and political standpoint for responsible and effective future industry growth.

INTRODUCTION

Wireless connected health is the most current and inclusive phrase to describe personal healthcare that incorporates mobile devices and/or wireless technologies. It represents one of the fastest growing sectors in the mobile and wireless ecosystem, with extraordinary change occurring daily.

Today, simply by using their smartphones, people around the world can watch their diets, track their exercise, log their sleep, access motivational fitness coaches and/or programs, and do myriad other things to support lifelong wellness.

Rapid advances in technology now make possible a global initiative to ultimately eradicate disease by empowering individuals to make informed, positive lifestyle choices that are disease-preventive. This bodes well not only for increased human longevity, but also, and perhaps even more importantly, for improved quality of life (WHO, 2013a).

DOI: 10.4018/978-1-4666-6166-0.ch017

The early literature and focus concerning the advent of wireless connected health addressed the successful linkage of systems and networks so machines could communicate with machines, and data – specifically, a patient's electronic health record (EHR) – could be accessed ubiquitously (Chronaki et al., 2007).

Current literature (Sejdić et al., 2013) focuses on the engineering and software behind new and innovative wireless medical devices and mobile medical applications (MMAs).

Furthermore, now that a variety of wireless medical devices and MMAs have entered into progressive clinical practices and mainstream teaching centers, a growing body of literature has begun to examine the clinical performance of products already on the market. Weight loss, weight management and weight gain all have mobile-inclusive protocols (Turner-McGrievy and Tate, 2011; Patrick et al., 2013; Carter et al., 2013; and Cardi, Clarke & Treasure, 2013).

Behavioral sciences and mental health have noteworthy studies and pilot programs that address everything from bipolar disorder (do patients prefer using mobile phones for charting moods or traditional pen-&-paper diaries?) (Depp et al., 2012), to psychosis (Palmier-Claus et al., 2013), to depression, anxiety and stress (Proudfoot et al., 2013), to the effectiveness of a suicide prevention app for indigenous Australian youths (Shand et al., 2013).

Two excellent studies (Spyridonis, Ghinea and Frank, 2013; Kristjánsdóttir et al., 2013) address pain as a significant component of many chronic illnesses and how wireless and mobile technologies can help alleviate this aspect of noncommunicable chronic disease (NCD).

Unfortunately, too few papers focus on ways the changing healthcare landscape affects the end-user as a patient, beyond becoming a personal source of biological data points. One that does is "Telecare, Surveillance, and the Welfare State" (Sorrell & Draper, 2012), which examines whether or not bringing a variety of health and wellness monitoring devices into the sanctity of the home strips individuals living there of autonomy, depersonalizes their care and, as an unintended consequence, actually increases their isolation.

Another excellent paper, entitled "How places matter: Telecare technologies and the changing spatial dimensions of healthcare," (Oudshoorn, 2012, p.124) argues that "Places are not only important because assumptions about the contexts of use are inscribed in technologies.... They also matter because places shape how technological devices are used, or not, and (de)stabilize the specific identities of technologies. Equally important, technologies participate in redefining the meaning and practices of the spaces in which they are used and...introduce new spaces in which people and objects interact." The author goes onto say, "The idea that places matter thus provides an important point of departure for an investigation of how reciprocal relationships between people, places and technologies enable or constrain the identities of users, places and technologies (Oudshoorn, 2012, p. 124)."

Clearly, wireless connected health already has begun to help many people with chronic illnesses. Four chronic illnesses typically used to illustrate its benefits – diabetes, hypertension, heart disease and asthma – are such that if a patient ignores unusual, fluctuating symptoms, results can prove catastrophic or fatal. New wireless medical devices and apps make regular surveillance of signs and symptoms remarkably easy, and they empower these patients to enjoy better health as a result of having better tracking systems to prevent undesired health crises.

However, not enough has been written about whether or how wireless connected health can bring significant, positive outcomes to resistant patients and those with lesser known and rarely fatal chronic illnesses...patients who tax the global healthcare system with a never-ending variety of very real, perplexing ailments that defy easy diagnosis or standard remedy.

And it remains to be seen, despite all arguments to the contrary, whether regular people outside the healthcare industry, without any special interest in science, technology, medicine, or illness prevention, will adopt the plethora of new and future behavior-changing technologies.

This paper provides a current overview of the global wireless connected health sector and offers observations and recommendations from a socio-economic standpoint for responsible industry growth going forward.

A PubMed literature review conducted in Q4/2013 at the Louise M. Darling Biomedical Library at UCLA David Geffen School of Medicine, using search terms *mobile*, *wireless, mobile health*, *healthcare*, *patients*, and *patient perceptions*, returned over 150 papers related to wireless connected health. Abstracts were read for all, eliminating those that dealt exclusively with systems design, engineering, software and strict technology issues. Those abstracts selected for further review broke out as 11 papers about mobile and physical activity; eight (8) about mobile and behavior modification; seven (7) about mobile and mental health; six (6) about mobile and older adults; five (5) about chronic disease; four (4) about mobile and pain management; three (3) about mobile and diabetes management; two (2) papers about mobile and asthma; and, numerous papers that contained an amalgam of relevant material. In addition, there were individual papers about wearables (wireless technology built into clothing or wearable accessories); social networks; games; alcoholism; access to care; medical residents' education; privacy; digital communication failure; and, cancer.

The wireless connected health sector has seen unprecedented growth over the last five years, and major news organizations have given it increased coverage accordingly, in print and online. As it changes significantly on an almost daily basis, a wide variety of industry-specific websites, newsletters, and blogs also served as reliable information sources. Finally, two major governing and regulatory bodies – the World Health Organization (WHO) and the United States Food and Drug Administration (USFDA or FDA) – provided invaluable statistical, regulatory and practical information essential to understanding some of the reasons why the wireless connected health industry, and specifically, mobile medical devices and mobile medical applications, have developed in the initial trajectories they have.

TECHNOLOGIES AND HEALTHCARE

The practice of medicine has a long tradition of incorporating new scientific knowledge through responsible applications. Within the last 200 years, many new technologies have been developed and taken their place in the annals of medicine. In 1816, Laennec introduced the stethoscope to better hear the beating heart. In 1884, the autoclave was introduced to sterilize tools used in the operating theatre. Nine years later, in 1895, Roentgen introduced the radiograph x-ray. French surgeon Alexis Carrel and American aviator Charles Lindbergh introduced the external perfusion pump in 1935. By 1957, the first artificial heart was implanted in a dog; the first artificial heart was placed in a human for 64 hours a mere twelve years later, in 1969. Artificial vertebral discs were introduced at the end of the 20th century and their successful implantation record has greatly minimized the need for more restrictive spinal procedures.

Today, thanks to advances in mobile and wireless technologies, smartphones also can function as medical devices, such as stethoscopes, ultrasound machines, cardiac monitors, and more. In the United States, any medical device must receive approval from the Food and Drug Administration (FDA) prior to use by physicians or the general public (U.S. Food and Drug Administration, 2012a). This law includes mobile and wireless devices intended for diagnostic or therapeutic use, or disease management.

The FDA defines a medical device as "an instrument, apparatus, implement, machine, contrivance, implant, in vitro reagent, or other similar or related article, including a component part, or accessory which is recognized in the official National Formulary, or the United States Pharmacopoeia, or any supplement to them; intended for use in the diagnosis of disease or other conditions, or in the cure, mitigation, treatment, or prevention of disease, in man or other animals; or intended to affect the structure or any function of the body of man or other animals, and which does not achieve its primary intended purposes through chemical action within or on the body of man or other animals and which is not dependent upon being metabolized for the achievement of any of its primary intended purposes (FDA, 2013a)."

Many wireless medical devices already have received clearance for use and are on the market. The GE Vscan ultrasound is a smartphone-based professional medical ultrasound with a probe attached by USB. The AliveCor Heart Monitor is a single-channel smartphone-based electrocardiogram for prescribed patients and their physicians. A number of glucose monitors are on the market. The Infrascanner Model 2000 is a new and exciting addition to medical diagnostics. It is a portable screening device that uses Near-Infrared (NIR) technology to screen patients for intracranial bleeding, identifying those who would most benefit from immediate referral to a CT scan and neurosurgical intervention (InfraScan, 2013).[1] The RP-VITA™ Remote Presence Robot – with audio/video capabilities – can rove throughout a hospital and permit remote telepresence between doctors and patients.

Wireless medical devices should not be confused with mobile medical applications (MMAs), some of which are intended solely for health and wellness, while others are intended for managing or treating a condition (or pre-condition). The FDA has draft guidances for MMAs, with fairly clear distinctions between the two.

"Apps that have the following functionalities will not be regulated by FDA:

- Mobile apps that are solely used to log, record, track, evaluate, or make decisions or suggestions related to developing or maintaining general health and wellness, if not intended to cure, treat, diagnose, or mitigate a specific disease, disorder, patient state, or any specific, identifiable health condition.
- Mobile apps that are used as dietary tracking logs and appointment reminders, or provide dietary suggestions based on a calorie counter, posture suggestions, exercise suggestions, or similar decision tools that generally relate to a healthy lifestyle and wellness and are not intended to cure, mitigate, diagnose, or treat a disease or condition.

Examples of software products intended for use by consumers that the draft guidance says may be subject to "device" regulation include:

- Apps that allow users to input their health information and through the application of formulas, data comparisons, or processing algorithms, issue a diagnosis or treatment recommendation that is specific to that person, such as his or her risk for colon cancer or heart disease, or recommend that the patient take a certain medication or seek a particular treatment.
- Software that is intended to be used to physically or wirelessly connect to and download information from a diagnostic device like a glucose meter to allow the user to display, store, analyze, and/or keep track of his or her medical data values (Hyman, Phelps & McNamara, 2013)."[2]

While the FDA historically has stayed away from MMAs whose intent has been restricted to

Figure 1. Health apps by general categories
Graphics courtesy of Amit Goel (Used with permission)

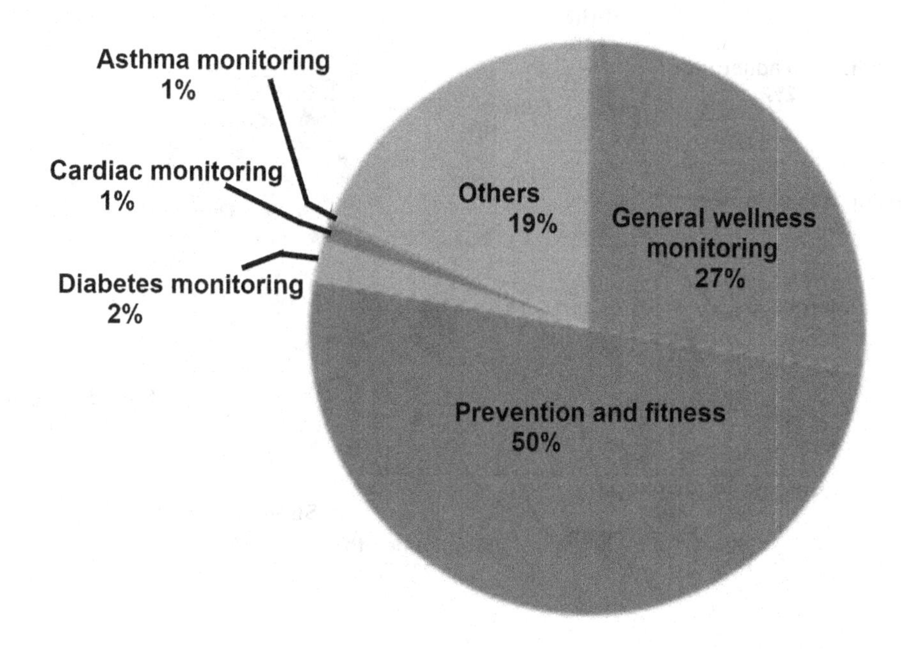

general health and wellness, it always has maintained regulatory power with respect to MMAs that are more integrally involved in disease diagnosis, treatment and management. However, the FDA also has the prerogative to eliminate the regulatory burden on certain MMAs, such as those whose intent is to help patients better manage their own disease(s).

"The following is a list of mobile apps that FDA believes fall within the medical device definition, but that FDA would like not to regulate:

1. Medication reminder apps for therapy adherence.
2. App for tabulating an Apgar score.
3. App for calculating drip rate for IV solution.
4. BMI calculator apps for use by patients and physicians.
5. Apps that help flag drug-drug interactions for physicians as they prescribe.
6. Diabetes management guide apps such as nutritional guides or pre-diabetes risk assessments.
7. Apps that offer behavior guides to help, for example, wean off smoking.
8. Calorie counters that would be specifically marketed to obese people or other people with health conditions trying to manage weight.
9. Cancer management apps manage medication schedules, and allow the patient to diary side effects and symptoms for reporting to their doctor.
10. Asthma management apps to assess symptoms, medication use and breathing data entered by the user to tell the user when their risk is changing.
11. Hypertension apps to help users log and chart their blood pressure, set medication reminders, record medicine taken, and share data with their doctor.

Figure 2. Detailed segmentation of health apps
Graphics courtesy of Amit Goel (Used with permission)

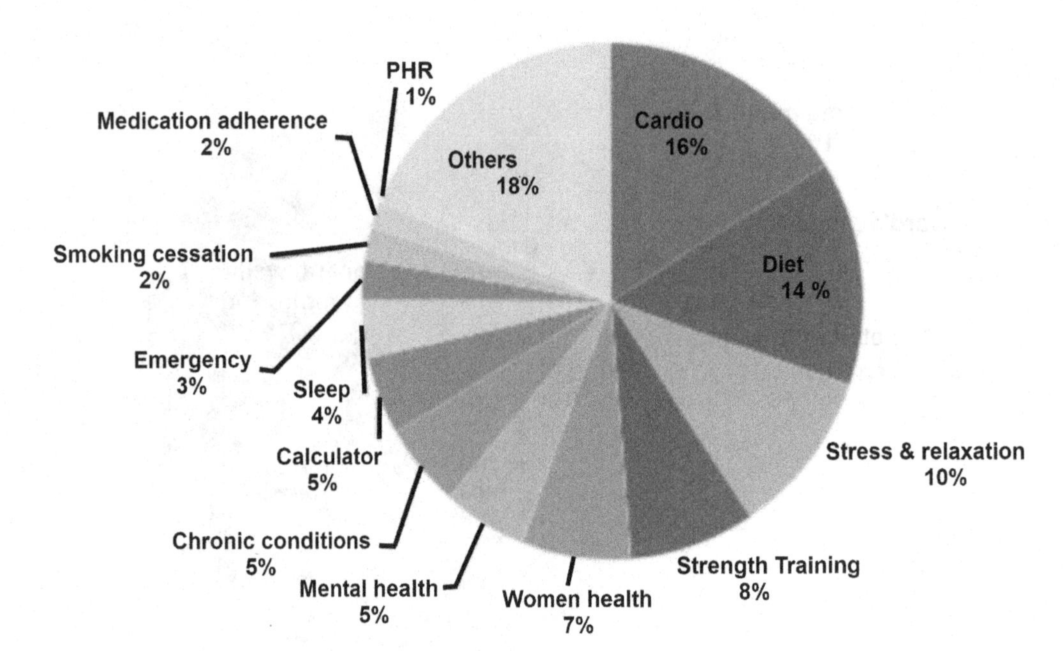

12. Arthritis management apps containing screening tools and questionnaires to help users determine the type of arthritis, provide treatment strategies and medication information, as well as information on diet and nutrition.
13. Chronic pain (fibromyalgia, headaches) management apps to help users track symptoms and triggers; weather conditions (humidity, pressure); photo attachments to document swelling, rash, discoloration; and interactive graphs of symptoms vs. weather to be shared with their doctor.
14. An app to help users who suffer from Chronic Kidney Disease (CKD) or End Stage Renal Disease (ESRD) make better decisions about their diet, by tracking their daily intake of certain nutrients and comparing their consumption to guidelines their nephrologist and nutritionist have set for them.
15. Digestive disease (Crohn's disease) management apps to help users record and track their food and fluid intake along with when symptoms arise and the time they take pain medication, all logged so a report may be generated for the doctor to review and analyze (Thompson, 2013)."[3]

As of August 2013, the FDA had approved just 75 mobile medical apps (MMAs), although many others are in various stages of development or approvals processes (FDA, 2013b).

The current global spend on MMAs is US$1.3 billion (as of 2012). The U.S. is the major contributor with $700 million. Right now, over 100,000 health and wellness apps are available and this number is expected to double by 2020. As more regulations come into play concerning security issues, future focus most likely will be on the quality of the apps. Also, there will be a significant increase in the number of downloads (Goel, 2013).[4]

Clearly, opportunities exist for creating apps that support health and wellness for practically every biological function, and innovators around the world are competing in a robust market for the attention of global consumers. New offerings show up almost daily in the various platforms' app stores. These are some of them: Tictrac, Fitocracy, Fitbit, UP by Jawbone, Zombies, Run!, and, Withings.

Physicians and patients both have a vast array of new mobile medical devices and MMAs to help diagnose disease, treat it, and begin, where possible, to promote wellness and even prevent certain illnesses.

One might easily assert that technology, itself, is the driving force behind the practice of medicine and the direction of healthcare in the 21st century. By using an array of digital and wireless technologies, modern healthcare providers can deliver first-world medical care to people who previously would not have had access to any healthcare at all.

They do this through eHealth, or electronic health, defined by the World Health Organization (WHO) as "the use of information and communication technologies (ICT) for health." eHealth is a very broad term – which is good, according to three authors of a compelling editorial in the *Journal of the International Society for Telemedicine and eHealth:* "...how do you define something that is incomplete? Do you know where – or what – eHealth will be in five, 20, 50 years' time? The WHO definition is simple, powerful, clear, descriptive, and flexible enough to accommodate future areas of application (Scott et al., 2013, p.53)."

If ICTs are used (at all) to provide healthcare to patients any time, by physicians almost anywhere, then that is colloquially known as connected health care. If wireless and mobile technologies are used, it's known as wireless connected health or mobile connected health, and sometimes expressed as mHealth. All are subsets of eHealth.

As previously stated, the FDA wields considerable power in the approval of pharmaceuticals, medical devices, and medical software applications. Without FDA approval, no drug, device or software application can be put on the market for sale in the United States, even if it already has gained approval elsewhere by other internationally recognized regulatory organizations.

According to the FDA, connected health "refers to electronic methods of health care delivery that allow users to deliver and receive care outside of traditional health care settings. Examples include mobile medical apps, medical device data systems, software, and wireless technology. The FDA's role in connected health continues to evolve along with medical device technology. The FDA's Center for Devices and Radiological Health (CDRH) plays an important role in enabling a connected health environment while assuring that patients stay safe and the new technologies work as intended. Currently, CDRH is focusing its efforts in several different areas including the convergence of wireless technologies with medical devices (in partnership with the Federal Communications Commission (FCC); medical devices used in a home environment; mobile medical apps; medical device data systems; and, the role of software in medical devices (FDA, 2013c)."

Another subset of eHealth is wireless medical telemetry, which is generally used "to monitor a patient's vital signs (e.g. pulse, and respiration) using radio frequency (RF) communication. These devices have the advantage of allowing patient movement without restricting patients to a bedside monitor with a hard-wired connection (FDA, 2013d)." Wireless medical telemetry deserves credit for reducing the number of hospital re-admissions post-30 days because it allows physicians the ability to better monitor their patients after they originally leave the hospital.

Finally, telemedicine is another type of eHealth. Telemedicine allows patients to have remote access to physicians and medical/healthcare resources via a variety of means, including telephone, video monitoring, video chat, etc. It can bring the world to patients in the most remote locations, and likewise, it can bring some of the most difficult and fascinating patients' cases to world-renowned physicians, wherever they practice.

The single-most important vehicle for facilitating all the above listed forms of eHealth to work optimally is an electronic health record (EHR) – sometimes known as an electronic medical record (EMR). Just as each person has a unique fingerprint by which to be identified, so, too, each person will have a unique EHR that will include information about allergies, pharmaceutical history, lab results, doctors' notes, digital imaging and reports, and (eventually) full genetic blueprints. Authorized healthcare providers anywhere should have digital access to this vital information in order to deliver both *symptom*-appropriate and *patient*-appropriate treatment, irrespective of either the healthcare provider's or the patient's location.

This long-awaited convergence of technology, innovation and delivery makes it possible for eHealth to begin to tackle, on a global basis, the most prevalent noncommunicable diseases (NCDs) and chronic health conditions that account for the bulk of healthcare costs. "Of the 57 million deaths globally, NCDs contribute to an estimated 36 million deaths every year, including 14 million people dying between the ages of 30 and 70. Using mobile telephone technology mHealth practices can help save lives, reduce illness and disability, and reduce healthcare costs significantly (WHO, 2012a)."

The WHO has targeted five of the most pervasive NCDs (and two contributing factors) for global reduction or eradication. They are listed alphabetically below. It is important to understand the direction of the wireless connected health sector's development, including research and development funds for wireless medical devices and mobile medical applications, has taken its cue from this WHO global initiative. Many other diseases and illnesses deserve attention, but they will not garner significant funds for wireless R&D until these first priorities are met.

Cancer

About 30 percent of cancer deaths are due to the five leading behavioral and dietary risks: high body mass index; low fruit and vegetable intake; lack of physical activity; tobacco use; and/or alcohol use. Tobacco use is the most important risk factor for cancer causing 22 percent of global cancer deaths and 71 percent of global lung cancer deaths. About 70 percent of all cancer deaths in 2008 occurred in low- and middle-income countries. Deaths from cancer worldwide are projected to continue rising, with an estimated 13.1 million deaths in 2030 (Globocan 2008, IARC, 2010). Tobacco use, alcohol use, unhealthy diet and physical inactivity are the main cancer risk factors worldwide. Cervical cancer, which is caused by HPV, is a leading cause of cancer death among women in low-income countries (WHO, 2013b).

Cardiovascular Diseases (CVDs)

CVDs are the number one cause of death globally: more people die annually from CVDs than from any other cause. An estimated 17.3 million people died from CVDs in 2008, representing 30 percent of all global deaths (WHO, 2011a). The number of people who die from CVDs, mainly from heart disease and stroke, will increase to reach 23.3 million by 2030 (WHO, 2011a; Mathers & Loncar, 2006). Most cardiovascular diseases can be prevented by addressing risk factors such as tobacco use, unhealthy diet and obesity, physical inactivity, high blood pressure, diabetes and raised lipids (WHO, 2013c).

Chronic Respiratory Diseases

Some 235 million people currently suffer from asthma. It is the most common chronic disease among children. The strongest risk factors for developing asthma are inhaled substances and particles that may provoke allergic reactions or irritate the airways. These would include indoor and outdoor allergens, tobacco smoke, chemical irritants in the workplace, and/or air pollution. Other triggers can include cold air, extreme emotional arousal, such as fear or anger, and/or physical exercise (WHO, 2011b).

Chronic Obstructive Pulmonary Dysfunction (COPD) is a lung ailment that is characterized by a persistent blockage of airflow to the lungs. It is an under-diagnosed, life-threatening lung disease that interferes with normal breathing and is not fully reversible. Symptoms are breathlessness, abnormal sputum, and a chronic cough. COPD is more than a 'smokers cough. (WHO, 2012b).'

An estimated 64 million people have COPD worldwide in 2004 (WHO, 2008). More than 3 million people died of COPD in 2005, which is equal to 5 percent of all deaths globally that year. The primary cause of COPD is tobacco smoke (through tobacco use or second-hand smoke). The disease now affects men and women almost equally, due in part to increased tobacco use among women in high-income countries. Total deaths from COPD are projected to increase by more than 30 percent in the next 10 years without interventions to cut risks, particularly exposure to tobacco smoke. Because COPD develops slowly, it is frequently diagnosed in people aged 40 or older (WHO, 2012b).

Diabetes

Diabetes is a chronic disease that occurs either when the pancreas does not produce enough insulin or when the body cannot effectively use the insulin it produces. Insulin is a hormone that regulates blood sugar. More than 80 percent of diabetes deaths occur in low- and middle-income countries. (WHO, 2013d) 347 million people worldwide have diabetes (Danaei et al., 2011).

WHO projects that diabetes will be the 7th leading cause of death in 2030 (WHO, 2011). Diabetes increases the risk of heart disease and stroke. 50 percent of people with diabetes die of cardiovascular disease, including primary heart disease and stroke (Morrish et al., 2001). Combined with reduced blood flow, neuropathy (nerve damage) in the feet increases the chance of foot ulcers, infection and eventual need for limb amputation (WHO, 2013d). One percent of global blindness can be attributed to diabetes (WHO, 2012c). The overall risk of dying among people with diabetes is at least double the risk of their peers without diabetes (Roglic et al., 2005).

Obesity

While obesity is categorized as a chronic illness with high priority for improving global health, it has many different components and causes. As such, it commands an arsenal of health and wellness/fitness apps geared toward education as well as behavioral changes to promote healthier living. Some of these include the following: weight charts, caloric intake diaries, caloric output (exercise) monitors, and nutritional values of foods and food groups, etc.

There is a medical distinction between overweight and obesity. The WHO defines a person with a body mass index (BMI) greater than or equal to 25 as overweight. A person with a BMI greater than or equal to 30 is obese (WHO, 2013e). BMI is calculated as weight in kilograms divided by height (in meters) squared.

Worldwide obesity has nearly doubled since 1980. In 2008, more than 1.4 billion adults, 20 and older, were overweight. Of these over 200 million men and nearly 300 million women were obese. More than 40 million children under the age of five were overweight in 2011. Overweight and obesity are the fifth leading risk for global deaths. At least 2.8 million adults die each year as a result of being overweight or obese. In addition, 44 percent of the diabetes burden, 23 percent of the ischaemic heart disease burden, and between seven percent and 41 percent of certain cancer burdens are attributable to overweight and obesity. Obesity is preventable (*emphasis by author*) (WHO, 2013e).

Hypertension (high blood pressure) is one of the key risk factors for cardiovascular disease. Hypertension is a silent, invisible killer that rarely causes symptoms. Many developing countries are seeing growing numbers of people who suffer from heart attacks and strokes due to undiagnosed and uncontrolled risk factors such as hypertension. Researchers have estimated that raised blood pressure currently kills nine million people every year. Preeclampsia is hypertension that occurs in some women during pregnancy. Women who experience preeclampsia are more likely to have hypertension in later life (WHO, 2013f).

Tobacco is an indisputable contributing factor to cancer and to chronic respiratory diseases. It kills up to half of its users. Nearly six million people die annually: over five million deaths result from direct tobacco use; over 600,000 non-smokers perish from second-hand smoke from cigarettes and water pipes. Unless urgent action is taken, the annual death toll could rise to more than eight million by 2030 (WHO, 2013g).

"Tobacco users who die prematurely deprive their families of income, raise the cost of health care and hinder economic development. In some countries, children from poor households are frequently employed in tobacco farming to provide family income. These children are especially vulnerable to 'green tobacco sickness,' caused by nicotine that is absorbed through the skin from the handling of wet tobacco leaves.

Tobacco smoke has over 4000 chemicals in it, with 250 of them known to be harmful and 50 known to cause cancer. Simply put, there is no safe level of exposure to second-hand smoke. In adults, second-hand smoke causes serious cardiovascular and respiratory diseases, including coronary heart disease and lung cancer. In infants, it causes sudden death. Over 40 percent of children have at least one smoking parent. In 2004, children accounted for 28 percent of the deaths attributable to second-hand smoke (WHO, 2013g)."

Today the world has one billion smokers, of which almost 80 percent live in low- and middle-income countries. This fact plays an important role in the disparity between a particular population's health (or lack thereof) and healthcare distribution. Those people with the greatest medical need typically have the least means with which to get it. Wireless connected health can help level the playing field.

Mobile telephony and mobile technologies have jumpstarted the economic growth of five specific populations that merit attention. They are Brazil, Russia, India, China, and South Africa. Colloquially, they are known as BRICS. Their vast geographies and unique cultures play an integral role in, and present enormous challenges to, providing adequate healthcare to their people living in remote locations.

For this reason, BRICS also offer some of the greatest opportunities for improving healthcare and quality of life through existing and emerging eHealth technologies, individually tailored to each place. All these populations face the WHO-identified global health challenges; each one also faces unique health challenges. Wireless connected health can address all of them right now.

In Brazil, the greatest health risk factors for adults come from diabetes, hypertension, obesity and tobacco use, according to the World Health Organization (WHO, 2013h). Brazil currently has five of Latin America's 23 mobile health projects running live, and one of its 10 pilots (Goel, 2013).[5]

According to a 2012 PriceWaterhouseCoopers (PwC) report[6], "U.S. mobile health development will trail developing countries like Brazil and India in the near future (Comstock, 2013)."[7]

In Russia, different sources give differing statistics, all based on data compiled over different periods of time. Essentially, the top health issues, in no particular order, are smoking, cancer, CVDs, chronic respiratory conditions, HIV/AIDS, and alcoholism. Alcoholism takes an enormous toll on Russia and its people, contributing to alcohol poisoning, road traffic injuries, other unintentional injuries, as well as suicides (WHO, 2009; WHO, 2006; Curtis, 1996). Furthermore, "a very high proportion of decedents whose death was attributed to 'other' or 'not classified' cardiovascular diseases had lethal or potentially lethal concentrations of ethanol in blood (Zaridze et al, 2009 p. 149)."

Cancer, CVDs, diabetes, tuberculosis, undernutrition, malaria, and maternal/infant mortality account for the most serious national healthcare challenges in India (WHO, 2013i). Out of 80 live- and 37 pilot-stage mobile health projects running in the entire Asia Pacific region, India has more (24 live and 10 pilot) than any other single country (Goel, 2013).[8]

As of 2009, the major causes of death from noncommunicable diseases in China were cancer, CVDs, respiratory ailments, and injuries and poisoning (WHO, 2011d). "Studies show that few people understand the specific health risks of tobacco use. For example, a 2009 survey in China revealed that only 38 percent of smokers knew that smoking causes coronary heart disease and only 27 percent knew that it causes stroke (WHO, 2013g)."

Finally, the "S" in BRICS stands for South Africa, geographically smaller than Brazil, Russia, India or China, but economically vibrant and progressive in terms of its adoption of mobile. However, to focus only on South Africa, which must reduce the devastation of HIV/AIDS among its people, and exclude thinking about other African nations would be a grave mistake when considering wireless technologies as means for improving healthcare and quality of life. This paper focuses on sub-Saharan and southern African countries.

Kenya, South Africa, Uganda and Nigeria represent sub-Saharan Africa's healthiest economies, but they, too, have health burdens to overcome. Out of a total of 106 live and 46 pilot mobile health projects in place in sub-Saharan Africa, Kenya has the most: 20 live and eight in pilot stage (Goel, 2013).[9]

CHANGES, CHALLENGES, AND CHANCES IN WIRELESS CONNECTED HEALTH

Daily advances in wireless medical devices and software make it increasingly possible for physicians and patients to benefit from easier and/or earlier diagnosis of adverse medical conditions. As previously described, the GE Vscan ultrasound is a striking example of positive and possible change in wireless connected health. AliveCor Heart Monitor is another, allowing patients to monitor adverse cardiac events and then relay the information through a secure website to their doctor, or store less critical information for later.

In addition to wireless medical devices used only by professionals or in tandem with prescribed patients, numerous new wireless medical devices are currently in the pipeline either for diagnosis by physicians or for prevention through personal monitoring by patients. One, called Eclipse, is a mobile breast imaging tool for home use in between regular mammograms. Still in the early stages of development, its team has chosen to use social media as a tool for generating further public awareness, building interest, and seeking funding for additional development. Another device, currently awaiting final FDA clearance, is QardioCore, a lightweight wearable heart monitor that comfortably can be worn continuously and frequently as required or desired, and for years (if needed). It sends clinical-grade, continuous monitoring of ECG, heart rate, heart rate variability, intensity of physical activity and skin temperature up to a secure cloud, where the data can be downloaded to practitioners who can monitor patients remotely in real-time (Qardio, 2013).[10]

For someone with a family history of breast or heart disease, both of these devices (along with myriad others), if approved by the FDA, may provide patients with self-empowering tools to regularly monitor their own health.

Finally, the very well publicized (and very well funded) Qualcomm X-PRIZE Tricorder global competition aims to foster development of a handheld diagnostic device, ultimately intended for consumer use, which will read for a minimum of 15 separate conditions. The expected date of the prize award is sometime in mid-2015, to be followed by the necessary FDA approval process, with consumer rollout as soon as possible after that.[11]

Prevention and wellness require knowledge, which can be gained with ease or effort depending where one lives in the world. While developed nations may have an edge in established healthcare and medical protocols, they can be rather entrenched and somewhat resistant to change. Developing nations and regions have the advantage of leapfrogging over such structure (and restriction) and going straight to eHealth through a mashup of means of delivery. Some places use telemedicine. Some places depend on portable kiosks and clinics as first lines of healthcare, with referrals to specialists after an initial evaluation from a certified healthcare professional.

Gamification is a purposely-broad umbrella term used to encompass the process of using 'gaming' elements to motivate and engage people in non-game contexts (Deterding et al., 2011). It is an especially valuable tool since it engages young people – and *very* young people – through two forms of learning and entertainment they already validate: interactive games with a competitive aspect, plus incentives to win prizes or ranking; and, interactive storytelling games with role playing game (RPG) mechanics that keep players coming back to play more. "The 'gamification' of health care is the latest strategy for motivating pediatric patients and their parents to make efforts to adopt a healthier lifestyle. Gamification is driven by data collection and interpretation. Patients use applications and monitoring devices to document compliance with treatment regimens and to visualize progress and goals achieved (Schuman, 2013, p.33)."[12]

Games have been used in a feasibility study in Sweden and the United States to teach high school students how to learn cardiopulmonary resuscitation (CPR) using avatars in massively multiplayer virtual worlds (MMVWs). Conclusion? "A high level of appreciation was reported among these adolescents and their self-efficacy increased significantly. The described training is a novel and interesting way to learn CPR teamwork, and in the future could be combined with psychomotor skills training (Creutzfeldt et al., 2013, abstract)."

"Games have also been created for specific health conditions. An example is Bant, a mobile app targeted at adolescents with diabetes that has successfully used incentives to improve the frequency of glucose monitoring (Cafazzo, 2012)." "If we show patients that we feel mobile devices are accurate and reliable enough for office use, we encourage patients and parents to consider using mobile health technology at home when indicated (Schuman, 2013)."

Vaccinations remain one of the most important tools for success in the global fight to eradicate communicable diseases. Unfortunately, many people fear inoculating their children because they mistakenly believe a relationship exists between vaccines and autism, among many other health issues. TiltFactor, a game company/research lab out of Dartmouth, has created a board game with a free app called ZombiePox. (Apparently, anything with zombies seems to motivate kids!) The app is a learning tool for children so they, themselves, will understand the importance of vaccinations. "When people see that the only way to win the game is to focus on vaccinations rather than trying to cure a rapidly spreading disease, the developers believe they will internalize the lesson and hopefully will be motivated to get their shots (Comstock, 2012)."[13]

An enormous number of MMAs on the global market target health, fitness, nutrition, hygiene and disease prevention, among a whole host of other

Figure 3. Graphic illustration of the Fogg behavior model (Fogg, 2011)
BJ Fogg. (2007). [Graphic illustration of the Fogg Behavior Model]. *BJ Fogg's Behavior Model.* Retrieved from http://behaviormodel.org/. Used with permission.

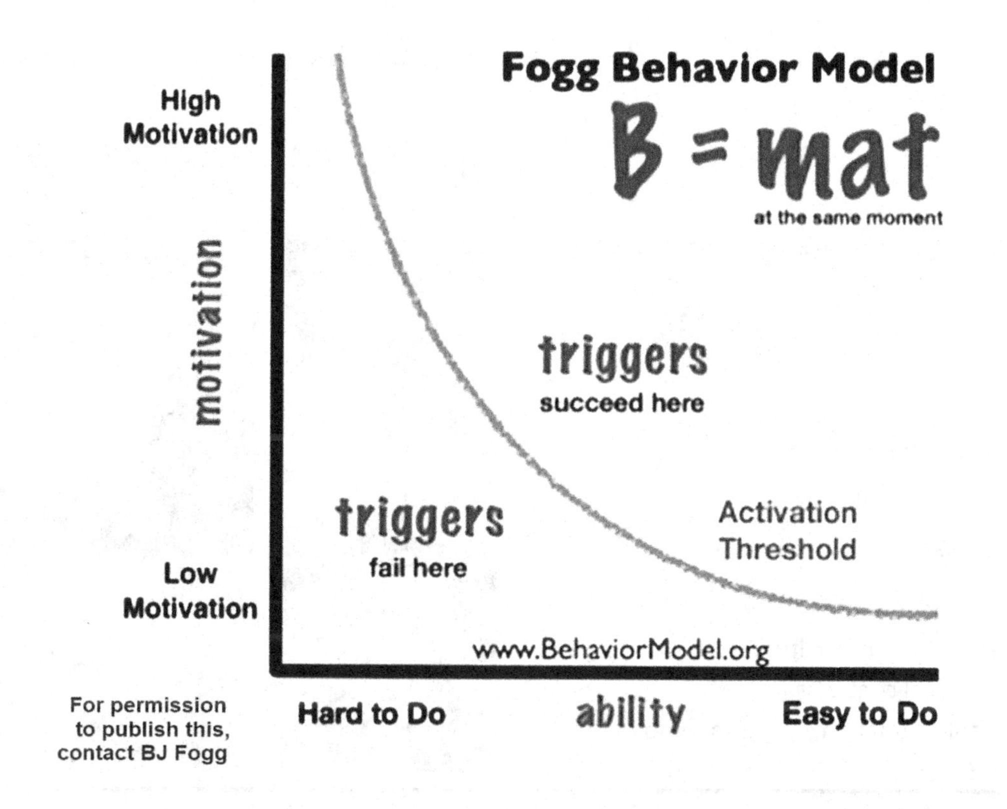

health and wellness topics. Notably, African life and health experiences bear little to no resemblance to life outside of Africa. Thus, in order to deliver region- and experience-appropriate health care and wellness/prevention resources, providers must use different models than elsewhere.

In South Africa, Anne Githuku-Shongwe, the 2013 Schwab Social Entrepreneur of the Year and founder and CEO of Afroes Transformational Games, has created a company whose *raison d'être* is to "build games, simulations and interactive engagements to inform, inspire and challenge young Africans using the mobile phone as an educational platform.... With over 450,000 users, we have built a series of mobile games designed to shape new choices and conversations. We built MORABA, an award-winning mobile game addressing difficult questions of gender-based violence and challenging the user to contemplate what he or she believes about sexual relations and sexual violence (Githuku-Shongwe, 2013)."[14]

While these games seem, perhaps, to be more socially driven than health-centric, they actually play a vital role in preventive healthcare individually, and societal wellness as a whole. Domestic and gender-based sexual violence and their consequences place incalculable economic burdens on healthcare systems around the world due to physical injuries, emotional trauma, psychosocial disruption, and the fact that many survivors of sexual violence never admit the true nature of their injuries to healthcare providers.

Figure 4. Background for Minas Gerais State, Brazil
Alkmim, M.B. et al. Used with permission.

Minas Gerais State, Brazil

The biggest push for disease prevention and wellness in Africa has just begun, with the announcement that Samsung has created the Smart Health Hub, a pan-African mobile health platform. Initially, it will be featured on all new Samsung smartphones and tablets sold and distributed across the African countries in which Samsung operates.[15] It will be free to anyone who already has a Samsung smartphone or tablet, and available for free through Google Play and the Samsung Store for travelers to the region. Tapping on the Smart Health Hub button will take you into the following sub-platform areas (Simon, 2013):

1. Pre- and post-natal care.
2. Nutrition (providing guidelines for healthy African nutrition, since Africa has its own food groups and fruit groups pertinent to the African experience).
3. HIV-AIDS.[16]
4. Tuberculosis.[3]
5. Malaria.[3]
6. Symptom checker with colloquial relevance for Africa.
7. Aerobics and general exercise (games custom-designed in Ireland with motion sensor capture for interaction with a low-cost touch screen smartphone or a Samsung).
8. SafePoint Single-Use Syringe Safety and Information Campaign.[17] (This sub-platform area shows what happens if one needle is

Figure 5. Telecardiology in Minas Gerais, Brazil
Ribeiro, A.L., Alkmim, M.B., et al. Implementation of a telecardiology system in the state of Minas Gerais: the Minas Telecardio Project. *Arq Bras Cardiol 2010, **95(1)**: 70-78.* Used with permission

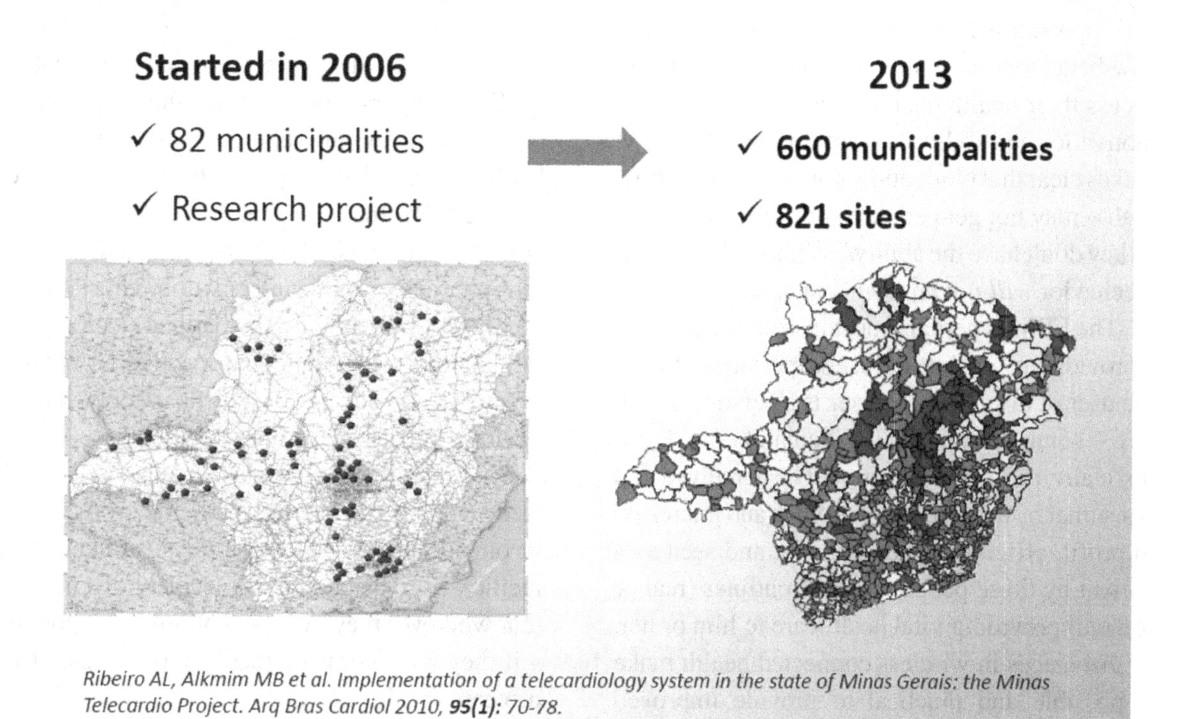

used on a number of patients, and educates how secondary infection can be curbed and HIV-AIDS can be prevented by not using any needle twice. Videos and other tools teach good patient and clinical safety regarding injections.)

Wellness also depends upon motivation. Much of the thought behind mHealth revolves around, and depends upon, the theory that, given the ability and opportunity, most people gladly will wear a variety of body sensors to monitor everything that's quantifiable about their physiology, moment-to-moment. Enough people already have started doing this to earn the moniker "Quantified Self." Trendsetters in the Quantified Self movement have been, predominantly, athletes and dieters, according to Alexandra Sifferlin in a March 14, 2013 article she wrote for *Time.com* (Sifferlin, 2013).[18]

Convincing people to purchase and wear a wireless monitoring device or sensor, as well as to perform different and specific behaviors related to it, may not be as easy as it seems.

Renowned Stanford University behavioral psychologist Dr. B.J. Fogg has written extensively about the three specific conditions necessary for an individual to make a change in behavior: motivation, ability, and a trigger.

His Fogg Behavior Model (www.behavior-model.org) asserts, "For a target behavior to happen, a person must have sufficient motivation, sufficient ability, and an effective trigger. All three factors must be present at the same instant for the behavior to occur (Fogg, 2009)."[19]

In other words, it will take more than ability and opportunity for most people to adopt radically new behaviors to secure their own wellness and access their health data. According to Dr. Fogg, motivation and ability can trade off, but "the FBM makes clear that motivation alone – no matter how high – may not get people to perform a behavior if they don't have the ability." Without the trigger, a behavior *will not happen* (Fogg, 2009).

The FBM is perhaps most instructive because it provides insight into the user experience. Who is the user of eHealth? The user is a person. A real, live – sometimes quite ill – human being. Paradoxically, this same person will be perceived as a consumer by those selling a product and interested in profit, efficiency and statistics, and seen as a patient by those people on the frontlines (real or remote) providing vital healthcare to him or her.

Advances in wireless connected health make it possible and practical to provide improved healthcare delivery to rural and outlying distances through the portability of technology and digital communications. Telemedicine has proved itself to be especially effective in this realm. The study "1,000,000 Electrocardiograms by Distance: An Outstanding Milestone for Telehealth in Minas Gerais, Brazil," provides an excellent example (Alkmim et al., 2013).

The telehealth model developed to support primary healthcare in Minas Gerais has produced good clinical and economic results. As a consequence, it is now a regular health service in the State, covering 660 of the 853 municipalities, and integrated to the healthcare system. The model and technology characteristics permit the replication in other parts of the world (Alkmim et al., 2013).

The 'connected' in wireless connected health refers to the ability to electronically link multiple departments within the same hospital; multiple hospitals in the same network; and, ideally, different hospital networks and clinics, locally and globally. Interoperability (to be addressed later) will be mandatory for this last linkage to happen.

With the ability to digitally input all of a patient's data – her protected health information (PHI) – doctors and other healthcare providers can create a unique, comprehensive electronic health record (EHR), or electronic medical record (EMR), that also should include allergies, complete pharmaceutical records, and, eventually, pharmacogenetic information. This then can be accessed directly by the patient, doctors, and other authorized people, irrespective of location or mobile device. This will absolutely have a positive impact on healthcare globally.

In other words, wireless connected health frees doctors to practice medicine from within or outside the confines of a physical healthcare facility. It gives people the security of knowing that wherever they may be – at home or globally – if they fall ill or need medical assistance, their protected health information can be retrieved from their electronic health record for a local doctor to attend to them, or for any specialist in the world to connect via eHealth.

Mobile communication has tilted toward text; medical tests have drifted toward digital; and impatient patients now insist on instant. The effects these changes will have on private practitioners cannot be overstated.

Some practices will upgrade to mobile and wireless ICTs and remain relevant. The up-front costs of doing business will increase, but they can be mitigated by the age of the medical practitioner and the potential number of years he or she likely will continue to practice.

Those people who do not shift to incorporate mobile and wireless ICTs into their practices likely may find themselves with fewer and fewer patients since they will be perceived as having lost relevancy in a digital age. Older physicians who foolishly think they can stick with their ways may find themselves forced into unanticipated early retirement.

Before we can examine how these changes will impact patients, we must address the greatest challenge, by far, to the success or failure of the entire eHealth initiative: interoperability. This refers to the ability for all healthcare IT systems around the world to be able to 'talk' to each other. It is not enough that all the healthcare facilities in one country run on the same operating system and software. "Health information stored in one IT system must be retrievable by others, including doctors and hospitals that are a part of other health systems. This is particularly important in emergency situations (Kellerman & Jones, 2013a)." If a patient from one country winds up in another needing medical assistance, his or her local doctor(s) must have meaningful access to that patient's complete health record. If there is incompatibility between or among systems that prevents this, then meaningful access to a patient's EHR does *not* exist and the patient is ill-served, indeed.

So far, patients have greater ability than ever before to access information in order to understand their own medical issues. They have improved opportunity to actively participate in their own healthcare decisions. They have better access to physicians, relevant medical information, and advanced technology anywhere around the world. And with myriad MMAs and games to educate, motivate and monitor compliance, as well as encourage wellness and illness prevention, patients have an arsenal of support to help them live longer, healthier and, hopefully, happier lives.

Changing over to connected health presents daunting behavioral challenges – for patients, physicians and other healthcare providers, plus institutions. As previously mentioned, motivation will play a key role in determining whether or not healthy individuals and existing patients successfully integrate wireless technologies as a fundamental part of a new medical paradigm. Older patients, many of whom are either not tech savvy or not tech inclined, may find this especially challenging.

Perhaps the most perplexing obstacle to the gospel of mHealth preached by its zealots comes from age-old human behaviors such as denial, exhaustion, and hedonism, to name but a few. There may, indeed, be excellent wireless medical devices and mobile medical apps available to help corral people into better health...provided they truly want it. People commonly say, "Without good health, nothing else matters." But in reality, people's actions speak louder than their words.

As long as someone with a bad knee is in denial, then he won't acknowledge that playing one more game of soccer could sideline him forever. As long as someone with hypertension regularly skips exercise in favor of a hot bath and a hot toddy, then no amount of body sensors and data tracking can incent that person toward better health. And if a clearly overweight person has a lousy day at work and seeks comfort later with "just a single piece of pie...maybe with a little ice cream on top," then obesity will eventually win. Healthcare industry entrepreneur and angel investor, Esther Dyson, refers to this paradox as one of "the most interesting unsolved problems in health care and human behavior (Regalado, 2013)."[20]

Consumers will make any number of impulse purchases based on an incalculable number of different factors at any given time. Yet, for all the hoopla surrounding the ever-increasing number of MMAs available to help folks "become 'Quantified Self' types who can never have too much data on themselves (Wieners, 2013)"[21] as they track their vital signs, count calories, log their footsteps (and the list goes on), "most of the apps are merely part of the entertainment industry," comments Robert B. McCray, president and CEO of the influential Wireless-Life Sciences Alliance (www.wirelesslifesciences.org).

Unless the apps receive FDA certification as to the science behind their design, and the genuine health and wellness benefits they can confer, they're likely to become one-hit wonders. Still, there's still a benefit in these apps in that they make personal health and quantification fashionable and, in my view, cultural shifts have the potential to produce more health benefits, more efficiently, than the health care system (McCray, 2013).[22]

Health clearly fascinates doctors, medical device makers, and MMA developers. Presumably, that's why they choose the work they do. The mistake they make is in assuming the rest of the world is as focused on health, body sensors, data monitoring, fitness tracking, etc., as they are. Cars take up most of an auto mechanic's attention during the day. A fashion designer concentrates on fabric, texture and color. A chef wants to know who has the freshest ingredients and how to use them in an original recipe, and a political journalist cares most about getting the story first and getting the story right.

Someone who already *is* healthy and is *not* a fitness buff may not necessarily have an interest in taking up the sport of 24/7 health monitoring. And people with non-fatal, yet painful chronic illnesses (e.g., chronic migraine, osteoarthritis, fibromyalgia, Ehlers-Danlos syndrome[23]) typically don't want to think about their health on days when they feel good. They want to live for that day and enjoy it completely without thinking at all about when they're going to feel bad next.

This is actually when people with non-fatal chronic illnesses are at risk for acting outside their own best interests medically. To people living with chronic pain, sometimes having a bit of fun *is* worth the price of pain later, since almost everyone with chronic pain knows the pain will inevitably return, no matter what. "The development and maintenance of chronic widespread pain and fibromyalgia involve a complex dynamic process with biological, cognitive, and psychosocial factors.... Maladaptive thoughts and feelings seem to play an important part in the negative spiral resulting in the maintenance of chronic pain (Kristjánsdóttir et al., 2013; Flor, 2011).

Change never happens without resistance, particularly in medicine, according to Dr. Eric Topol, who wears many distinguished hats. He is director of Scripps Translational Science Institute; chief academic officer for Scripps Health; the Gary & Mary West Chair of Innovative Medicine; professor of genomics at The Scripps Research Institute; and senior consultant for the division of cardiovascular diseases at Scripps Clinic. "Of all the professions represented on the planet, perhaps none is more resistant to change than physicians. If there were ever a group defined by lacking plasticity, it would first apply to doctors (Topol, 2012, p. 177)."

Not all doctors or healthcare workers like the changes they have begun to see as a direct result of connected healthcare. "The typical 15-minute office visit is rarely enough time to fully address the clinical needs of patients with multiple chronic illnesses, and the onerous documentation demands of electronic medical records ensure that doctors spend most of that visit interacting with the computer rather than with the patient (Ofri, 2013)."[24]

The costs of re-tooling a large health system for eHealth can be staggering. Memorial Healthcare System (MHS) (www.mhs.net) in Hollywood, Florida, is the second-largest public healthcare system in the United States. It has five hospitals (not including Joe DiMaggio Children's Hospital), numerous outpatient facilities throughout South Broward County, and over 1800 in-patient beds. Although MHS provides more than $200 million in direct costs for uncompensated care, it receives only $15 million (one percent) of its $1.5 billion annual budget from local taxes. The System's revenue is generated primarily by fees for healthcare services. A relatively small, but meaningful, amount also comes from donors through the System's two foundations (Janser, 2013).

MHS recently finished installing EPIC software across its entire health system (clinical data, financial data, EHRs, MyChart portal for patients, etc.) to achieve true connected healthcare among its different hospitals and facilities, and to begin empowering every MHS patient with meaningful access to his/her own health record. EPIC also offers secure messaging and scheduling apps that run on smartphones and tablets for those doctors who choose to use them.

The entire transition took three years, at a cost of approximately $130 million. MHS has invested extraordinary capital to keep the healthcare system at the top in its field, not only in terms of technology and innovation, but also in its commitment to patient care by engaging patients earlier and more comprehensively in making medical decisions for themselves. The greatest resistance to change came from…the doctors (Blanton, 2013).

In commentary co-written by Arthur Kellermann and Spencer S. Jones, appearing on *Project-Syndicate.org*, they, too, address physician resistance to change: "…a newly hired neurosurgeon with 27 years of education may have to read a thick user manual, attend tedious classes, and accept periodic tutoring from a 'change champion' to master the various steps required to use his hospital's IT system. Not surprisingly, despite its theoretical benefits, health IT has few fans among health-care providers. In fact, many complain that it slows them down (Kellerman & Jones, 2013b)."[25] In a recent PricewaterhouseCoopers report in which a number of health industry leaders were interviewed, the difference between doctors and patients, according to Steinar Pedersen[26], "is the centre of the battlefield" over mHealth. Misha Chellam[27] adds that such technology "changes the balance of power. It is not surprising that doctors would be concerned (PricewaterhouseCoopers, 2012)."

Educating new doctors and other healthcare providers, as well as re-educating older ones, present additional challenges for realizing the promise of eHealth. Many medical schools (and nursing schools) already have begun to incorporate mobile and wireless tools for teaching, diagnosis and therapeutics. Students can use mobile technology for reference tools, for virtual anatomy practice in place of cadavers, and as stethoscopes and ultrasound devices, for example. While these schools train future caregivers to use and trust data provided by spectacular wireless technologies, they also must teach students to trust their patients, too.

Today's medical schools must educate for a new "connected bedside manner." Students weaned on smartphones and wireless technologies are at risk, themselves, of forgetting to make eye contact with patients. Doctors cannot practice good medicine without this.

Skillful, compassionate and well-aligned care takes time. It goes slowly. It requires face-time, not computer time. We have to listen to the fellow human in our midst, examine her, go over both relative and absolute risks and benefits of treatment options, and then be clear about expectations. You don't really think an EMR is capable of removing fear and ignorance from medical decisions, do you? And the 6-page office note…this helps align care with a patient's goals (Mandrola, 2013)?[28]

Dr. Eric Topol is uniquely poised to envision and describe the future of medicine, not only because he is brilliant and leads an extraordinary team in the field of human genetics research, but also because he literally sees the future become reality each time another genetic mystery unfolds. He refers to the ability to unlock every detail of human DNA as the ability to 'digitize humans.' "Digitizing a human being is determining all of the letters ('life codes') of his or her genome – there are six billion letters in a whole genome sequence. It is about being able to remotely and continuously monitor each heart beat, moment to moment blood pressure readings, the rate and depth of breathing…all the things that make us tick (Topol, 2012, p.vi)."

With all due respect to Dr. Topol, the things that make us tick go well beyond mere genetic data. In Act IV of Shakespeare's *The Tempest*, Prospero says, "…We are such stuff / As dreams are made on, and our little life / Is rounded with a sleep." The phrase "digitizing humans" speaks to a dispassionate, clinical approach to issues that go to the very figurative heart of who we are as people. As a person *and* as a patient, I take offense at it. Dr. Topol concedes, with the various technology convergences that make digitizing humans possible, "there will be legitimate worries about depersonalization, about treating the digital information instead of the individual (Topol, 2012, p.xi)."

There can be little argument left against the benefits wireless connected health can bring. It will facilitate better prevention of illness or disease, thus engendering sustained wellness of most people. People want to feel well and they want to feel good getting there. Mobile games and other mobile means of education allow individuals to learn and process at their own pace and according to their own schedules. Peer pressure advocating for healthy living that promotes wellness will increase, just as peer pressure to stop smoking, reduce drinking and driving, etc., has had a positive effect on human behavior. As good healthcare becomes available ubiquitously, excuses for poor health will no longer be socially acceptable.

Wireless connected health will bring improved diagnostic and treatment opportunities for people living in rural and distant communities. Telemedicine has already proved itself a viable and desirable option for first line healthcare in remote areas. Lab work and other diagnostics performed on site can be interpreted at distant locations by transmitting data/images via mobile and wireless devices. Patients need not wait days and weeks for results, which benefits them if they have conditions meriting immediate treatment. As smaller mobile medical devices become standard tools for both health providers and patients, alike, problems can be detected sooner, therapeutic treatment can begin more expediently, and people with chronic illnesses have greater opportunity for extended periods of wellness through active health maintenance.

Another clear benefit of wireless connected health comes from using mobile and wireless media for faster containment of disease outbreaks. Upon discovery of the presence of any highly communicable disease, such as an influenza outbreak, healthcare providers can utilize many forms of instant communication to prevent epi- and pandemics. Mobile social media can alert local and global populations at highest risk. Mobile apps (such as GPS) can direct these same populations to safety zones. And mobile devices such as Propeller Health (www.propellerhealth.com formerly known as Asthmapolis) can be used to detect poor breathing conditions for at-risk asthmatics or people with COPD, both chronic respiratory conditions among the top global NCDs to address.

Opportunities for innovation will increase as more people become stakeholders in the healthcare process itself. Mobile social media makes it possible for online support groups and communities to connect patients with the same disease or chronic illness living anywhere in the world. People often innovate in those areas where they have a personal interest or need, and by joining forces and funds with mutually motivated people, they potentially can fast-track research, new products, therapeutics, and more.

SOCIAL, POLITICAL, AND ECONOMIC IMPLICATIONS OF WIRELESS CONNECTED HEALTH

Health and wellness, illness and disability all contribute to self-image and how others perceive us. Based on our health status, we may gain or lose social standing, political power, and/or economic opportunity. The expectation of connected health is that, individually, people will be empowered to monitor and safeguard their health or improve it by using an ever-increasing number of mobile medical devices and mobile medical apps (and general health and wellness apps) powered by wireless technologies. On a larger scale, entire populations will eventually be enabled to shift from poor health toward wellness.

Psychology, sociology and culture play important and interconnected roles in the social implications of connected health. The extraordinary popularity of social media such as Facebook and Twitter prove (if ever there were doubt) that humans are social creatures.

Chatting online or via mobile social platforms about personal experience validates toward personal empowerment, and experience sharing among groups via social networks and online support groups further helps to validate an individual's own experience.

James Fowler, PhD, is professor of medical genetics and political science at the University of California, San Diego. His work lies at the intersection of the natural and social sciences, with a focus on social networks, behavioral science, evolution, politics, genetics, and big data. A panelist at the 8th Annual Convergence Conference of the Wireless-Life Sciences Alliance in San Diego, California, he asserted, "Friends will be critical over the next ten years." He discussed friends as data ("Data from your friends can predict your political party."); friends as sensors ("Data can predict, academically, by Twitter, about where there are H1N1 virus outbreaks."); and friends as motivators[29], referring to a Brown University paper called "The Social Context of Dietary Behaviors: The Role of Social Relationships and Support on Dietary Fat and Fiber Intake" (Dube, 2010). Not surprisingly, among the authors' conclusions: social support improved results.

Researchers repeatedly have shown the health benefits of belonging to a community (Helliker, 2005). Communities contribute to wellness. When it comes to connected health, the definition of community, itself, expands beyond geographic boundaries (it can exist online or locally), social similarities, or religious or moral values. Community can derive from people who share the same chronic illness, genetic expression, or other health-related similarity, and it also can develop around non-professional people who provide care to others with health issues.

In the United States, the National Multiple Sclerosis Society offers community not only to people with MS, but also to their families and friends. Alzheimer's support groups provide community for an increasing number of people who personally understand the challenges of caring for a loved one with Alzheimer's dementia. Perhaps the best-known example, globally, of healthful benefits from belonging to a community comes from Alcoholics Anonymous.

Dr. Fowler wants to "unlock the 'socio' for bio-pharmaceutical prediction and prevention," describing his own work as "tapping the social side for prediction and prevention. We need to uncover the emotional drivers of crazy decisions regarding our own healthcare (Fowler, 2013)." We need his expertise now.

Distracted driving, by using mobile media for texting, talking, or otherwise not paying complete attention to the road-train tracks-bus lane-bicycle lane-or-boating conditions (or others) invites potential injury or death for the driver-pedestrians-passengers-cyclists- swimmers-oncoming vehicles…or all of them. Still, people continue making a crazy decision regarding their own well being, and the sanctity of others' lives, every time they consciously make the choice to do this.

Cultural will figure enormously in the ultimate success or failure of global eHealth initiatives. "Culture is a term that refers to the inherited set of implicit and explicit rules guiding how a group's members view, feel about, and interact with the world. Cultural expressions and, to a lesser extent, cultural values change over time and are influenced by others (García, 2006)."

We already know wireless medical devices and MMAs work. They can help patients more precisely self-monitor chronic health conditions to enjoy extended periods of wellness, and they can help people learn to adopt new and healthy behaviors. This holds true for children as well as adults. However, the cultural drivers behind Western medicine do not necessarily have relevance for other cultures holding differing values, beliefs, and perspectives on health.

A commonly-held Western assumption that all people everywhere strive for excellent health in order to have the longest life possible is just that...an assumption. Many Eastern cultures give higher importance to living a balanced life, rather than achieving longevity. Some cultures are more present- than future-oriented, so trying to 'sell' the long-term benefits of changing behaviors now for better health in the future may have little impact. Still other cultures are group-oriented. Anything that places sustained focus on individual change – be it diet, exercise, or lifestyle – can seem threatening to the group as a whole (García, 2006).

Poverty is its own culture, or rather, a subculture within nearly every known culture around the world, and a subculture most every member would happily abandon if only he or she knew how. Poverty affects attitude and expectations. If a person grows up with nothing, is told to expect nothing, gets nothing, and finds no visible means of improving her lot in life, then by being told that simply changing certain behaviors will dramatically improve her life, she probably won't believe it.

eHealth can make enormous inroads to improve the quality of life for the world's most disadvantaged or at-risk populations if it first can crack the mindset that says, essentially, "Life is hard and suffering is inevitable."

In many countries other than the United States, healthcare is viewed as a human right and basic costs are covered by a national health plan. For these people, eHealth already has begun to take hold and likely will continue to grow. eHealth also has caught on quickly in developing nations without a pre-existing medical infrastructure. According to a 2013 PricewaterhouseCoopers report, "Patients and doctors in emerging markets are much more likely to use mHealth than those in developed countries – and more payers in emerging markets cover the cost of mHealth than in developed countries. Why? Existing healthcare is scarce — in many cases, mobile technology is the only (rather than alternative) affordable tool to reach people. The lack of existing infrastructure means fewer entrenched interests, so lower barriers. Change is more welcome (PricewaterhouseCoopers, 2013, overview)."[30]

As time passes and technology advances, savvy healthcare providers will lean more and more on eHealth innovations to accurately and efficiently manage chronic diseases, ameliorate or eradicate non-chronic medical conditions, and prevent illness in the first place. Prevention – wellness – will become the focus and new mindset in medicine.

It is good public policy to emphasize population wellness. However, population wellness cannot be achieved if people cannot afford the tools necessary for their health, such as a basic eHealth tool kit (for instance, a thermometer, blood pressure cuff and reader, ECG device, pulse oximeter[31], and glucometer). How will patients pay for additional devices when they already can't adequately cover the costs of their existing health burdens?

In the U.S., it remains to be seen how long it will take for eHealth to become the standard of care (beyond the federal mandate to implement EHRs by 2015). In our current political climate, private insurers dictate the economics of healthcare. Will insurers penalize patients who do not use eHealth tools because they can't afford to? And if so, will this stratify our society into a post-modern "caste system" based on health status, with healthy people being 'Brahmins' and chronically ill and/or low-income patients being 'untouchables?' As much as these may seem like economic questions, they are actually highly political in nature.

In our near future, wireless and mobile medical devices and MMAs will help us to avoid adverse health events (heart attack, brain attack/stroke, asthma attack) in the first place. Soon, using an ever-increasing variety of wireless medical devices and mobile apps, everyone (who can afford them, that is) should be able, theoretically, to prevent anything unmanageable from happening. Theoretically, then, no one would ever get sick.

Preventive medicine using eHealth technologies would make unnecessary, or greatly reduced, the social and medical "safety nets" governments have struggled to sustain in the face of shifting political winds and economic realities. It sounds almost too good to be true, and there's a reason.

Over 80 percent of the world's deaths from CVDs occur in low- and middle-income countries. People in low- and middle-income countries who suffer from CVDs and other noncommunicable diseases have less access to effective and equitable health care services that respond to their needs (including early detection services) (WHO, 2013c).

The promise of wireless connected health is just that – a promise – if it's out of reach for the very people who need preventive healthcare the most. Furthermore, women in low-income countries suffer disproportionately, simply for lack of access to early detection services. A woman who dies young often leaves behind young children, stunting their growth for loss of a provider, especially if the woman was nursing, and crippling their economic opportunities should they survive into adulthood.

Many other obstacles exist (besides the loss of one or both parents) to threaten a person's well being. All it takes is a vehicle crash, or tick bite, or fire, for our health status to permanently change in an instant.

Fortunately, technology is in place and enough data already exists for us to know on a day-to-day basis, and even on an hourly basis, for example, where air quality exceeds safe breathing conditions for people with chronic respiratory diseases; or, where climate conditions have created breeding grounds for mosquitoes. The truth is mosquitoes can actually drive public health policy.

In a 2013 broadcast on National Public Radio in the U.S., Beenish Ahmed reported on a two-year concerted effort by the local government in Lahore, Pakistan, to stop the spread of dengue fever, a deadly tropical disease caused by mosquitoes. Mosquitoes can breed just about anywhere there's standing water. In 2011, one of the world's worst epidemics of dengue fever hit Punjab (Ahmed, 2013).[32]

The government hired Umar Saif, a Cambridge-educated computer scientist who developed a smartphone app called Clean Lahore "to track all efforts to prevent the disease." Local investigators first canvassed the city to identify trouble spots and take a photo using the app. City workers then went about doing their jobs. If their jobs happened to involve prevention activities at known trouble spots, an investigator would follow, taking pictures of the workers in action.

With 'before' and 'after' documentation, workers knew they could not be accused of skipping work or shirking their full responsibility. "If Punjab averted another epidemic in 2012, then it didn't happen by accident," said Saif. "There were 67,000 different prevention activities [that] were performed and photo-logged by the smartphones." Saif went on to develop a Google map that correlates the locations of dengue cases and mosquito larvae for "a clear pattern of disease outbreak that corresponds to reports of positive dengue larvae (Ahmed, 2013)."

Sending patients across national borders for critical care transcends local politics and can sometimes be good for international relations. Almost everyone knows the story of Malala Yousafzai, the young, outspoken Pakistani schoolgirl who advocated for girls' education. The politically powerful and highly feared Taliban targeted her for assassination, but failed. Nevertheless, she suffered life-threatening injuries. Local Pakistani doctors and resources could not treat her severe wounds, so they airlifted her to a hospital in Birmingham, England, for critical care and long-term rehabilitation. What began as a gesture of good will between two nations turned an extraordinary young women into an international symbol of hope and a U.N. advocate for worldwide access to education for children.

When a crisis of almost any nature occurs anywhere, brave men and women show up to help, often traveling great distances and sometimes at their own expense: doctors, medics, engineers, trained rescuers, and others with the expertise needed to solve immediate problems on the ground. Telemedicine can play an increasing role as an international "partner" and "global ambassador" when disaster strikes. As long as a wireless signal can be accessed, ICTs can connect caregivers on the ground with expert doctors and caregivers all around the world. Having access to EHRs will help treat injured people promptly and *appropriately* to reduce serious casualties and possibly avert fatalities.

Politically, there is one very compelling reason to encourage and facilitate the adoption of eHealth to support population wellness locally and nationally, and that is to *sustain* population wellness. Illness and communicable diseases, such as influenza, cross borders without discrimination. Today, the global mobility and migration of people mean disease can spread rapidly. One infected passenger on an international flight can scatter sickness around the world. The best defense against illness is a foundation of good health and good public health policy.

Wireless medical devices and MMAs are quickly becoming "go to" tools for diagnosing, monitoring and/or management of a variety of health conditions that affect people worldwide. Costs and benefits go along with this wireless makeover – for the healthcare industry, for local and global economies, and for individuals. Most all of it is good.

Initially, healthcare providers, especially institutional ones, will face significant upfront costs to transition their facilities into eHealth entities. Not only will a variety of ICTs be involved, but also, new hardware and software may need to be installed or upgraded, and if pre-existing systems aren't replaced, then new systems will need to be compatible with them, at least for some time. Some of the costs were addressed previously in this chapter. The benefits show up as added jobs in the work force: real people sell the systems and must oversee every installation.

There are costs affiliated with training end users, too. These include doctors, nurses and other hospital personnel, who must actively train anywhere from eight to 16 hours per person (anesthesiologists are required to train the maximum time) – time and billing which is taken away from regular patient care (Blanton, 2013). For some providers, the high costs and complexities may force them out of business altogether, or to align with larger medical groups that can absorb them.

Hospitals and large medical groups also will need to spend money on community outreach to market and promote the new, user-friendly EHR systems that allow patients to access their entire medical record. Not all people will recognize or understand EHRs as a benefit, at first. However, most systems have been created to be highly intuitive, so users learn quickly how to navigate them once they try.

People who join the "Quantified Self" movement, whether out of vanity or for medical necessity, will face increased costs for the privilege or necessity of non-stop monitoring. Currently, it is not clear who will pay for these expenditures. As of October 2013, a pulse oximeter can cost approximately $250, a blood pressure reader and cuff, $129, and a heart monitor that can generate an ECG runs about $200. A high-end baby monitor costs $250, while the same company's baby scale is $180, and its fitness scale costs $150.[33] These costs are out of reach for most low- and many middle-income families.

Most MMAs currently range in price from free to a few dollars, while medical devices cost much more. One new market has emerged as a clear economic winner in the race toward connected health: "wearables." Until recently, "wearables" went by another, more clinical name: body sensors. "Wearables" is definitely cooler. Last year, nearly 30 million wearable wireless medical devices were shipped, according to ABI Research (Slabodkin, 2012).[34] Wearables are the *sine qua non* of any bona fide "Quantified Self." Some people sport multiple types of wearables simultaneously. UP by

Jawbone, for one example, currently lists at $130. Clearly, paying for the ability to "know (about) thyself" is not for the faint of heart or wallet.

Wearables allow for 24/7 monitoring, collection, and/or transmission of a garden variety of physiological data, ranging from body temperature, skin temperature, amount and quality of sleep, to number of steps taken in a day, heart rate during exercise and rest, and more. They have become a new fashion statement, taking the form of bracelets, rings, watches, and even 'smart' tattoos. Fabric has been explored for its wearability for several years already, which says everything about its greater potential and absolutely nothing about how it looks or feels.

As people become healthier through the benefits available via connected health, the social cost of poor health should decrease. More households' incomes can be allocated for education and other opportunities previously unavailable due to financial resources being usurped by healthcare.

One unstoppable vehicle for health education is mobile games. Within the mobile games market, educational games are referred to as "game-based learning" or "serious games." Games that are used for such things as corporate training are referred to as "simulation-based learning." If there is any lesson to be learned, it's this: games make serious money.

According to Sam Adkins, chief research officer at Ambient Insight, "Mobile educational games are now outselling PC educational games. And the entrepreneurs are incredibly passionate about what they are doing (Takahashi, 2013)."[35]

In a report from research firm Ambient Insight, forecasters predict the serious games market will grow from $1.5 billion in 2012 to $2.3 billion in 2017, with the total market (including simulation-based learning) to grow to $8.9 billion in 2017 (Takahashi, 2013). The global mobile health market is estimated to generate $56.6 billion[36] in revenue by 2020, with the number of mobile apps on the market at approximately 200,000. The number of smartphone users by 2020 will grow to over four billion. Approximately half of these users are likely to use health and wellness apps, which indicates there will be numerous downloads and great usage of these apps (Goel, 2013).

While games-based learning offers many choices for youngsters and adults, it doesn't cater well to learners after the 4th grade because teachers seem to prefer using other methods to teach older learners. "There's a lot of debate still about the effectiveness of game-based learning, but it has long since proved its worth," Adkins asserted. "I don't know why we are still having that debate (Takahashi, 2013)."

Wireless connected health augurs economic growth locally and globally. As the field expands, it will create a need for new wireless medical devices, wearables, and apps. Every product will go through research and development, manufacturing and sales, and any aspect of the process can take place anywhere in a connected world. Globally, one source of growth will come through greater dispersion of population centers since adequate, even excellent, healthcare will no longer be restricted to major metropolitan areas. As the cost of healthcare delivery goes down, more people can enter into the system to access it. As the benefits of healthcare delivery go up, and an emphasis on illness prevention becomes the norm, overall population wellness will help maintain economic stability.

CONCLUSION

The high penetration rates of smartphones around the world now make the promise of wireless connected health a reality. Wireless medical devices and mobile medical apps have become the primary means for bringing mainstream medicine and healthcare to previously overlooked or underserved populations, especially in remote locations around the world. People unable to access regular healthcare, most notably in developing nations, now have another option for improving their lives through better health. They adopt eHealth quite readily.

While the road ahead remains uncharted in some ways, it also offers the opportunity for industry leaders to establish a connected health ecosphere that truly responds to the needs of end-users, taking into consideration not only best medical practices, but also the social, cultural, educational, economic, and political issues that contribute to a patient's experience of the world.

In the following paragraphs, I will provide my observations, insights and recommendations regarding people behavior, community, culture, education, entertainment/games, economics, public policy, and international relations in a wireless connected health world.

Behavior

Observations: In a wireless connected health world, a plethora of medical devices and MMAs can monitor conditions and provide massive data about one's health status, moment-to-moment, to the patient, the doctor, and others. From this data, preventive or corrective recommendations can be made and appropriate action taken, accordingly. Wireless connected health technologies allow us to create tailored communication to support patients and encourage compliance. Data alone does not persuade people to change their behavior, and patient non-compliance remains one of the most vexing issues in the fight against illness.

Insights: We need to better understand why patients do crazy things and don't comply with sound medical advice. We have the ability to track people's moods. This could be a part of every tool used in service to patient compliance. An empowered patient is one who will tend to do what is beneficial, not detrimental. If current technologies can predict which patients likely will not comply, then we have the choice of doing nothing and, ultimately, letting that patient fail, or else helping to provide whatever support is needed. If we can prevent or cure almost everything, does this mean we could theoretically live forever? Will it be a question of who can afford it?

Recommendations:

- Societal transformation begins by learning how to treat our bodies with respect and admiration for the extraordinary machines they are.
- Begin teaching children their bodies must last them a very long time and they must take good care of them.
- Organize a well-thought out, international clinical trial of known, non-compliant patients to search for insight that will lead to better medical outcomes with less patient resistance.

Community

Observations: Our notion of what defines a community has radically changed in light of social media and mobile social media. Improving health or sustaining wellness has a better chance of success with the support of one's social networks. Fowler tells us we can gather data from a patient's typical activity on social media and track it to see variances, which can indicate when that patient is reaching out to (or isolating from) friends. Evidence shows a well-supported person will be more inclined to do what is good for him/herself.

Insights: There has been much talk about healthcare solutions, but less about patient support, which takes different forms for different problems in different communities. The understanding of community has expanded to mean almost any group of people held together by at least one common interest. Since we know that community support improves patient compliance and health outcomes, helping people find – or create – the right support as soon as possible would benefit patients and caregivers alike. If illness and disease were colloquially (not medically) reframed as "conditions," and support groups were reframed as "communities," then perhaps the disempowerment associated with 'needing support' could be turned into positive participation in a communal group activity. Everyone has something to offer. This could occur either locally or online/mobile. In addition, the concept of rape hotlines and suicide hotlines is well established: chronic and serious health conditions can precipitate crises for patients and caregivers, too.

Recommendations: The various medical societies/sub-specialties should take the lead to create guided frameworks for support groups – caring communities – for the types of patients they treat.

- Create separate support groups/communities for patients from their families/caregivers and/or other support team members to encourage candor. Have at least one facilitator, to prevent the group from devolving into a 'pity party.'
- Engage another co-facilitator who is a trained healthcare professional to answer psychosocial or medical FAQs.
- Facilitators may need training in cultural sensitivity/cultural relevancies.
- Offer additional, trusted resources for reliable information.
- Every country should create these, so each community is language-appropriate. Translation capabilities still may be a digital necessity.

Culture

- **Observations:** Cultural awareness helps when bringing Western-based medicine into developing nations. Working knowledge of customary and traditional local remedies for medical conditions also can prove instructive in building bridges that will integrate best Western medical practices with complementary healing regimens.

Insights: Culture plays a much larger role than possibly we have accounted for in the prescription for health and wellness. Many people around the world have ancient remedies and healing methods indigenous to their culture. We must find a way to bring in 'best practices' of modern medicine while, at the same time, honoring culturally respected traditions, provided they do no known harm.

Furthermore, when we talk about health, we also are *not* talking about death. Cultural beliefs about health, illness, life and death must be known in order to implement best modern medical practices that also respect cultural views surrounding the end of life. This even may extend to death rites of passage. Anthropologists and ethnologists can offer not only insight, but also directed pathways, to identify leaders of cultural communities, who can convey this information back and forth to build trust among all stakeholders. In this way, entire communities can be enlisted all together in the quest for wellness.

Recommendations:

- Identify colloquial relevancies and cultural absolutes.
- Ensure all recommendations are realistic by paying attention to the economic standing of patients and any other considerations that could impede compliance in spite of best intentions.
- Find something of value in each culture and introduce it to another, especially promoting cross-cultural exchange and conversation.
- Pay attention to religious mandates such as gender role expectations, dietary restrictions, and separation of genders so modesty (among other issues) can be observed/preserved as required.
- Above all else: ask leaders from every culture what is important to them!

Education

Observations: Everything about healthcare is always about education and learning for someone.

Healthcare professionals' education and learning never stops. People may have rudimentary knowledge about health in general, but they only start serious learning on an as-needed basis when a particular health condition manifests personally.

- **Insights:** Once people discover they "have" something, they become patients. Fear is often their primary emotion that drives everything else, whether it's fast-track learning of latest information, thorough investigation of complementary medicine, prayer/spirituality, or denial. Knowledge can mitigate fear. As our understanding of science and medicine increases daily, we gain greater choices in how to deliver newer, better, more customized care to patients, especially through advances in wireless connected health. One day, genetic mapping may make it possible to prevent NCDs or other conditions by interventional therapeutics. How might this affect our understanding of not "blaming" the patient for the disease?

Recommendations: This could be an extraordinary opportunity to change the entire conversation about health and wellness.

- Teach age-specific and demographic-specific audiences about a variety of health issues well before they would typically occur, including risk factors and known behaviors that can lead to disease.
- Teach how positive behaviors can help prevent disease onset.
- Teach how to recognize important physiological symptoms not to ignore.
- Empower patients, through early and ongoing education, not to deny pain or symptoms, but rather, to recognize them as important messages that our bodies need prompt care, like fixing a flat tire on a car, or watering a wilting plant.
- Start teaching health and wellness in preschool or grammar school. If we teach good habits in a positive way early on, then kids in the schoolyard (another type of community) can support each other in making good food choices and other health-positive decisions.

Entertainment/Games

Observations: MMAs developed as games are being used successfully to educate people while simultaneously entertaining them. They work especially well with young people and pediatric patients, who already play games and validate their design structure of using incentives or rewards to a) encourage repeat play, or b) drive competition with others. Apps for chronic conditions risk boring the very people who need to master what they teach and then maintain what they've learned, which often involves repeating one or more activities indefinitely.

Insights: Games have proved worthy vehicles for edutainment in many different fields. It's still too early to know definitively how effective game-based learning will be for instruction in health and wellness. Time will tell; initial results seem positive. If developers work in an interdisciplinary, collaborative fashion not only with epidemiologists, psychologists, and sociologists, but also with experts in artificial intelligence (AI), then they can develop agile games with compelling player/avatar interaction that not only can educate, but also continue to entertain and challenge players. Structuring games 'learning' in stages would allow players to complete one level of play and then graduate up the learning ladder.

Recommendations:

- Healthcare providers should seek to connect motivated patients with games designer/developers.
- Gamers with chronic diseases can explain, first-hand, what it's like to live with their

particular condition and whether or not (or how) it affects their ability to play games being designed potentially for them (e.g., people with fibromyalgia might experience muscle exhaustion or pain from repetitive movements; people with diabetes might have neuropathy in their hands, plus loss of fine motor skills, both of which make using hand controls an issue; people with extreme hypertension might need to avoid stressful games or competitions with elevated emotional involvement; and, people with anything-plus-a-hearing impairment may need on-screen captions).

- Knowing the particulars of a specific condition, games designers could actually create a massive multi-player game for patients all around the world with similar (or same) conditions. This would create entirely new communities of people initially connected by their condition, but united over time by having fun and feeling better.

Economics

Observations: The complexity of networking different ICT systems to each other takes significant time, expertise, and capital. The costs for healthcare providers to transition their facilities into eHealth operations can be staggering, if not lethal barriers. The necessity for interoperability of networks and systems around the world is an imperative for eHealth to live up to its promise of providing excellent, personalized healthcare to any patient anywhere any time.

Much has been made of the costs to providers. Less has focused on the economics of eHealth implementation, especially for people in low- to middle-income countries. Eighty percent of NCDs occur in such places. Poverty, cultural traditions and reduced expectations for quality of life conspire to keep down those populations with the greatest need and least access to modern tools that can change those dynamics.

Insights: A smartphone or tablet is just the first of many costly devices and apps necessary for monitoring one's own vital signs and managing any number of chronic ailments. Who will pay for all the technologies necessary for poor, or even mid-income people to access the benefits of connected health? Will governments or insurers look to the models of telco operators and subsidize the cost of smartphones, wireless medical devices and MMAs to reduce the social cost of poor health? Would there be a value proposition in subsidizing the cost of a smartphone or medical device, and if so, for whom, and in what ways? If governments, for example, spend less on the cost of patient healthcare because of effective medical ICTs, would they pump such newly 'available' resources back into technology development?

As patients learn to better manage their NCDs, remaining healthier for longer, and as preventive healthcare increasingly keeps more people from developing NCDs in the first place, hospitals should see a downturn in admissions. What will this mean in terms of their own budgets, and more so, their fundamental identity? If people don't get sick in the first place, what new purpose might hospitals serve?

Recommendations:

- Hospitals and other qualified care providers should learn from each other about best practices for transitioning large healthcare systems and smaller, independent hospitals and clinics to eHealth.
- Consider alternative ways, besides full payment up front, for patients to afford devices and apps, such as rent-to-own, or micro lending to a small group of people who can share equipment, where feasible.
- As more and more people avoid illness, they will be able, theoretically, to remain productive members of society and enter, or re-enter, the job marketplace. Now is the time to begin thinking about how to maximize this future asset, including creating

new job categories, such as interdisciplinary wellness mentors.

Public Policy

Observations: Data from wireless and mobile ICTs can help track disease, recognize adverse environmental and health conditions, and send alerts to people at risk. Wireless technologies can play a vital, positive role in public health policies and initiatives, which can benefit people whose lives are often at the mercy of elements beyond their control.

Insights: Good health doesn't happen by accident. Public policies promoting health and wellness need to be seen as relevant and beneficial by the constituents affected by them.

Recommendations: Leaders in connected health should seek out municipal (or higher) politicos to learn what kind of data sets already exist about their own populations.

- Examine existing data for new insights that specifically relate to health and wellness as they relate to communicable- and NCD patterns.
- In low- and middle-income nations, help frontline healthcare workers validate the efficacy and benefits of wireless connected health by implementing a massive campaign that educates and entertains, showing wireless and mobile as essential and welcome tools in the arsenal of healthcare, prevention and wellness.
- Communicate that effective public policy for healthcare, such as vaccinating school children against a host of illnesses, has contributed globally to the near elimination of smallpox, polio, and other dreaded diseases.
- Clearly communicate tangible benefits to individuals, as well as the community, of every policy initiated and approved.
- If public policy requires individuals to download apps to comply with the policy, then they should be not only free of charge, but also understandable to the broadest diversity of individuals.

International Relations

Observations: An old aphorism says, "Good fences make good neighbors." A newer one might say, "Good connected healthcare keeps connected neighbors healthy."

Wireless connected health already has shown results as a vehicle for positive international relations. Telemedicine, especially, builds bridges between nations, benefits patients who might otherwise not receive adequate treatment, and now can turn these patients into lifelong ambassadors of goodwill for their "adopted" nation that helped them.

Insights: Physicians routinely rotate through the various specialties of medicine to attain broad-based knowledge, plus a sense of what it would be like to pursue a particular career direction. As a direct result of this sampling and exploration, they're highly versatile, broadly speaking. Those doctors who opt to specialize after general training often relocate in order to gain specific expertise.

Carrying out effective international relations requires knowledge about a broad range of subjects, too. So does connected healthcare. In a highly mobilized world that increasingly will rely upon wireless connected health, physicians and other qualified care providers could conceivably be called upon at any time, anywhere, to substitute briefly in the absence of a patient's regular doctor or qualified care provider. In essence, they would be conducting international relations *without* regard to nationality or politics because blood is always red no matter where it drips. Wireless connected health could hold a key to healing the world.

Recommendations: A new form of practicing medicine – via wireless connected health – may require a new twist in medical students' education and post-graduate training.

- Medical schools, nursing schools and training programs for other allied healthcare professionals should consider instituting a mandatory pre-graduation 'study abroad' clinical experience in a low- or middle-income country. This will give students even greater insight and connection to patients from different cultures and vastly different circumstances. While this recommendation would certainly entail complicated global negotiations regarding financial underwriting, legal liability, and other considerations, it would challenge the ingenuity of innovation that is the hallmark of exceptional and productive international relations.
- International physician exchange programs should be encouraged and cultivated through all available means.
- Interoperability of systems almost necessitates some level of interoperability among globally connected physicians and qualified healthcare providers, too, so they not only have a wealth of medical knowledge with the ability to collaborate, but they also carry within themselves a breadth of cultural and geographic awareness if called into service anytime anyone, anywhere needs medical help.

REFERENCES

Ahmed, B. (2013, September 16). How Smartphones Became Vital Tools Against Dengue In Pakistan. *NPR: National Public Radio*. Retrieved from http://www.npr.org/blogs/health/2013/09/16/223051694/how-smartphones-became-vital-tool-against-dengue-in-pakistan

AliveCor. (2013). *AliveCor Heart Monitor*. Retrieved from http://www.alivecor.com/en

Alkmim, M. B., Marcolino, M., Figueira, R., Maia, J., Cardoso, C., & Abreu, M. … Ribeiro, A. (2013, April 10). 1,000,000 electrocardiograms by distance: An outstanding milestone for telehealth in Minas Gerais, Brazil. In Presentation to 2013 Med@Tel/Luxembourg. International Society for Telemedicine and eHealth.

Cafazzo, J. A., Casselman, M., Hamming, N., Katzman, D. K., & Palmert, M. R. (2012, May 8). Design of an mHealth app. for the self-management of adolescent type 1 diabetes: A pilot study. *Journal of Medical Internet Research*, *14*(3), e70. doi:10.2196/jmir.2058 PMID:22564332

Cardi, V., Clarke, A., & Treasure, J. (2013, May 16). The Use of Guided Self-help Incorporating a Mobile Component in People with Eating Disorders: A Pilot Study. *European Eating Disorders Review*, *21*, 315–322. doi:10.1002/erv.2235 PMID:23677740

Carter, M. C., Burley, V. J., Nykjaer, C., & Cade, J. E. (2013, April 15). Adherence to a Smartphone Application for Weight Loss Compared to Website and Paper Diary: Pilot Randomized Controlled Trial. *Journal of Medical Internet Research*, *15*(4), e32. doi:10.2196/jmir.2283 PMID:23587561

Chronaki, C., Kontoyiannis, V., Mytaras, M., Aggoutakis, N., Kostomanolakis, S., & Roumeliotaki, T. … Tsiknakis, M. (2007). Evaluation of shared EHR services in primary healthcare centers and their rural community offices: the twister story. In *Proceedings - IEEE Engineering in Medicine and Biology Society*. Retrieved from http://www.ncbi.nim.nih.gov/pubmed/18003492

Comstock, J. (2012, December 4). Real games for health and the trouble with gamification. *MobiHealthNews*. Retrieved from http://mobihealthnews.com/19323/real-games-for-health-and-the-trouble-with-gamification/

Comstock, J. (2013, May 28). Brazil: An up-and-coming mobile health market. *MobiHealthNews*. Retrieved from http://mobihealthnews.com/22654/brazil-an-up-and-coming-mobile-health-market/

Creutzfeldt, J., Hedman, L., Heinrichs, L., Youngblood, P., & Felländer-Tsai, L. (2013, January 14). Cardiopulmonary Resuscitation Training in High School Using Avatars in Virtual Worlds: An International Feasibility Study. *Journal of Medical Internet Research*. doi:10.2196/jmir.1715 PMID:23318253

Curtis, G. E. (Ed.). (1996). *Russia: A Country Study*. Washington, DC: GPO for the Library of Congress. Retrieved from http://countrystudies.us/russia/53.htm

Danaei, G., Finucane, M. M., Lu, Y., Singh, G. M., Cowan, M. J., & Paciorek, C. J. et al. (2011, July 2). National, regional, and global trends in fasting plasma glucose and diabetes prevalence since 1980: Systematic analysis of health examination surveys and epidemiological studies with 370 country-years and 2.7 million participants. *Lancet*, *378*(9785), 31–40. doi:10.1016/S0140-6736(11)60679-X PMID:21705069

Depp, C. A., Kim, D. H., de Dios, L. V., Wang, L., & Ceglowski, J. (2012, January 1). A Pilot Study of Mood Ratings Captured by Mobile Phone Versus Paper-and-Pencil Mood Charts in Bipolar Disorder. *Journal of Dual Diagnosis*, *8*(4), 326–332. doi:10.1080/15504263.2012.723318 PMID:23646035

Deterding, S., Miguel, S., Nacke, L., O'Hara, K., & Dixon, D. (Eds.). (2011). Gamification: Using game-design elements in non-gaming contexts. In *Proceedings of the 2011 Annual Conference Extended Abstracts on Human Factors in Computing Systems* (pp. 2425–2428). doi: 10.1177/0141076813480996

Dube, A. R., & Stanton, C. A. (2010). The Social Context of Dietary Behaviors: The Role of Social Relationships and Support on Dietary Fat and Fiber Intake. *Modern Dietary Fat Intakes in Disease Promotion. Nutrition and Health (Berkhamsted, Hertfordshire)*. doi: doi:10.1007/978-1-60327-571-2_2

Flor, H., & Turk, D. C. (2011). *Chronic Pain: An Integrated Biobehavioral Approach. International Association for the Study of Pain Press*. Retrieved from.

Fogg, B. J. (2009). A Behavior Model for Persuasive Design. In *Proceedings of the 4th International Conference on Persuasive Technology*. Retrieved from http://bjfogg.com/fbm_files/page4_1.pdf

Fogg, B. J. (2011). BJ Fogg's Behavior Model. *BJ Fogg 6d*. Retrieved from http://behaviormodel.org/

García, A. (2006). Is health promotion relevant across cultures and the socioeconomic spectrum? *Family & Community Health*, *29*(1), 20S–27S. doi:10.1097/00003727-200601001-00005 PMID:16344633

General Electric Healthcare. (2013). *Vscan Ultrasound*. Retrieved from www.vscanultrasound.gehealthcare.com

Githuku-Shongwe, A. (2013, March 6). The Greatest Return of Investment Is Investing in the Mindsets of the Future Generation of Leaders. *The Huffington Post*. Retrieved from http://www.huffingtonpost.com/anne-githukushongwe/afroes-anne-githuku-shongwe_b_2819045.html?view=print&comm_ref=false

Globocan. (2008). Retrieved from http://globocan.iarc.fr/factsheets/populations/factsheet.asp?uno=900

Helliker, K. (2005, May 3). Body and Spirit: Why Attending Religious Services May Benefit Health. *Wall Street Journal*. Retrieved from http://online.wsj.com/article/0,SB111507405746322613-email,00.html

Hyman, Phelps, & McNamara. (2013, March 12). Are All Health and Wellness Mobile Apps Exempt from FDA Regulatory Requirements? *FDA Law Blog*. Retrieved from http://www.fdalawblog.net/fda_law_blog_hyman_phelps/2013/03/are-all-health-and-wellness-mobile-apps-exempt-from-fda-regulatory-requirements.html

InfraScan, Inc. (2013). *The Infrascanner Model 2000*. Retrieved from www.infrascanner.com

Kellermann, A. L., & Jones, S. S. (2013a, Jan). What It Will Take to Achieve the As-Yet-Unfulfilled Promises of Health Information Technology? *Health Affairs*, *32*(1), 63–68. doi:10.1377/hlthaff.2012.0693 PMID:23297272

Kellermann, A. L., & Jones, S. S. (2013b, February 26). The Delayed Promise of Health-Care IT. *Project Syndicate*. Retrieved from http://www.project-syndicate.org/commentary/the-delayed-promise-of-health-care-it-by-art-kellermann-and-spencer-jones

Kristjánsdóttir, O. B., Fors, E. A., Eide, E., Finset, A., Stensrud, T. L., & van Dulmen, S. et al. (2013, January 7). A Smartphone-Based Intervention With Diaries and Therapist-Feedback to Reduce Catastrophizing and Increase Functioning in Women With Chronic Widespread Pain: Randomized Controlled Trial. *Journal of Medical Internet Research*, *15*(1), e5. doi:10.2196/jmir.2249 PMID:23291270

Mandrola, J. (2013, August 19). *Compassionate, well-aligned healthcare takes time, a luxury few doctors have*. Retrieved from http://medcitynews.com/2013/08/compassionate-well-aligned-healthcare-takes-time-a-luxury-few-doctors-have/

Mathers, C. D., & Loncar, D. (2006). Projections of global mortality and burden of disease from 2002 to 2030. *PLoS Medicine*, *3*(11), e442. doi:10.1371/journal.pmed.0030442 PMID:17132052

Morrish, N. J., Wang, S. L., Stevens, L. K., Fuller, J. H., & Keen, H. (2001). Mortality and Causes of Death in the WHO Multinational Study of Vascular Disease in Diabetes. *Diabetologia*, *22*(2), S14–S21. doi:10.1007/PL00002934 PMID:11587045

National Institutes of Health. (2010). *Traditional Chinese Medicine: An Introduction*. Retrieved from http://nccam.nih.gov/health/whatiscam/chinesemed.htm

National Institutes of Health. (2013). *Ayurvedic Medicine: An Introduction*. Retrieved from http://nccam.nih.gov/health/ayurveda/introduction.htm

Ofri, D. (2013, August 14). Why Doctors are Reluctant to Take Responsibility for Rising Medical Costs. *The Atlantic.com*. Retrieved from http://www.theatlantic.com/health/archive/2013/08/why-doctors-are-reluctant-to-take-responsibility-for-rising-medical-costs/278623/

Oudshoorn, N. (2012). How places matter: Telecare technologies and the changing spatial dimensions of healthcare. *Social Studies of Science*, *42*(1), 121–142. doi:10.1177/0306312711431817 PMID:22530385

Paiz, J. M., Angeli, E., Wagner, J., Lawrick, E., Moore, K., & Anderson, M. … Keck, R. (2013, March 1). *General Format*. Retrieved from http://owl.english.purdue.edu/owl/resource/560/01/

Palmier-Claus, J. E., Rogers, A., Ainsworth, J., Machin, M., Barrowclough, C., & Laverty, L. et al. (2013). Integrating mobile-phone based assessment for psychosis into people's everyday lives and clinical care: A qualitative study. *BioMed Central Psychiatry*, *13*, 34. doi:10.1186/1471-244X-13-34 PMID:23343329

Patrick, K., Marshall, S. J., Davila, E. P., Kolodziejczyk, J. K., Fowler, J. H., & Calfas, K. J. et al. (2013, November 9). Design and implementation of a randomized controlled social and mobile weight loss trial for young adults (project SMART). *Contemporary Clinical Trials*, *37*, 10–18. doi:10.1016/j.cct.2013.11.001 PMID:24215774

PricewaterhouseCoopers. (2012, June 7). *Emerging mHealth: Paths for growth.* Retrieved from http://www.pwc.com/us/en/press-releases/2012/consumers-are-ready-to-adopt-mobile-health.jhtml

PricewaterhouseCoopers. (2013). *mHealth implementation in emerging markets: PwC*. Retrieved from http://www.pwc.com/gx/en/healthcare/mhealth/opportunities-emerging-markets.jhtml

Proudfoot, J., Clarke, J., Birch, M. R., Whitton, A. E., Parker, G., & Manicavasagar, V. et al. (2013, November 15). Impact of a mobile phone and web program on symptom and functional outcomes for people with mild-to-moderate depression, anxiety and stress: A randomised controlled trial. *BioMed Central Psychiatry*, *13*, 312. doi:10.1186/1471-244X-13-312 PMID:24237617

Qardio. (2013). *QardioCore*. Retrieved from http://www.getqardio.com

Regalado, A. (2013, August 18). Esther Dyson: We Need to Fix Health Behavior. *MIT Technology Review*. Retrieved from http://www.technologyreview.com/news/518901/esther-dyson-we-need-to-fix-health-behavior/

Roglic, G., Unwin, N., Bennett, P. H., Mathers, C., Tuomilehto, J., & Nag, S. et al. (2005). The burden of mortality attributable to diabetes: Realistic estimates for the year 2000. *Diabetes Care*, *28*(9), 2130–2135. doi:10.2337/diacare.28.9.2130 PMID:16123478

Schuman, A. J. (2013). Improving patient care: Smartphones and mobile medical devices. *Contemporary Pediatrics*, *30*(6), 33.

Scott, R. E., Mars, M., & Jordanova, M. (2013). Would a Rose By Any Other Name - Cause Such Confusion?. *Journal of the International Study for Telemedicine and eHealth, 1*(2).

Sejdić, E., Rothfuss, M. A., Stachel, J. R., Franconi, N. G., Bocan, K., Lovell, M. R., & Mickle, M. H. (2013). Innovation and translation efforts in wireless medical connectivity, telemedicine and eMedicine: A story from the RFID Center of Excellence at the University of Pittsburgh. *Annals of Biomedical Engineering*, *41*(9), 1913–1925. doi:10.1007/s10439-013-0873-8 PMID:23897048

Shand, F. L., Ridani, R., Tighe, J., & Christensen, H. (2013). The effectiveness of a suicide prevention app. for indigenous Australian youths: study protocol for a randomized controlled trial. *Trials*, *14*, 396. doi:10.1186/1745-6215-14-396 PMID:24257410

Sifferlin, A. (2013, March 14). South By Southwest (SXSW), Will Collecting Data on Your Body Make You Healthier? *TIME.com.* Retrieved from http://healthland.time.com/2013/03/14/south-by-southwest-sxsw-will-collecting-data-on-your-body-make-you-healthier/

Slabodkin, G. (2012, December 11). Wearable mHealth device shipments to hit 30 million by year's end. *FierceMobileHealthcare.com*. Retrieved from http://www.fiercemobilehealthcare.com/story/wearable-mhealth-device-shipments-hit-30-million-years-end/2012-12-11

Sorrell, T., & Draper, H. (2012, Aug 10). Telecare, Surveillance, and the Welfare State. *The American Journal of Bioethics*, *12*(9), 36–44. doi:10.1080/15265161.2012.699137 PMID:22881854

Spyridonis, F., Ghinea, G., & Frank, A. O. (2013, April 10). Attitudes of Patients Toward Adoption of 3D Technology in Pain Assessment: Qualitative Perspective. *Journal of Medical Internet Research*, *15*(4), e55. doi:10.2196/jmir.2427 PMID:23575479

Takahashi, D. (2013, August 16). With a mobile boom, learning games are a $1.5B market headed toward $2.3B by 2017 (exclusive). *VentureBeat*. Retrieved from http://venturebeat.com/2013/08/16/with-a-mobile-boom-learning-games-are-a-1-5b-market-headed-toward-2-3b-by-2017-exclusive

Thompson, B. (2013, August 27). Set the FDA mobile medical app. guidance free! *Mobihealth News*. Retrieved from http://mobihealthnews.com/25040/set-the-fda-mobile-medical-app-guidance-free/

Topol, E. (2012). *The Creative Destruction of Medicine: How the Digital Revolution Will Create Better Health Care*. New York: Basic Books.

Turner-McGrievy, G., & Tate, D. (2011, December 20). Tweets, Apps, and Pods: Results of the 6-month Mobile Pounds Off Digitally (Mobile POD) Randomized Weight-Loss Intervention Among Adults. *Journal of Medical Internet Research*. doi:10.2196/jmir.1841 PMID:22186428

U.S. Food and Drug Administration. (2012, January 24). *Medical Devices: Premarket Approval (PMA)*. Retrieved from http://www.fda.gov/medicaldevices/deviceregulationandguidance/howtomarketyourdevice/premarketsubmissions/premarketapprovalpma/

U.S. Food and Drug Administration. (2013a, February 8). *Medical Devices: Is the Product a Medical Device?* Retrieved from http://www.fda.gov/medicaldevices/deviceregulationandguidance/overview/classifyyourdevice/ucm051512.htm

U.S. Food and Drug Administration. (2013b, June 6). *Mobile Medical Applications: Examples of MMAs the FDA Has Cleared or Approved*. Retrieved from http://www.fda.gov/MedicalDevices/ProductsandMedicalProcedures/ConnectedHealth/MobileMedicalApplications/ucm368784.htm

U.S. Food and Drug Administration. (2013c, August 13). *Medical Devices: Connected Health*. Retrieved from http://www.fda.gov/MedicalDevices/ProductsandMedicalProcedures/ConnectedHealth/default.htm

U.S. Food and Drug Administration. (2013d, August 13). *Medical Devices: Wireless Medical Telemetry Systems*. Retrieved from http://www.fda.gov/MedicalDevices/ProductsandMedicalProcedures/ConnectedHealth/WirelessMedicalDevices/ucm364308.htm

Wieners, B. (2013, August 8). Dude, My Testosterone's Pushing 1290: How About Yours? *Bloomberg Businessweek*. Retrieved from http://www.businessweek.com/printer/articles/142612-dude-my-testosterone-s-pushing-1290-dot-how-about-yours

World Health Organization. (2006). *World Health Statistics 2006*. Retrieved from http://www.cdc.gov/globalhealth/countries/russia/pdf/russia.pdf

World Health Organization. (2008). *The global burden of disease: 2004 update*. Retrieved from http://www.who.int/healthinfo/global_burden_disease/GBD_report_2004update_full.pdf

World Health Organization. (2009). *Public Health and the Environment, Russian Federation*. Retrieved from http://www.who.int/quantifying_ehimpacts/national/countryprofile/russianfederation.pdf

World Health Organization. (2011a). *Global status report on noncommunicable diseases 2010*. Retrieved from http://www.who.int/mediacentre/factsheets/fs317/en/

World Health Organization. (2011b, May). *Asthma fact sheet No.307*. Retrieved from http://www.who.int/mediacentre/factsheets/fs307/en

World Health Organization. (2011c). *Country Health Information Profiles: China*. Retrieved from http://www.wpro.who.int/countries/chn/5CHNpro2011_finaldraft.pdf

World Health Organization. (2012a, October 17). *ITU and WHO launch mHealth initiative to combat noncommunicable diseases*. Joint ITU/WHO news release. Retrieved from http://www.who.int/mediacentre/news/releases/2012/mHealth_20121017/en/

World Health Organization. (2012b, November). *Chronic obstructive pulmonary disease fact sheet No.315*. Retrieved from http://www.who.int/mediacentre/factsheets/fs315/en/

World Health Organization. (2012c). *Global data on visual impairments 2010*. Retrieved from http://www.who.int/mediacentre/factsheets/fs312/en/index.html

World Health Organization. (2012d). *China Health Service Delivery Profile*. Retrieved from http://www.wpro.who.int/health_services/service_delivery_profile_china.pdf

World Health Organization. (2013a, October). *Millennium Development Goals (MDGs) fact sheet No.290*. Retrieved from http://www.who.int/mediacentre/factsheets/fs290/en/index.html

World Health Organization. (2013b, January). *Cancer fact sheet No.297*. Retrieved from http://www.who.int/mediacentre/factsheets/fs297/en/index.html

World Health Organization. (2013c, March). *Cardiovascular Diseases (CVDs) fact sheet No.317*. Retrieved from http://www.who.int/mediacentre/factsheets/fs317/en/index.html

World Health Organization. (2013d, March). *Diabetes fact sheet No.312*. Retrieved from http://www.who.int/mediacentre/factsheets/fs312/en/index.html

World Health Organization. (2013e, March). *Obesity and overweight fact sheet No.311*. Retrieved from http://www.who.int/mediacentre/factsheets/fs311/en

World Health Organization. (2013f, April). *A Global Brief on Hypertension*. Retrieved from http://www.who.int/cardiovascular_diseases/publications/global_brief_hypertension/en/

World Health Organization. (2013g, July). *Tobacco fact sheet No.339*. Retrieved from http://www.who.int/mediacentre/factsheets/fs339/en/

World Health Organization. (2013h). *Brazil: Health profile*. Retrieved from http://www.who.int/gho/countries/bra.pdf

World Health Organization. (2013i, May). *India: Country Cooperation Strategy*. Retrieved from http://www.who.int/countryfocus/cooperation_strategy/ccsbrief_ind_en.pdf

World Health Organization. (n.d.). *Health topics: eHealth*. Retrieved from http://www.who.int/topics/ehealth/en

X Prize Foundation. (2013). *Life Sciences Prize Group | XPRIZE*. Retrieved from http://www.xprize.org/prize-development/life-sciences

Zaridze, D., Maximovitch, D., Lazarev, A., Igitov, V., Boroda, A., & Boreham, J. et al. (2008). Alcohol poisoning is a main determinant of recent mortality trends in Russia: Evidence from a detailed analysis of mortality statistics and autopsies. *International Journal of Epidemiology*, *38*(1), 143–153. doi:10.1093/ije/dyn160 PMID:18775875

KEY TERMS AND DEFINITIONS

eHealth: Electronic health; a broad, general term referring to health care that uses information communication technologies (ICTs) to transmit medical data and healthcare information among patients, qualified caregivers, doctors and healthcare systems. The term eHealth is frequently interchanged with the terms 'connected health,' 'wireless connected health,' and 'mobile connected health (mHealth).'

Electronic Health Record (EHR): The unique digital health record for each person. Ideally it will contain all medical data across the entire healthcare spectrum, allowing any doctor or qualified caregiver to access all relevant medical information about the person they treat.

Health and Wellness Apps: Applications that can be downloaded (or pre-loaded) onto smartphones for the purpose of maintaining or achieving fitness and wellness and would include apps that help count calories, measure footsteps taken in a day, record amount and quality of nightly sleep, measure pulse, etc. Health and wellness apps are not regulated by the USFDA.

Interoperability: Refers to the ability for all healthcare IT systems around the world to be able to 'talk' to each other. Interoperability includes not only text, such as doctors' notes and lab reports, but also imaging, such as x-rays and sophisticated diagnostic radiological scans.

Mobile Medical Apps (MMAs): Typically, these are applications that can be downloaded (or pre-loaded) onto smartphones for the purpose of helping patients manage an existing medical condition (such as through diet, medication scheduling, awareness of exacerbation triggers, etc.) and/or helping qualified caregivers correctly administer medications, obtain vital statistics, or synch with a wireless medical device for performance. In the US, the Food and Drug Administration regulates MMAs.

Noncommunicable Chronic Diseases (NCDs): Long-lasting illnesses that are not spread via person-to-person contact, such as cancer, cardiovascular diseases (heart disease or stroke), chronic respiratory diseases (asthma or chronic obstructive pulmonary disease), diabetes, and overweight/obesity.

Quantified Self: The name of a movement that began in the late 2000s, whereby people collect, analyze and store (through MMAs) data on an increasing number of their biological functions and daily physical activities. People do this, currently, through the use of 'wearables' (see definition below).

Telemedicine: Refers to the transfer of patient information from one place to another via electronic means (e.g., email, video conferencing, telephones, smartphones) for the purpose of direct patient care. It is not a subspecialty of medicine, but rather, an augmented means of communication allowing medicine to be practiced when a patient and physician (and/or the patient's personal health information) are geographically distant.

Wearables: A broad term referring to a category of wireless data-gathering devices that can be worn on the body – either as clothing made from 'smart fabric,' or as accessories, such as wristbands, headbands, etc. – and then synchronized with other wireless devices where that data can be analyzed.

Wireless Medical Devices: Any wireless device that can diagnose disease or a health condition, or is intended for the monitoring, treatment or cure of such. This would include, but is not limited to, devices that monitor/track biological functions such as blood glucose levels, blood oxygenation, or blood pressure. It also would include smartphones that can function as an ultrasound machine, or record an ECG, or function as a stethoscope, among many other purposes. Wireless medical devices are regulated by the USFDA.

Wireless Medical Telemetry: The ability to connect patients to data measuring/monitoring/recording devices via radio frequency rather than cables. This untethering allows today's hospital patient to be freer to ambulate short distances without physical connection to vital equipment.

ENDNOTES

1. ©2014 InfraScan. Used with permission.
2. © Hyman, Phelps & McNamara, P.C. Used with permission.
3. © 2013 Chester Street Publishing, Inc. All rights reserved. Used with permission.
4. Used with permission.
5. Used with permission.
6. ©2012 PricewaterhouseCoopers LLP, a Delaware limited liability partnership. All rights reserved. Used with permission. PwC refers to the United States member firm, and may sometimes refer to the PwC network. Each member firm is a separate legal entity. Please see www.pwc.com/structure for further details. This content is for general purposes only and should not be used as a substitute for consultation with professional advisors. ©2012 The Economist Intelligence Unit Ltd. All rights reserved. Used with permission. Whilst efforts have been taken to verify the accuracy of this information, neither The Economist Intelligence Unit Ltd. nor its affiliates can accept responsibility or liability for reliance by any person on this information.
7. ©2013 Chester Street Publishing, Inc. All rights reserved. Used with permission.
8. Used with permission.
9. Used with permission.
10. Used with permission.
11. The dire need to improve healthcare and health in the U.S. is a problem whose solution has evaded the brightest minds. The Qualcomm Tricorder XPRIZE is a $10 million competition to stimulate innovation and integration of precision diagnostic technologies, making definitive health assessment available directly to "health consumers." These technologies on a consumer's mobile device will be presented in an appealing, engaging way that brings a desire to be incorporated into daily life. Advances in fields such as artificial intelligence, wireless sensing, imaging diagnostics, lab-on-a-chip, and molecular biology will enable better choices in when, where, and how individuals receive care, thus making healthcare more convenient, affordable, and accessible. The winner will be the team that most accurately diagnoses a set of diseases independent of a healthcare professional or facility and that provides the best consumer user experience. Visit the competition website to learn more. This prize is made possible by a generous grant from the Qualcomm Foundation. TRICORDER is a trademark of CBS Studios, Inc. Used under license. (X Prize Foundation, 2013)"
12. COPYRIGHT NOTICE: Adapted, displayed and reprinted with permission from *Contemporary Pediatrics*, June 2013. *Contemporary Pediatrics* is a copyrighted publication of Advanstar Communications Inc. All rights reserved.
13. ©2012 Chester Street Publishing, Inc. All rights reserved. Used with permission.
14. ©2013 Anne Githuku-Shongwe, Afroes Transformational Games. Used with permission.
15. As of September 2013, Samsung operates out of all but six African nations.
16. The United Nations Global Fund for AIDS, TB and Malaria is an official participant in Smart Health Hub.
17. SafePoint is the inventor of the single-use disposable syringe.
18. ©2013 Time Inc. All rights reserved.
19. ©2009 ACM. Used with permission.
20. ©2014 MIT Technology Review. (www.technologyreview.com) All rights reserved. This quote has been reproduced with permission.

[21] ©2013 Bloomberg L.P. Used with permission.

[22] Used with permission.

[23] An exception to this would be Vascular Type EDS. http://ghr.nlm.nih.gov/condition/ehlers-danlos-syndrome

[24] ©2013 The Atlantic Monthly Group. All rights reserved. Used with permission.

[25] ©2013 Project Syndicate. All rights reserved. Used with permission.

[26] Steinar Pedersen, Chief Executive Officer, Tromsø Telemedicine Consult

[27] Misha Chellam, Chief Operating Officer, Scanadu

[28] ©2013 John Mandrola, MD. All rights reserved. Used with permission.

[29] Used with permission.

[30] ©2013 PricewaterhouseCoopers LLP, a Delaware limited liability partnership. All rights reserved. Used with permission. PwC refers to the United States member firm, and may sometimes refer to the PwC network. Each member firm is a separate legal entity. Please see www.pwc.com/structure for further details. This content is for general purposes only and should not be used as a substitute for consultation with professional advisors.

[31] "According to Vancouver-based LionsGate Technologies (LGT Medical), the World Health Organization has recognized the importance of making pulse oximetry available to the developing world, where 64 percent of mobile phone users are found."

[32] ©2013 Beenish Ahmed. All rights reserved. Used with permission.

[33] All costs in US dollars.

[34] ©2012 FierceMarkets and FierceMobileHealthcare.com. All rights reserved. Used with permission.

[35] ©2013 VentureBeat. All rights reserved. Used with permission.

[36] All costs in US dollars.

Chapter 18
Mobile Usability:
State of the Art and Implications

Linda M. Gallant
Emerson College, USA

Gloria Boone
Suffolk University, USA

Christopher S. LaRoche
Massachusetts Institute of Technology, USA

ABSTRACT

Context and the pervasive environment play a much greater role in mobile technology usage than stationary technology for which usability standards and methods were traditionally developed. The examination of mobile usability shows complex issues due to the ubiquitous and portable nature of mobile devices. This chapter presents the current state of mobile usability testing. More specially, topics covered are various usability testing methods, contextual complexity, audio interfaces, eye and hands-free interactions, augmented reality, and recommendation systems.

INTRODUCTION

Despite a decade of mobile device usability testing, technical challenges such as bandwidth, physical features, and financial issues slowed the progress of mobile communication (Zhang & Adipat, 2005). With recent advances in technology infrastructures, mobile communication is now faster and geographically more flexible. The removal of technological challenges such as network reliability, bandwidth, device in-puts and screen resolution necessitates the need for analyzing the multi-layered complexities of mobile usability. These include contextual factors, multimodal input and output choices, geographic locality, physical movement, and social interaction. Studies on mobile usability can be divided between studies of mobile phones with tiny screens, the arrival of smartphones such as the early Black Berry phones and the more recent full screen phones like iPhones, Android and Windows Phones (Nielsen & Budiu, 2013). Kjeldskov and Graham (2003) in a review of 42 mobile evaluations found that only 19 percent employ field studies and that 71 per-

DOI: 10.4018/978-1-4666-6166-0.ch018

cent employ lab-based experiments. Kaikkonen, A., et. el. (2005) and Kjeldskov (2004) found few differences between a usability test in the lab or a usability tests in the field. Garzonis (2005) stated that field evaluation provide insights with the everyday use of technology, divided attention spans, and the dynamic context of use. Coursaris & Kim (2006) offer a review of 45 empirical mobile usability studies based on variables like user, task, environment and technology. The majority of these studies were done in a lab-setting. Only eleven studies were done in the field or in the lab and the field. These early studies examined effectiveness (62%) or the accuracy or completeness of achieving a goal; efficiency (33%) of task performance using a particular device; or the degree of satisfaction (20%) a product gives a user (Coursaris & Kim, 2007). "Less than 7 percent of studies explored dynamic factors, i.e. lighting and noise levels. Hence there is a lack of research on physical, psychosocial, and other environment-specific factors (Coursaris & Kim, 2007, p. 2343). Kjeldskov & Paay (2010) provide evidence of using indexicality, specific contexts, to understanding mobile human-computer interaction in different contexts. Indexes help users to interpret their environment or the interaction potential with other people in the spatial context, the physical context, or the social context.

Rahmadi & Zhong (2013) offer a study of teens mobile usage evolution based on initial opinion, knowledge and skills, context dependence, boredom, and personalization. They believe a field study is needed to observe change in usage after people become familiar with their mobile device. Another approach is to conduct a longitudinal study testing interaction with users during a prolonged period of time. This could be repeated several times until redundancy is reached or no new patterns emerge.

Jacob Nielsen and Raluca Budiu in their book on Mobile Usability (2013), explained many of the findings from their 124 phone and 35 tablet in- lab usability tests using think-aloud methodology: keep text copy short (25), enlarge interface elements (20), a limited screen space should employ limited features (26), mobile apps can perform better for some websites (34), mobile task completion averaged 62% (45), do not waste space for visuals or content that is secondary (52), consistently use of gestures, home, or back buttons (59-66), too many images can hurt download time (80), avoid early registration (81), test and retest page designs (95), reading comprehension is lower on mobile (102), speak the user's language (111), cut the fluff (110), the average user reads only 120 words per page (113), and consider what your users know when organizing content by usage, task, subtopics, one long page or alphabetical sorting (124-129).

While a small body of research is emerging in the development of techniques and standards for mobile usability evaluation, the current state of mobility usability testing is still evolving from creatively applying traditional usability methods to the specific challenges of mobile contexts.

This chapter discusses how personal mobile communication devices are evaluated for mobile usability based on technological and methodological challenges. Specifically in technological challenges, issues like contextual complexity, audio interfaces, eye and hand free interactions, augmented reality, and recommendation systems are examined. For methodological challenges, we consider expert reviews, empirical research, and methods of data collection. The social and economic impacts of mobile usability are considered before conclusions are drawn.

TECHNOLOGICAL CHALLENGES IN MOBILE USABILITY

While laboratory usability testing may be viable for particular inquiries into cognitive and visual functions; the major challenge of mobile usability is often the environmental context. This is beyond traditional usability laboratory testing and more closely follows a field methods approach to usability testing. Testing must take into account the physical and environmental situations in which mobile communication occurs in daily life. Legacy and stationary computing, even laptops, can have better simulation in laboratory settings given the less complex physical and social environments in which the human computer interaction occurs. As mobile communication becomes more socially interactive and databases driven, usability testing should be more context or field-based. Mobile usability testing needs to find better ways to capture the real experience of users in their natural settings. A possible way to develop such methods and natural use patterns for mobile usability would be to test users with their personal devices instead of new produces like with traditional usability tests.

Several topics, central to the current state of the art of mobile usability, are covered in this chapter. First, the impact of contextual complexity is reviewed. Secondly, ongoing work in multimodal mobile usability is presented. Third, accessibility challenges of mobile devices are explored.

Contextual Complexity

Like any general assessment of usability, a key step to providing user insights is designing the proper research plan for the situation. This is especially true when fields are new and there is less disciplinary knowledge to draw upon. One recognized crucial point for mobile usability testing is context. As such, the planning phase of usability in testing mobile technology must be heavily tied to context of use (Roto, et al, 2011). Mobile technology is inherently contextual complex, which impacts the practicality of developing standards of mobile usability testing efforts.

As a general rule, people, place, and general objects are identified as fundamental to consider in designing usability assessments (Biancalana et al. 2013). The itinerant nature of mobile technology centralizes contexts of use as crucial assessing product usability (see, Racadio, Rose, & Boyd, 2012). In a more specific consideration of mobile usability, Jumisko-Pyykkö and Vainio (2010) provide a five component framework which focuses on content and context for mobile human computer interaction. These five are physical content, task context, social content, temporal context, and technical and information context. Other advice on the importance of context to mobile usability testing emphasizes that in planning participant testing user tasks must be designed in close examination of use environment to best determine context characteristics (Biancalana et al. 2013). After the initial planning, methods and tools need to be combined with a budget in mind. A pilot test can help to determine the correctness of the methods for the context and its characteristics.

Real-time location-based services utilizing databases queries add to the complexity of mobile usability. Knowledge databases queries must be experienced as functional by users. While users input information into databases, these systems must analyze user requests within the parameters of social and location-based data specific to each individual user. These types of mobile applications with e (e.g., by GPS) need evaluation in real would settings (Cheng, 2011) in order to confirm high levels of usability.

Multimodal Mobile Usability

Comparative testing between older more traditional user interactions and new forms of user interactions with products help to highlight implementation problems and usability concerns in innovative mobile products (Lai, Mitchell, & Pavlovski, 2007). Using this comparative testing method, one study found that even with little or no prior experience with systems users of primarily text-only systems preferred to use a speech based interface for in-put and navigation (Lai, Mitchell, & Pavlovski, 2007). Thus, comparing a text input with a speech-based input interface, result found that the less traditional speech-based interaction was preferred by users. Multimodal interactions are heavily involved in audio interfaces, eyes and hand free interaction, and augmented reality interfaces.

Audio Interfaces

Audio interfaces are prominent in mobile devices. Context awareness is important to audio interfaces usability (Kaila, et. al., 2009). For instance, speech recognition systems must capture users' speech in order to process information properly. If an environment has too much ambient or background sounds, the usability of a speech-based interface can be negatively impacted.

Beyond traditional usability concerns, several specific concerns surround the usability of audio interfaces such as time and effort to complete tasks in a mobile environment, the flexibility of usage in mobile environments, and user attitudes (Gu, Gu & Laffey, 2011). More specifically, differing environmental contexts such as room size, ambient noise, or number of voices in a space can impact usability. Testing microphones and audio processing in differing environmental context in which a device maybe typically used is important (Kaila, et. al., 2009).

Eyes and Hand Free Interaction

Williamson, Crossan, and Brewster (2011) tested users' access of news feeds in a hand free environment using audio, speech, and wearable sensors. They found that participants interacted well with the system and as time went system usage increased. A benefit of testing in the real world environment was discovering that users were more comfortable making task gesture on a street rather than on public transportation. This adds social knowledge to how people will engage with mobile products in real world environments. Laboratory testing would fall short in revealing this natural interaction condition that is revealed in-situ with other people.

Augmented Reality

Laboratory experiments are still an important part of mobile multimedia usability testing. In these controlled environments, researchers can test fine cognitive, visual, and motor skills for best user outcomes. For instance, Choi, Jang, and Kim (2011) used a laboratory setting for testing the organizing and presentation of geospatial tags in location-based augmented reality. They found that an "automatic but-less-accurate" routine is more preferred by users than a "precise-but-manual" approach. The researchers reason that this user preference is based on the "dynamic nature of the mobile interaction and less than perfect sensing," (Choi, Jang, & Kim, 2011). Additionally, laboratory experiments help reveal visually sensitive interaction such as with augmented reality overlays on mobile devices. For instance, some research has shown that real-time interactive feedback provides a better user interaction with augmented reality on smart phones than standard textual and graphical instructions (Liu, et. al., 2012).

Recommendation Systems

Web-based ubiquitous mobile media technologies, which are becoming more heavily tied to intelligent multi-sensory agents such as location-based systems must come to terms with a multiplicity of user preferences that can rapidly change based on location sensing and individual user profiling preferences. Central to the usability evaluation for this type of mobile user environment is log tracking as a participant moves through real life situations in potentially diverse physical environments (see, Kim, Kwon, & Hong, 2010). This can provide remote ethnographic insights that can be triangulated with other data such as observation, survey, and interviews. While usability of the mobile interface central in testing user experience, databases must also provide a high level of usability for recommendation systems to provide appropriate levels of user interaction.

Interactive Data and Training Sets

Training sets for knowledge databases, like social recommendation systems or location-based recommendation systems, can be an important part of mobile usability testing (Biancalana et al. 2013). Databases serving mobile technology often have the purpose of providing predictive relationships between the user and environment. This is seen in location-based services such as visitor guides, maps, and restaurants recommendations that provide users with information on their surrounding physical environments. As a result, these intelligent systems or knowledge databases are trained to provide appropriate response values based on user inputs. A location-based recommendation system providing information on user inquires needs this capability. As users move through physical environments and interacting with databases, recommendation systems need to integrate user input and environmental cues. Training sets help model typical user behavior in a test environment.

Accessibility Challenges

The evolution of technological changes required in mobile usability has increased the awareness and need of accessibility so that all users, including those with disabilities, can access mobile applications. The main areas of disability include visual, hearing/auditory, motor, and cognitive impairments. With the increased awareness and recognition of the need for improved usability, particularly in mobile devices, this has in turn increased the awareness and requirements for accessibility within products. With legal force, such as the American with Disabilities Act several decades old, now behind many accessibility initiatives, the twin thrust of legal force and a desire to expand to a wider market has made accessibility a crucial area within mobile usability. The WC3 (2013) does provide some information and resources on mobile accessibility. The WC3 (2013) plans to include more guidance as professional knowledge and research provides more insights.

Often products that are made accessible are often pared down to eliminate any unnecessary information, content, or barriers to the end goal of the user and this ties into the need for mobile usability to be minimalist in what it shows to accomplish the task at hand. So the intersection of mobile usability and accessibility not only helps the widest community possible but it also can help with the idea of designing products in the most minimalist way, that will benefit all users of mobile products. The increasing influence of universal and inclusive design has also assisted with the increasing usability of products, including mobile devices. The next section lays out current data collection and analysis methods being used in mobile usability studies.

METHODOLOGICAL CHALLENGES IN MOBILE USABILITY

Inspection Methods

Standardized and generally agreed upon set of design and usability principles can be used to evaluate an artifact's usability. Usability inspection methods, which do not require user testing like heuristic evaluations and expert reviews for mobile usability, are in a nascent state. Thus, a well-developed body of general knowledge on mobile usability testing does not exist.

In general, heuristic evaluation, a commonly known technique has often been criticized, especially when not used by usability and content experts with sufficient knowledge to correctly apply design and usability principles. A second important inspection method is an expert review. An expert evaluator analyzes an artifact by means of user scenarios and tasks or examines the usability and interactivity in a self-guided exploration of important core user tasks.

The knowledge bases of heuristic evaluation and expert review specifically for mobile devices must be built over time by usability professionals. This knowledge should include developing general usability principles plus industry and device specific usability criteria for mobile usability testing.

Empirical Research

As empirical research for usability is often done in laboratory and field settings, the design of any empirical mobile usability testing and assessment needs to be tied closely to the artifact's use environment. Methodological choices depend on questions to be answered and situations to be studied. Laboratory experiments can produce fine-tuned understandings of user interaction under controlled environments; however mobile technologies are by nature used in differing contexts offering little predictable control. Data triangulation is a viable way to counterbalance the weaknesses inherent in laboratory settings. In such cases, laboratory and field testing can be done on the same product to compare and integrate findings. Hybrid approaches using data triangulation should be considered in a testing plan if its implementation can optimize user experience for the mobile products.

Developing a set of design principles to guide analysis that uses data triangulation to produce a findings based on multiple data points could overcome the current weakness of having few industry standards. This approach could also bolster testing solely relying on controlled laboratory testing for mobile usability. Triangulation can involve multiple data collection processes. This can included setting design principles then conducting expert evaluations such as cognitive walkthroughs as well as focus groups and think aloud session to investigate the usability of mobile communication technologies (Gu, Gu & Laffey, 2011). This could also include field studies. What is important at the end of the data collection is the cross-case analysis of data to search for patterns and correlations related to usability.

Methods of Data Collection and Analysis

Methods of data collection variety among observational analysis, automatic capture data, and user reporting of data. Heat maps, eye-tracking, video and audio recording, and field observation are useful methods of data collection in mobile usability testing.

Heat maps for multi-touch interaction in mobile usability studies are limited (Lettner & Holzmann, 2012). This is due to limitations imposed by mobility and hardware. Lettner and Holzmann (2012) devised a method using heat maps for finger-based multi-touch interaction with mobile applications. Their central concern was to develop a usability testing methods to collect data from people using the applications in real world settings. While some research casts limits on usability tools for visualizing eye gaze data (Lettner & Holzmann, 2012), Cheng (2011) has developed a usability testing model described as context comprehensive.

Using a remote eye-tracker along with an on-screen simulator in a lab setting, Cheng (2011) collects quantitative measure of eye gaze on static material. Remote eye tracking provides the strength of quantifiable precision in usability performance evaluation. This method cannot be used in field testing. To solve this problem, Cheng (2011) supplemented usability testing with a portable eye-tracker which can record user interaction with mobile devices in the lab and field settings. The remote tracker records eye-movements, finger operations and surroundings. This captures observation data to analyze the interaction process between users, interfaces and environments. This approach brings together discrete performance measures captured on remote eye-tracker in lab settings with observational data showing user interaction with devices in lab and real world environments.

Along with communicating and interacting with other people on social media platforms, users will increasingly be part of the internet of everything in which near-field communication (NFC) and RFID (radio frequency identification) are increasingly connecting people and devices. This makes user field studies for mobile devices even more important. Recording user actions in the field is an essential data collection method for mobile usability (Racadio, Rose, & Boyd, 2012). As tag and location-based services are becoming common place, there is a compelling need for developing specific field-based methods for mobile usability testing so that a user has positive experiences with tag-based services (Isomursu, 2008). When tag and location-based services are limited or too expensive to set-up a beta system, a robust field testing of their usability is unlikely to succeed.

Researchers such as Kaasinen (2005) have designed field studies which combine shorter and longer testing intervals in which a cross-case analysis of user performance with devices produced insights on usability, utility, and user trust of a location-based mobile guide. The field study required the user be accompanied by a primary evaluator, who traveled with the participant and conducted the test. While the test was being audio recorded, another evaluator followed the test subject and primary evaluator. The secondary evaluator took photos interaction at pivotal usage points. The triangulated analysis combined recorded data, photos, user reactions, and evaluators' observations.

SOCIAL AND ECONOMIC IMPLICATIONS OF MOBILE USABILITY

The ubiquitous nature of mobile communication impacts our social and economic worlds. With easy to use devices that can access 24/7 communication, people will be able to engage more in social platforms, upload photos, post comments, and check on friends and family from anytime and from anywhere. In fact, Cisco (2013) estimates that "by the end of 2013, the number of mobile-connected devices will exceed the number of people on earth....There will be over 10 billion mobile-connected devices in 2017." The spread and speed of mobile communication continues to accelerate. Cisco (2013) predicts that "mobile network connection speeds will increase 7-fold by 2017 with 4G accounting for 45 percent of total traffic" and that "two-thirds of the world's mobile data traffic will be video by 2017."

As mobile communication devices' ease of use improves more segments of the population will be able to increasingly use mobile technology for social purposes. Designing low income and low literacy platforms and devices for mobile communication, software and applications developers will level the social playing field for mobile usage and will allow more users to quickly learn and use mobile communication technologies. This increases economic impacts of mobile communication. The economic implication of improved mobile usability allow customers to quickly purchase items online, obtain coupons and sales information, to comparison shop in brick and mortar stores, check stocks, weather, traffic, market or currency rates or to enjoy entertainment or search for information.

Connecting to social media by mobile devices is becoming commonplace. In March of 2013, IDC did a sentiment analysis of how users' feel about Facebook's mobile application. Users expressed feeling of being connected, excited, curious/interested, and productive when using their mobile phone (Meeker & Lu, 2013). Since 68% of Facebook users often connect via mobile devices, mobile usability is a necessary concern for social media platforms (Meeker & Lu, 2013). Social media and location-based services are becoming more integrated as a social and economic power. People use recommendation systems on multiple social media platforms which also heavily expose users to advertising and marketing campaigns.

The economic impact of mobile communication through online marketing is growing. eMarketer (2013) asserts that more money will continue to flow into mobile advertising. Many more ads will be on smartphones and tablets as advertisers shift to correlate with the amount of time consumers are spending on mobile devices. Currently, Google has more than 53% and Apple has nearly 16% of the $16.65 billion mobile global advertising market (eMarketer, 2013). In North American by 2017, digital advertising is expected to increase to 37% or nearly half of all digital advertising spending (eMarketer, 2013). This is up from 18.8% in 2013 (eMarketer, 2013). Additionally, by 2017 almost 40% of all worldwide advertising will be on mobile devices (eMarketer, 2013). To keep up with these changes, usability testing of advertising on different mobile platforms is likely to increase in the next several years. Platform differences will drive the need for more usability tests as well.

Given the increased speed and usability of mobile devices, mobile search and information seeking is a familiar user routine. The National Technical Transfer Network (NTTN) reported in 2012 that mobile search queries have surpassed search queries on personal computers in South Korea. Boland (2012) calculates that local searches on mobile will surpass local searches on personal computers in the United States in 2015. Besides search, people are using their smartphones for messaging, voice calls, clock, music, gaming, social media, alarms, camera, news/ alerts, and calendar functions (TomiAhonen Almanac, 2013). Each of these usages allow for different tasks and interactions and ultimately different usability measures for both mobile devices and mobile websites.

CONCLUSION

Testing for mobile usability as a specialty field is still taking shape. The technological and methodological challenges are being overcome by researchers creatively reconfiguring traditional usability methods for stationary technology. Specific challenges of mobile usability are being addressed in areas such as contextual complexity, audio interfaces, eye and hand free interactions, augmented reality, and recommendation systems. The mobile usability methods found to be used in these areas are numerous and range from expert reviews to field research. Since standard methods for usability testing of mobile devices is still in its early stages, having an understanding and working knowledge of how to choose an appropriate methodology for specific testing conditions will produce the best usability testing for mobile communication devices.

Platform differences will drive the need for more usability tests as well. For example, what factors account for why online shoppers are three times more likely to make a purchase on a tablet then a smartphone (Adobe Digital Index). Mobile usability can impact the trust that users have in transactional websites. People may be more reluctant to purchase items from poorly working website that are perceived as less credible due to bad design. Ultimately, increased mobile usability testing is likely to surge as a result of the rapid increase in the number of smartphones around the globe. According to IDC, the worldwide mobile phone market is "forecast to grow 7.3% in 2013 driven by 1 billion smartphone shipments" with over 75% using the Android platform and over 16% using the Apple's iOS.

Given mobile communication's rapid growth and its wide range of devices, general principles specifically for mobile communication products will continue to emerge. For the future, beyond traditional concerns of usability (learnability, efficiency, memorability, errors, satisfaction, effectiveness, simplicity, comprehensibility, learning performance), inspection methods for mobile usability need particular attention to touch screen, hands free interaction audio, and location-based services, and changing environmental spaces. Testing in the contextual environment is key to the future development of mobile usability testing.

REFERENCES

Adobe Digital Index. (2013). *How tablets are catalyzing brand website engagement*. Retrieved from http://success.adobe.com/assets/en/downloads/whitepaper/13926.tablets-brand-engagement-v5.pdf

Biancalana, C., Gasparetti, F., Micarelli, A., & Sansonetti, G. (2013). An approach to social recommendation for context-aware mobile services. *ACM Trans. Intell. Syst. Technol.*, *4*(1). doi:10.1145/2414425.2414435 PMID:24883228

Boland, M. (2012, April 20). When Will Mobile Local Searches Eclipse Desktop? *BIA/Kelsey*. Retrieved from http://blog.biakelsey.com/index.php/2012/04/20/when-will-mobile-local-searches-eclipse-desktop/

Cheng, S. (2011). The research framework of eye-tracking based mobile device usability evaluation. In *Proceedings of the 1st international workshop on pervasive eye tracking & mobile eye-based interaction* (PETMEI '11). ACM. DOI: 10.1145/2029956.2029964

Cisco. (2013, February 6). *Cisco Visual Networking Index: Global Mobile Data Traffic Forecast Update, 2012–2017*. Retrieved from http://www.cisco.com/en/US/solutions/collateral/ns341/ns525/ns537/ns705/ns827/white_paper_c11-520862.html

Coursaris, C. K., & Kim, D. J. (2006). A Qualitative Review of Empirical Mobile Usability Studies. In *Proceedings of the Twelfth Americas Conference on Information Systems*. Acapulco, Mexico: Academic Press.

Coursaris, C. K., & Kim, D. J. (2011). A meta-analytical review of empirical mobile usability studies. *Journal of Usability Studies*, *6*(3), 117–171.

eMarketer. (2013, August 28). *Facebook Sees Big Gains in Global Mobile Ad Market Share*. Retrieved from http://www.emarketer.com/Article/Facebook-Sees-Big-Gains-Global-Mobile-Ad-Market-Share/1010171

Garzonis, S. (2005). Usability evaluation of context-aware mobile systems: A review, 3rd UK-UbiNet Workshop, Bath, UK.

Gu, X., Gu, F., & Laffey, J. M. (2011). Designing a mobile system for lifelong learning on the move. *Journal of Computer Assisted Learning*, *27*, 204–215. doi:10.1111/j.1365-2729.2010.00391.x

IDC. (2013, September 4). *Worldwide Mobile Phone Market Forecast to Grow 7.3% in 2013 Driven by 1 Billion Smartphone Shipments*. Retrieved from http://www.idc.com/getdoc.jsp?containerId=prUS24302813

Isomursu, M. (2008). Tags and the city. *Psychology Journal*, *6*(2), 131–156.

Jumisko-Pyykkö, S., & Vainio, T. (2010). Framing the context of use for mobile HCI. *International Journal of Mobile Human Computer Interaction*, *2*(4), 1–18. doi:10.4018/jmhci.2010100101

Kaikkonen, A. et al. (2005). Usability testing of mobile applications: A comparison between laboratory and field testing. *Journal of Usability Studies*, *1*(1), 4–17.

Kjeldskov, J., & Graham, C. (2003). A Review of Mobile HCI Research Methods. In *Proceeding of ACM Int'l Conf. Human Computer Interaction with Mobile Devices and Services* (MobileHCI). ACM.

Kjeldskov, J., & Paay, J. (2010). Indexicality: Understanding mobile human-computer interaction in context. *ACM Transactions on Computer-Human Interaction, 17*(4). doi:10.1145/1879831.1879832

Kjeldskov, J., Skov, M. B., Als, B. S., & Høegh, R. T. (2004). Is it Worth the Hassle? Exploring the Added Value of Evaluating the Usability of Context-Aware Mobile Systems in the Field. In *Proceedings of the 6th International Mobile HCI 2004 Conference*. Glasgow, Scotland: Springer-Verlag. DOI: 10.1007/978-3-540-28637-0_6

Kjeldskov, J., Skov, M. B., Nielsen, G. W., Thorup, S., & Vestergaard, M. (2013). Digital Urban Ambience: Mediating Context on Mobile Devices in the City. *Journal of Pervasive and Mobile Computing, 19*(5), 738–749. doi:10.1016/j.pmcj.2012.05.002

Lai, J., Mitchell, S., & Pavlovski, C. (2007). Examining modality usage in a conversational multimodal application for mobile e-mail access. *International Journal of Speech Technology, 10*(1), 17–30. doi:10.1007/s10772-009-9017-9

Lettner, F., & Holzmann, C. (2012a). Heat maps as a usability tool for multi-touch interaction in mobile applications. In *Proceedings of the 11th International Conference on Mobile and Ubiquitous Multimedia* (MUM '12). ACM. DOI: 10.1145/2406367.2406427

Lettner, F., & Holzmann, C. (2012b). Automated and unsupervised user interaction logging as basis for usability evaluation of mobile applications. In *Proceedings of the 10th International Conference on Advances in Mobile Computing & Multimedia* (MoMM '12). ACM. DOI: 10.1145/2428955.2428983

Lettner, F., & Holzmann, C. (2012c). Sensing mobile phone interaction in the field. In *Proceedings of percomw, 2012 IEEE International Conference on Pervasive Computing and Communications Workshops*. Lugano, Switzerland: IEEE.

Meeker, M., & Lu, L. (2013, May 29). *Internet Trends D11 Conference*. KPCB Report. Retrieved from http://www.kpcb.com/insights/2013-internet-trends

Nielsen, J., & Budiu, R. (2013). *Mobile Usability*. Berkeley, CA: New Riders.

Racadio, R., Rose, E., & Boyd, S. (2012). Designing and evaluating the mobile experience through iterative field studies. In *Proceedings of the 30th ACM international conference on Design of communication* (SIGDOC '12). ACM. DOI: 10.1145/2379057.2379095

Rahmati, A., & Zhong, L. (2013). Studying Smartphone Usage: Lessons from a Four-Month Field Study. *IEEE Transactions on Mobile Computing, 12*(7), 1417–1427. doi:10.1109/TMC.2012.127

Roto, V., Väätäjä, H., Jumisko-Pyykkö, S., & Väänänen-Vainio-Mattila, K. (2011). Best Practices for Capturing Context in User Experience Studies in the Wild. In *Proceedings of MindTrek 2011*. ACM. doi:10.1145/2181037.2181054

Ryan, C., & Gonsalves, A. (2005). *The effect of context and application type on mobile usability: An empirical study TomiAhonen Almanac*. Retrieved from http://communities-dominate.blogs.com/brands/2013/03/the-annual-mobile-industry-numbers-and-stats-blog-yep-this-year-we-will-hit-the-mobile-moment.html

WC3. (2013). *Mobile Accessibility*. Retrieved from http://www.w3.org/WAI/mobile/

Williamson, J. R., Crossan, A., & Brewster, S. (2011). Multimodal mobile interactions: Usability studies in real world settings. In *Proceedings of the 13th international conference on multimodal interfaces* (ICMI '11). ACM. DOI: 10.1145/2070481.2070551

Zhang, D., & Adipat, B. (2005). Challenges, methodologies, and issues in the usability testing of mobile applications. *International Journal of Human-Computer Interaction*, *18*(3), 293–308. doi:10.1207/s15327590ijhc1803_3

KEY TERMS AND DEFINITIONS

Field Studies: A methodology, originating in social science fields such as anthropology and sociology, utilized by researchers seeking to understand how people interact in their natural environments.

Geographic Locality: When a specific physical position to a point on earth is provided.

Inspection Methods: An evaluation method using a standard set of heuristics for the evaluation of user interfaces by usability experts.

Laboratory Studies: Research studies conducted in a controlled environment so that variables can be observed and manipulated.

Multimodal: When the existence of several modes of expression such as text, audio and visual are present.

Qualitative: A paradigm for inquiry in which methods focus on understanding how people produce meaning in social interaction.

Usability: When a created object, tool, or product has a high level of ease of use and learnability for a user.

Compilation of References

Aaltonen, A., Huuskonen, P., & Lehikoinen, J. (2005). Context Awareness Perspectives for Mobile Personal Media. *Information Systems Management*, *22*(4), 43–55. doi:10.1201/1078.10580530/45520.22.4.20050901/90029.5

Aaltonen, A., & Lehikoinen, J. (2005). Refining Visualization Reference Model for Context Information. *Personal and Ubiquitous Computing*, *9*(1), 381–394. doi:10.1007/s00779-005-0349-4

Aarseth, E. (2000). Allegories of Space: The Question of Spatiality in Computer Games. In R. Koskimaa (Ed.), *Cybertext Yearbook*. Jyvaskyla: University of Jyvaskyla.

Abbink, J. (2004). Being young in Africa: The politics of despair and renewal. In *Vanguard or Vandals: Youth, Politics and Conflict in Africa* (pp. 1–34). Leiden: Brill Inc.

Abdelhay, N. (2012). The Arab uprising 2011: New media in the hands of a new generation in North Africa. *Aslib Proceedings: New Information Perspectives*, *64*(5), 529–539. doi:10.1108/00012531211263148

Adey, P. (2010). Vertical Security in the Megacity: Legibility, Mobility and Aerial Politics. *Theory, Culture & Society*, *27*(6), 51–67. doi:10.1177/0263276410380943

Adobe Digital Index. (2013). *How tablets are catalyzing brand website engagement*. Retrieved from http://success.adobe.com/assets/en/downloads/whitepaper/13926.tablets-brand-engagement-v5.pdf

Afolayan, G. E. (2011). *Widowhood Practices and the Rights of Women: The Case of South-Western Nigeria*. (Unpublished MA dissertation). Erasmus University, Rotterdam, The Netherlands.

Afolayan, G. E. (2014). Critical Perspectives of E-government in developing World: Insights from Emerging Issues and Barriers. In *Technology Development and Platform Enhancements For Successful Global E-government Design* (pp. 395–414). Hershey, PA: IGI Global.

Agonia, A. (2012). *Qatar students view mobile tech as an aid to education* (p. 17). Qatar Tribune.

Aguilar, D. D. (1998). *Toward a nationalist feminism*. Quezon City, Philippines: Giraffe Books.

Aguilar, F. V. Jr. (2009). *Maalwang buhay: Family, overseas migration, and cultures of relatedness in Barangay Paraiso*. Quezon City: Ateneo de Manila University Press.

Ahlers, D. (2006). News consumption and the new electronic media. *The Harvard International Journal of Press/Politics*, *11*(1), 29–52. doi:10.1177/1081180X05284317

Ahmed, B. (2013, September 16). How Smartphones Became Vital Tools Against Dengue In Pakistan. *NPR: National Public Radio*. Retrieved from http://www.npr.org/blogs/health/2013/09/16/223051694/how-smartphones-became-vital-tool-against-dengue-in-pakistan

Al Aamri, K. S. (2011). The use of mobile phones in learning English language by Sultan Qaboos University students: Practices, attitudes and challenges. *Canadian Journal on Scientific & Industrial Research*, *2*(3), 143–152.

Albright, J., & Luke, A. (Eds.). (2008). *Pierre Bourdieu and literacy education*. New York: Routledge.

Alessi, S., & Trollip, S. (2001). Multimedia for learning: Methods and development (3rd Ed.). Allyn & Bacon, A Pearson Education Company.

Al-Fahad, F. N. (2009). Students' attitudes and perceptions towards the effectiveness of mobile learning in King Saud University, Saudi Arabia. *The Turkish Online Journal of Educational Technology*, *8*(2), 111–119.

AliveCor. (2013). *AliveCor Heart Monitor*. Retrieved from http://www.alivecor.com/en

Al-Khalidi, S. (2013). *Interview: Zain Iraq sees double-digit growth this year*. Retrieved from http://uk.reuters.com/article/2013/06/06/iraq-zain-telecoms-idUKL5N0EH11Q20130606

Alkmim, M. B., Marcolino, M., Figueira, R., Maia, J., Cardoso, C., & Abreu, M. … Ribeiro, A. (2013, April 10). 1,000,000 electrocardiograms by distance: An outstanding milestone for telehealth in Minas Gerais, Brazil. In Presentation to 2013 Med@Tel/Luxembourg. International Society for Telemedicine and eHealth.

Allan, S. (1999). *News culture*. Buckingham, UK: Open University Press.

Allthings, D. (2013). *CEO Dennis Crowley on Foursquare's biggest mistake*. Retrieved on May 24, 2013 from http://allthingsd.com/20130311/ceo-dennis-crowley-on-foursquares-biggest-mistake/

Ally, M., Samaka, M., Impagliazzo, J., & Abu-Dayya, A. (2012). Use of emerging mobile computer technology to train the Qatar workforce. *Qatar Foundation Annual Research Forum Proceedings: 2012, CSP6*. Retrieved from http://www.qscience.com/doi/abs/10.5339/qfarf.2012.CSP6

Almarwani, A. (2011). ML for EFL: Rationale for mobile learning. In *Proceeding of the International Conference ICT Language learning* (4th ed.). Florence, Italy: Academic Press.

Almutawwa, K. (2012). *An international expert: Mobile phone penetration in Saudi Arabia reaches 200 percent*. Retrieved from http://www.alsharq.net.sa/2012/06/05/324934

Alsanaa, B. (2012). Students' acceptance of incorporating emerging communication technologies in higher education in Kuwait. *World Academy of Science. Engineering and Technology*, *64*, 1412–1419.

Al-Shehri, S. (2012). *Contextual language learning: The educational potential of mobile technologies and social media*. (Unpublished doctoral dissertation). The University of Queensland, Brisbane, Australia.

Althaus, S. L., & Tewksbury, D. (2000). Patterns of Internet and traditional news media use in a networked community. *Political Communication*, *17*(1), 21–45. doi:10.1080/105846000198495

Alunan-Melgar, G., & Borromeo, R. (2002). The Plight of Children of OFWs. In E. Dizon-Añonuevo, & A. Añonuevo (Eds.), *Coming home: Women, migration, and reintegration*. Balikbayani Foundation, Inc., & Atikha Overseas Workers and Communities Initiatives, Inc.

Alzaza, N. S. (2012). Opportunities for utilizing mobile learning services in the Palestinian higher education. *International Arab Journal of e-Technology*, *2*(4), 216-222.

Al-Zoubi, A. Y., Jeschke, S., & Pfeiffer, O. (2010). Mobile learning in engineering education: The Jordan example. In *Proceedings of the International Conference on E-Learning in the Workplace 2010* (pp. 8-15). New York: Academic Press.

Amin, S. (2013, March 20). PKR Launches Mobile App. As Election Campaign Tool. *Malaysian Digest*.

Analysys International. (2011). *China LBS Accumulate Account Exceeded 10 Million in Q2, 2011*. Retrieved July 26, 2012, from http://english.analysys.com.cn/article.php?aid=112335

Anderson, C. (2006). *The Long Tail: Why the Future of Business is Selling More of Less*. New York: Hyperion.

Angeles, L. C. (2001). The Filipino male as macho-Machunurin: Bringing Men and Masculinities into Gender and Development Studies. *Kasarinlan*, *16*(1), 9–30.

Añonuevo, A. (2002). Migrant Women's Dream for a Better Life: At What Cost? In E. Dizon-Añonuevo, & A. Añonuevo (Eds.), *Coming home: Women, migration, and reintegration*. Balikbayani Foundation, Inc., & Atikha Overseas Workers and Communities Initiatives, Inc.

Anstey, M., & Bull, G. (2000). *Reading the visual: Written and illustrated children's literature*. Sydney: Harcourt.

App. Annie, & IDC. (2013). *The Future of Mobile & Portable Gaming*. App. Annie IDC.

Aquiling-Dalisay, G., Nepomuceno-Van Heugten, M. L., & Sto. Domingo, M. R. (1995). Ang Pagkalalaki ayon sa mga lalaki: Pagaaral sa Tatlong Grupong Kultural sa Pilipinas. *Philippine Social Sciences Review, 52*(1-4), 143–166.

Archambault, J. S. (2011). Breaking up 'because of the phone' and the transformative potential of information in Southern Mozambique. *New Media & Society, 13*(3), 444–456. doi:10.1177/1461444810393906

Archambault, J. S. (2013). Cruising through uncertainty: Cell Phones and the politics of display and disguise in Inhambane, Mozambique. *American Ethnologist, 40*(1), 88–101. doi:10.1111/amet.12007

Arikawa, M., Konomi, S., & Ohnishi, K. (2007). Navitime: Supporting Pedestrian Navigation in the Real World. *IEEE Pervasive Computing / IEEE Computer Society [and] IEEE Communications Society, 6*(3), 21–29. doi:10.1109/MPRV.2007.61

Arminen, I. (2006). Social Functions of Location in Mobile Telephony. *Personal and Ubiquitous Computing, 10*(5), 319–323. doi:10.1007/s00779-005-0052-5

Arnold, M. (2003). On the phenomenology of technology: The 'Janus-faces' of mobile phones. *Information and Organization, 13*(4), 231–256. doi:10.1016/S1471-7727(03)00013-7

Asis, M. M. B. (2000). Imagining the future of migration and families in Asia. *Asian and Pacific Migration Journal, 9*(3), 255–272.

Asis, M. M. B. (2008). The Social Dimensions of International Migration in the Philippines: Findings from Research. In M. M. B. Asis, & F. Baggio (Eds.), *Moving out, back and up: International migration and development prospects in the Philippines* (pp. 77–108). Quezon City: Scalabrini Migration Center.

Asis, M. M. B., Huang, S., & Yeoh, B. S. A. (2004). When the light of the home is abroad: Unskilled female migration and the Filipino family. *Singapore Journal of Tropical Geography, 25*(2), 198–215. doi:10.1111/j.0129-7619.2004.00182.x

Asiwaju, A. I. (1976). *Western Yorubaland under European Rule, 1889–1945: A Comparative Analysis of French and British Colonialism*. London: Longman and Highland.

Asiwaju, A. I. (2001). *West African Transformations: Comparative Impacts of French and British Colonialism. Lagos*. Malthouse Press Ltd.

Asiwaju, A. I. (2004). Frontier in Egba [Yoruba] History: Abeokuta, Dahomey and Yewaland in the 19th Century. *Lagos Historical Review, 4*(1), 18–48.

Atkin, D. J. (1994). Newspaper readership among college students in the information age: The influence of telecommunication technology. *Journal of the Association for Communication Administration, 2*, 95–103.

Atkin, D. J., & Jeffres, L. W. (1998). Understanding Internet adoption as telecommunications behavior. *Journal of Broadcasting & Electronic Media, 42*(4), 475–490. doi:10.1080/08838159809364463

Attorney-General's Chambers. (2012). *Laws of Malaysia, Peaceful Assembly Act 2012*. Retrieved August 28, 2013, from http://www.federalgazette.agc.gov.my

Australian Communications and Media Authority. (2008). *Telecommunications today: Report 3: Farming sector attitudes to take-up and use*. Canberra, Australia: Australian Communications and Media Authority (Commonwealth of Australia). Retrieved on 26 December 2010, from www.qrwn.org.au/pdfs/technologytoday/2007_telecomms_today_report.pdf

Austria, F. (2007). Gays, the Internet, and Freedom. *Plaridel, 4*(1), 47–76.

Azizuddin Mohd Sani, M. (2008). Freedom of speech and democracy in Malaysia. *Asian Journal of Political Science, 16*(1), 85–104. doi:10.1080/02185370801962440

Azuma, R. (1997). *A Survey of Augmented Reality Presence: Teleoperators and Virtual Environments*. Retrieved from http://www.cs.unc.edu/~azuma/ARpresence.pdf

Babrow, A. S., & Swanson, D. L. (1988). Disentangling antecedents of audience exposure levels: Extending expectancy-value analyses of gratifications sought from television news. *Communication Monographs, 55*(1), 1–21. doi:10.1080/03637758809376155

Baike. (2013). *Digu*. Retrieved on June 1, 2013 from http://baike.baidu.com/link?url=pW9Mf-Oj0SMJlRkwg76CBC8jMxIyDVAMrK-i9gxZfb-wJJjHGdp9RbTDbuEy0Ew2lCaFxkQJOqTCyo5Kluct-GQ0l2cj8YzicuyUnLlO8Rq-SopP5pdweKBdtD_U1wIdr

Baike. (2013). *Momo*. Retrieved on June 1, 2013 from http://baike.baidu.com/link?url=j7M3z1E-i9fUKrBmS-BwkmgDnWQdlhRvrnSb74Udy70t4QdQG1JIA19Ii-HfGorPx6

Baker, A. J. (2009). Mick or Keith: Blended Identity of Online Rock Fans. *Identity in the Information Society*, *2*(1), 7–21. doi:10.1007/s12394-009-0015-5

Bakhtin, M. (1941). *Rabelais and His World* (H. Iswolsky, Trans.). Bloomington, IN: Indiana University Press.

Balakrishnan, V., & Shamim, A. (2013). Malaysian Facebookers: Motives and addictive behaviours unraveled. *Computers in Human Behavior*, *29*(4), 1342–1349. doi:10.1016/j.chb.2013.01.010

Ball-Rokeach, S. J., & DeFleur, M. L. (1976). A dependency model of mass-media effects. *Communication Research*, *3*(1), 3–21. doi:10.1177/009365027600300101

Banerjee, I. (2002). The locals strike back? *Gazette: The International Journal for Communications Studies*, *64*(6), 517–535. doi:10.1177/17480485020640060101

Banerjee, S. A. (2008). Does location based advertising work. *International Journal of Mobile Marketing*, *3*(2), 68–75.

Barkuss, L., & Dey, A. (2003, September). *Location-Based Services for Mobile Telephony: A study of user's privacy concerns.* Paper presented at the 9th IFIP TC13 International Conference on Human-Computer Interaction. Zurich, Switzerland.

Barreras, A., & Mathur, A. (2007). Wireless Location Tracking. In K. R. Larson, & Z. A. Voronovich (Eds.), *Convenient or Invasive - The Information Age* (pp. 176–186). Boulder, CO: Ethica Publishing.

Basil, M. D., Brown, W. J., & Bocarnea, M. C. (2002). Differences in univariate values versus multivariate relationships: Findings from a study of Diana, Princess of Wales. *Human Communication Research*, *28*(4), 501–514. doi: doi:10.1111/j.1468-2958.2002.tb00820.x

Battistella, G., & Conaco, M. C. G. (1998). The impact of labour migration on the children left behind: A study of elementary school children in the Philippines. *Sojourn: Journal of Social Issues in Southeast Asia*, *13*(2), 220–241. doi:10.1355/SJ13-2C

Baumbach, D. (2009). Web 2.0 and you. *Knowledge Quest*, *37*(4), 12–19.

BBC. (2013). *Nigerian Texters to take on the drug counterfeiters*. Retrieved from http://www.bbc.com/news/world-africa-20976277

Beavis, C. (1999). Magic or mayhem? New texts and new literacies in technological times. In *Proceedings of Annual Conference of the Australian Association for Research in Education*. University of Queensland.

Beck, U., & Beck-Gernsheim, M. E. (1995). *The Normal Chaos of Love*. Cambridge, MA: Polity Press.

Beer, D. (2012). The Comfort of Mobile Media: Uncovering Personal Attachments with Everyday Devices. *Convergence: The International Journal of New Media*, *18*(4), 361–367. doi:10.1177/1354856512449571

Bellavista, P., Kupper, A., & Helal, S. (2008). Location-Based Services: Back to the Future. *IEEE Pervasive Computing / IEEE Computer Society [and] IEEE Communications Society*, *7*(2), 85–89. doi:10.1109/MPRV.2008.34

Beltran, R. P., Samonte, E. L., & Walker, L. (1996). Filipino women migrant workers: Effects on family life and challenges for intervention. In R. P. Beltran, & G. F. Rodriguez (Eds.), *Filipino women migrant workers: At the crossroads and beyond Beijing* (pp. 15–45). Quezon City: Giraffe Books.

Benford, S., Giannachi, G., Koleva, B., & Rodden, T. (2009). From Interaction to Trajectories: Designing Coherent Journeys Through User Experiences. In *Proceedings of Conference on Human Factors in Computing Systems*. Retrieved from http://www.mrl.nott.ac.uk/~sdb/research/research-and-publications.htm

Benford, S., Seager, W., Flintham, M., Anastasi, R., Rowland, D., Humble, J., et al. (2004). The error of our ways: The experience of self-reported position in a location-based game. In N. Davies, E. D. Mynatt, & I. Siio (Eds.), *Proceedings of the UbiComp 2004* (pp. 70-87). Nottingham, UK: Springer.

Benisch, M. K. (2010). Capturing location-privacy preferences: Quantifying accuracy and user-burden tradeoffs. *Journal of Personal and Ubiquitous Computing, 15*(7), 679–694. doi:10.1007/s00779-010-0346-0

Benson, R., & Brack, C. (2009). Developing the scholarship of teaching: What is the role of e-teaching and learning? *Teaching in Higher Education, 14*(1), 71–80. doi:10.1080/13562510802602590

Beresford, A., & Stanjano, F. (2003). Location Privacy in Pervasive Computing. *IEEE Pervasive Computing / IEEE Computer Society [and] IEEE Communications Society, 2*(1), 45–55. doi:10.1109/MPRV.2003.1186725

Berlant, L. (Ed.). (2000). *Intimacy*. Chicago: University of Chicago Press.

Berry, C., Martin, F., & Yue, A. (Eds.). (2003). *Mobile Cultures: New Media in Queer Asia*. Duke University Press. doi:10.1215/9780822384380

Bertel, T., & Stald, G. (2013). From SMS to SNS: The use of the internet on the mobile phone among young Danes. In K. Cumiskey, & L. Hjorth (Eds.), *Mobile media practices, presence and politics: The challenge of being seamlessly mobile* (pp. 198–213). New York: Routledge.

Bhargav-Spantzel, A., Squicciarini, A. C., & Bertino, E. (2006). Establishing and Protecting Digital Identity in Federation Systems. *Journal of Computer Security, 14*(3), 269–300.

Biancalana, C., Gasparetti, F., Micarelli, A., & Sansonetti, G. (2013). An approach to social recommendation for context-aware mobile services. *ACM Trans. Intell. Syst. Technol., 4*(1). doi:10.1145/2414425.2414435 PMID:24883228

Biddix, P. J., & Park, H. W. (2008). Online networks of student protest: The case of the living wage campaign. *New Media & Society, 10*(6), 871–891. doi:10.1177/1461444808096249

Bimber, B. (1999). The Internet and citizen communication with government: Does the medium matter? *Political Communication, 16*(4), 409–429. doi:10.1080/105846099198569

Bin Yahya, E. (2013). *Ipsos's study about the internet penetration and usage in the Arab world*. Retrieved from http://www.tech-wd.com/wd/2013/03/15/ipsos-report-2012

Bivens, R. K. (2008). The Internet, mobile phones and blogging. *Journalism Practice, 2*(1), 113–129. doi:10.1080/17512780701768568

Blackbox. (2012, May). *Smartphones in Singapore – A whitepaper release*. Retrieved from http://www.blackbox.com.sg/wp/wp-content/uploads/2012/05/Blackbox-YKA-Whitepaper-Smartphones.pdf

Blast Theory. (2008). *You Get Me*. London: Deloitte Ignite.

Boal, A. (1992). *Theatre of the Oppressed*. London: Pluto Press.

Boberg, M. (2008). *Mobile Phone and Identity: A Comparative Study of the Representations of Mobile Phone among French and Finnish Adolescents* (Dissertation). Joensuu: University of Joensuu. Retrieved from http://epublications.uef.fi/pub/urn_isbn_978-952-219-103-8/urn_isbn_978-952-219-103-8.pdf

Boesen, J., Rode, J. A., & Mancini, C. (2010, September). *The domestic panopticon: Location tracking in families*. Paper presented at the 12th ACM International Conference on Ubiquitous Computing. Copenhagen, Denmark.

Boland, M. (2012, April 20). When Will Mobile Local Searches Eclipse Desktop? *BIA/Kelsey*. Retrieved from http://blog.biakelsey.com/index.php/2012/04/20/when-will-mobile-local-searches-eclipse-desktop/

Bond, E. (2010). Managing mobile relationships: children's perceptions of the impact of the mobile phone on relationships in their everyday lives. *Childhood, 17*(4), 514–529. doi:10.1177/0907568210364421

Borden, I. (2001). *Skateboarding, Space and the City: Architecture and the body*. Oxford, UK: Berg.

Borneo Post. (2012, December 28). *Registration for 3G smartphone rebate start from next Tuesday –MCMC*. Retrieved September 27, 2013, from http://www.skmm.gov.my/Media/Press-Clippings/Registration-for-3G-smartphone-rebate-start-from-n.aspx

Bourdieu, P. (1977). *Outline of a theory of practice*. Cambridge, UK: Cambridge University Press. doi:10.1017/CBO9780511812507

Bourdieu, P. (1980). *The logic of practice*. Stanford, CA: Stanford University Press.

Bourdieu, P. (1986). The Forms of Capital. In J. G. Richardson (Ed.), *Handbook of theory and research for the sociology of education* (pp. 241–258). New York: Greenwood Press.

Bourdieu, P. (1991). *Language and Symbolic Power*. Cambridge, MA: Polity Press.

Bowen, T. (2013). Graffiti as Spatializing Practice and Performance. *Rhizomes, 25*.

boyd, d. (2009). Friendship. In M. Ito, et al. (Eds.), *Hanging Out, Messing Around, and Geeking Out: Kids Living and Learning with New Media* (pp. 79-115). Cambridge, MA: MIT Press.

boyd, d., & Ellison, N. (2007). Social Network Sites: Definition, History, and Scholarship. *Journal of Computer-Mediated Communication, 13* (1), 210-230.

Braga, D., & Busnardo, J. (2004). Digital Literacy for Autonomous Learning. In I. Snyder, & C. Beavis (Eds.), *Doing Literacy Online: Teaching, Learning and Playing in an Electronic World* (pp. 45–68). Cresskill, NJ: Hampton Press.

Bratton, M. (2013). Citizens and Cell Phones in Africa. *African Affairs, 112*(447), 304–319. doi:10.1093/afraf/adt004

Brickell, C. (2012). Sexuality, power and the sociology of the internet. *Current Sociology, 60*(28), 28–44. doi:10.1177/0011392111426646

Brimicombe, A., & Li, C. (2009). *Location-based services and geo-information engineering*. Chichester, UK: Wiley-Blackwell.

Brooks-Young, S. (2008, May). *Web tools: The second generation*. District Administration, Professional Media Group.

Brown, B., Green, N., & Harper, R. (Eds.). (2001). *Wireless world: Social and interactional aspects of the mobile age*. New York, NY: Springer-Verlag.

Brown, B., Harper, R., & Green, N. (2002). *Wireless World: Social, Cultural, and Interactional Issues in Mobile Communications and Computing*. London: Springer Verlag. doi:10.1007/978-1-4471-0665-4

Browne, M. W., & Cudeck, R. (1993). Alternative ways of assessing model fit. In K. A. Bollen, & J. S. Long (Eds.), *Testing structural equation models* (pp. 445–455). Newbury Park, CA: Sage.

Bruck, P. A., & Rao, M. (Eds.). (2013). *Global mobile: Applications and innovations for the worldwide mobile ecosystem*. Information Today, Inc.

Bruner, G. A. (2007). Attitude towards location based advertising. *Journal of Interactive Advertising, 7*(2), 3–15. doi:10.1080/15252019.2007.10722127

Bryan, A., & Vavrus, F. (2005). The promise and peril of education: The teaching of in/tolerance in an era of globalization. *Globalisation, Societies and Education, 3*(2), 183–202. doi:10.1080/14767720500167033

Buckingham, D. (2007). *Beyond Technology: Children's Learning in the Age of Digital Culture*. Cambridge, UK: Polity Press.

Buerkle, R. (2008). *Of Worlds and Avatars: A Playercentric Approach to Videogame Discourse*. (Unpublished PhD thesis). University of Southern California.

Bull, M. (2005). The Intimate Sounds of Urban Experience: An Auditory Epistemology of Everyday Mobility. In K. Nyíri (Ed.), *Sense of Place: The Global and the Local in Mobile Communication* (pp. 169–178). Vienna, Austria: Passagen Verlag.

Burke, A. (2006). Literacy as Design. *Orbit (Amsterdam, Netherlands), 36*(1), 15–17.

Burmann, C., & Arnould, U. (2009). *User-generated branding, state of the art research*. Münster: Lit.

Burrell, J. (2008). Problematic empowerment: West African Internet scams as strategic misrepresentation. *Information Technologies and International Development, 4*(4), 15–30. doi:10.1162/itid.2008.00024

Burrell, J. (2010). Evaluating shared access: Social equality and the circulation of mobile phones in rural Uganda. *Journal of Computer-Mediated Communication, 15*(2), 230–250. doi:10.1111/j.1083-6101.2010.01518.x

Bursa Malaysia. (n.d.). Retrieved September 21, 2013, from http://announcements.bursamalaysia.com/edms/hsubweb.nsf/1c0706d8c060912d48256c6f0017b41c/48256aaf0027302c48256bcd001175b4/$FILE/Maxis-IndustryOverview-Material%20Indebtedness-Conditions-RPT-Additional%20Info%20(780KB).pdf

Business Wire. (2012). Research and markets: Mobile LBS - for 2017 in the EU27, mobile LBS penetration of mobile subscribers will reach 35% - or 255 million users. *Business Wire*. Retrieved September 5, 2013, from http://search.proquest.com/docview/1015995647?accountid=13876

Buss, A., & Strauss, N. (2009). *Online Communities Handbook*. Berkeley, CA: New Riders.

Butler, R. (2005). *Cell phones may help 'save' Africa*. Retrieved from http://newsmongabay.com/2005/0712-rhet_butler.html

Butler, J. (1993). *Bodies that matter: On the discursive limits of sex*. London: Routledge.

Cabanes, J. V., & Acedera, K. (2012). Of mobile phones and mother-fathers: Calls, text messages, and conjugal power relations in mother-away Filipino families. *New Media & Society*, *14*(6), 916–930. doi:10.1177/1461444811435397

Cabrera-Balleza, M. (2005). Gendered, wired and globalized: Gender and globalization issues in the new information and communication technologies. *Review of Women's Studies*, *15*(2), 140–156.

Cafazzo, J. A., Casselman, M., Hamming, N., Katzman, D. K., & Palmert, M. R. (2012, May 8). Design of an mHealth app. for the self-management of adolescent type 1 diabetes: A pilot study. *Journal of Medical Internet Research*, *14*(3), e70. doi:10.2196/jmir.2058 PMID:22564332

Caillois, R. (2001). *Man, Play and Games* (M. Barash, Trans.). Urbana, IL: University of Illinois Press.

Calabrese, F., Crowcroft, J., Di, L. G., Lathia, N., & Quercia, D. (2010, December). *Recommending Social Events from Mobile Phone Location Data*. Paper presented at the IEEE International Conference on Data Mining. Sydney, Australia.

Calantone, R. J., Griffith, D. A., & Yalcinkaya, G. (2006). An empirical examination of a technology adoption model for the context of China. *Journal of International Marketing*, *14*(4), 1–27. doi:10.1509/jimk.14.4.1

Calleja, G. (2011). *In-Game: From Immersion to Incorporation*. Cambridge, MA: MIT Press.

Calleja, G. (2012). Erasing the Magic Circle. In J. R. Sageng, H. Fossheim, & T. M. Larsen (Eds.), *The Philosophy of Computer Games*. London: Springer Press. doi:10.1007/978-94-007-4249-9_6

Camacho, M. S. T. (2010). The public transcendence of intimacy: The social value of recogimiento. In *More Pinay Than We Admit: The Social Construction of the Filipina* (pp. 295–317). Manila: Vibal.

Cammaerts, B. (2012). Protest logics and the mediation opportunity structure. *European Journal of Communication*, *27*(2), 117–134. doi:10.1177/0267323112441007

Campbell, S. W. (2013). Mobile media and communication: A new field, or just a new journal? *Mobile Media & Communication*, *1*(1), 8–13. doi:10.1177/2050157912459495

Campbell, S. W., & Kwak, N. (2010). Mobile communication and civic life: Linking patterns of use to civic and political engagement. *The Journal of Communication*, *60*, 536–555. doi:10.1111/j.1460-2466.2010.01496.x

Campbell, S. W., & Kwak, N. (2010). Mobile communication and social capital: An analysis of geographically differentiated usage patterns. *New Media & Society*, *12*(3), 435–451. doi:10.1177/1461444809343307

Cardi, V., Clarke, A., & Treasure, J. (2013, May 16). The Use of Guided Self-help Incorporating a Mobile Component in People with Eating Disorders: A Pilot Study. *European Eating Disorders Review*, *21*, 315–322. doi:10.1002/erv.2235 PMID:23677740

Carling, J., Menjívar, C., & Schmalzbauer, L. (2012). Central themes in the study of transnational parenthood. *Journal of Ethnic and Migration Studies*, *38*(2), 191–217. doi:10.1080/1369183X.2012.646417

Caron, A., & Caronia, L. (2007). *Moving cultures*. Montreal, Canada: McGill-Queen's University Press.

Caronia, L. (2005). Mobile Culture: An Ethnography of Cellular Phone Uses in Teenagers' Everyday Life. *Convergence*, *11*(3), 96–103. doi:10.1177/13548565050l100307

Carr, D., Buckingham, D., Burn, A., & Schott, G. (2006). *Computer Games: Text, Narrative and Play*. Cambridge, MA: Polity.

Carrington, V. (2005). Txting: The end of civilization (again)? *Cambridge Journal of Education*, *35*(2), 161–175. doi:10.1080/03057640500146799

Carrington, V., & Luke, A. (1997). Literacy and Bourdieu's Sociological Theory: A Reframing. *Language and Education*, *11*(2), 96–112. doi:10.1080/09500789708666721

Carrington, V., & Marsh, J. (2005). Editorial Overview: Digital Childhood and Youth: New texts, new literacies. *Discourse: Studies in the Cultural Politics of Education*, *26*(3), 279–285.

Carr, J., Herman, N., & Harris, D. (2005). *Creating dynamic schools through mentoring, coaching, and collaboration*. Association for Supervision and Curriculum Development.

Carroll, N. (2003). *Engaging the Moving Image*. New Haven, CT: Yale University Press. doi:10.12987/yale/9780300091953.001.0001

Carter, M. C., Burley, V. J., Nykjaer, C., & Cade, J. E. (2013, April 15). Adherence to a Smartphone Application for Weight Loss Compared to Website and Paper Diary: Pilot Randomized Controlled Trial. *Journal of Medical Internet Research*, *15*(4), e32. doi:10.2196/jmir.2283 PMID:23587561

Casado, E. (2002). *La construccio'n sociocognitiva de las identidades de ge'nero de las mujeres espan˜olas (1975–1995)*. Madrid: Universidad Complutense de Madrid.

Castelli, M. (2003). *Mogi: Item Hunt*. Tokyo, Japan: Newt Games.

Castells, M., et al. (2005/6). Electronic communication and socio-political mobilization. *Global Civil Society*, 266-285.

Castells, M., Fernandez-Ardevol, M., Qiu, J., & Sey, A. (2004). *The mobile communication society: A cross-cultural analysis of available evidence on the social uses of wireless communication technology*. Annenberg: Annenberg Research Network on International Communication.

Castells, M. (1997). *The information age: Economy, society and culture: The power of identity* (Vol. 2). Oxford, UK: Blackwell Publishing.

Castells, M. (2000). *End of Millennium*. Oxford, UK: Blackwell.

Castells, M. (2000). *The rise of the network society*. Oxford, UK: Blackwell.

Castells, M., Fernández-Ardèvol, M., Qiu, J. L., & Sey, A. (2007). *Mobile communication and society: A global perspective*. Cambridge, MA: MIT Press.

Castronova, E. (2005). *Synthetic worlds: The business and culture of online games*. Chicago, IL: University of Chicago Press.

Caulfield, B. (2011). *Apple now selling more iPads than Macs, iOS eclipses Dell and HP's PC Businesses*. Retrieved March 17, 2013 from http://www.forbes.com/sites/briancaulfield/2011/07/19/apple-didnt-just-sell-more-ipads-than-macs-ios-has-now-eclipsed-dell-and-hps-pc-business-too/

Cavanaugh, C., Hargis, J., Munns, S., & Kamali, T. (2012). iCelebrate teaching and learning: Sharing the iPad experience. *Journal of Teaching and Learning with Technology*, *1*(2), 1–12.

Center for Mobile Communication Studies. (2013). *What is mobile communication studies?* Retrieved from http://www.bu.edu/com/academics/emerging-media/cmcs/

Centre for New Media. (2004). *CitiTag*. Centre for New Media.

Chaffee, S., & Metzger, M. (2001). The end of mass communication? *Mass Communication & Society*, *4*(4), 365–379. doi:10.1207/S15327825MCS0404_3

Chanchary, F. H., & Islam, S. (2011). *Mobile learning in Saudi Arabia: Prospects and challenges*. Retrieved from http://www.nauss.edu.sa/acit/PDFs/f2535.pdf

Chan-Olmsted, S., Rim, H., & Zerba, A. (2013). Mobile news adoption among young adults: Examining the roles of perceptions, news consumption, and media usage. *Journalism & Mass Communication Quarterly*, *90*(1), 126–147. doi:10.1177/1077699012468742

Chan, T. W., Roschelle, J., & Hsi, S., Kinshuk, Sharples, M., & Brown, T., et al. (2006). One-to-one Technology Enhanced Learning: An Opportunity for Global Research Collaboration. *Research and Practice in Technology-Enhanced Learning*, *1*(1), 3–29. doi:10.1142/S1793206806000032

Chau, P. Y. K., & Lai, V. S. K. (2003). An empirical investigation of the determinants of user acceptance of Internet banking. *Journal of Organizational Computing and Electronic Commerce*, *13*(2), 123–145. doi:10.1207/S15327744JOCE1302_3

Cheng, S. (2011). The research framework of eye-tracking based mobile device usability evaluation. In *Proceedings of the 1st international workshop on pervasive eye tracking & mobile eye-based interaction* (PETMEI '11). ACM. DOI: 10.1145/2029956.2029964

Chen, R. (2012). *Ubiquitous Positioning and Mobile Location-Based Services in Smartphones*. Hershey, PA: Information Science Reference. doi:10.4018/978-1-4666-1827-5

Cheverst, K., Fitton, D., Rouncefield, M., & Graham, C. (2004). *Smart mobs and technology probes: Evaluating texting at work*. Academic Press.

Chew, C. (2009, August 15). Young crusaders. *The Straits Times*, pp. D1-9.

Chib, A., Malik, S., Aricat, R. G., & Kadir, S. Z. (2014). Migrant mothering and mobile phones: Negotiations of transnational identity. *Mobile Media & Communication*, *2*(1), 73–93. doi:10.1177/2050157913506007

Cho, H. (2011). Theoretical intersections among social influences, beliefs, and intentions in the context of 3G mobile services in Singapore: Decomposing perceived critical mass and subjective norms. *The Journal of Communication*, *61*, 283–306. doi:10.1111/j.1460-2466.2010.01532.x

Christodoulides, G., & Jevons, C. (2011). The voice of the consumer speaks forcefully in brand identity. *Journal of Advertising Research*, *51*, 101–108. doi:10.2501/JAR-51-1-101-111

Chronaki, C., Kontoyiannis, V., Mytaras, M., Aggoutakis, N., Kostomanolakis, S., & Roumeliotaki, T. ... Tsiknakis, M. (2007). Evaluation of shared EHR services in primary healthcare centers and their rural community offices: the twister story. In *Proceedings - IEEE Engineering in Medicine and Biology Society*. Retrieved from http://www.ncbi.nim.nih.gov/pubmed/18003492

Chu, H. C., Deng, D. J., & Chao, H. C. (2011). Potential Cyberterrorism via a Multimedia Smart Phone Based on a Web 2.0 Application via Ubiquitous Wi-Fi Access Points and the Corresponding Digital Forensics. *Multimedia Systems*, *17*(4), 341–349. doi:10.1007/s00530-010-0216-7

Chye, K. T. (2013, February 15). *Gross Encounters with the Zin Kind*. Retrieved August 17, 2013, from http://www.malaysiandigest.com/opinion/256182-gross-encounters-of-the-zin-kind.html

Cisco. (2013, February 6). *Cisco Visual Networking Index: Global Mobile Data Traffic Forecast Update, 2012–2017*. Retrieved from http://www.cisco.com/en/US/solutions/collateral/ns341/ns525/ns537/ns705/ns827/white_paper_c11-520862.html

Clark, L. S. (2013). *The parent app*. Oxford, UK: Oxford University Press.

CNNIC. (2013). *The 32nd statistical report on internet development in China*. Retrieved December 5, 2013, from http://www1.cnnic.cn/IDR/ReportDownloads/201310/P020131029430558704972.pdf

Cohen, L. (2007, October). Information literacy in the age of social scholarship. *Library 2.0: An Academic's Perspective*.

Colbert, M. (2005). Usage and user experience of communication before and during rendezvous. *Behaviour & Information Technology*, *24*(6), 449–469. doi:10.1080/01449290500043991

Coleman, S. (2007). How democracies have disengaged from young people. In B. D. Loader (Ed.), *Young citizens in the digital age: Political engagement, younger people and new media*. London: Routledge.

Collins, J., & Bold, R. K. (2003). *Literacy and Literacies: Texts, Power, and Identity*. Cambridge, UK: Cambridge University Press. doi:10.1017/CBO9780511486661

Comaroff, J., & Comaroff, J. L. (1999). Occult Economies and the Violence of Abstraction: Notes from the South African Postcolony. *American Ethnologist*, *26*(2), 279–303. doi:10.1525/ae.1999.26.2.279

Comstock, J. (2012, December 4). Real games for health and the trouble with gamification. *MobiHealthNews*. Retrieved from http://mobihealthnews.com/19323/real-games-for-health-and-the-trouble-with-gamification/

Comstock, J. (2013, May 28). Brazil: An up-and-coming mobile health market. *MobiHealthNews*. Retrieved from http://mobihealthnews.com/22654/brazil-an-up-and-coming-mobile-health-market/

Consalvo, M. (2009). There is No Magic Circle. *Games and Culture*, *4*(4), 408–417. doi:10.1177/1555412009343575

Consolvo, S., Smith, I. E., Matthews, T., LaMarca, A., Tabert, J., & Powledge, P. (2005, April). *Location disclosure to social relations: Why, when, & what people want to share*. Paper presented at the Human Factors in Computing Systems. Portland, OR.

Cooley, B. (2013). *CBS This Morning, Gadgets and gizmos galore*. Retrieved Jan. 8 from http://www.cbsnews.com/video/watch/?id=50138517n

Cope, B., & Kalantzis, M. (2000). *Multiliteracies: Literacy Learning and the Design of Social Futures*. London: Routledge.

Copier, M. (2009). Challenging the magic circle: How online role-playing games are negotiated by everyday life. In M. v. Boomen, S. Lammes, A.-S. Lehmann, J. Raessens, & M. T. Schafer (Eds.), *Digital Material: Tracing New Media in Everyday Life and Technology*. Amsterdam: Amsterdam University Press.

Corporation, V. F. (2010). *VF Corporation Annual Report 2009*. Retrieved March 2, 2010, from http://reporting.vfc.com/2009/pdfs/vfc_ar09.pdf

Cotten, S. R., Anderson, W. A., & Tufekci, Z. (2009). Old wine in a new technology, or a different type of digital divide? *New Media & Society*, *11*(7), 1163–1186. doi:10.1177/1461444809342056

Couldry, N. (2003). *Media Rituals: A Critical Approach*. London: Routledge.

Coursaris, C. K., & Kim, D. J. (2006). A Qualitative Review of Empirical Mobile Usability Studies. In *Proceedings of the Twelfth Americas Conference on Information Systems*. Acapulco, Mexico: Academic Press.

Coursaris, C. K., & Kim, D. J. (2011). A meta-analytical review of empirical mobile usability studies. *Journal of Usability Studies*, *6*(3), 117–171.

Crampton, J. (2003). *The Political Mapping of Cyberspace*. Chicago: University of Chicago Press.

Crang, M., Crosbie, T., & Graham, S. (2006). Variable Geometrics of Connection: Urban Digital Divides and the Uses of Information Technology. *Urban Studies (Edinburgh, Scotland)*, *42*(13), 2551–2570. doi:10.1080/00420980600970664

Crato, N. (2010). How GPS Works. In N. Crato (Ed.), *Figuring It Out: Entertaining Encounters with Everyday Math* (pp. 49–52). Berlin: Springer. doi:10.1007/978-3-642-04833-3_12

Creutzfeldt, J., Hedman, L., Heinrichs, L., Youngblood, P., & Felländer-Tsai, L. (2013, January 14). Cardiopulmonary Resuscitation Training in High School Using Avatars in Virtual Worlds: An International Feasibility Study. *Journal of Medical Internet Research*. doi:10.2196/jmir.1715 PMID:23318253

Crowley, D., & Selvadurai, N. (2009). *Foursquare*. New York: Foursquare.

Crystal, D. (2008). *Txtng: The Gr8 Db8*. Melbourne, Australia: Oxford University Press.

Culler, J. (2001). *The Pursuit of Signs: Semiotics, Literature, Deconstruction*. New York: Routledge.

Curtis, G. E. (Ed.). (1996). *Russia: A Country Study*. Washington, DC: GPO for the Library of Congress. Retrieved from http://countrystudies.us/russia/53.htm

Cyzone. (2012). *Dilemma of Jiepang's check-in*. Retrieved on May 18, 2013 from http://www.cyzone.cn/a/20120425/226227.html

Daft, R. L., & Lengel, R. H. (1984). Information richness: A new approach to managerial behavior and organizational design. In B. M. Staw, & L. L. Cummings (Eds.), *Research in organizational behavior: An annual series of analytical essays and critical reviews* (Vol. 6, pp. 191–233). Greenwich, CT: JAI Press Inc.

Dahlstrom, E., & Warraich, K. (2012). Student mobile computing practices, 2012: Lessons learned from Qatar. *A report by EDUCAUSE*. Retrieved from http://www.educause.edu/library/resources/student-mobile-computing-practices-2012-lessons-learned-qatar

Damásio, M. J., Henriques, S., Teixeira-Botelho, I., & Dias, P. (2013). Social activities and mobile Internet diffusion: A search for the Holy Grail? *Mobile Media & Communication, 1*(3), 335–355. doi:10.1177/2050157913495690

Danaei, G., Finucane, M. M., Lu, Y., Singh, G. M., Cowan, M. J., & Paciorek, C. J. et al. (2011, July 2). National, regional, and global trends in fasting plasma glucose and diabetes prevalence since 1980: Systematic analysis of health examination surveys and epidemiological studies with 370 country-years and 2.7 million participants. *Lancet, 378*(9785), 31–40. doi:10.1016/S0140-6736(11)60679-X PMID:21705069

Daniel, M. S., & de Souza e Silva, A. (2011). Location-aware mobile media and urban sociability. *New Media & Society, 13*(5), 807–823. doi:10.1177/1461444810385202

Davidson, J. (2010). Cultivating Knowledge: Development, Dissemblance, and Discursive Contradictions among the Diola of Guinea-Bissau. *American Ethnologist, 37*(2), 212–226. doi:10.1111/j.1548-1425.2010.01251.x

Davies, J. (2007). Display, Identity and the Everyday: Self-presentation through Digital Image Sharing. *Discourse. Studies in the Cultural Politics of Education, 28*(4), 549–564. doi:10.1080/01596300701625305

Davis, F. D. (1989). Perceived usefulness, perceived ease of use, and user acceptance of information technology. *Management Information Systems Quarterly, 13*(3), 318–340. doi:10.2307/249008

Davis, F. D. (1993). User acceptance of information technology: System characteristics, user perceptions and behavioral impacts. *International Journal of Man-Machine Studies, 38*(3), 475–487. doi:10.1006/imms.1993.1022

Davis, F. D., Bagozzi, R. P., & Warshaw, P. R. (1989). User acceptance of computer technology: A comparison of two theoretical models. *Management Science, 35*(8), 982–1003. doi:10.1287/mnsc.35.8.982

de Bruijn, M., Nyamnjoh, F., & Brinkman, I. (2009). Introduction. In *Mobile Phones: The New Talking Drums of Africa* (pp. 11–22). Leiden: Langaa & African Studies Centre.

De Castro, L. (1995). Pagiging Lalaki, Pagkalalaki at Pagkamaginoo. *Philippine Social Science Review, 52*(1-4), 127–142.

de Cristofaro, E., Manulis, M., & Poettering, B. (2011/2013). Private Discovery of Common Social Contacts. *International Journal of Information Security, 12*(1), 49–65. doi:10.1007/s10207-012-0183-4

de Souza e Silva, A. (2006). From cyber to hybrid: Mobile technologies as interfaces of hybrid spaces. *Space and Culture, 9*(3), 261–278. doi:10.1177/1206331206289022

de Souza e Silva, A. (2013). Location-aware mobile technologies: Historical, social and spatial approaches. *Mobile Media & Communication, 1*(1), 116–121. doi:10.1177/2050157912459492

de Souza e Silva, A., & Delacruz, G. C. (2006). Hybrid reality games reframed: Potential uses in educational contexts. *Games and Culture, 1*(3), 231–251. doi:10.1177/1555412006290443

de Souza e Silva, A., & Frith, J. (2010). Locational privacy in public spaces: Media discourses on location-aware mobile technologies. *Communication, Culture & Critique, 3*(4), 503–525. doi:10.1111/j.1753-9137.2010.01083.x

de Souza e Silva, A., & Frith, J. (2010). Locative mobile social networks: Mapping communication and location in urban spaces. *Mobilities, 5*(4), 485–505. doi:10.1080/17450101.2010.510332

de Souza e Silva, A., & Hjorth, L. (2009). Playful urban spaces: A historical approachto mobile games. *Simulation & Gaming, 40*(5), 602–625. doi:10.1177/1046878109333723 de Souza e Silva, A., & Sutko, D. M. (2008). Playing life and living play: How hybridreality games reframe space, play, and the ordinary. *Critical Studies in Media Communication, 25*(5), 447–465. doi:10.1080/15295030802468081

de Souza e Silva, A., & Sutko, D. M. (2011). Theorizing locative technologies through philosophies of the virtual. *Communication Theory, 21*(1), 23–42. doi:10.1111/j.1468-2885.2010.01374.x

de Souza e Silva. A. (2006). Interfaces of Hybrid Spaces. In A. Kavoori & N. Arceneaux (Eds.), The Cell Phone Reader: Essays in Social Transformation. New York, NY: Peter Lang.

deCerteau, M. (1984). *The Practice of Everyday Life.* Berkeley, CA: University of California Press.

Dela Cruz, P. (1988). *Images of Women in Philippine Media: From Virgin to Vamp.* Manila: Asian Social Institute in cooperation with the World Association for Christian Communication.

Dennen, V. P. (2009). Constructing Academic Alter-Egos: Identity Issues in a Blog-Based Community. *Identity in the Information Society, 2*(1), 23–38. doi:10.1007/s12394-009-0020-8

Denzin, N. K. (1997). *Interpretive Ethnography: Ethnographic Practices for the 21st Century.* Thousand Oaks, CA: Sage Publications. doi:10.4135/9781452243672

Depp, C. A., Kim, D. H., de Dios, L. V., Wang, L., & Ceglowski, J. (2012, January 1). A Pilot Study of Mood Ratings Captured by Mobile Phone Versus Paper-and-Pencil Mood Charts in Bipolar Disorder. *Journal of Dual Diagnosis, 8*(4), 326–332. doi:10.1080/15504263.2012.723318 PMID:23646035

Deterding, S., Miguel, S., Nacke, L., O'Hara, K., & Dixon, D. (Eds.). (2011). Gamification: Using game-design elements in non-gaming contexts. In *Proceedings of the 2011 Annual Conference Extended Abstracts on Human Factors in Computing Systems* (pp. 2425–2428). doi: 10.1177/0141076813480996

Dey, A. K. (2001). Understanding and Using Context. *Personal and Ubiquitous Computing Journal, 5*(1), 4–7. doi:10.1007/s007790170019

Diani, M. (2000). Social movement networks virtual and real. *Information Communication and Society, 3*(3), 386–401. doi:10.1080/13691180051033333

Dibdin, P. (2001). *Where are Mobile Location Based Services.* Retrieved on April 3, 2013 from http://mms.ecs.soton.ac.uk/mms2002/papers/4.pdf?origin=publication_detail

Diddi, A., & LaRose, R. (2006). Getting hooked on news: Uses and gratifications and the formation of news habits among college students in an internet environment. *Journal of Broadcasting & Electronic Media, 50*(2), 193–210. doi:10.1207/s15506878jobem5002_2

Digu. (2010). *Digu and Starbucks.* Retrieved on April 3, 2013 from http://wenku.baidu.com/view/1f943749f7ec4afe04a1df53.html

Dimmick, J., Feaster, J. C., & Hoplamazian, G. J. (2011). News in the interstices: The niches of mobile media in space and time. *New Media & Society, 13*(1), 23–39. doi:10.1177/1461444810363452

Dishaw, M. T., & Strong, D. M. (1999). Extending the technology acceptance model with task-technology fit constructs. *Information & Management, 36*(1), 9–21. doi:10.1016/S0378-7206(98)00101-3

Dobashi, S. (2005). The gendered use of *ketai* in domestic contexts. In M. Ito et al. (Eds.), *Personal, Portable, Pedestrian. Mobile Phones in Japanese Life* (pp. 219–236). MIT Press.

Dockter, J., Haug, D., & Lewis, C. (2010, February). Redefining Rigor: Critical engagement, digital media, and the new English/Language Arts. *Journal of Adolescent & Adult Literacy.* doi:10.1598/JAAL.53.5.7

Donner, J. (2008). Research approaches to mobile use in the developing world: A review of the literature. *The Information Society, 24*(3), 140–159. doi:10.1080/01972240802019970

Donner, J., Gitau, S., & Marsden, G. (2011). Exploring mobile-only internet use: Results of a training study in urban South Africa. *International Journal of Communication, 5*, 574–597.

Doron, A. (2012). Mobile Persons: Cell phones, Gender and the Self in North India. *The Asia Pacific Journal of Anthropology*, *13*(5), 414–433. doi:10.1080/14442213.2012.726253

Dourish, P., & Anderson, K. (2006). Collective information practice: Exploring privacy and security as social and cultural phenomena. *Human-Computer Interaction*, *21*(3), 319–342. doi:10.1207/s15327051hci2103_2

Dourish, P., & Bell, G. (2011). *Divining a Digital Future*. Cambridge, MA: MIT Press. doi:10.7551/mitpress/9780262015554.001.0001

Downes, S. (2007, February 6). *Msg. 30, Re: What Connectivism Is. Connectivism Conference*. University of Manitoba. Retrieved from http://ltc.umanitoba.ca/moodle/mod/forum/discuss.php?d=12

Dube, A. R., & Stanton, C. A. (2010). The Social Context of Dietary Behaviors: The Role of Social Relationships and Support on Dietary Fat and Fiber Intake. *Modern Dietary Fat Intakes in Disease Promotion. Nutrition and Health (Berkhamsted, Hertfordshire)*. doi: doi:10.1007/978-1-60327-571-2_2

Duckham, M., & Kulik, L. (2006). Location Privacy and Location-Aware Computing. In J. Drummond, R. Billen, E. Joao, & D. Forrest (Eds.), *Dynamic & Mobile GIS: Investigating Change in Space and Time*. Boca Raton, FL: CRC Press.

Dumdum, O. (2010). *Jerks without faces: The XTube spectacle and the modernity of the Filipino bakla*. Paper presented at the Annual Meeting of the International Communication Association. Suntec, Singapore.

Duncum, P. (2004). Visual Culture Isn't Just Visual: Multiliteracy, Multimodality and Meaning. *Studies in Art Education*, *45*(3), 252–264.

Dupagne, M. (1999). Exploring the characteristics of potential high-definition television adopters. *Journal of Media Economics*, *12*(1), 35–50. doi:10.1207/s15327736me1201_3

Dürager, A., & Livingstone, S. (2012). *How can parents support children's Internet Safety?* London: EU Kids Online. Retrieved December 15, 2013 from http://www.eukidsonline.net

Duran, A. (2006). Flash mobs: Social influence in the 21st century. *Social Influence*, *1*(4), 301–315. doi:10.1080/15534510601046569

Dutta-Bergman, M. J. (2004). Complementarity in consumption of news types across traditional and new media. *Journal of Broadcasting & Electronic Media*, *48*(1), 41–60. doi:10.1207/s15506878jobem4801_3

Economics. (2010). *Itisalat supports learning via mobile phones*. Retrieved from http://www.albayan.ae/economy/1277246021486-2010-10-22-1.295970

Edge Malaysia. (2013, May 16). *New home minister tells unhappy Malaysians to emigrate*. Retrieved August 5, 2013, from http://www.theedgemalaysia.com/political-news/239111-new-home-minister-tells-unhappy-malaysians-to-emigrate.html

Eid, N. (2013). *Egypt's ICT ministry: 3.7% decline in mobile penetration march-end*. Retrieved from http://www.amwalalghad.com/en/investment-news/technology-news/19204-egypts-ict-ministry-37-decline-in-mobile-penetration-march-end.html

Eisenberg, M. (2008). The Parallel Information Universe. *Library Journal*, *133*(8).

Electronic Times. (1998). How Bluetooth works. *Electonics Times*, 46.

Elegbeleye, O. S. (2005). Prevalent Use of Global System of Mobile Phone (GSM) for Communication in Nigeria: A Breakthrough in Interactional Enhancement or a Drawback? *Nordic Journal of African Studies*, *14*(2), 193–207.

El-Gamal, H. R. (2012). The power of e-learning for Egypt: A spot light on e-learning. *A report by GNSE Group*. Retrieved from http://www.gnsegroup.com/news/The_Power_of_elearning_for_Egypt_ver01.pdf

eMarketer. (2013, August 28). *Facebook Sees Big Gains in Global Mobile Ad Market Share*. Retrieved from http://www.emarketer.com/Article/Facebook-Sees-Big-Gains-Global-Mobile-Ad-Market-Share/1010171

Engels, F. (2001). *The Condition of the Working Class in England*. London: ElecBook.

Episcopal Commission for Pastoral Care of Migrants and Itinerant People-Catholic Bishops Conference of the Philippines/Apostleship of the Sea-Manila, Scalabrini Migration Center, and Overseas Workers Welfare Administration. (2004). *Hearts apart: Migration in the eyes of Filipino children*. Manila: Author.

Escobar, A. (2005). Imagining a post-development era. In *The Anthropology of Development and Globalization: From Classical Political Economy to Contemporary Neoliberalism* (pp. 341–351). Oxford, UK: Blackwell.

Eshet-Alkalai, Y. (2004). Digital literacy: A conceptual framework for survival skills in the digital era. *Journal of Educational Multimedia and Hypermedia*, *13*(1), 93–106.

Ethos Interactive. (2012). *Most popular mobile application in Saudi Arabia*. Retrieved from http://blog.ethosinteract.com/2012/09/27/most-popular-mobile-applications-in-saudi-arabia/

Ettema, J. S. (1984). Three phases in the creation of information inequities: An empirical assessment of a prototype videotex system. *Journal of Broadcasting*, *28*(4), 383–395. doi:10.1080/08838158409386548

Ettema, J. S. (1989). Interactive electronic text in the United States: Can videotex ever go home again? In J. L. Salvaggio, & J. Bryant (Eds.), *Media use in the information age: Emerging patterns of adoption and consumer use* (pp. 105–123). Hillsdale, NJ: Lawrence Erlbaum Associates.

Etzo, S., & Collender, G. (2010). The Mobile Phone 'Revolution' in Africa: Rhetoric or Reality? *African Affairs*, *109*(437), 659–668. doi:10.1093/afraf/adq045

Evans-Cowley, J. (2010). Planning in the Real-Time City: The Future of Mobile Technology. *Journal of Planning Literature*, *25*(2), 136–149. doi:10.1177/0885412210394100

Evans, E. (2011). *Transmedia Television: Audiences, New Media and Daily Life*. London: Routledge.

Evans, L. (2011). Location-based services: Transformation of the experience of space. *Journal of Location Based Services*, *5*(3-4), 242–260. doi:10.1080/17489725.2011.637968

Evans, M. (2003). The relational oxymoron and personalisation pragmatism. *Journal of Consumer Marketing*, *20*(7), 665–685. doi:10.1108/07363760310506193

Evans, P., & Wurster, T. S. (1999). Getting real about virtual commerce: Strategies for success in electronic and physical retail commerce. *Harvard Business Review*, *77*(6), 84–94. PMID:10662007

Eversmann, P. (2004). The Experience of the Theatrical Event. In V. A. Cremona et al. (Eds.), *Theatrical Events: Borders Dynamics Frames* (pp. 139–174). Amsterdam: Rodopi.

Faigley, L. (1999). Beyond Imagination: The Internet and Global Digital Literacy. In G. Hawisher, & C. Selfe (Eds.), *Passions and Pedagogies and 21st Century Technologies* (pp. 129–139). Logan, UT: Utah State University Press.

Farkas, M. (2007). *Social software in libraries: Building collaboration, communication, and community online*. Information Today, Inc.

Ferguson, J. (2002). Of mimicry and membership: Africans and the 'New World Society'. *Cultural Anthropology*, *17*(4), 551–569. doi:10.1525/can.2002.17.4.551

Firat, A. F., & Venkatesh, A. (1995). Liberatory postmodernism and the reenchantment with consumption. *The Journal of Consumer Research*, *22*(December), 239–267. doi:10.1086/209448

Flanagan, M. (2009). *Critical Play: Radical game design*. Cambridge, MA: MIT Press.

Flanagin, A. J., Stohl, C., & Bimber, B. (2006). Modeling the structure of collective action. *Communication Monographs*, *73*(1), 29–54. doi:10.1080/03637750600557099

Flor, H., & Turk, D. C. (2011). *Chronic Pain: An Integrated Biobehavioral Approach. International Association for the Study of Pain Press*. Retrieved from.

Fogg, B. J. (2009). A Behavior Model for Persuasive Design. In *Proceedings of the 4th International Conference on Persuasive Technology*. Retrieved from http://bjfogg.com/fbm_files/page4_1.pdf

Fogg, B. J. (2011). BJ Fogg's Behavior Model. *BJ Fogg 6d*. Retrieved from http://behaviormodel.org/

Forsyth, A. (2006). Economic Literacy – An Essential Dimension in the Social Education Curriculum for the Twenty-First Century. *Science Educator*, *24*(2), 29–33.

Fortunati, L. (2002). The mobile phone: Towards new categories and social relations. *Information Communication and Society*, *5*(4), 513–528. doi:10.1080/13691180208538803

Fortunati, L. (2005). Mobile Telephone and the Presentation of the Self. In R. Ling, & P. E. Pederson (Eds.), *Mobile communications: Re-negotiation of the social sphere*. London: Springer. doi:10.1007/1-84628-248-9_13

Fortunati, L. (2009). Gender and the Mobile Phone. In G. Goggin, & L. Hjorth (Eds.), *Mobile Technologies: From Telecommunications to Media* (pp. 23–34). New York: Routledge.

Fortunati, L., & Manganelli, A. (2002). Young people and the mobile telephone. *Estudios de Juventud*, *57*(2), 59–78.

Foursquare. (2013). In *Wikipedia*. Retrieved on April 5, 2013 from http://en.wikipedia.org/wiki/Foursquare

Fredrick, K. (2010). In the driver's seat: Learning and library 2.0 tools. *School Library Monthly*, *26*(6).

Frith, J. (2012). Splintered Space: Hybrid Spaces and Differential Mobility. *Mobilities*, *7*(1), 131–179. doi:10.1080/17450101.2012.631815

Frith, J. (2013). Turning Life into a Game: Foursquare, Gamification, and Personal Mobility. *Mobile Media & Communication*, *1*(2), 248–262. doi:10.1177/2050157912474811

Fuchs, C. (2012). The Political Economy of Privacy on Facebook. *Television & New Media*, *13*(2), 139–159. doi:10.1177/1527476411415699

Fusco, S. J., Michael, K., Michael, M. G., & Abbas, R. (2010). *Exploring the Social Implications of Location Based Social Networking: An Inquiry into the Perceived Positive and Negative Impacts of Using LBSN between Friends*. Paper presented at the Ninth International Conference on Mobile Business/2010 Ninth Global Mobility Roundtable. Piscataway, NJ.

Gall, J., Gall, M., & Borg, W. (1999). *Applying educational research: A practical guide* (4th ed.). Addison Wesley Longman, Inc.

Garcia, L. (2008). *Manila beams with pride, despite debut of anti-gay protesters*. Retrieved from http://www.fridae.asia/newsfeatures/2008/12/08/2168.manila-beams-with-pride-despite-debut-of-anti-gay-protesters#sthash.rOWaDqSM.dpuf

García, A. (2006). Is health promotion relevant across cultures and the socioeconomic spectrum? *Family & Community Health*, *29*(1), 20S–27S. doi:10.1097/00003727-200601001-00005 PMID:16344633

Garcia, J. N. C. (2000). Performativity, the bakla and the orientalizing gaze. *Inter-Asia Cultural Studies*, *1*(2), 265–281. doi:10.1080/14649370050141140

Garcia, J. N. C. (2008). *Philippine gay culture: Binabae to bakla, silahis to MSM* (2nd ed.). Quezon City: University of the Philippines Press.

Garg, V. K. (2007). Wireless Local Area Networks. In V. K. Garg (Ed.), *Wireless Communications & Networking* (pp. 713–776). Burlington, MA: Morgan Kaufmann. doi:10.1016/B978-012373580-5/50055-7

Garzonis, S. (2005). Usability evaluation of context-aware mobile systems: A review, 3rd UK-UbiNet Workshop, Bath, UK.

Gauntlett, D., & Hill, A. (1999). *TV Living: Television, Culture and Everyday Life*. London: Routledge.

Gazzard, A. (2011). Location, location, location: Collecting space and place in mobile media. *Convergence*, *17*(4), 405–417. doi:10.1177/1354856511414344

Gee, J. P. (2003). *What video games have to teach us about literacy and learning*. New York: Palgrave Macmillan.

Gee, J. P. (2004). *Situated Language and Learning: A critique of traditional schooling*. New York: Routledge.

Gee, J. P. (2008). *Social Linguistics and Literacies: Ideology in Discourse* (3rd ed.). New York: Routledge.

General Electric Healthcare. (2013). *Vscan Ultrasound*. Retrieved from www.vscanultrasound.gehealthcare.com

George, C. (2011, June 25). Alternative media: New era in ties. *The Straits Times*, p. A32.

George, C. (2003). The Internet and the narrowing tailoring dilemma for Asian democracies. *Communication Review*, *6*, 247–268. doi:10.1080/10714420390226270.002

Gergen, K. (2002). Cell phone technology and the realm of absent presence. In J. E. Katz, & M. Aakhus (Eds.), *Perpetual contact* (pp. 227–241). New York, NY: Cambridge University Press.

Gergen, K. (2002). The challenge of absent presence. In J. Katz, & M. Aakhus (Eds.), *Perpetual contact: Mobile communication, private talk, public performance* (pp. 227–241). Cambridge, UK: Cambridge University Press.

Gergen, K. J. (2002). The Challenge of Absent Presence. In J. E. Katz, & M. Aakhus (Eds.), *Perpetual Contact: Mobile Communication, Private Talk, Public Performance* (pp. 227–243). Cambridge, UK: Cambridge University Press.

Geser, H. (2004). Towards a Sociological Theory of the Mobile Phone. *Sociology in Switzerland: Sociology of the Mobile Phone*. Retrieved 22 December 2010, from http://socio.ch/mobile/t_geser1.pdf

GfK Retail and Technology Asia. (2012, April 26). *Boom Times Continue as Southeast Asia's Smartphone Market Value Expands by 62 Percent in Quarter One 2012: GfK Asia*. Retrieved September 27, 2013, from http://www.gfkrt.com/asia/news_events/news/news_single/009744/index.en.html

Giaglis, G. M., Kourouthanassis, P., & Tsamakos, A. (2003). Towards a Classification Framework for Mobile Location Services. In B. E. Mennecke & T. J. Strader (Eds.), Mobile Commerce: Technology, Theory and Applications (pp. 64-81). Hershey, PA: Idea Group Inc (IGI).

Gibson, M. (2008). Beyond literacy panics: Digital literacy and educational optimism. *Media International Australia: Culture and policy (Digital Literacies), 128*, 73-79.

Gibson, E. J., & Pick, A. D. (2000). *An ecological approach to perceptual learning and development*. Oxford, UK: Oxford University Press.

Gillis, E. K. (2005). *Singapore civil society and British power*. Singapore: Talisman.

Gilster, P. (1997). *Digital literacy*. New York: John Wiley & Sons.

Githuku-Shongwe, A. (2013, March 6). The Greatest Return of Investment Is Investing in the Mindsets of the Future Generation of Leaders. *The Huffington Post*. Retrieved from http://www.huffingtonpost.com/anne-githukushongwe/afroes-anne-githuku-shongwe_b_2819045.html?view=print&comm_ref=false

Globocan. (2008). Retrieved from http://globocan.iarc.fr/factsheets/populations/factsheet.asp?uno=900

Goffman, E. (1967). *Interaction Ritual: Essays on Face to Face Behavior*. New York: Pantheon.

Goffman, E. (1972). *Relations in public*. Transaction Books.

Goggin, G. (2006). *Cell Phone Culture: Mobile technology in everyday life*. London: Routledge.

Goggin, G. (2010). *Global mobile media*. New York, NY: Routledge.

Goggin, G. (2013). Youth culture and mobiles. *Mobile Media & Communication*, *1*(1), 83–88. doi:10.1177/2050157912464489

Goggin, G., & Hjorth, L. (Eds.). (2009). *Mobile technologies: From telecommunications to media*. London: Routledge.

Goh, D. H.-L., Lee, C. S., & Low, G. (2012). I played games as there was nothing else to do: Understanding motivations for using mobile content sharing games. *Online Information Review*, *36*(6), 784–806. doi:10.1108/14684521211287891

Goldstein, T. (2008). The capital of attentive silence and its impact on English language and literacy education. In J. Albright, & A. Luke (Eds.), *Pierre Bourdieu and Literacy Education* (pp. 209–232). New York: Routledge.

Gonzales-Rosero, M. A. P. (2000). The Household as a Workplace: Articulation of Class and Gender in Filipino Middle Class Households. *Review of Women's Studies*, *10*(1-2), 41–68.

Gordon, E., & de Souza e Silva, A. (2011). *Net Locality: Why Location Matters in a Networked World*. Boston: Blackwell-Wiley.

Gordon, J. (2007). The mobile phone and the public sphere: Mobile phone usage in three critical situations. *Convergence*, *13*(3), 307–319. doi:10.1177/1354856507079181

Gow, G. A. (2005). Information Privacy and Mobile Phones. *Convergence: The International Journal of Research into New Media Technologies*, *11*(2), 76–87. doi:10.1177/135485650501100208

Graft, K. (2011). *DICE 2011: EA's Boatman Busts Five Mobile Gaming Myths*. Retrieved 27th June, 2013, from http://www.gamasutra.com/view/news/32977/DICE_2011_EAs_Boatman_Busts_Five_Mobile_Gaming_Myths.php

Grant, A. E., & Wilkinson, J. (2009). *Understanding media convergence: The state of the field*. New York: Oxford University Press.

Green, E., & Singleton, C. (2007). Mobile Selves: Gender, ethnicity and mobile phones in the everyday lives of young Pakistani-British men and women. *Information. Cultura e Scuola*, *10*(4), 506–526.

Greenfield, A. (2006). *Everyware: The dawning age of ubiquitous computing*. Berkley, CA: New Riders.

Greenhow, C. (2009). Social scholarship: Applying social networking technologies to research practices. *Knowledge Quest*, *37*(4), 42–47.

Green, N., & Haddon, L. (2009). *Mobile communications: an introduction to new media*. Oxford, UK: Berg.

Gregg, M. (2011). *Work's Intimacy*. Cambridge, MA: Polity Press.

Griffiths, A. (2008). *Shivers Down Your Spine: Cinemas, Museums and the Immersive View*. New York: Columbia University Press.

Gross, L. (2003). The Gay Global Village in Cyberspace. In N. Couldry, & J. Curran (Eds.), *Contesting Media Power: Alternative Media in a Networked World* (pp. 259–272). New York: Routledge.

GSMA & NTT Docomo. (2012). *Children's use of mobile phones: An international comparison 2012*. Retrieved December 15, 2013 from http://www.gsma.com/publicpolicy/wp-content/uploads/2012/03/GSMA-ChildrenES_English2012WEB.pdf

Gurumurthy, A. (2010). *Understanding gender in a digitally transformed world*. Retrieved from www.itforchange.net/

Gu, X., Gu, F., & Laffey, J. M. (2011). Designing a mobile system for lifelong learning on the move. *Journal of Computer Assisted Learning*, *27*, 204–215. doi:10.1111/j.1365-2729.2010.00391.x

Haddon, L. (2000). The social consequences of mobile telephony: Framing questions. In R. Ling & K. Thrane (Eds.), *Sosiale Konsekvenser av mobiltelefoni: Proceedings fra et seminar om samfunn, barn og mobiltelefoni* (pp. 2-7). Kjeller: Telenor FoU N 38.

Haddon, L. (2004). *Information and Communication Technologies in Everyday Life*. Oxford, UK: Berg.

Haddon, L. (2013). Mobile media and children. *Mobile Media & Communication*, *1*(1), 89–95. doi:10.1177/2050157912459504

Haddon, L., & Vincent, J. (2009). Children's broadening use of mobile phones. In G. Goggin, & L. Hjorth (Eds.), *Mobile technologies: from telecommunications to media* (pp. 37–49). London: Routledge.

Hagel, J., & Armstrong, A. G. (1997). *Net gain: Expanding markets through virtual communities*. Boston: Harvard Business School Press.

Hahn, H. P. (2012). Mobile phones and the transformation of the society: Talking about criminality and the ambivalent perception of new ICT in Burkina Faso. *African Identities*, *10*(2), 181–192. doi:10.1080/14725843.2012.657862

Hahn, H. P., & Kibora, L. (2008). The domestication of the mobile phone: Oral society and new ICT in Burkina Faso. *The Journal of Modern African Studies*, *46*(1), 87–109. doi:10.1017/S0022278X07003084

Håkansson, H., & Ford, D. (2002). How Should Companies Interact in Business Networks? *Journal of Business Research*, *55*(2), 133–139. doi:10.1016/S0148-2963(00)00148-X

Hampton, K. N., Sessions, L. F., & Her, E. J. (2011). Core networks, social isolation, and new media. *Information Communication and Society*, *14*(1), 130–155. doi:10.1080/1369118X.2010.513417

Hargittai, E., & Kim, S. J. (2011). *The Prevalence of Smartphone Use Among a Wired Group of Young Adults*. Northwestern University. Retrieved December 15, 2013 from http://www.ipr.northwestern.edu/publications/docs/workingpapers/2011/IPR-WP-11-01.pdf

Haythornthwaite, C. (2005). Social networks and Internet connectivity effects. *Information Communication and Society*, *8*(2), 125–147. doi:10.1080/13691180500146185

Heath, S. B., & Street, B. V. (2008). *On Ethnography: Approaches to Language and Literacy Research*. New York: Teachers College Press.

Heller, M. (2008). Bourdieu and literacy education. In J. Albright, & A. Luke (Eds.), *Pierre Bourdieu and Literacy Education* (pp. 50–67). New York: Routledge.

Helliker, K. (2005, May 3). Body and Spirit: Why Attending Religious Services May Benefit Health. *Wall Street Journal*. Retrieved from http://online.wsj.com/article/0,SB111507405746322613-email,00.html

Henke, L. L. (1985). Perceptions and use of news media by college students. *Journal of Broadcasting & Electronic Media*, *29*(4), 431–436. doi:10.1080/08838158509386598

Herbert, M. (2004). Parenting across the lifespan. In M. Hoghughi, & N. Long (Eds.), *Theory and research for practice* (pp. 55–71). Handbook of parentingLondon: Sage Publications.

Hermanns, H. (2008). Mobile Democracy: Mobile Phones as Democratic Tools. *Politics*, *28*(2), 74–82. doi:10.1111/j.1467-9256.2008.00314.x

Hermida, A. (2010). Twittering the news. *Journalism Practice*, *4*(3), 297–308. doi:10.1080/17512781003640703

He, Z. (2008). SMS in China: A major carrier of the nonofficial discourse universe. *The Information Society*, *24*, 182–190. doi:10.1080/01972240802020101

Hill, M. L. (2008). Towards a pedagogy of the popular: Bourideu, hip-hop, and out-of-school literacies. In J. Albright, & A. Luke (Eds.), *Pierre Bourdieu and Literacy Education* (pp. 136–161). New York: Routledge.

Himfr. (2008). *Himfr.com reports that China's outdoor sports brandshave enormous market opportunities*. Retrieved November 26, 2009, from http://www.reuters.com/article/2008/11/17/idUS146919+17-Nov-2008+PRN20081117

Hjorth, L. (2003). Pop and Ma: The Landscape of Japanese Commodity Characters and Subjectivity. In Mobile Cultures: New Media in Queer Asia (pp. 158-179). Duke University Press.

Hjorth, L. (2009). *Mobile Media in the Asia-Pacific: Gender and the art of being mobile*. New York: Routledge.

Hjorth, L. (2011). Mobile@game Cultures: The Place of Urban Mobile Gaming. *Convergence: The International Journal of Research into New Media Technologies*, *17*(4), 357–371. doi:10.1177/1354856511414342

Hjorth, L., & Arnold, M. (2013). The place of intimate visualities: Ba ling hou, LBS and camera phones (Shanghai). In *Online@AsiaPacific: Mobile, social and locative media in the Asia-Pacific*. London: Routledge.

Hjorth, L., Burgess, J., & Richardson, I. (2012). *Studying mobile media: Cultural technologies, mobile communication, and the iPhone*. New York, NY: Routledge.

Hjorth, L., Burgess, J., & Richardson, I. (2012). Studying the Mobile. Locating the Field. In L. Hjorth, J. Burgess, & I. Richardson (Eds.), *Studying mobile Media: Cultural technologies, Mobile Communication and the iPhone* (pp. 1–7). London: Routledge.

Hjorth, L., & Kim, H. (2005). Being there and being here: Gendered customising of mobile 3G practices through a case study on Seoul. *Convergence Journal*, *11*(2), 49–55.

Hobbs, R. (1998). The Seven Great Debates in the Media Literacy Movement. *The Journal of Communication*, *48*(1), 16–32. doi:10.1111/j.1460-2466.1998.tb02734.x

Hochschild, A. (2001). *The Time Bind: When Work becomes Home and Home becomes Work*. Holt Press.

Hochschild, A. (2003). *The Commercialization of Intimate Life: Notes From Home and Work*. Berkeley, CA: University of California Press.

Hodson, H. (2012). Why Google's *Ingress* game is a data gold mine. *New Scientist*, 2893.

Hofheinz, A. (2011). Nextopia? Beyond revolution 2.0. *International Journal of Communication*, *5*, 1417–1434.

Höflich, J. R. (2005). The mobile phone and the dynamic between private and public communication: Results of an international exploratory study. In P. Glotz, S. Bertschi, & C. Locke (Eds.), *Thumb Culture: The Meaning of Mobile Phones for Society* (pp. 123–135). London: Transaction Publishers.

Hoflich, J. R. (2010). Mobile media and the change of everyday life: A short introduction. In J. R. Hoflich, G. F. Kircher, C. Linke, & I. Schlote (Eds.), *Mobile Media and the Change of Everyday Life*. Frankfurt: Peter Lang.

Hoflich, J., & Linke, C. (2011). Mobile Communication and Intimate Relationships. In *Mobile Communication: Bringing Us Together and Tearing Us Apart*. New Brunswick, NJ: Transaction Publishers.

Hoghughi, M. (2004). Parenting: An introduction. In M. Hoghughi, & N. Long (Eds.), *Theory and research for practice* (pp. 1–18). Handbook of parentingLondon: Sage Publications.

Ho, K. C., Baber, Z., & Khondker, H. (2002). 'Sites' of resistance: Alternative web sites and state-society relations. *The British Journal of Sociology*, *53*(1), 127–148. doi:10.1080/00071310120109366 PMID:11958682

Holbert, R. L. (2005). Intramedia mediation: The cumulative and complementary effects of news media use. *Political Communication*, *22*(4), 447–461. doi:10.1080/10584600500311378

Holbert, R. L., & Stephenson, M. T. (2002). Structural equation modeling in the communication sciences, 1995-2000. *Human Communication Research*, *28*(4), 531–551. doi: doi:10.1111/j.1468-2958.2002.tb00822.x

Hondagneu-Sotelo, P., & Avila, E. (2003). I'm here, but I'm there: The meanings of Latina transnational motherhood. In P. Hondagneu-Sotelo (Ed.), *Gender and U.S. Immigration: Contemporary Trends* (pp. 317–340). Berkeley, CA: University of California Press. doi:10.1525/california/9780520225619.003.0015

Hong, R., & Chen, V. H.-H. (2013). Becoming and ideal co-creator: Web materiality and intensive laboring practices in game modding. *New Media & Society*, 1–16.

Horst, H. A., & Miller, D. (2006). *The Cell Phone: An Anthropology of Communication*. Oxford, UK: Berg.

Horst, H., & Miller, D. (2006). *The cell phone: An anthropology of communication*. Oxford, UK: Berg.

Hughes, N., & Lonie, S. (2007). M-PESA: Mobile Money for the unbanked Turning Cellphones into 24-Hour Tellers in Kenya. *Innovations*, *2*(1–2), 63–81. doi:10.1162/itgg.2007.2.1-2.63

Huizinga, J. (1955). *Homo Ludens*. Boston: Beacon Press.

Hu, L.-T., & Bentler, P. M. (1999). Cutoff criteria for fit indexes in covariance structure analysis: Conventional criteria versus new alternatives. *Structural Equation Modeling*, *6*(1), 1–55. doi:10.1080/10705519909540118

Hulme, M., & Truch, A. (2005). Social identity: The new sociology of the mobile phone. In K. Nyíri (Ed.), *Sense of place: The global and the local in mobile communication* (pp. 459–466). Vienna: Passagen Verlag.

Humphreys, L., Von Pape, T., & Karnowski, V. (2013). Evolving mobile media: Uses and conceptualizations of the mobile internet. *Journal of Computer-Mediated Communication*, *18*(4), 491–507. doi:10.1111/jcc4.12019

Hurst, M. (2007). *Bit Literacy*. New York: Good Experience Press.

Hurt, H. T., Joseph, K., & Cook, C. (1977). Scales for the measurement of innovativeness. *Human Communication Research*, *4*(1), 58–65. doi:10.1111/j.1468-2958.1977.tb00597.x

Hussain, M. M., & Howard, P. N. (2013). What best explains successful protest cascades? ICTs and the fuzzy causes of the Arab Spring. *International Studies Review*, *15*, 48–66. doi:10.1111/misr.12020

Huyer, S., Hafkin, N., Ertl, H., & Dryburgh, H. (2006). Women in the Information Society. In G. Sciadas (Ed.), *From the Digital Divide to Digital Opportunities: Measuring Infostates for Development*. Montreal, Canada: Orbicom.

Hyman, Phelps, & McNamara. (2013, March 12). Are All Health and Wellness Mobile Apps Exempt from FDA Regulatory Requirements? *FDA Law Blog*. Retrieved from http://www.fdalawblog.net/fda_law_blog_hyman_phelps/2013/03/are-all-health-and-wellness-mobile-apps-exempt-from-fda-regulatory-requirements.html

IAB. (2008). *IAB platform status report: A mobile advertising overview*. Retrieved October 29, 2008, from http://www.iab.net/media/file/moble_platform_status_report.pdf

Ibrahim, Y. (2009). Textual and symbolic resistance: Remediating politics through the blogosphere in Singapore. In A. Russell, & N. Echchaibi (Eds.), *International blogging* (pp. 173–198). New York, NY: Peter Lang Publishing.

IDC. (2013, September 4). *Worldwide Mobile Phone Market Forecast to Grow 7.3% in 2013 Driven by 1 Billion Smartphone Shipments*. Retrieved from http://www.idc.com/getdoc.jsp?containerId=prUS24302813

Idhe, D. (2003). Auditory Imagination. In *The Auditory Culture Reader*. Oxford, UK: Berg.

ifeng. (2010). *The enlightenment from Foursquare*. Retrieved on May 6, 2013 from http://tech.ifeng.com/magazine/local/detail_2010_12/21/3615387_0.shtml

Igbaria, M., Guimaraes, T., & Davis, G. B. (1995). Testing the determinants of microcomputer usage via a structural equation model. *Journal of Management Information Systems*, *11*(4), 87–114.

iiMedia. (2011). *What lies behind LBS*. Retrieved on June 20, 2013 from http://www.iimedia.cn/16056_2.html

Infocomm Development Authority of Singapore. (2010). *Annual survey on infocomm usage in households and by individuals for 2010*. Retrieved on September 23, 2012 from http://www.ida.gov.sg/doc/Publications/Publications_Level3/Survey2010/HH2010ES.pdf

Infocomm Development Authority of Singapore. (2013). *Annual survey on infocomm usage in households and by individuals for 2011*. Retrieved from http://www.ida.gov.sg/~/media/Files/Infocomm%20Landscape/Facts%20and%20Figures/SurveyReport/2011/2011%20HH%20mgt%20rpt%20public%20final.pdf

InfraScan, Inc. (2013). *The Infrascanner Model 2000*. Retrieved from www.infrascanner.com

Intelligence, G. S. M. A. (2013). *Global mobile penetration — Subscribers versus connections*. Retrieved September 3, 2013, from https://gsmaintelligence.com/analysis/2012/10/global-mobile-penetration-subscribers-versus-connections/354/

International Herald Tribune. (2013, August 19). GPS and the law. *The International Herald Tribune*.

International Organization for Migration. (2013). *Facts and figures infographics*. International Organization for Migration.

International Telecommunication Union (ITU). (2013). *Key Global Telecom Indicators for the World Telecommunication Service Sector*. Retrieved from http://www.itu.int/ITU-D/ict/statistics/at_glance/KeyTelecom.html

International Telecommunication Union. (2009). *The World in 2009, ICT Facts and Figures*. Retrieved from http://www.itu.int/ITU-D/ict/material/Telecom09_flyer.pdf

International Telecommunications Union (ITU). (2008). *Mobile cellular subscription*. Retrieved from http://www.itu.int/ITU-D/icteye/Reporting/ShowReportFrame.aspx?ReportName=/WTI/CellularSubscribersPublic&RP_intYear=2008&RP_intLanguageID=1

Isomursu, M. (2008). Tags and the city. *Psychology Journal*, *6*(2), 131–156.

Ito, M. (2004). *Personal Portable Pedestrian: Lessons from Japanese Mobile Phone Use*. Paper presented at Mobile Communication and Social Change, the 2004 International Conference on Mobile Communication. Seoul, Korea.

Ito, M., & Okabe, D. (2005). *Intimate visual co-presence*. Paper presented at theUbicamp 2005. Retrieved January 3, 2010, from http://www.itofisher.com/mito/archives/ito.ubicomp05.pdf

Ito, M., Okabe, D., & Matsuda, M. (Eds.). Personal, portable, pedestrian: Mobile phones in Japanese life. Cambridge, MA: MIT Press.

Ito, M., & Okabe, D. (2005). Mobile Phones, Japanese Youth and the Replacement of Social Contact. In R. Ling, & P. Pedersen (Eds.), *Mobile communications: Renegotiation of the social sphere*. London: Springer. doi:10.1007/1-84628-248-9_9

Ito, M., Okabe, D., & Matsuda, M. (Eds.). (2005). *Personal, portable, pedestrian: Mobile phones in Japanese life*. Cambridge, MA: MIT Press.

Ito, M., Okabe, D., & Matsuda, M. (Eds.). (2005). *Personal, Portable, Pedestrian: Mobile Phones in Japanese Life*. Cambridge, MA: MIT Press.

It's Alive!. (2001). *BotFighters*. It's Alive!.

ITU. (2013). *The World in 2013: ICT Facts and Figures*. Retrieved December 15, 2013 from http://www.itu.int/en/ITU-D/Statistics/Documents/facts/ICTFactsFigures2013.pdf

Jamieson, L. (1998). *Intimacy: Personal Relationship in Modern Societies*. Cambridge, MA: Polity Press.

Jeffres, L., & Atkin, D. (1996). Predicting use of technologies for communication and consumer needs. *Journal of Broadcasting & Electronic Media*, *40*(3), 318–330. doi:10.1080/08838159609364356

Jenkins, H. (2006). *Convergence Culture: When New and Old Media Collide*. New York: New York University Press.

Jenkins, H. (2006). *Convergence Culture: Where Old and New Media Collide*. New York: New York University Press.

Jensen, K. B. (2010). *Media convergence: The three degrees of network, mass, and interpersonal communication*. London: Routledge.

Jensen, M. (2007). The new metrics of scholarly authority. *The Chronicle*, *53*(41).

Jensen, O. B. (2006). 'Facework', Flow and the City: Simmel, Goffman, and Mobility in the Contemporary City. *Mobilities*, *1*(2), 143–165. doi:10.1080/17450100600726506

Jewitt, C. (2006). *Technology, Literacy and Learning: A multimodal approach*. New York: Routledge.

Jiow, H. J., & Lin, J. (2013). The influence of parental factors on children's receptiveness towards mobile phone location disclosure services. *First Monday*, *18*(1). doi:10.5210/fm.v18i1.4284

Ji, P., & Skoric, M. M. (2013). Gender and social resources: Digital divides of social network sites and mobile phone use in Singapore. *Chinese Journal of Communication*, *6*(2), 221–239. doi:10.1080/17544750.2013.785673

Joffe, B. (2005). Mogi: Location and presence in a Pervasive Community Game. In *Proceedings of Ubicomp Workshop on Ubiquitous Gaming and Entertainment*. Academic Press.

Johnsen, E. (2003). The Social Context of the Mobile Phone Use of Norwegian Teens. In J. Katz (Ed.), *Machines that Become Us: The Social Context of Personal Communication Technology* (pp. 161–169). New Brunswick, NJ: Transaction Publishers.

Johnsen, T. E. (2003). The Social Context of the Mobile Phone Use of Norwegian Teens. In J. E. Katz (Ed.), *Machines That Become Us: The Social Context of Personal Communication Technology* (pp. 161–169). New Brunswick: Transaction Publishers.

Johnson, D. G. (2003). Reflections on campaign politics, the Internet and ethics. In D. M. Anderson, & M. Cornfield Johnson (Eds.), *The civic web: Online politics and democratic values* (pp. 19–34). Lanham, MD: Rowman & Littlefield.

Jones, D. (2005). iPod, Therefore I Am: Thinking Inside the White Box. New York: Bloomsbury.

Jöreskog, K. (1999). *Advanced topics on Lisrel by Professor Karl Jöreskog: How large can a standardized coefficient be?* Retrieved from http://www.ssicentral.com/lisrel/techdocs/HowLargeCanaStandardizedCoefficientbe.pdf

Jumisko-Pyykkö, S., & Vainio, T. (2010). Framing the context of use for mobile HCI. *International Journal of Mobile Human Computer Interaction*, *2*(4), 1–18. doi:10.4018/jmhci.2010100101

Junglas, I. A., & Watson, R. T. (2008). Location-based services. *Communications of the ACM*, *51*, 65–69. doi:10.1145/1325555.1325568

Jungwon, M., Byung, K., & Shu, W. (2008). *Location Based Services for Mobiles: Technologies and Standards.* Retrieved on May 12, 2013 from http://blue-penguin.org/cache/location-based-services-for-mobiles.pdf

Juul, J. (2004). *Half-Real: Video Games between Real Rules and Fictional Worlds.* Cambridge, MA: MIT Press.

Kabisch, E. (2010). Mobile after-media: Trajectories and points of departure. *Digital Creativity, 21*(1), 46–54. doi:10.1080/14626261003654996

Kahney, L. (2005). *The Cult of iPod.* San Francisco: No Starch Press.

Kaikkonen, A. et al. (2005). Usability testing of mobile applications: A comparison between laboratory and field testing. *Journal of Usability Studies, 1*(1), 4–17.

Kaiser, T. (2012). *Location Privacy Protection Act Passed by Senate Committee.* Retrieved August 27, 2013, from http://www.dailytech.com/Location+Privacy+Protection+Act+Passed+by+Senate+Committee/article29428.htm

Kaminer, N. (1997). Scholars and the use of the Internet. *Library & Information Science Research, 19*(4), 329–345. doi:10.1016/S0740-8188(97)90024-4

Kang, M.-H. (2002). Digital cable: Exploring factors associated with early adoption. *Journal of Media Economics, 15*(3), 193–207. doi:10.1207/S15327736ME1503_4

Karahanna, E., & Straub, D. W. (1999). The psychological origins of perceived usefulness and ease-of-use. *Information & Management, 35*(4), 237–250. doi:10.1016/S0378-7206(98)00096-2

Kato, F., Okabe, D., Ito, M., & Uemoto, R. (2005). Uses and possibilities of the Keitai camera. In M. Ito, D. Okabe, & M. Matsuda (Eds.), *Personal, portable, pedestrian: Mobile phones in Japanese life.* Cambridge, MA: The MIT Press.

Katz, J. (Ed.). (2003). *Machines that Become Us: The Social Context of Personal Communication Technology.* New Brunswick, NJ: Transaction Publishers.

Katz, J. E. (Ed.). (2003). *Machines That Become Us: The Social Context of Personal Communication Technology.* New Brunswick: Transaction Publishers.

Katz, J. E. (2005). Mobile communication and the transformation of daily life: The next phase of research on mobiles. In P. Glotz, S. Bertschi, & C. Locke (Eds.), *Thumb Culture: The Meaning of Mobile Phones for Society* (pp. 171–182). London: Transaction Publishers.

Katz, J. E. (2006). *Magic in the air: Mobile communication and the transformation of social life.* New Brunswick, NJ: Transaction Publishers.

Katz, J. E. (Ed.). (2008). *Handbook of mobile communication studies.* Cambridge, MA: The MIT Press. doi:10.7551/mitpress/9780262113120.001.0001

Katz, J. E. (Ed.). (2011). *Mobile communication: Dimensions of social policy.* New Brunswick, NJ: Transaction Publishers.

Katz, J. E., & Aakhus, M. (2002). Conclusion: Making Meaning of Mobiles—A Theory of *Apparatgeist.* In *Perpetual Contact: Mobile Communication, Private Talk, Public Performance* (pp. 301–318). Cambridge, UK: Cambridge University Press. doi:10.1017/CBO9780511489471.023

Katz, J. E., & Aakhus, M. (Eds.). (2002). *Perpetual Contact: Mobile Communication, Private Talk, Public Performance.* Cambridge, UK: Cambridge University Press. doi:10.1017/CBO9780511489471

Kaupins, G., & Minch, R. (2005). *Legal and Ethical Implications of Employee Location Monitoring.* Paper presented at the 38th Annual Hawaii International Conference on System Sciences. Hawaii, HI.

Kayany, J. M., & Yelsma, P. (2000). Displacement effects of online media in the socio-technical contexts of households. *Journal of Broadcasting & Electronic Media, 44*(2), 215–229. doi:10.1207/s15506878jobem4402_4

Kellermann, A. L., & Jones, S. S. (2013b, February 26). The Delayed Promise of Health-Care IT. *Project Syndicate.* Retrieved from http://www.project-syndicate.org/commentary/the-delayed-promise-of-health-care-it-by-art-kellermann-and-spencer-jones

Kellermann, A. L., & Jones, S. S. (2013a, Jan). What It Will Take to Achieve the As-Yet-Unfulfilled Promises of Health Information Technology? *Health Affairs, 32*(1), 63–68. doi:10.1377/hlthaff.2012.0693 PMID:23297272

Kennedy, G., Dalgarno, B., Bennett, S., Judd, T., Gray, K., & Chang, R. (2008). Immigrants and natives: Investigating differences between staff and students' use of technology. In *Hello! Where are you in the landscape of educational technology? Proceedings ascilite Melbourne 2008*. Retrieved on 7 January, 2011, from www.ascilite.org.au/conferences/melbourne08/procs/kennedy.pdf

Kennedy, G., Dalgarno, B., Gray, K., Judd, T., Waycott, J., Bennett, S., et al. (2007). The net generation are not big users of Web 2.0 technologies: Preliminary findings. In *ICT: Providing choices for learners and learning: Proceedings Ascilite Singapore 2007*. Ascilite. Retrieved on 7 January, 2011, from: www.ascilite.org.au/conferences/singapore07/procs/kennedy.pdf

Kennedy, G. E., Judd, T. S., Churchward, A., Gray, K., & Krause, K. L. (2008). First year students' experiences with technology: Are they really digital natives? *Australasian Journal of Educational Technology*, *24*(1), 108–122.

Kenway, J., & Bullen, E. (2001). *Consuming Children: Education-entertainment-advertising*. Buckingham, UK: Open University Press.

Kenway, J., Kraack, A., & Hickey-Moody, A. (2006). *Masculinity Beyond the Metropolis*. Houndsmills, UK: Palgrave Macmillan. doi:10.1057/9780230625785

Khondker, H. H. (2011). Role of the new media in the Arab Spring. *Globalizations*, *8*(5), 675–679. doi:10.1080/14747731.2011.621287

Kim, J., & Lee, J. E. R. (2011). The Facebook Paths to Happiness: Effects of the Number of Facebook Friends and Self-Presentation on Subjective Well-Being. *Cyberpsychology, Behavior, and Social Networking*, *14*(6), 359–364. doi:10.1089/cyber.2010.0374 PMID:21117983

Kist, W. (2005). *New Literacies in Action: Teaching and Learning in Multiple Media*. New York: Teacher's College Press.

Kjeldskov, J., & Graham, C. (2003). A Review of Mobile HCI Research Methods. In *Proceeding of ACM Int'l Conf. Human Computer Interaction with Mobile Devices and Services* (MobileHCI). ACM.

Kjeldskov, J., Skov, M. B., Als, B. S., & Høegh, R. T. (2004). Is it Worth the Hassle? Exploring the Added Value of Evaluating the Usability of Context-Aware Mobile Systems in the Field. In *Proceedings of the 6th International Mobile HCI 2004 Conference*. Glasgow, Scotland: Springer-Verlag. DOI: 10.1007/978-3-540-28637-0_6

Kjeldskov, J., & Paay, J. (2010). Indexicality: Understanding mobile human-computer interaction in context. *ACM Transactions on Computer-Human Interaction*, *17*(4). doi:10.1145/1879831.1879832

Kjeldskov, J., Skov, M. B., Nielsen, G. W., Thorup, S., & Vestergaard, M. (2013). Digital Urban Ambience: Mediating Context on Mobile Devices in the City. *Journal of Pervasive and Mobile Computing*, *19*(5), 738–749. doi:10.1016/j.pmcj.2012.05.002

Klabbers, J. H. (2009). *The magic circle: Principles of gaming & simulation*. Rotterdam, The Netherlands: Sense Publishers.

Koh, G., & Ooi, G. L. (2004). Relationship between state and civil society in Singapore: Clarifying the concept, assessing the ground. In *Civil society in Southeast Asia* (pp. 167–197). Singapore: ISEAS Publications.

Komisar, L. (1987). *Corazon Aquino: The Story of a Revolution*. New York: George Braziller, Inc.

Korea Times. (2004, July 3). Location-Based Information Service Due Next Year. *Korea Times*. Retrieved from http://www.koreatimes.co.kr

Kozinets, R. (2010). *Netnography*. London: Sage.

Krajewski, B. (1999, October). Enhancing Character Education through Experiential Drama and Dialogue. *NASSP Bulletin*, 40–45. doi:10.1177/019263659908360906

Kreimer, S. F. (2001). Technologies of protest: Insurgent social movements and the First Amendment in the era of the Internet. *University of Pennsylvania Law Review*, *150*(1), 119–171. doi:10.2307/3312914

Kress, G. (2000a). A Curriculum for the Future. *Cambridge Journal of Education*, *30*(1), 133–145. doi:10.1080/03057640050005825

Kress, G. (2000b). Multimodality. In B. Cope, & M. Kalantzis (Eds.), *Multiliteracies: Literacy Learning and the Design of Social Futures* (pp. 182–202). London: Routledge.

Kress, G. (2003). *Literacy in the New Media Age*. London: Routledge. doi:10.4324/9780203164754

Kress, G., & Pachler, N. (2007). Thinking about the 'm' in m-learning. In N. Pachler (Ed.), *Mobile learning: towards a research agenda* (pp. 7–32). London: The WLE Centre, Institute of Education, University of London.

Kress, G., & Van Leeuwen, J. (2001). *Multimodal Discourse: The modes and media of contemporary communication*. London: Arnold.

Kristjánsdóttir, O. B., Fors, E. A., Eide, E., Finset, A., Stensrud, T. L., & van Dulmen, S. et al. (2013, January 7). A Smartphone-Based Intervention With Diaries and Therapist-Feedback to Reduce Catastrophizing and Increase Functioning in Women With Chronic Widespread Pain: Randomized Controlled Trial. *Journal of Medical Internet Research*, *15*(1), e5. doi:10.2196/jmir.2249 PMID:23291270

Kshetri, N. (2009). The evolution of the Chinese online gaming industry. *Journal of Technology Management in China*, *4*(2), 158–179. doi:10.1108/17468770910965019

Kücklich, J. (2005). Precarious Playbour: Modders and the Digital Games Industry. *The Fibreculture Journal, 5*.

Kuo, C. Y. (1995). The making of a new nation: Cultural construction and national identity. In B. H. Chua (Ed.), *Communitarian ideology and democracy in Singapore* (pp. 101–123). London: Routledge.

Kuwait Telecommunications Report Q4. (2012). Retrieved from http://www.companiesandmarkets.com/Market/Telecommunications/Market-Research/Kuwait-Telecommunications-Report-Q4-2012/RPT1108596

Kwak, N., Campbell, S. W., Choi, J., & Bae, S. Y. (2011). Mobile communication and public affairs engagement in Korea: An examination of non-linear relationships between mobile phone use and engagement across age groups. *Asian Journal of Communication*, *21*(5), 485–503. doi:10.1080/01292986.2011.587016

Lacohee, H., & Wakeford, N. (2003). A Social History of the Mobile Telephone with a View of its Future. *BT Technology Journal*, *21*(3), 203–211. doi:10.1023/A:1025187821567

Lai, J., Mitchell, S., & Pavlovski, C. (2007). Examining modality usage in a conversational multimodal application for mobile e-mail access. *International Journal of Speech Technology*, *10*(1), 17–30. doi:10.1007/s10772-009-9017-9

Lai, V., & Guynes, J. (1994). A model of ISDN (integrated services digital network) adoption in U.S. corporations. *Information & Management*, *26*(2), 75–84. doi:10.1016/0378-7206(94)90055-8

Lane, G., Thelwall, S., Angus, A., Peckett, V., & West, N. (2006). *Urban tapestries: Public authoring, place & mobility*. Retrieved March 26, 2009, from http://socialtapestries.org/outcomes/reports/UT_Report_2006.pdf

Langman, L. (2005). From virtual public spheres to global justice: A critical theory of internetworked social movements. *Sociological Theory*, *23*(1), 42–74. doi:10.1111/j.0735-2751.2005.00242.x

LaRose, R., & Atkin, D. (1992). Audiotext and the re-invention of the telephone as a mass medium. *The Journalism Quarterly*, *69*(2), 413–421. doi:10.1177/107769909206900215

LaRose, R., & Atkin, D. J. (1988). Satisfaction, demographic, and media environment predictors of cable subscription. *Journal of Broadcasting & Electronic Media*, *32*(4), 403–413. doi:10.1080/08838158809386712

Lasén, A., & Casado, E. (2012). Mobile Telephony and the Remediation of Couple Intimacy. *Feminist Media Studies*, *12*(4), 550–559. doi:10.1080/14680777.2012.741871

Latham, K. (2007). SMS, communication, and citizenship in China's information society. *Critical Asian Studies*, *39*(2), 295–314. doi:10.1080/14672710701339493

Laurila, J. K., Gatica-Perez, D., Aad, I., Blom, J., Bornet, O., Do, T., et al. (2012). The Mobile Data Challenge: Big Data for Mobile Computing Research. In *Proc. Mobile Data Challenge Workshop (MDC) in conjunction with Int. Conf. on Pervasive Computing*. Retrieved from https://research.nokia.com/files/public/MDC2012_Overview_LaurilaGaticaPerezEtAl.pdf

Laursen, B., & Collins, W. A. (2004). Parent-child communication during adolescence. In A. L. Vangelisti (Ed.), *Handbook of Family Communication* (pp. 333–348). Mahwah, NJ: Lawrence, Erlbaum, Associates, Inc.

Le Poire, B. A. (2006). *Family communication: Nurturing and control in a changing world.* Thousand Oaks, CA: Sage Publications, Inc.

Ledbetter, A. M., Heiss, S., Sibal, K., Lev, E., Battle-Fisher, M., & Shubert, N. (2010). Parental Invasive and Children's Defensive Behaviors at Home and Away at College: Mediated Communication and Boundary Management. *Communication Studies, 61*(2), 184–204. doi:10.1080/10510971003603960

Lee, D. H. (2013). Smartphones, mobile social space, and new sociality in Korea. *Mobile Media & Communication, 1*(3), 269–284. doi:10.1177/2050157913486790

Lee, K. J., & Choi, D. J. (2012). Mobile junk message filter reflecting user preference. *Transactions on Internet and Information Systems (Seoul), 6*(11), 2849–2865.

Leek, S., & Christodoulides, G. (2009). Next-generation mobile marketing: How young consumers react to Bluetooth-enabled advertising. *Journal of Advertising Research, 49*(1), 44–53. doi:10.2501/S0021849909090059

Lefebvre, H. (1991). *The Production of Space* (D. Nicholson-Smith, Trans.). Oxford, UK: Blackwell.

Lehmann, D. R., & Ostlund, L. E. (1974). Consumer perceptions of product warranties: An exploratory study. *Advances in Consumer Research. Association for Consumer Research (U. S.), 1*(1), 51–65.

Lemish, P., & Cohen, A. (2005). On the gendered nature of mobile phone culture in Israel. *Sex Roles, 52*(7/8), 511–521. doi:10.1007/s11199-005-3717-7

Lemos, A. (2010). Post-mass media functions, locative media, and informational territories: New ways of thinking about territory, place, and mobility incontemporary society. *Space and Culture, 13*(4), 403–420. doi:10.1177/1206331210374144

Lenhart, A. (2009). *Teens and Sexting: How and why minor teens are sending sexually suggestive nude or nearly nude images via text messaging*. Washington, DC: Pew Research Center. Retrieved December 15, 2013 from http://pewresearch.org/assets/pdf/teens-and-sexting.pdf

Lenhart, A., Ling, R., Campbell, S., & Purcell, K. (2010). *Teens and Mobile Phones.* Washington, DC: Pew Research Center.

Leong, L. Y., Hew, T. S., Ooi, K. B., & Lin, B. (2011). Influence of gender and English proficiency on Facebook mobile adoption. *International Journal of Mobile Communications, 9*(5), 495–521. doi:10.1504/IJMC.2011.042456

Leonhard, G. (2014, February 26). *Big data, big business, big brother?* Retrieved from http://Ed.cnn.com/2014/02/26/business/big-data-big-business/index.html

Leppäniemi, M. S. (2006). A review of mobile marketing research. *International Journal of Mobile Marketing, 1*(1), 30–40.

Lettner, F., & Holzmann, C. (2012a). Heat maps as a usability tool for multi-touch interaction in mobile applications. In *Proceedings of the 11th International Conference on Mobile and Ubiquitous Multimedia* (MUM '12). ACM. DOI: 10.1145/2406367.2406427

Lettner, F., & Holzmann, C. (2012b). Automated and unsupervised user interaction logging as basis for usability evaluation of mobile applications. In *Proceedings of the 10th International Conference on Advances in Mobile Computing & Multimedia* (MoMM '12). ACM. DOI: 10.1145/2428955.2428983

Lettner, F., & Holzmann, C. (2012c). Sensing mobile phone interaction in the field. In *Proceedings of percomw, 2012 IEEE International Conference on Pervasive Computing and Communications Workshops.* Lugano, Switzerland: IEEE.

Leung, L., & Wei, R. (1999). Seeking news via the pager: An expectancy-value study. *Journal of Broadcasting & Electronic Media, 43*(3), 299–315. doi:10.1080/08838159909364493

Leung, L., & Wei, R. (2000). More than just talk on the move: Uses and gratifications of the cellular phone. *Journalism & Mass Communication Quarterly*, *77*(2), 308–320. doi:10.1177/107769900007700206

Levy, M. R. (1978). The audience experience with television news. *Journalism Monographs (Austin, Tex.)*, *55*, 1–29.

Levy, M. R. (1987). VCR use and the concept of audience activity. *Communication Quarterly*, *35*(3), 267–275. doi:10.1080/01463378709369689

Levy, S. (2006). *The Perfect Thing: How the iPod Shuffles Commerce, Culture, and Coolness*. New York: Simon & Schuster.

Levy, S., & Zaltman, G. (1975). *Marketing, Society, and Conflict*. Englewood Cliffs, NJ: Prentice Hall.

Libed, B. P. C. (2010). *Dekada '70 and activist mothers: A new look at mothering, militarism, and Philippine martial law*. (MA Thesis). San Diego State University, San Diego, CA.

Li, C. (2007). Online chatters' Self-Marketing in Cyberspace. *Cyberpsychology & Behavior*, *10*(1), 131–132. doi:10.1089/cpb.2006.9982 PMID:17305459

Licoppe, C. (2003). The Modes of Maintaining Interpersonal Relations Through the Telephone: From the Domestic to the Mobile Phone. In J. Katz (Ed.), *Machines that Become Us: The Social Context of Personal Communication Technology* (pp. 171–185). New Brunswick, NJ: Transactions.

Licoppe, C. (2004). 'Connected presence': The emergence of a new repertoire for managing social relationships in a changing communication technoscape. *Environment and Planning. D, Society & Space*, *22*(1), 135–156. doi:10.1068/d323t

Licoppe, C., & Inada, Y. (2006). Emergent uses of a location aware multiplayer game: The interactional consequences of mediated encounters. *Mobilities*, *1*(1), 39–61. doi:10.1080/17450100500489221

Lim, J. B. Y. (2013a). *Youth Participation: Social Media and East Asian Cultures in Malaysia*. Paper presented at the International Symposium - The Korean Wave in Southeast Asia: Consumption and Cultural Production. Kuala Lumpur, Malaysia.

Lim, J. B. Y. (2013b). Videoblogging and Youth Activism in Malaysia. *International Communication Gazette*, *75*(3), 300–321. doi:10.1177/1748048512472947

Lim, S. S. (2009). Home, school, borrowed, public or mobile: Variations in young Singaporeans' Internet access and their implications. *Journal of Computer-Mediated Communication*, *14*, 1228–1256. doi:10.1111/j.1083-6101.2009.01488.x

Lim, S. S., & Soon, C. (2010). The influence of social and cultural factors on mothers' domestication of household ICTs –Experiences of Chinese and Korean women. *Telematics and Informatics*, *27*(3), 205–216. doi:10.1016/j.tele.2009.07.001

Lim, S., Xue, L. S., Yen, C. C., Chang, L., Chan, H. C., & Tai, B. C. et al. (2011). A study on Singaporean women's acceptance of using mobile phones to seek health information. *International Journal of Medical Informatics*, *80*(12), E189–E202. doi:10.1016/j.ijmedinf.2011.08.007 PMID:21956003

Lin, C. A. (1993a). Adolescent viewing and gratifications in a new media environment. *Mass Communication Review*, *20*(1/2), 39–50.

Lin, C. A. (1993b). Modeling the gratification-seeking process of television viewing. *Human Communication Research*, *20*(2), 224–244. doi:10.1111/j.1468-2958.1993.tb00322.x

Lin, C. A. (1998). Exploring personal computer adoption dynamics. *Journal of Broadcasting & Electronic Media*, *42*(1), 95–112. doi:10.1080/08838159809364436

Lin, C. A. (2001). Audience attributes, media supplementation, and likely online service adoption. *Mass Communication & Society*, *4*(1), 19–38. doi:10.1207/S15327825MCS0401_03

Lin, C. A. (2004). Webcasting adoption: Technology fluidity, user innovativeness, and media substitution. *Journal of Broadcasting & Electronic Media*, *48*(3), 446–465. doi:10.1207/s15506878jobem4803_6

Lindtner, S., & Douris, P. (2011). The promise of play: A new approach to productive play. *Games and Culture*, *6*(5), 453–478. doi:10.1177/1555412011402678

Ling, R. (2007). Mobile Communication and Mediate Ritual. In *Communications in the 21st century*. Budapest: MTA – T-Mobile. Retrieved from http://www.richardling.com/papers/2007_Mobile_communication_and_mediated_ritual.pdf

Ling, R. S. (2004). *The mobile connection: The cell phone's impact on society*. Retrieved from http://libezp.lib.lsu.edu/login?url=http://www.netLibrary.com/urlapi.asp?action=summary&v=1&bookid=114180

Ling, R., Karnowski, V., von Pape, T., & Jones, S. (2013). About the title. *Mobile Media & Communication*. Retrieved from http://www.sagepub.com/journals/Journal202140

Ling, R. (2004). *The Mobile Connection: The Cell Phone's Impact on Society*. San Francisco: Morgan Kaufmann.

Ling, R. (2004). *The mobile connection: The cell phone's impact on society*. San Francisco, CA: Morgan Kaufmann.

Ling, R. (2007). Mobile communication and mediated ritual. In K. Nyiri (Ed.), *Communications in the 21st Century*. Budapest: Academic Press.

Ling, R. (2008). *New Tech, New Ties. How Mobile Communication is Reshaping Social Cohesion*. Cambridge, MA: MIT Press.

Ling, R. (2008). The mediation of ritual interaction via the mobile telephone. In J. E. Katz (Ed.), *Handbook of Mobile Communication Studies* (pp. 165–176). Cambridge, MA: MIT Press. doi:10.7551/mitpress/9780262113120.003.0013

Ling, R. (2012). *Taken for grantedness: The embedding of mobile communication into society*. Cambridge, MA: MIT Press.

Ling, R., & Bertel, T. (2013). Mobile communication culture among children and adolescents. In D. Lemish (Ed.), *The Routledge international handbook of children, adolescents and media* (pp. 127–133). London: Routledge.

Ling, R., Bertel, T., & Sundsøy, P. R. (2012). The socio-demographics of texting: An analysis of traffic data. *New Media & Society*, *14*(2), 281–298. doi:10.1177/1461444811412711

Ling, R., & Donner, J. (2010). *Mobile communication in everyday life: New choices, new challenges. Mobile Communication* (pp. 75–106). Malden, MA: Polity.

Ling, R., & Haddon, L. (2003). Mobile telephony, mobility, and the coordination of everyday life. In J. E. Katz (Ed.), *Machines that become us: The social context of personal communication technology* (pp. 245–265). New Brunswick, NJ: Transaction Publishers.

Ling, R., & Haddon, L. (2008). Children, Youth and the Mobile Phone. In K. Drotner, & S. Livingstone (Eds.), *The International Handbook of Children, Media and Culture* (pp. 137–151). London: Sage. doi:10.4135/9781848608436.n9

Ling, R., & Pedersen, P. E. (Eds.). (2005). *Mobile communications: Re-negotiation of the social sphere*. New York, NY: Springer.

Ling, R., & Yttri, B. (2002). Hyper-coordination via mobile phones in Norway. In J. E. Katz, & M. Aakhus (Eds.), *Perpetual Contact: Mobile Communication, Private Talk, Public Performance* (pp. 139–169). Cambridge, UK: Cambridge University Press.

Linke, C. (2013). Mobile media and communication in everyday life: Milestones and challenges. *Mobile Media & Communication*, *1*(1), 32–37. doi:10.1177/2050157912459501

Lips, M. (2010). Rethinking Citizen – Government Relationships in the Age of Digital Identity. *Information Polity*, *15*(4), 273–289. DOI: 10.3233/IP-2010-0216

Li, S. S.-C. (2004). Exploring the factors influencing the adoption of interactive cable television services in Taiwan. *Journal of Broadcasting & Electronic Media*, *48*(3), 466–483. doi:10.1207/s15506878jobem4803_7

Liu, Y., & Li, H. (2011). Exploring the impact of use context on mobile hedonic services adoption: An empirical study on mobile gaming in China. *Computers in Human Behavior*, *27*, 890–898. doi:10.1016/j.chb.2010.11.014

Livingstone, S., Haddon, L., Görzig, A., & Ólafsson, K. (2011). *Risks and safety on the internet: The perspective of European children: Full findings*. London: LSE. Retrieved December 15, 2013 from http://lse.ac.uk/EUKidsOnlineReports

Livingstone, S., & Bovill, M. (2001). *Children and their changing media environment: A European comparative study*. Mahwah, NJ: Lawrence Erlbaum.

Locher, D. A. (2002). *Collective behaviour*. Upper Saddle River, NJ: Prentice Hall.

Loudon, M. (2010). ICTs as an opportunity structure in southern social movements. *Information Communication and Society*, *13*(8), 1069–1098. doi:10.1080/13691180903468947

Luke, C. (1997). Technological Literacy. Melbourne, Australia: Adult Research Literacy Network (Language Australia Limited).

Lvren.cn. (2009). *Photolist of The North Face Campaign*. Retrieved November 10,2010, from http://active.lvren.cn/chuqizhisheng-0909/photolist.php

Lynch, K. (1960). *The Image of the City*. Cambridge, MA: MIT Press.

Madden, M., Lenhart, A., Cortesi, S., & Gasser, U. (2013). *Teens and Mobile Apps Privacy*. Washington, DC: Pew Research Center.

Madden, M., Lenhart, A., Duggan, M., Cortesi, S., & Gasser, U. (2013). *Teens and Technology 2013*. Washington, DC: Pew Research Center.

Madianou, M., & Miller, D. (2011). Mobile phone parenting: Reconfiguring relationships between Filipina migrant mothers and their left-behind children. *New Media & Society*, *13*(3), 457–470. doi:10.1177/1461444810393903

Madianou, M., & Miller, D. (2012). *Migration and new media: Transnational families and polymedia*. Oxon, UK: Routledge.

Madison, D. S. (2005). *Critical Ethnography: Methods, Ethics, and Performance*. Thousand Oaks, CA: Sage.

Mainwaring, S. D., Anderson, K., & Chang, M. F. (2005). *Living for the Global City: Mobile Kits, Urban Interfaces, and Ubicomp*. Paper presented at UbiComp. New York, NY.

Maisel, R. (1973). The decline of mass media. *Public Opinion Quarterly*, *37*(2), 159–170. doi:10.1086/268075

Malaby, T. M. (2007). Beyond play: A new approach to game. *Games and Culture*, *2*(2), 95–113. doi:10.1177/1555412007299434

Malaysian Communications and Multimedia Commission. (2012). *Hand Phone Users Survey 2012*. Retrieved September 20, 2013, from http://www.skmm.gov.my/skmmgovmy/media/General/pdf/130717_HPUS2012.pdf

Malaysian Digest. (2013, February 19). *Mohd Zin Defends Selangor BN's SMS Campaign*. Retrieved September 5, 2013, from http://www.malaysiandigest.com/sports/259902-mohd-zin-defends-selangor-bns-sms-campaign.html

Malaysian Insider. (2013, April 8). *Get 'Undi PRU13' apps for free on your Androids or Apple smartphones*. Retrieved August 10, 2013, from http://www.themalaysianinsider.com/malaysia/article/get-undi-pru13-apps-for-free-on-your-androids-or-apple-smartphones/

Malaysian Mobile Content Provider Association. (2012). *Home*. Retrieved September 4, 2013, from http://www.mmcp.org.my/

Malaysian Mobile Content Providers Association. (n.d.). Retrieved September 4, 2013, from http://www.mmcp.org.my/

Mandrola, J. (2013, August 19). *Compassionate, well-aligned healthcare takes time, a luxury few doctors have*. Retrieved from http://medcitynews.com/2013/08/compassionate-well-aligned-healthcare-takes-time-a-luxury-few-doctors-have/

Manovich, L. (2001). *The language of new media*. Cambridge, MA: MIT Press.

Marcus, G. E. (1998). *Ethnography Through Thick and Thin*. Princeton, NJ: Princeton University Press.

Marmasse, N., & Schmandt, C. (2003). *Safe & sound: A wireless leash*. Paper presented at the Human Factors in Computing Systems. Ft. Lauderdale, FL.

Mascheroni, G. (2013). Parenting the mobile internet in Italian households: Parents' and children's discourses. *Journal of Children and Media*. DOI: 10.1080/17482798.2013.830978

Mascheroni, G., & Ólafsson, K. (2013). *Mobile internet access and use among European children: Initial findings of the Net Children Go Mobile project*. Milan, Italy: Educatt. Retrieved December 15, 2013 from www.netchildrengombile.eu

Mascheroni, G., & Ólafsson, K. (2014). *Net Children Go Mobile: Risks and opportunities*. Milan, Italy: Educatt. Retrieved April 22, 2014 from www.netchildrengombile.eu

Mathers, C. D., & Loncar, D. (2006). Projections of global mortality and burden of disease from 2002 to 2030. *PLoS Medicine*, *3*(11), e442. doi:10.1371/journal.pmed.0030442 PMID:17132052

Matsuda, M. (2005). Mobile Communications and Selective Sociality. In M. Ito, D. Okabe, & M. Matsuda (Eds.), *Personal, Portable, Pedestrian: Mobile Phones in Japanese Life* (pp. 123–142). Cambridge, MA: MIT Press.

Mattoni, A. (2013). Repertoires of communication in social movement processes. In B. Cammaerts, A. Mattoni, & P. McCurdy (Eds.), *Mediation and protest movements* (pp. 39–56). Bristol: Intellect.

Mazzarella, W. (2010). Beautiful Balloon: The Digital Divide and the Charisma of New Media in India. *American Ethnologist*, *37*(4), 783–804. doi:10.1111/j.1548-1425.2010.01285.x

McAdam, D., Tarrow, S., & Tilly, C. (2001). *Dynamics of contention*. New York, NY: Cambridge University Press. doi:10.1017/CBO9780511805431

McCloskey, D. (2003). Evaluating electronic commerce acceptance with the technology acceptance model. *Journal of Computer Information Systems*, *44*(2), 49–57. doi: doi:10.4018/978-1-59140-066-0.ch101

McConnell, B., & Huba, J. (2006). *Citizen marketers: When people are the message*. Chicago: Kaplan Business.

McGonigal, J. (2011). *Reality is Broken: Why games make us better and how they can change the world*. London: Penguin.

McLuhan, M. (1994). *Understanding media: The extensions of man*. Cambridge, MA: MIT Press.

McVeigh, B. (2004). *Nationalisms of Japan: Managing and Mystifying Identity*. Oxford, UK: Rowman and Littlefield.

Medina, B. T. G. (2001). *The Filipino family* (2nd ed.). Quezon City: The University of the Philippines Press.

Meeker, M., & Lu, L. (2013, May 29). *Internet Trends D11 Conference*. KPCB Report. Retrieved from http://www.kpcb.com/insights/2013-internet-trends

Mele, C. (2000). The Materiality of Urban Discourse: Rational Planning in the Restructuring of the Early Twentieth-Century Ghetto. *Urban Affairs Review*, *35*(5), 628–648. doi:10.1177/10780870022184570

Melkote, D. R., & Steeves, H. L. (2004). Information and communication technologies for rural development. In *Development and Communication in Africa* (pp. 165–173). Oxford, UK: Rowman & Littlefield.

Mendes, S., Alampay, E., & Soriano, E. et al. (2007). *The Innovative Use of Mobile Applications in the Philippines: Lessons for Africa*. Stockholm: Swedish International Development Cooperation Agency.

Menkhoff, T., & Bengtsson, M. L. (2010). Engaging students in higher education through mobile learning. *Communications in Computer and Information Science*, *112*, 471–487. doi:10.1007/978-3-642-16324-1_56

Meuli, P. G., & Richard, J. E. (2013). Exploring and modellling digital natives' intention to use permission-based location-aware mobile advertising. *Journal of Marketing Management*, *29*(5/6), 698–719.

Meyrowitz, J. (1985). *No sense of place: The impact of electronic media on social behavior*. New York: Oxford University Press.

Meyrowitz, J. (2005). The Rise of Glocality: New Senses of Place and Identity in the Global Village. In K. Nyìri (Ed.), *The Global and the Local in Mobile Communication* (pp. 21–30). Wien: Passagen Verlag.

Midgley, D. F., & Dowling, G. R. (1978). Innovativeness: The concept and its measurement. *The Journal of Consumer Research*, *4*(4), 229–242. doi:10.1086/208701

Miles, M. B., & Huberman, A. M. (1994). *Qualitative data analysis*. Thousand Oaks, CA: Sage.

Mischler, E. (1986). *Research interviewing: Context and narrative*. Cambridge, MA: Harvard University Press.

Mobile Studies International. (2014). *Mobile campus*. Retrieved from http://www.mobile2studies.com/mobile-campus.html

Mohammad, S., & Anil Job, M. (2013). Adaption of M-Learning as a Tool in Blended Learning - A Case Study in AOU Bahrain. *International Journal of Science and Technology*, *3*(1), 14–20.

Mohd Azam Osman, A. Z.-Y. (2012). A Study of the Trend of Smartphone and its Usage Behavior in Malaysia. *International Journal on New Computer Architectures and Their Applications*, 2(1), 274–285.

Molnár, V. (2010). Reframing public space through digital mobilization: Flash mobs and the futility (?) of contemporary urban youth culture. *Theory, Culture & Society*.

Molony, T. (2008). Non-developmental uses of mobile communication in Tanzania. In *Handbook of Mobile Communication Studies* (pp. 339–351). Cambridge, MA: The MIT Press. doi:10.7551/mitpress/9780262113120.003.0025

Montgomery, J. (1998). Making a City: Urbanity, Vitality, and Urban Design. *Journal of Urban Design*, *3*(1), 93–116. doi:10.1080/13574809808724418

MoPub. (2012). *Mobile Games Played Most Often at Home*. San Francisco, CA: MoPub Miniclip.

Morley, D. (1986). *Family Television: Cultural Power and Domestic Leisure*. London: Routledge.

Morrish, N. J., Wang, S. L., Stevens, L. K., Fuller, J. H., & Keen, H. (2001). Mortality and Causes of Death in the WHO Multinational Study of Vascular Disease in Diabetes. *Diabetologia*, *22*(2), S14–S21. doi:10.1007/PL00002934 PMID:11587045

Morse, M. (1990). An Ontology of Everyday Distraction: The Freeway, the Mall, and Television. In *Logics of Television: Essays in Cultural Criticism*. London: BFI.

Mubasher. (2012). *UAE ranks among highest mobile penetration countries worldwide – SHUAA*. Retrieved from http://english.mubasher.info/DFM/news/2189413/UAE-ranks-among-highest-mobile-penetration-countries-worldwide-SHUAA

Muchie, M., & Baskaran, A. (2006). Introduction. In *Bridging the Digital Divide: Innovation Stems for ICT in Brazil, China, India, Thailand and Southern Africa* (pp. 23–50). London: Adonis & Abbey Publishers Ltd.

Muniandy, P. (2012). Malaysias Coming Out! Critical Cosmopolitans, Religious Politics and Democracy. *Asian Journal of Social Science*, *40*(5-6), 5–6. doi:10.1163/15685314-12341263

Muse, J. H. (2010). Flash Mobs and the Diffusion of Audience. *Theater*, *40*(3), 9–23. doi:10.1215/01610775-2010-005

Nabeth, T. (2006, May). Understanding the Identity Concept in the Context of Digital Social Environment. In T. Nabeth (Ed.), Del 2.2 Set of Use Cases and Scenarios, FIDIS Deliverable, (pp. 74–91). FIDIS.

Nagasaka, I. (2007). Cellphones in Rural Philippines. In *The Social Construction and Usage of Communication Technologies Asian and European Experiences* (pp. 100–125). University of the Philippines Press.

Nash, R. (2009). *The active classroom: Practical strategies for involving students in the learning process*. Corwin Press.

Nassuora, A. B. (2013). Students acceptance of mobile learning for higher education in Saudi Arabia. *International Journal of Learning Management Systems*, *1*(1), 1–9. doi:10.12785/ijlms/010101

National Institutes of Health. (2010). *Traditional Chinese Medicine: An Introduction*. Retrieved from http://nccam.nih.gov/health/whatiscam/chinesemed.htm

National Institutes of Health. (2013). *Ayurvedic Medicine: An Introduction*. Retrieved from http://nccam.nih.gov/health/ayurveda/introduction.htm

Nations, D. (2013). *How many iPads have been sold?* Retrieved March17, 2013 from http://ipad.about.com/od/iPad-FAQ/a/How-Many-iPads-Have-Been-Sold.htm

Ndukwe, E. C. (2006). *Three Years of GSM Revolution in Nigeria*. Retrieved from http://www.ncc.gov.ng/speeches_presentations/EVC's%20Presentation/GSM%20REVOLUTION%20IN%20NIGERIA%20%20-140504.pdf

Neal, A. G. (1998). *National trauma and collective memory: Major events in the American century*. Armonk, NY: M.E. Sharpe.

Networks Asia. (2013, June 18). *Singapore smartphone and tablet penetration on the rise, app. usage increasing*. Retrieved from http://www.networksasia.net/content/singapore-smartphone-and-tablet-penetration-rise-app-usage-increasing?page=0%2C0

New London Group. (2000). A Pedagogy of Multiliteracies: Designing social futures. In B. Cope, & M. Kalantzis (Eds.), *Multiliteracies: Literacy Learning and the Design of Social Futures* (pp. 9–37). London: Routledge.

Ng, T.Y. (2011, May 17). Facebook trolls can be the most valuable fans. *The Straits Times*, p. A2.

Ng, T. P., Lim, M. L., Niti, M., & Collinson, S. (2012). Long-term digital mobile phone use and cognitive decline in the elderly. *Bioelectromagnetics*, *33*(2), 176–185. doi:10.1002/bem.20698

Niantic Labs. (2012). *Ingress*. Google.

Nicholson, J. A. (2005). Flash! mobs in the age of mobile connectivity. *Fibreculture Journal, 6*.

Nicol, D. (2013). *Mobile strategy: How your company can win by embracing mobile technologies*. IBM Press & Pearson Education, Inc.

Nielinger, O. (2006). *Information and Communication Technologies (ICT) for Development in Africa*. Frankfurt, Germany: Peter Lang.

Nielsen, J., & Budiu, R. (2013). *Mobile Usability*. Berkeley, CA: New Riders.

Nie, N. H. (2001). Sociability, interpersonal relations, and the Internet: Reconciling conflicting findings. *The American Behavioral Scientist*, *45*, 420–435. doi:10.1177/00027640121957277

Nip, J. Y. M. (2004). The Queer Sisters and its electronic bulletin board: A study of the Internet for social movement mobilization. *Information Communication and Society*, *7*(1), 23–49. doi:10.1080/13691180420000208889

Nitsche, M. (2008). *Video Game Spaces: Image, Play, and Structure in 3D Worlds*. Cambridge, MA: MIT Press. doi:10.7551/mitpress/9780262141017.001.0001

Norris, P. (2002). *Democratic phoenix*. Cambridge, UK: Cambridge University Press. doi:10.1017/CBO9780511610073

Nyamba, A. (2000). La parole du téléphone: Significations sociales et individuelles du téléphone chez les Sanan du Burkina Faso. In *Enjeux des Technologies de la Communication en Afrique: Du Téléphone à Internet* (pp. 193–210). Paris: Karthala.

Nyíri, K. (2005). *A sense of place: The global and the local in mobile communication*. Vienna, Austria: Passagen Verlag.

Nyíri, K. (Ed.). (2007). *Mobile studies: Paradigms and perspectives*. Vienna, Austria: Passagen Verlag.

Nystedt, D. (2005). *Online gaming growing fast in China, study says*. Retrieved August 6, 2010, from http://www.macworld.com/article/44065/2005/04/chinagaming.html

Nystrom, P. C., Ramamurthy, K., & Wilson, A. L. (2002). Organizational context, climate and innovativeness: Adoption of imaging technology. *Journal of Engineering and Technology Management*, *19*(3/4), 221–247. doi:10.1016/S0923-4748(02)00019-X

Nysveen, H., Pedersen, P. E., Thorbjornsen, H., & Berthon, P. (2005). Mobilizing the brand - The effects of mobile services on brand relationships and main channel use. *Journal of Service Research*, *7*(3), 257–276. doi:10.1177/1094670504271151

O' Riordan, K. (2007). Queer Theories and Cybersubjects: Intersecting Figures. In K. O'Riordan, & D. Phillips (Eds.), *Queer Online* (pp. 13–30). New York: Peter Lang.

O'Brien, D., & Scharber, C. (2010). Teaching old dogs new tricks: The luxury of digital abundance. *Journal of Adolescent & Adult Literacy*, *53*(7).

O'Neil, J. (2008). *Remix Identity: Cultural Mash-Ups and Aesthetic Violence in Digital Media.* Retrieved from mcluhanremix.com

O'Dell, T. (2009). My Soul for a Seat: Commuting and the Routines of Mobility. In R. Wilk, & E. Shove (Eds.), *Time, Consumption and Everyday Life*. Oxford, UK: Berg.

Odigie, V. I., Yusufu, L. M. D., Dawotola, D. A., Ejagwulu, F., Abur, P., & Mai, A. et al. (2012). The mobile phone as a tool in improving cancer care in Nigeria. *Psycho-Oncology*, *21*(3), 332–335. doi:10.1002/pon.1894 PMID:22383275

Ofcom. (2013).*Communications market report: United Kingdom.* London: Ofcom. Retrieved from http://stakeholders.ofcom.org.uk/market-data-research/market-data/communications-market-reports/cmr13/uk/

Ofri, D. (2013, August 14). Why Doctors are Reluctant to Take Responsibility for Rising Medical Costs. *The Atlantic.com.* Retrieved from http://www.theatlantic.com/health/archive/2013/08/why-doctors-are-reluctant-to-take-responsibility-for-rising-medical-costs/278623/

Ohiagu, O. P. (2010). Influence of information & communication technologies on the Nigerian society and culture. In *Indigenous societies and cultural globalization in the 21st century*. Germany: Muller Aktiengesellschaft & Co.

Okazaki, S. (2004). How do Japanese Consumers Perceive Wireless Ads? A Multivariate Analysis. *International Journal of Advertising*, *23*(4), 429–454.

Okazaki, S., & Taylor, C. R. (2008). What is SMS advertising and why do multinationals adopt it? Answers from an empirical study in European markets. *Journal of Business Research*, *61*(1), 4–12. doi:10.1016/j.jbusres.2006.05.003

Oksman, V. (2005). MMS and Its 'Early Adopters' in Finland. In K. Nyíri (Ed.), *A Sense of Place: The Global and the Local in Mobile Communication* (pp. 349–362). Vienna: Passagen Verlag.

Olatokun, M. W., & Bodunwa, I. O. (2006). GSM usage at the University of Ibadan. *The Electronic Library*, *24*(4), 530–547. doi:10.1108/02640470610689214

Orlich, D., Harder, R., Callahan, R., Trevisan, M., & Brown, A. (2007). *Teaching strategies: A guide to effective Instruction* (8th ed.). Houghton Mifflin Company.

Ornebring, H. (2007). Alternate reality gaming and convergence culture: The case of Alias. *International Journal of Cultural Studies*, *10*(4), 445–462. doi:10.1177/1367877907083079

Osborn, M. (2008). Fuelling the flames: Rumour and politics in Kibera. *Journal of Eastern African Studies*, *2*(2), 315–327. doi:10.1080/17531050802094836

Osman, M. A., Talib, A. Z., Sanusi, Z. A., Shiang-Yen, T., & Alwi, A. S. (2012). A Study of the Trend of Smartphone and its Usage Behavior in Malaysia. *International Journal of New Computer Architectures and their Applications*, *2*(1), 274-285.

Oudshoorn, N. (2012). How places matter: Telecare technologies and the changing spatial dimensions of healthcare. *Social Studies of Science*, *42*(1), 121–142. doi:10.1177/0306312711431817 PMID:22530385

Outdoor Industry Association. (2010). *China's outdoor market spawns intense competition*. Retrieved December 16, 2010, from http://www.outdoorindustry.org/news.ceo.php?newsId=13116&newsletterId=158&action=display

Paavilainen & Jouni. (2002). *Mobile Business Strategies*. London: IT Press.

Pahl, K., & Rowsell, J. (2005). *Literacy and Education: Understanding the New Literacy Studies in the Classroom.* Los Angeles, CA: Sage.

Paine, R. (1967). What is gossip about? An alternative hypothesis. *New Series*, *2*(2), 278–285.

Paiz, J. M., Angeli, E., Wagner, J., Lawrick, E., Moore, K., & Anderson, M. … Keck, R. (2013, March 1). *General Format*. Retrieved from http://owl.english.purdue.edu/owl/resource/560/01/

Pajitnov, A. (1984). *Tetris*. Academic Press.

Palloff, R., & Pratt, K. (2005). *Collaborating online: Learning together in community*. Jossey-Bass.

Palmgreen, P., & Rayburn, J. D. (1982). Gratifications sought and media exposure: An expectancy value model. *Communication Research*, *9*(4), 561–580. doi:10.1177/009365082009004004

Palmgreen, P., & Rayburn, J. D. (1985). An expectancy-value approach to media gratifications. In K. E. Rosengren, L. Wenner, & P. Palmgreen (Eds.), *Media gratification research* (pp. 61–72). Beverly Hills, CA: Sage Publications.

Palmier-Claus, J. E., Rogers, A., Ainsworth, J., Machin, M., Barrowclough, C., & Laverty, L. et al. (2013). Integrating mobile-phone based assessment for psychosis into people's everyday lives and clinical care: A qualitative study. *BioMed Central Psychiatry*, *13*, 34. doi:10.1186/1471-244X-13-34 PMID:23343329

Panagakos, A. N., & Horst, H. A. (2006). Return to cyberia: Technology and the social worlds of transnational migrants. *Global Networks*, *6*(2), 109–124. doi:10.1111/j.1471-0374.2006.00136.x

Papacharissi, Z. (Ed.). (2010). *A Networked Self: Identity, Community, and Culture on Social Network Sites*. New York: Routledge.

Paradiso, M. (2013). The role of information and communications technologies in migrants from Tunisia's Jasmine Revolution. *Growth and Change*, *44*(1), 168–182. doi:10.1111/j.1468-2257.2012.00603.x

Paragas, F. (2005). Migrant mobiles: Cellular telephony, transnational spaces, and the Filipino diaspora. In K. Nyiri (Ed.), *A sense of place: The global and the local in mobile communication* (pp. 241–249). Vienna: Die Deutsche Bibliothek.

Paragas, F. (2008). Migrant workers and mobile phones: Technological, temporal, and spatial simultaneity. In R. Ling, & S. W. Campbell (Eds.), *The reconstruction of space and time: Mobile communications practices* (pp. 39–65). New Brunswick, NJ: Transaction Publishers.

Parikka, J., & Suominen, J. (2006). Victorian Snake? Towards a Cultural History of Mobile Gaming and the Experience of Movement. *Game Studies*, *6*(1). Retrieved from http://gamestudies.org/0601/articles/parikka_suominen

Parikka, J., & Suominen, J. (2006). Victorian Snakes? Towards a Cultural History of Mobile Games and the Experience of Movement. *Game Studies*, *6*(1).

Park, H. W. (2002). Examining the determinants of who is hyperlinked to whom: A survey of webmasters in Korea. *First Monday*, *7*(11). doi:10.5210/fm.v7i11.1005

Park, W. (2005). Mobile Phone Addiction. In R. Ling, & P. E. Pederson (Eds.), *Mobile Communications* (Vol. 31, pp. 253–272). London: Springer. doi:10.1007/1-84628-248-9_17

Parreñas, R. (2001). Mothering from a distance: Emotions, gender, and intergenerational relations in Filipino transnational families. *Feminist Studies*, *27*(2), 361–390. doi:10.2307/3178765

Parreñas, R. (2005). Long distance intimacy: Class, gender and intergenerational relations between mothers and children in Filipino transnational families. *Global Networks*, *5*(4), 317–336. doi:10.1111/j.1471-0374.2005.00122.x

Parreñas, R. S. (2001). *Servants of globalization: Women, migration, and domestic work*. Stanford, CA: Stanford University Press.

Parreñas, R. S. (2005a). *Children of global migration: Transnational families and gendered woes*. Stanford, CA: Stanford University Press.

Parreñas, R. S. (2008). Transnational fathering: Gendered conflicts, distant disciplining and emotional gaps. *Journal of Ethnic and Migration Studies*, *34*(7), 1057–1072. doi:10.1080/13691830802230356

Patrick, K., Marshall, S. J., Davila, E. P., Kolodziejczyk, J. K., Fowler, J. H., & Calfas, K. J. et al. (2013, November 9). Design and implementation of a randomized controlled social and mobile weight loss trial for young adults (project SMART). *Contemporary Clinical Trials*, *37*, 10–18. doi:10.1016/j.cct.2013.11.001 PMID:24215774

Pearce, C. (2006). Productive play: Game culture from the bottom up. *Games and Culture*, *1*(1), 17–24. doi:10.1177/1555412005281418

Penman, R., & Turnbull, S. (2007). Media Literacy—Concepts, Research and Regulatory Issues. Canberra, Australia: Australian Communications and Media Authority (Australian Government).

Perez, S. (2012, July 2). *comScore: In U.S. Mobile Market, Samsung, Android Top The Charts, Apps Overtake Web Browsing*. Retrieved from http://www.techcrunch.com

Pernia, E. M., Pernia, E. E., Ubias, J. L., & San Pascual, M. R. S. (2013). *International Migration, Remittances, and Economic Development in the Philippines*. National Research Council of the Philippines. Unpublished.

Perse, E. M., & Dunn, D. G. (1998). The utility of home computers and media use: Implications of multimedia and connectivity. *Journal of Broadcasting & Electronic Media*, *42*(4), 435–456. doi:10.1080/08838159809364461

Perse, E. M., & Ferguson, D. A. (1994). The impact of the newer television technologies on television satisfaction. *The Journalism Quarterly*, *70*(4), 843–853. doi:10.1177/107769909307000410

Pertierra, R. Ugarte, E., Pingol, A., et al. (2012). Txt-ing selves: Cellphones and Philippine modernity. Manila: De La Salle University Press.

Pertierra, R. (2005). Mobile phones, identity and discursive intimacy. *Human Technology*, *1*(1), 23–44.

Pertierra, R. (2006). *Transforming Technologies, Altered Selves. Mobile Phone and Internet Use in the Philippines*. Manila: DLSU Press.

Philippine LGBT Crime Watch. (2012). *A Database of Killed Lesbian, Gay, Bisexual and Transgendered Filipinos*. Retrieved from http://thephilippinelgbthate-crimewatch.blogspot.com/

Philippine Overseas Employment Administration. (2006). *OFW global presence: A compendium of overseas employment statistics 2006*. Philippine Overseas Employment Administration.

Philippine Overseas Employment Administration. (2011). *2007-2011 Overseas employment statistics*. Philippine Overseas Employment Administration.

Philippine Overseas Employment Administration. (2012). *2008-2012 Overseas employment statistics*. Philippine Overseas Employment Administration.

Philippine Overseas Employment Agency (POEA). (2012). *OFW Deployment Statistics*. Retrieved from http://www.poea.gov.ph

Piotrowski, C., & Armstrong, T. R. (1998). Mass media preference in disaster: A study of Hurricane Danny. *Social Behavior & Personality: An International Journal*, *26*(4), 341–346. doi:10.2224/sbp.1998.26.4.341

Plant, S. (2001). *On the mobile: The effects of mobile telephones on social and individual life*. Motorola.

Pollara, P. (2011). *Mobile learning in higher education: A glimpse and a comparison of student and faculty readiness, attitudes and perceptions*. (Unpublished doctoral dissertation). Louisiana State University, Baton Rouge, LA.

Prahalad, C. K., & Ramaswamy, V. (2003). The new frontier of experience innovation. *MIT Sloan Management Review*, *44*(4), 12–18.

Prahalad, C. K., & Ramaswamy, V. (2004). Co-creation experiences: The next practice in value creation. *Journal of Interactive Marketing*, *18*(3), 5–14. doi:10.1002/dir.20015

Prahalad, C. K., & Ramaswamy, V. (2005). *The future of competition: Co-creating unique value with customers*. Boston: Harvard Business School Press.

Prensky, M. (2001a). Digital Natives, Digital Immigrants. *Horizon*, *9*(5). doi:10.1108/10748120110424816

Prensky, M. (2001b). Digital Natives, Digital Immigrants, Part II: Do They Really Think Differently? *Horizon*, *9*(6). doi:10.1108/10748120110424843

PricewaterhouseCoopers. (2012). *The evolution of video gaming and content consumption*. Pricewaterhouse Coopers.

PricewaterhouseCoopers. (2012, June 7). *Emerging mHealth: Paths for growth*. Retrieved from http://www.pwc.com/us/en/press-releases/2012/consumers-are-ready-to-adopt-mobile-health.jhtml

PricewaterhouseCoopers. (2013). *mHealth implementation in emerging markets: PwC*. Retrieved from http://www.pwc.com/gx/en/healthcare/mhealth/opportunities-emerging-markets.jhtml

Proudfoot, J., Clarke, J., Birch, M. R., Whitton, A. E., Parker, G., & Manicavasagar, V. et al. (2013, November 15). Impact of a mobile phone and web program on symptom and functional outcomes for people with mild-to-moderate depression, anxiety and stress: A randomised controlled trial. *BioMed Central Psychiatry*, *13*, 312. doi:10.1186/1471-244X-13-312 PMID:24237617

Pullen, C. (2010). The Murder of Lawrence King and LGBT Online Stimulations of Narrative Copresence. In C. Pullen, & M. Cooper (Eds.), *LGBT Identity and Online New Media* (pp. 17–36). New York: Routledge.

Puro, J. P. (2002). Finland: A mobile culture. In J. Katz, & M. Aakhus (Eds.), *Perpetual contact: Mobile communication, private talk, public performance* (pp. 19–29). Cambridge, UK: Cambridge University Press.

Pyramid Research. (2010). *The Impact of Mobile Services in Nigeria: How Mobile Technologies are Transforming Economic and Social Activities*. Author.

Qardio. (2013). *QardioCore*. Retrieved from http://www.getqardio.com

Qiu, J.L. (2008). Mobile civil society in Asia: A comparative study of People Power II and the Nosamo Movement. *Javnost – The Public, 15*(3), 39-58.

Qiu, J. L. (2007). The wireless leash: Mobile messaging service as a means of control. *International Journal of Communication, 1*(1), 74–91.

Quindoza-Santiago, L. (2010). Roots of Feminist Thought in the Philippines. In *More Pinay Than We Admit: The Social Construction of the Filipina* (pp. 105–119). Manila: Vibal.

Quinn, C. (2012). *The mobile academy mlearning for higher education*. Jossey-Bass.

Racadio, R., Rose, E., & Boyd, S. (2012). Designing and evaluating the mobile experience through iterative field studies. In *Proceedings of the 30th ACM international conference on Design of communication* (SIGDOC '12). ACM. DOI: 10.1145/2379057.2379095

Radbourne, J., Johanson, K., & Glow, H. (2010). Empowering Audiences to measure Quality. *Participations: International Journal of Audience Research, 7*(2).

Rahmati, A., & Zhong, L. (2013). Studying Smartphone Usage: Lessons from a Four-Month Field Study. *IEEE Transactions on Mobile Computing, 12*(7), 1417–1427. doi:10.1109/TMC.2012.127

Rainie, L., & Wellman, B. (2012). *Networked: The new social operating system*. Cambridge, MA: MIT Press.

Rakow, L. F. (1992). *Gender on the line*. Urbana, IL: University of Illinois Press.

Rakow, L., & Navarro, V. (1993). Remote mothering and the parallel shift: Women meet the cellular telephone. *Critical Studies in Mass Communication, 10*(2), 144–157. doi:10.1080/15295039309366856

Ramaprasad, U., & Robert, H. (2003). Location-Based Services: Models for Strategy Development in M-Commerce. In Proceedings of Management of Engineering and Technology (pp. 416-424). IEEE.

Ramaprasad, U., & Robert, H. (2007). Perceived Effectiveness of Push vs. Pull Mobile Location Based Advertising. *Journal of Interactive Advertising, 7*(2), 28–40. doi:10.1080/15252019.2007.10722129

RandomAlphabets.com. (2010, May 15). *Glee Flashmob Dance 2010*. Retrieved September 2, 2013, from http://randomalphabets.com/2010/05/glee-flashmob-dance-2010/

Rao. (2011). *Discovering the distance to discount ratio*. Retrieved on July 8, 2013 from http://techcrunch.com/2011/02/04/distance-discount-ratio/

Rao, B., & Minakakis, L. (2003). Evolution of mobile location-based services. *Communications of the ACM, 46*(12), 61–65. doi:10.1145/953460.953490

Raper, J., Gartner, G., Karimi, H., & Rizos, C. (2007). Applications of location-based services: A selected review. *Journal of Location Based Services, 1*(2), 89–111. doi:10.1080/17489720701862184

Rayburn, J. D., & Palmgreen, P. (1984). Merging uses and gratifications and expectancy-value theory. *Communication Research, 11*(4), 537–562. doi:10.1177/009365084011004005

Reason, M. (2004). Theatre Audiences and Perceptions of Liveness in Performance. *Participations: International Journal of Audience Research, 1*(2).

Regalado, A. (2013, August 18). Esther Dyson: We Need to Fix Health Behavior. *MIT Technology Review*. Retrieved from http://www.technologyreview.com/news/518901/esther-dyson-we-need-to-fix-health-behavior/

Reiser, R., & Dempsey, J. (2007). *Trends and issues in instructional design and technology* (2nd ed.). Pearson, Merrill Prentice Hall.

Rheingold, H. (2002). *Smart mobs*. New York: Basic Books.

Rheingold, H. (2002). *Smart Mobs: The Next Social Revolution*. Cambridge, MA: Perseus Books.

Rice, R. E., & Rogers, E. M. (1980). Reinvention in the innovation process. *Science Communication*, *1*(4), 499–514. doi:10.1177/107554708000100402

Richardson, I. (2007). Pocket technospaces: The bodily incorporation of mobile media. *Continuum: Journal of Media & Cultural Studies*, *21*(2), 205–215. doi:10.1080/10304310701269057

Richardson, I. (2010). Ludic Mobilities: The Corporealities of Mobile Gaming. *Mobilities*, *5*(4), 431–447. doi:10.1080/17450101.2010.510329

Rivera, M., Walton, M., & Sreekumar, T. T. (2012). *ICTs, Gender, and Leisure: Soft Forms of Subversion*. Paper presented at the International Association of Media and Communications Research. Durban, South Africa.

Roberts, K. (2005). *Lovemarks*. New York: PowerHouse Books.

Robins, K., & Webster, F. (1999). *Times of the Technoculture*. London: Routledge.

Rodan, G. (1998). The Internet and Political Control in Singapore. *Political Science Quarterly*, *113*(1), 63–89. doi:10.2307/2657651

Rogers, E. M. (1986). *Communication technology: The new media in society*. New York: Free Press.

Rogers, E. M. (2003). *Diffusion of innovations* (5th ed.). New York: Free Press.

Roglic, G., Unwin, N., Bennett, P. H., Mathers, C., Tuomilehto, J., & Nag, S. et al. (2005). The burden of mortality attributable to diabetes: Realistic estimates for the year 2000. *Diabetes Care*, *28*(9), 2130–2135. doi:10.2337/diacare.28.9.2130 PMID:16123478

Rojas, H., & Puig-i-Abril, E. (2009). Mobilizers mobilized: Information, expression, mobilization and participation in the digital age. *Journal of Computer-Mediated Communication*, *14*, 902–927. doi:10.1111/j.1083-6101.2009.01475.x

Rook, D. (1985). The Ritual Dimension of Consumer Behavior. *The Journal of Consumer Research*, *12*, 251–264. doi:10.1086/208514

Rosenberg, M. J. (2000). *E-Learning: Strategies for delivering knowledge in the digital age*. McGraw-Hill.

Roto, V., Väätäjä, H., Jumisko-Pyykkö, S., & Väänänen-Vainio-Mattila, K. (2011). Best Practices for Capturing Context in User Experience Studies in the Wild. In *Proceedings of MindTrek 2011*. ACM. doi:10.1145/2181037.2181054

Rozario, K. (2013, April 14). *BN Spam SMS: Why it's such a bad idea*. Retrieved September 4, 2013, from http://www.keithrozario.com/2013/04/barisan-nasional-bn-spam-sms.html

Rubin, A. M. (1983). Television uses and gratifications: The interactions of viewing patterns and motivations. *Journal of Broadcasting*, *27*(1), 37–51. doi:10.1080/08838158309386471

Rule, J. B. (2002). From mass society to perpetual contact: Models of communication technologies in social context. In J. E. Katz, & M. Aakhus (Eds.), *Perpetual Contact: Mobile Communication, Private Talk, Public Performance* (pp. 242–254). Cambridge, UK: Cambridge University Press.

Ryan, C., & Gonsalves, A. (2005). *The effect of context and application type on mobile usability: An empirical study TomiAhonen Almanac*. Retrieved from http://communities-dominate.blogs.com/brands/2013/03/the-annual-mobile-industry-numbers-and-stats-blog-yep-this-year-we-will-hit-the-mobile-moment.html

Sabq. (2013). *Launching the iPad project (Limitless Knowledge) at Saudi Schools*. Retrieved from http://www.sabq.org/AcCfde

Salen, K., & Zimmerman, E. (2004). *Rules of Play: Game Design Fundamentals*. Cambridge, MA: MIT Press.

SamMobile. (2013). *Apple losing market share among non-English speakers (Apple vs. Samsung)*. Retrieved November 15, 2013, from http://www.sammobile.com/2013/02/07/apple-losing-market-share-among-non-english-speakers-apple-vs-samsung/

San Pascual, M. R. S. (2012). *Communicated Parenting: Singapore-Based Filipino Migrant Mothers and Their Long-Distance Parenting of Their Teenage Children in the Philippines*. (Unpublished Master's Thesis). National University of Singapore, Singapore.

Sarrab, M., & Elgamel, L. (2013). Contextual m-learning system for higher education providers in Oman. *World Applied Sciences Journal, 22*(10), 1412–1419.

Satchell, C., & Graham, C. (2010). Conveying Identity with Mobile Content. *Personal and Ubiquitous Computing, 14*(3), 251–259. doi:10.1007/s00779-009-0254-3

Satchell, C., Shanks, G., Howard, S., & Murphy, J. (2011). Identity Crisis: User Perspectives on Multiplicity and Control in Federated Identity Management. *Behaviour & Information Technology, 30*(1), 51–62. doi:10.1080/01449290801987292

Saudi Arabia Consumer Electronic Report Q3. (2013). Retrieved from http://www.marketresearch.com/Business-Monitor-International-v304/Saudi-Arabia-Consumer-Electronics-Q3-7691595/

Sawsaa, A., Lu, J., & Meng, Z. (2012). Using an application of mobile and wireless technology in Arabic learning system. In Z. J. Lu (Ed.), *Learning with Mobile Technologies, Handheld Devices, and Smart Phones* (pp. 171–186). Hershey, PA: IGI Global. doi:10.4018/978-1-4666-0936-5.ch011

Scharl, A., Dickinger, A., & Murphy, J. (2005). Diffusion and success factors of mobile marketing. *Electronic Commerce Research and Applications, 4*, 159–173. doi:10.1016/j.elerap.2004.10.006

Schawbel, D. (2009). *Me 2.0*. New York: Kaplan.

Schlozman, K. L. (2002). Citizen participation in America: What do we know? Why do we care? In I. Katznelson, & H. V. Milner (Eds.), *Political science: The state of the discipline* (pp. 433–461). New York: W. W. Norton & Company, Inc.

Schrøder, K. C., & Steeg Larsen, B. (2010). The shifting cross-media news landscape. *Journalism Studies, 11*(4), 524–534. doi:10.1080/14616701003638392

Schuman, A. J. (2013). Improving patient care: Smartphones and mobile medical devices. *Contemporary Pediatrics, 30*(6), 33.

Schumate, M., & Lipp, J. (2007, May). *Connective collective action online: An examination of the network structure of the English speaking Islamic Resistance Movement*. Paper presented at International Communication Association Conference. San Francisco, CA.

Scifo, B. (2005). The Domestication of Camera-Phone and MMS Communication: Early Experiences of Young Italians. In K. Nyìri (Ed.), *The Global and the Local in Mobile Communication* (pp. 363–373). Wien: Passagen Verlag.

Scott, R. E., Mars, M., & Jordanova, M. (2013). Would a Rose By Any Other Name - Cause Such Confusion?. *Journal of the International Study for Telemedicine and eHealth, 1*(2).

Sejdić, E., Rothfuss, M. A., Stachel, J. R., Franconi, N. G., Bocan, K., Lovell, M. R., & Mickle, M. H. (2013). Innovation and translation efforts in wireless medical connectivity, telemedicine and eMedicine: A story from the RFID Center of Excellence at the University of Pittsburgh. *Annals of Biomedical Engineering, 41*(9), 1913–1925. doi:10.1007/s10439-013-0873-8 PMID:23897048

Seliaman, M. E., & Al-Turki, M. S. (2012). Mobile learning adoption in Saudi Arabia. *World Academy of Science, Engineering, and Technology, 69*, 391–393.

Shadbolt, N. (2008). A Crisis of Identity. *Engineering & Technology, 7*(6), 20. doi: doi:10.1049/et:20081000

Sha, L., Looi, C. K., Chen, W., Seow, P., & Wong, L. H. (2012). Recognizing and measuring self-regulated learning in a mobile learning environment. *Computers in Human Behavior, 28*(2), 718–728. doi:10.1016/j.chb.2011.11.019

Sha, L., Looi, C. K., Chen, W., & Zhang, B. H. (2012). Understanding mobile learning from the perspective of self-regulated learning. *Journal of Computer Assisted Learning, 28*(4), 366–378. doi:10.1111/j.1365-2729.2011.00461.x

Shand, F. L., Ridani, R., Tighe, J., & Christensen, H. (2013). The effectiveness of a suicide prevention app. for indigenous Australian youths: study protocol for a randomized controlled trial. *Trials*, *14*, 396. doi:10.1186/1745-6215-14-396 PMID:24257410

Shapiro, C., & Varian, H. R. (1999). *Information rules: A strategic guide to the network economy.* Boston: Harvard Business School Press.

Shavelson, R. (1981). *Statistical reasoning for the behavioral sciences* (3rd ed.). Allyn and Bacon.

Sheth, J. N., & Uslay, C. (2007). Implications of the revised definition of marketing: From exchange to value creation. *Journal of Public Policy & Marketing*, *26*(2), 302–307. doi:10.1509/jppm.26.2.302

Shiode, N., Li, C., Batty, M., Longley, P., & Maguire, D. (2002). *The Impact and Penetration of Location-Based Services*. London: University College London.

Shirky, C. (2008). *Here Comes Everybody: The Power of Organizing Without Organizations*. New York: The Penguin Press.

Siapera, E. (2012). *Understanding New Media*. London: Sage.

Sifferlin, A. (2013, March 14). South By Southwest (SXSW), Will Collecting Data on Your Body Make You Healthier? *TIME.com.* Retrieved from http://healthland.time.com/2013/03/14/south-by-southwest-sxsw-will-collecting-data-on-your-body-make-you-healthier/

Silverstone, R., & Hirsch, E. (Eds.). (1992). *Consuming Technologies: Media and Information in Domestic Spaces*. London: Routledge. doi:10.4324/9780203401491

Silverstone, R., Hirsch, E., & Morley, D. (1992). Information and communication technologies and the moral economy of the household. In R. Silverstone, & E. Hirsch (Eds.), *Consuming Technologies: Media and Information in Domestic Space* (pp. 15–31). London: Routledge. doi:10.4324/9780203401491_chapter_1

Sina. (2003). *China Unicom to start new business.* Retrieved on July 1, 2013 from http://tech.sina.com.cn/it/t/2003-08-14/1420221020.shtml

Sinclair, I. (1997). *Lights out for the territory*. London: Penguin Books.

Singh, K. (2012, May 15). *Interesting insights into smartphone behavior.* Retrieved September 27, 2013, from http://www.digitalnewsasia.com/node/107

Singtel. (2013). *Locator Plus.* Retrieved September 5, 2013, from http://info.singtel.com/personal/phones-plans/mobile/vas/locator-plus/detail

Skuse, A., & Cousins, T. (2008). Getting connected: the social dynamics of urban telecommunications access and use in Khayelitsha, Cape Town. *New Media & Society*, *10*(1), 9–26. doi:10.1177/1461444807085319

Slabodkin, G. (2012, December 11). Wearable mHealth device shipments to hit 30 million by year's end. *FierceMobileHealthcare.com.* Retrieved from http://www.fiercemobilehealthcare.com/story/wearable-mhealth-device-shipments-hit-30-million-years-end/2012-12-11

Smala, S., & Al-Shehri, S. (2013). Privacy and identity management in social media: Driving factors for identity hiding. In J. Keengwe (Ed.), *Research Perspectives and Best Practices in Educational Technology Integration* (pp. 304–320). Hershey, PA: IGI Global.

Smala, S., & Al-Shehri, S. (2013). Privacy and Identity Management in Social Media: Driving Factors for Identity Hiding. In J. Keengwe (Ed.), *Research Perspectives and Best Practices in Educational Technology Integration* (pp. 304–320). Hershey, PA: IGI Global.

Smith, A. (2012). *17% of cell phone owners do most of their online browsing on their phone, rather than a computer or other device.* Retrieved September 5, 2013, from http://pewinternet.org/Reports/2012/Cell-Internet-Use-2012/Key-Findings.aspx

Smith, A. (2013). *Smartphone Ownership - 2013 Update.* Retrieved September 5, 2013, from http://pewinternet.org/Reports/2013/Smartphone-Ownership-2013/Findings.aspx

Smith, D. J. (2006). Cell Phones, Social Inequality, and Contemporary Culture in Nigeria. *Canadian Journal of African Studies*, *40*(3), 96–523.

Smith, M. (1995). *Engaging Characters: Fiction, Emotion and the Cinema*. Oxford, UK: Clarendon Press.

Snyder, I. (2008). *The Literacy Wars: Why teaching children to read and write is a battleground in Australia. Crow's Nest.* Australia: Allen & Unwin.

Sobritchea, C. I. (2007). Constructions of mothering: The experience of female Filipino overseas workers. In T. W. Devasahayam, & B. S. A. Yeoh (Eds.), *Working and mothering in Asia: Images, ideologies and identities* (pp. 173–194). Singapore: NUS Press.

So, H. J., Tan, E., & Tay, J. (2012). Collaborative mobile learning in situ from knowledge building perspectives. *Asia-Pacific Education Researcher, 21*(1), 51–62.

Soh, J., & Tan, B. (2008). Mobile gaming. *Communications of the ACM, 51*(3), 35–39. doi:10.1145/1325555.1325563

Solove, D. J. (2007). *The Future of Reputation: Gossip, Rumor, and Privacy on the Internet.* New Haven, CT: Yale University Press.

Soon, C., & Cho, H. (forthcoming). OMGs! Offline-based movement organizations, online-based movement organizations and network mobilization: A case study of political bloggers in Singapore. *Information Communication and Society.*

Soon, C., & Kluver, R. (forthcoming). Uniting political bloggers in diversity: Collective identity and web activism. *Journal of Computer-Mediated Communication.*

Soopramanien, D. G. R., & Robertson, A. (2007). Adoption and usage of online shopping: An empirical analysis of the characteristics of buyers? browsers? and non-Internet shoppers? *Journal of Retailing and Consumer Services, 14*(1), 73–82. doi:10.1016/j.jretconser.2006.04.002

Soriano, C. (2014). Constructing Collectivity in Diversity: Online Political Mobilization of a National LGBT Political Party. *Media Culture & Society, 36*(1), 20–36. doi:10.1177/0163443713507812

Soriano, C., Lim, S., & Rivera, M. (forthcoming). The Virgin Mother with a Mobile Phone: Ideologies of mothering and technology consumption in Philippine television advertisements. *Communication, Culture & Critique.*

Sorrell, T., & Draper, H. (2012, Aug 10). Telecare, Surveillance, and the Welfare State. *The American Journal of Bioethics, 12*(9), 36–44. doi:10.1080/15265161.2012.699137 PMID:22881854

Southern Daily. (2011). *Digu to cut off its 70% staff.* Retrieved on July 2, 2013 from http://epaper.nfdaily.cn/html/2011-10/19/content_7016439.htm

Spiekermann, S. (2004). General Aspects of Location-Based Services. In J. Schiller, & A. Voisard (Eds.), *Location-Based Services* (pp. 9–26). San Francisco, CA: Morgan Kaufmann Publishers. doi:10.1016/B978-155860929-7/50002-9

Spradley, J. P. (1979). *The Ethnographic Interview.* New York: Holt, Rinehart & Winston.

Sprenger, P. (1999). *Sun on Privacy: 'Get Over It!'.* Retrieved September 2, 2013, from http://www.wired.com/politics/law/news/1999/01/17538

Spyridonis, F., Ghinea, G., & Frank, A. O. (2013, April 10). Attitudes of Patients Toward Adoption of 3D Technology in Pain Assessment: Qualitative Perspective. *Journal of Medical Internet Research, 15*(4), e55. doi:10.2196/jmir.2427 PMID:23575479

Sreekumar, T. T. (2011). Mobile Phones and the Cultural Ecology of Fishing in Kerala, India. *The Information Society, 27*, 172–180. doi:10.1080/01972243.2011.566756

Stald, G., & Ólafsson, K. (2012). Mobile access – Different users, different risks, different consequences? In S. Livingstone, L. Haddon, & A. Görzig (Eds.), *Children, risk and safety online: Research and policy challenges in comparative perspective* (pp. 285–295). Bristol, MA: Policy Press. doi:10.1332/policypress/9781847428837.003.0022

Stattin, H., & Kerr, M. (2000). Parental Monitoring: A Reinterpretation. *Child Development, 71*(4), 1072–1085. doi:10.1111/1467-8624.00210 PMID:11016567

Stein, L. (2007, May). *National social movement organizations and the World Wide Web: A survey of web-based activities and attributes.* Paper presented at International Communication Association Conference. San Francisco, CA.

Steinbock, D. (2005). *The mobile revolution: The making of mobile services worldwide.* London: Kogan Page.

Stevens, Q. (2007). *The Ludic City.* London: Routledge.

Stigler, G. J. (1970). The Case, If Any, for Economic Literacy. *The Journal of Economic Education, 1*(2), 77–84. doi:10.1080/00220485.1970.10845301

Stockwell, G. (2008). Investigating learner preparedness for and usage patterns of mobile learning. *ReCALL, 20*(3), 253–270. doi:10.1017/S0958344008000232

Strathern, M. (1992). Foreword: The mirror of technology. In *Consuming Technologies*. London: Routledge.

Street, B. V. (1984). *Literacy in theory and practice*. Cambridge, UK: Cambridge University Press.

Student Monitor. (2005a). *College student readers of national newspapers gain*. Retrieved from http://www.studentmonitor.com/press/05.pdf

Student Monitor. (2005b). *Study finds record number of student cell phone owners*. Retrieved from http://www.studentmonitor.com/press/02.pdf

Study on Mobile Identity Management. (2005). In G. Müller & S. Wohlgemuth (Eds.), *Future of Identity in the Information Society by WP3, Albert-Ludwigs-Universität Freiburg*. Retrieved from http://www.fidis.net/fileadmin/fidis/deliverables/fidis-wp3-del3.3.study_on_mobile_identity_management.pdf

Swanson, D. L. (1979). Political communication research and the uses and gratifications model: A critique. *Communication Research, 6*(1), 37–53. doi:10.1177/009365027900600103

Symes, C. (2007). Coaching and Training: An Ethnography of Student Commuting on Sydney's Suburban Trains. *Mobilities, 2*(3), 443–461. doi:10.1080/17450100701597434

Tadeo-Pingol, A. (1999). Absentee Wives and Househusbands: Power, Identity & Family Dynamics. *Review of Women Studies, 9*(1-2).

Takahashi, D. (2013, August 16). With a mobile boom, learning games are a $1.5B market headed toward $2.3B by 2017 (exclusive). *VentureBeat*. Retrieved from http://venturebeat.com/2013/08/16/with-a-mobile-boom-learning-games-are-a-1-5b-market-headed-toward-2-3b-by-2017-exclusive

Takahashi, K., & Hatano, G. (1999). Recent Trends in Civic Engagement among Japanese Youth. In *Roots of civic identity: International perspectives on community service and activism in youth*. Academic Press.

Tan, W. (2008, August 9). Rise of online activists. *The Straits Times*, p. B11.

Tang, K., Lin, J., Hong, J., Siewiorek, D., & Sadeh, N. (2010). Rethinking location sharing: Exploring the implications of social-driven vs purpose-driven location sharing.[Copenhagen, Denmark: ACM Press.]. *Proceedings of UbiComp, 2010*, 85–94. doi:10.1145/1864349.1864363

Tan, K. P. (2007). New politics for a renaissance city? In K. P. Tan (Ed.), *Renaissance Singapore?* (pp. 17–36). Singapore: NUS Press.

Tarrow, S. (1998). *Power in movement: social movements and contentious politics*. New York, NY: Cambridge University Press. doi:10.1017/CBO9780511813245

Tassi, P. (2012). Is the Playstation Vita a Dinosaur Already? *Forbes*. Retrieved 27th July, 2013, from http://www.forbes.com/sites/insertcoin/2012/02/17/is-the-playstation-vita-a-dinosaur-already/

Taylor, C. A. (2011). *The Mobile Literacy Practices of Adolescents: An Ethnographic Study*. (Doctoral dissertation). Retrieved from Monash University Research Repository. (Identifier: ethesis-20110810-160841)

Taylor, C. (2009a). Choice, Coverage and Cost in the Countryside: A topology of adolescent rural mobile technology use. *Education in Rural Australia, 19*(1), 53–64.

Taylor, C. (2009b). Pre-paid literacy: Negotiating the cost of adolescent mobile technology use. *Engineers Australia, 44*(2), 26–34.

Telecommunication Development Bureau. (2013). *The World in 2013 - ICT Facts and Figures*. Retrieved September 5, 2013, from http://www.itu.int/en/ITU-D/Statistics/Pages/default.aspx

Telecompaper. (2012). *Egypt passes 100% mobile penetration*. Retrieved from http://www.telecompaper.com/news/egypt-passes-100-mobile-penetration--853147

Tencent. (2010). *Digu to transform from Weibo to LBS*. Retrieved on May 5, 2013 from http://tech.qq.com/a/20101104/000171.htm

Tenheunen, S. (2008). Mobile technology in the village: ICTs, culture, and social logistics in India. *The Journal of the Royal Anthropological Institute, 14*, 515–534. doi:10.1111/j.1467-9655.2008.00515.x

Tenhunen, S. (2011). Culture, conflict, and translocal communication: Mobile technology and politics in Rural West Bengal, India. *Ethnos*, *76*(3), 398–420. doi:10.1080/00141844.2011.580356

Teri, R. (2000). *Wireless Marketing is about Location, Location, Location*. Retrieved on May 14, 2013 from http://www.dimensiondata.com/Global/Downloadable%20Documents/Location%20Location%20Location%20-%20Using%20Wireless%20to%20Deliver%20Process%20Improvement%20and%20Innovation%20Opinion%20Piece.pdf

Thompson, B. (2013, August 27). Set the FDA mobile medical app. guidance free! *Mobihealth News*. Retrieved from http://mobihealthnews.com/25040/set-the-fda-mobile-medical-app-guidance-free/

Tilly, C. (2004). *Social movements, 1768-2004*. Boulder, CO: Paradigm Publishers.

Tisch School of the Arts. (2004). *Pac-Manhattan*. New York: Tisch School of the Arts.

Topol, E. (2012). *The Creative Destruction of Medicine: How the Digital Revolution Will Create Better Health Care*. New York: Basic Books.

Tornatzky, L. G., & Klein, K. J. (1982). Innovation characteristics and innovation adoption–implementation: A meta-analysis of findings. *IEEE Transactions on Engineering Management*, *29*(1), 28–45. doi:10.1109/TEM.1982.6447463

Totilo, S. (2012). PlayStation Vita Review. *Techspot*. Retrieved 27th June, 2013, from http://www.techspot.com/review/500-playstation-vita/

Touré, H. I. (2008). Welcome address. In *ICTs in Africa: A Continent on the Move*. ITU TELECOM Africa.

Traxler, J. (2009). Learning in a Mobile Age. *International Journal of Mobile and Blended Learning*, *1*(1), 1–12. doi:10.4018/jmbl.2009010101

Trinklein, E., & Parker, G. (2013). *Combining multiple GPS receivers to enhance relative distance measurements*. Paper presented at the Sensors Applications Symposium. Galveston, TX.

Tromski, D., & Doston, G. (2003). Interactive Drama: A Method for Experiential Multicultural Training. *Journal of Multicultural Counseling and Development*, *31*(1), 52–62. doi:10.1002/j.2161-1912.2003.tb00531.x

Tufekci, Z. (2008). Can You See Me Now? Audience and Disclosure Regulation in Online Social Network Sites. *Bulletin of Science, Technology & Society*, *28*(1), 20–36. doi:10.1177/0270467607311484

Turkle, S. (2011). *Alone together: Why we expect more from technology and less from each other*. New York: Basic Books.

Turner-McGrievy, G., & Tate, D. (2011, December 20). Tweets, Apps, and Pods: Results of the 6-month Mobile Pounds Off Digitally (Mobile POD) Randomized Weight-Loss Intervention Among Adults. *Journal of Medical Internet Research*. doi:10.2196/jmir.1841 PMID:22186428

Tuters, M. (2012). From mannerist situationism to situated media. *Convergence: The International Journal of Research into New Media Technologies*, *18*(3), 267–282. doi:10.1177/1354856512441149

Tybout, A. M., Sternthal, B., Malaviya, P., Bakamitsos, G. A., & Park, S.-B. (2005). Information accessibility as a moderator of judgments: The role of content versus retrieval ease. *The Journal of Consumer Research*, *32*(1), 76–85. doi:10.1086/426617

U.S. Food and Drug Administration. (2012, January 24). *Medical Devices: Premarket Approval (PMA)*. Retrieved from http://www.fda.gov/medicaldevices/deviceregulationandguidance/howtomarketyourdevice/premarketsubmissions/premarketapprovalpma/

U.S. Food and Drug Administration. (2013a, February 8). *Medical Devices: Is the Product a Medical Device?* Retrieved from http://www.fda.gov/medicaldevices/deviceregulationandguidance/overview/classifyyourdevice/ucm051512.htm

U.S. Food and Drug Administration. (2013b, June 6). *Mobile Medical Applications: Examples of MMAs the FDA Has Cleared or Approved*. Retrieved from http://www.fda.gov/MedicalDevices/ProductsandMedicalProcedures/ConnectedHealth/MobileMedicalApplications/ucm368784.htm

U.S. Food and Drug Administration. (2013c, August 13). *Medical Devices: Connected Health*. Retrieved from http://www.fda.gov/MedicalDevices/ProductsandMedicalProcedures/ConnectedHealth/default.htm

U.S. Food and Drug Administration. (2013d, August 13). *Medical Devices: Wireless Medical Telemetry Systems*. Retrieved from http://www.fda.gov/MedicalDevices/ProductsandMedicalProcedures/ConnectedHealth/WirelessMedicalDevices/ucm364308.htm

United Nations, Department of Economics and Social Affairs, Population Division. (2013). *UN Global Migration Statistics 2013*. Retrieved on 13 September 2013 from http://esa.un.org/unmigration/wallchart2013.htm

United States Postal Service. (2013). *Track & Confirm*. Retrieved August 26, 2013, from https://tools.usps.com/go/TrackConfirmAction!input.action

UNSTATS. (2008). *Millennium Development Goals Indicators*. Retrieved from http://mdgs.un.org/unsd/mdg/Data.aspx

Urdal, H. (2006). A clash of generations? Youth Bulges and Political Violence. *International Studies Quarterly*, *50*, 607–629. doi:10.1111/j.1468-2478.2006.00416.x

Urian, D. (1998). On Being an Audience: A Spectator's Guide. In *On the Subject of Drama* (pp. 133–150). (N. Paz, Trans.). London: Routledge.

Uy-Tioco, C. (2007). Overseas Filipino workers and text messaging: Reinventing transnational mothering. *Continuum: Journal of Media & Cultural Studies*, *21*(2), 253–265. doi:10.1080/10304310701269081

Valenza, J. K., & Johnson, D. (2009, October). Things that keep us up at night. *School Library Journal*.

Value of Our Identity. (2012). *Liberty Global, Inc. with permission of The Boston Consulting Group, Inc*. Retrieved from http://www.libertyglobal.com/PDF/public-policy/The-Value-of-Our-Digital-Identity.pdf

van Dijck, J. (2009). Users like you? Theorizing agency in user-generated content. *Media Culture & Society*, *31*(1), 41–58. doi:10.1177/0163443708098245

van House, N., & Davis, M. (2005). The Social Life of Cameraphone Images. In *Proceedings of Workshop on Pervasive Image, Capturing and Sharing: New Social Practices and Implications for Technology Workshop (PICS 2005) at the Seventh International Conference on Ubiquitous Computing (UbiComp 2005)*. Tokyo, Japan: UbiComp. Retrieved from http://people.ischool.berkeley.edu/~vanhouse/Van%20House,%20Davis%20-%20The%20Social%20Life%20of%20Cameraphone%20Images.pdf

Van Laer, J., & Van Aelst, P. (2010). Internet and social movement action repertoires. *Information Communication and Society*, *13*(8), 1146–1171. doi:10.1080/13691181003628307

Venkatesh, A. (1999). Postmodern perspectives for macromarketing: An inquiry into the global information and sign economy. *Journal of Macromarketing*, *19*(2), 153–169. doi:10.1177/0276146799192006

Verve. (2012). *Location Powered Mobile Advertising Report 2012 Annual Review*. Retrieved on June 15, 2013 from http://www.vervemobile.com/pdfs/LIR/LIR_web.pdf

Villi, M. (2010). *Visual Mobile Communication: Camera Phone Photo Messages as Ritual Communication and Mediated Presence*. Aalto University School of Art and Design.

Villi, M., & Stocchetti, M. (2011). Visual Mobile Communication, Mediated Presence and the Politics of Space. *Visual Studies*, *26*(2), 102–112. doi:10.1080/1472586X.2011.571885

Vincent, J. (2011). *Emotion in the Social Practices of Mobile Phone Users*. PhD Thesis at University of Surrey. Retrieved from http://epubs.surrey.ac.uk/770244/1/Vincent_2011.pdf

Vincent, R. C., & Basil, M. D. (1997). College students' news gratifications, media use, and current events knowledge. *Journal of Broadcasting & Electronic Media*, *41*(3), 380–392. doi:10.1080/08838159709364414

Virtual Logic Sdn Bhd. (2013, March 4). *JustUndi GE13*. Retrieved September 19, 2013, from https://play.google.com/store/apps/details?id=com.etoff.undilah&hl=en

Vishwanath, A., & Golohaber, G. M. (2003). An examination of the factors contributing to adoption decisions among late-diffused technology products. *New Media & Society, 5*(4), 547–572. doi:10.1177/146144480354005

von Hippel, E., & Katz, R. (2002). Shifting innovation to users via toolkits. *Management Science, 48*(7), 821–833. doi:10.1287/mnsc.48.7.821.2817

Vyas, R. S., Singh, N. P., & Shilpi, B. (2007). Media displacement effect: Investigating the impact of Internet on newspaper reading habits of consumers. *Vision: The Journal of Business Perspective, 11*(2), 29–40. doi: doi:10.1177/097226290701100205

Wajcman, J., Bittman, M., & Brown, J. (2008). Families without Borders: Mobile Phones, Connectedness, and Work-Home Divisions. *Sociology, 42*(4), 635–652. doi:10.1177/0038038508091620

Wajcman, J., Brown, J., & Bittman, M. (2009). Intimate connections: The impact of the mobile phone on work life boundaries. In *Mobile technologies* (pp. 9–22). Routledge.

Wang, Y., Yang, X., Zhao, Y., Liu, Y., & Cuthbert, L. (2013). *Bluetooth Positioning using RSSI and Triangulation Methods*. Paper presented at the Consumer Communicaitons and Networking Conference. Las Vegas, NV.

Wang, A. (2007). Branding over mobile and internet advertising: The cross-media effect. *International Journal of Mobile Marketing, 2*(1), 34–42.

Warf, B., & Arias, S. (2008). Spatial Turn, The: Interdisciplinary Perspectives. In *Routledge Studies in Human Geography*. New York: Taylor & Francis.

WC3. (2013). *Mobile Accessibility*. Retrieved from http://www.w3.org/WAI/mobile/

Wearing, B. (1999). *Leisure and feminist theory*. London: Sage Publications Ltd.

Weber, A. S. (2010). *Web-based learning in Qatar and the GCC states*. Georgetown University. Retrieved from http://cirs.georgetown.edu/publications/papers/120276.html

Weerakkody, N. N. (2007). *Framing the Discourses of Harm and Loss: A Case Study of Power Relations, Mobile Phones, and Children in Australia*. Paper presented at the Australia New Zealand Communication Association (ANZCA) 2007 Conference: Communications, Civics, Industry. Retrieved on 2 March, 2009, from http://www.latrobe.edu.au/ANZCA2007/proceedings/Weerakkody.pdf

Wei, C. (2007). *Mobile Hybridity: Supporting Personal and Romantic Relationships with Mobile Phones in Digitally Emergent Spaces*. (Unpublished doctoral dissertation). University of Washington, Seattle, WA.

Wei, C., & Lo, V.-H. (2013). Examining sexting's effects among adolescent mobile phone users. *International Journal of Communication, 11*(2), 176–193.

Weight, J. (2007). Living in the Moment: Transience, Identity and the Mobile Device. In G. Goggin & L. Hjorth (Eds.), *Mobile Media 2007: Proceedings of an International Conference on Social and Cultural Aspects of Mobile Phones, Convergent Media and Wireless Technologies*. Sydney: University of Sydney.

Weilenmann, A., & Larsson, C. (2001). Local Use and Sharing of the Mobile Phone. In *Wireless World: Social, Cultural and Interactional Issues in Mobile Communications and Computing* (pp. 92–107). London: Springer.

Wei, R. (2001). From luxury to utility: A longitudinal analysis of cell phone laggards. *Journalism & Mass Communication Quarterly, 78*(4), 702–719. doi:10.1177/107769900107800406

Wei, R. (2013). Mobile media: Coming of age with a big splash. *Mobile Media & Communication, 1*(1), 50–56. doi:10.1177/2050157912459494

Wei, R. (2014). Texting, tweeting, and talking: Effects of smartphone use on engagement in civic discourse in China. *Mobile Media & Communication, 2*(1), 3–19. doi:10.1177/2050157913500668

Wei, R., & Lo, V.-H. (2006). Staying connected while on the move: Cell phone use and social connectedness. *New Media & Society, 8*(1), 53–72. doi:10.1177/1461444806059870

Wellman & Rainie. (2013). If Romeo and Juliet Had Mobile Phones. *Mobile Media & Communication, 1*(1), 166–171. doi:10.1177/2050157912459505

Westlund, O. (2008). From mobile phone to mobile device: News consumption on the go. *Canadian Journal of Communication, 33*(3), 443–463.

Westlund, O. (2013). Mobile news: A review and model of journalism in an age of mobile media. *Digital Journalism, 1*(1), 6–26. doi:10.1080/21670811.2012.740273

Wieners, B. (2013, August 8). Dude, My Testosterone's Pushing 1290: How About Yours? *Bloomberg Businessweek*. Retrieved from http://www.businessweek.com/printer/articles/142612-dude-my-testosterone-s-pushing-1290-dot-how-about-yours

Wilhelm, L. (2005). Increasing Visual Literacy Skills With Digital Imagery. *T.H.E. Journal, 32*(7), 24–27.

Wilken, R. (2010). A Community of Strangers? Mobile Media, Art and Urban Encounters with the Other. *Mobilities, 5*(4), 449–468. doi:10.1080/17450101.2010.510330

Wilken, R., & Goggin, G. (Eds.). (2013). *Mobile technology and place*. New York, NY: Routledge.

Wilken, R., & Sinclair, J. (2009). 'Waiting for the kiss of life': Mobile media andadvertising. *Convergence: The International Journal ofResearch into Newmedia Technologies, 15*(4), 427–446. doi:10.1177/1354856509342343

Williamson, J. R., Crossan, A., & Brewster, S. (2011). Multimodal mobile interactions: Usability studies in real world settings. In *Proceedings of the 13th international conference on multimodal interfaces* (ICMI '11). ACM. DOI: 10.1145/2070481.2070551

Willinsky, J. (1991). Postmodern Literacy: A Primer. *Interchange, 22*(4), 56–76. doi:10.1007/BF01806966

Willis, K. S., Roussos, G., Chorianopoulos, K., & Struppek, M. (Eds.). (2010). *Sharedencounters*. London: Springer.

Willis, P. (2000). *The Ethnographic Imagination*. Cambridge, MA: Polity Press.

Wilzig, L. S., & Avigdor, C. N. (2004). The natural life cycle of new media evolution. *New Media & Society, 6*(6), 707–730. doi:10.1177/146144804042524

Wisniewski, D., Morton, D., Robbins, B., Welch, J., DeBenedictis, S., Dunin, E., et al. (2005). *2005 Mobile Games White Paper*. International Game Developers Association.

Wolpin, S. (2014, February 23). *7 new smartphone features that will help define your future*. Retrieved from http://Ed.cnn.com/2014/02/23/business/future-smartphone-mobile-world-congress/index.html?iref=allsearch

Wong, E. L. (2012, May 8). *More Malaysians Using Smartphone*. Retrieved September 27, 2013, from http://www.marketing-interactive.com/news/32749

Wong, K. (2004). Asian-Based Development Journalism and Political Elections Press Coverage of the 1999 General Elections in Malaysia. *Gazette, 66*(1), 25–40. doi:10.1177/0016549204039940

Woodill, G. (2011). *The mobile learning edge: Tools and technologies for developing your teams*. The McGraw-Hill Companies.

Woon, L. (2013, February 22). *DAP man sees red over BN SMS*. Retrieved September 5, 2013, from http://www.freemalaysiatoday.com/category/nation/2013/02/22/dap-man-sees-red-over-bn-sms/

World Health Organization. (2006). *World Health Statistics 2006*. Retrieved from http://www.cdc.gov/globalhealth/countries/russia/pdf/russia.pdf

World Health Organization. (2008). *The global burden of disease: 2004 update*. Retrieved from http://www.who.int/healthinfo/global_burden_disease/GBD_report_2004update_full.pdf

World Health Organization. (2009). *Public Health and the Environment, Russian Federation*. Retrieved from http://www.who.int/quantifying_ehimpacts/national/countryprofile/russianfederation.pdf

World Health Organization. (2011a). *Global status report on noncommunicable diseases 2010*. Retrieved from http://www.who.int/mediacentre/factsheets/fs317/en/

World Health Organization. (2011b, May). *Asthma fact sheet No.307*. Retrieved from http://www.who.int/mediacentre/factsheets/fs307/en

World Health Organization. (2011c). *Country Health Information Profiles: China*. Retrieved from http://www.wpro.who.int/countries/chn/5CHNpro2011_finaldraft.pdf

World Health Organization. (2012a, October 17). *ITU and WHO launch mHealth initiative to combat non-communicable diseases*. Joint ITU/WHO news release. Retrieved from http://www.who.int/mediacentre/news/releases/2012/mHealth_20121017/en/

World Health Organization. (2012b, November). *Chronic obstructive pulmonary disease fact sheet No.315*. Retrieved from http://www.who.int/mediacentre/factsheets/fs315/en/

World Health Organization. (2012c). *Global data on visual impairments 2010*. Retrieved from http://www.who.int/mediacentre/factsheets/fs312/en/index.html

World Health Organization. (2012d). *China Health Service Delivery Profile*. Retrieved from http://www.wpro.who.int/health_services/service_delivery_profile_china.pdf

World Health Organization. (2013a, October). *Millennium Development Goals (MDGs) fact sheet No.290*. Retrieved from http://www.who.int/mediacentre/factsheets/fs290/en/index.html

World Health Organization. (2013b, January). *Cancer fact sheet No.297*. Retrieved from http://www.who.int/mediacentre/factsheets/fs297/en/index.html

World Health Organization. (2013c, March). *Cardiovascular Diseases (CVDs) fact sheet No.317*. Retrieved from http://www.who.int/mediacentre/factsheets/fs317/en/index.html

World Health Organization. (2013d, March). *Diabetes fact sheet No.312*. Retrieved from http://www.who.int/mediacentre/factsheets/fs312/en/index.html

World Health Organization. (2013e, March). *Obesity and overweight fact sheet No.311*. Retrieved from http://www.who.int/mediacentre/factsheets/fs311/en

World Health Organization. (2013f, April). *A Global Brief on Hypertension*. Retrieved from http://www.who.int/cardiovascular_diseases/publications/global_brief_hypertension/en/

World Health Organization. (2013g, July). *Tobacco fact sheet No.339*. Retrieved from http://www.who.int/mediacentre/factsheets/fs339/en/

World Health Organization. (2013h). *Brazil: Health profile*. Retrieved from http://www.who.int/gho/countries/bra.pdf

World Health Organization. (2013i, May). *India: Country Cooperation Strategy*. Retrieved from http://www.who.int/countryfocus/cooperation_strategy/ccsbrief_ind_en.pdf

World Health Organization. (n.d.). *Health topics: eHealth*. Retrieved from http://www.who.int/topics/ehealth/en

X Prize Foundation. (2013). *Life Sciences Prize Group | XPRIZE*. Retrieved from http://www.xprize.org/prize-development/life-sciences

Xu, X. (2012). *Mobile studies*. Retrieved from http://www.mobile2studies.com/glossary.html

Xue, L. S., Yen, C. C., Chang, L., Chan, H. C., Tai, B. C., & Tan, S. B. et al. (2012). An exploratory study of ageing women's perception on access to health informatics via a mobile phone-based intervention. *International Journal of Medical Informatics*, *81*(9), 637–648. doi:10.1016/j.ijmedinf.2012.04.008 PMID:22658778

Yau, S. S., & Fariaz, K. (2004). A Context-Sensitive Middleware for Dynamic Integration of Mobile Devices with Network Infrastructures. *Journal of Parallel and Distributed Computing*, *64*(2), 301–317. doi:10.1016/j.jpdc.2003.10.007

Yee, N. (2006). The labor of fun: How video games blur the boundaries of work and play. *Games and Culture*, *1*(1), 68–71. doi:10.1177/1555412005281819

Yiannopoulus, M. (2010, August 25). Grindr: Combatting loneliness or a cruising ground for gays? Gay social networks remain controversial and iPhone app. Grindr is no exception. *The Telegraph*. Retrieved from http://www.telegraph.co.uk/technology/social-media/7964000/Grindr-combatting-loneliness-or-a-cruising-ground-for-gays.html

Yoon, K. (2003). Retraditionalizing the Mobile Young People's Sociality and Mobile Phone Use in Seoul, South Korea. *European Journal of Cultural Studies*, *6*(3), 327–343. doi:10.1177/13675494030063004

Yoon, K. (2006). Local Sociality in Young People's Mobile Communications: A Korean case study. *Childhood, 13*(2), 155–174. doi:10.1177/0907568206062924

Yuen, M. K. (2012, June 14). *Govt Denies Jamming Calls during Bersih Rally*. Kuala Lumpur, Malaysia: Star Publications (M) Bhd. Retrieved August 3, 2013, from http://www.thestar.com.my/News/Nation/2012/06/14/Govt-denies-jamming-calls-during-Bersih-rally.aspx

Yu, H. (2004). The power of thumbs: The politics of SMS in urban China. *Graduate Journal of Asia-Pacific Studies, 2*, 30–43.

Zainudeen, A., Iqbal, T., & Samarajiva, R. (2010). Who's got the phone? Gender and the use of the telephone at the bottom of the pyramid. *New Media & Society, 12*(4), 549–566. doi:10.1177/1461444809346721

Zaridze, D., Maximovitch, D., Lazarev, A., Igitov, V., Boroda, A., & Boreham, J. et al. (2008). Alcohol poisoning is a main determinant of recent mortality trends in Russia: Evidence from a detailed analysis of mortality statistics and autopsies. *International Journal of Epidemiology, 38*(1), 143–153. doi:10.1093/ije/dyn160 PMID:18775875

Zhang, D., & Adipat, B. (2005). Challenges, methodologies, and issues in the usability testing of mobile applications. *International Journal of Human-Computer Interaction, 18*(3), 293–308. doi:10.1207/s15327590ijhc1803_3

Zhang, J., & Mao, E. (2012). What's around me? Applying the theory of consumption values to understanding the use of location-based services (LBS) on smart phones. *International Journal of E-Business Research, 8*(3), 33–49. doi:10.4018/jebr.2012070103

Zhou, W. T., & Peng, Y. N. (2010). Breaking free from virtual reality and Web of boredom. *China Daily*. Retrieved March 2, 2011, from http://www.chinadaily.com.cn/china/2010-10/12/content_11396450.htm

About the Contributors

Xiaoge Xu, PhD, is the founder of Mobile Studies International, a global institute of mobile media and communication research. He has been passionately promoting and conducting Mobile Studies, an emerging and interdisciplinary discipline. Among his research and business activities are various projects ranging from mobile journalism to mobile government in collaboration with practitioners and professors of mobile media and communication around the world. In providing service to the mobile industry, he serves as the editor-in-chief of the Advances in Wireless Technologies and Telecommunication Book Series of IGI Global. He also serves as a consultant of China Mobile Labs. He is an associate professor in digital media and communication at the School of International Communications, the University of Nottingham, China Campus. His expertise lies mainly in comparative mobile studies, including mobile experience and mobile journalism.

* * *

Saleh Al-Shehri, PhD, is an Assistant Professor of Mobile Learning and Pedagogies at College of Education, King Khalid University, Saudi Arabia. He received his PhD in mobile and contextual learning from the University of Queensland, Australia in 2012. Al-Shehri has published several articles and book chapters and presented at different international conferences. His research interests include mobile learning, mobile language learning, mobile social media, behavior of mobile learners, design-based research, and connectivism. Saleh is currently focusing on prospects of mobile learning in Saudi Arabia and the Arab world.

Gloria Boone, PhD, is a professor of communication and graduate director at Suffolk University in Boston, Massachusetts. She has published articles in *TripleC: Cognition, Communication, Cooperation, Journal of Computer Mediated Communication, First Monday,* and *Public Relations Review* in social media, Websites, usability, and rhetoric. Gloria is a consultant on social media, marketing communication, and Website design and usability.

Florie Brizel is a strategic thinker who has spent the last 10 years studying mobile and wireless technologies as drivers of profound socio-economic change and innovation. Jim Luce profiled her on *The Huffington Post,* and she has contributed to CNN.com's Our Mobile World series. She is a keynote speaker and mentors a TED Fellow, a Global Leadership Fellow at World Economic Forum, and a global mobile industry leader, among others. Ms. Brizel serves on the advisory board to Mobile Studies International (www.mobile2studies.com), and belongs to Dutch think tank FreedomLab:*Future Studies*. She holds a BFA from NYU and is currently developing a new diagnostic medical device.

Afolayan Gbenga Emmanuel is currently a doctoral candidate in Public Administration. He holds both Diploma in Educational Management and Bachelor of Education in Counselling Psychology and Economics from University of Ibadan, Nigeria. He is a past recipients of the Netherlands Fellowship Programme for his MA in develepment Studies at ISS of Erasmus University Rotterdam, and the Commonwealth Scholarship Award for his MA in Public Policy and Management at the University of York, Heslington, UK. His research is focused on international development, public personnel management and administrative reforms, public service delivery, organisational theory, gender and conflict resolution, human rights and public policy related to health, education, sexuality and crime, governance and ICTs, and local government studies. He had previously worked as a Programme Assistant and Administrative/ Business officer for a local NGO and a business venture in Nigeria respectively for more than five years. He has several publications in joiHe and currently teaches at the Federal Polytechnic Ilaro, Nigeria.

Elizabeth Evans, PhD, is a Assistant Professor in Film and Television Studies in the Department of Culture, Film, and Media at the University of Nottingham, UK Campus. She is the author of *Transmedia Television: Audiences, New Media, and Daily Life* (2011) and has published articles in *Media, Culture and Society*, *Participations*, and a number of edited collections. Her research explores the relationship between audiences, technology, and screen narratives.

Katalin Feher, PhD, is a Senior Lecturer at Taylor's University, head of R&D at Digital Identity Agency Ltd. Member of the European Communication Research and Education Association, the European Society for Aesthetics, and the Hungarian Communication Studies Association, expert of Tempus Public Foundation. Current research projects: head of Digital Identity Strategy research project in international open source network (2013-), head of Systematic Analysis of Network Competences research project at Budapest Business School Research Centre, supported by the Hungarian Ministry for Human Resources (2013-), research fellow in International Mobile Studies hosted by University of Nottingham Ningbo China (2012-). Research interests: trends of digital and virtual media, network impact in new media. Current focus of interest: digital identity, digital culture, social media, network society, and business.

Linda M. Gallant, PhD, is an associate professor and the graduate program director for the MA in Communication Management at Emerson College in Boston, Massachusetts. Her research focuses on how a Web-based technology impacts communication. She is published in *Journal of Computer-Mediated Communication, Personal and Ubiquitous Computing, First Monday, Journal of Participatory Medicine, tripleC,* and *e-Service Journal*.

Janet Holland, PhD, is an Associate Professor at Emporia State University in Emporia, Kansas, teaching graduate students in Instructional Design and Technology. She has served as president of the Kansas Association for Educational and Communications Technology, and conference chair. Research and publication interests include human-computer interaction in the field of instructional design and technology, online learning, mobile learning, globalization, and wearable technologies. Dr. Holland has published many book chapters and journal articles and has presented internationally in Paris, London,

Rome, and many other locations. She has won awards for both teaching and research at Emporia State University. As an instructional designer, new technologies are continually examined in an effort to inspire innovative teaching and learning practices.

Hee Jhee Jiow, is a research scholar at the National University of Singapore investigating parental mediation of video gaming. His interests revolve around media's impact in the domestic realm and parents' response to media influences on their children. Recipient of the Graduates Students Teaching Award, he lectures on cybercrime, governance, advertising strategies and research methods at the university.

Christopher S. LaRoche works as a usability consultant at the Massachusetts Institute of Technology (MIT). He is also a Senior Lecturer at the College of Professional Studies (CPS) at Northeastern University, where he has taught for a dozen years. He teaches graduate courses in usability/UX and information architecture, and undergraduate courses in modern Irish and British history. Chris is active in the usability community and regularly presents at professional conferences. Chris is on the Board of Directors at the Boston chapter of the Usability Professionals' Association (UXPA).

Xigen Li, PhD, is an Associate Professor of Department of Media and Communication, City University of Hong Kong. Dr Li's research focuses on impact of communication technology on mass communication, media use and communication behaviors on the Internet, and social influence on media content. His publications have appeared in *Journal of Communication, Journal of Broadcasting and Electronic Media, Journal of Computer-Mediated Communication, New Media and Society,* and *Journalism and Mass Communication Quarterly*.

Joanne B.Y. Lim, PhD, is Associate Professor of Media and Cultural Studies and Deputy Head of the School of Modern Languages and Cultures, University of Nottingham, Malaysia. Her research focuses on areas of participatory media and New Communication Technologies, interculturality, youth identities and civic/political engagement within the Malaysian- Southeast Asian context. Her work has been published in books and academic journals including the International Communication Gazette and the International Journal of Cultural Studies. Joanne is currently leading on three research projects: 'Integrating New Communication Technology: A Study of Media Convergence in the Malaysian Democracy'; 'Social Media and The Agency of Youth in Malaysia' and 'Youth Theatre: Fostering Interculturality Through the Performing Arts in Malaysia'. Joanne was recently appointed Visiting Research Fellow at the Asia Research Institute, National University of Singapore.

Paul Martin, PhD, is an Assistant Professor in Digital Media and Communications at University of Nottingham Ningbo China. His background is in English literature, but he has since migrated to game studies via an interest in hypertext novels, reader response, and theories of performance. He has degrees in Psychology and English Literature, and his PhD was on space and place as means of expression in digital games. His work in the area of game studies focuses on textual analysis, expression in games, and the phenomenology of digital game play. He is currently expanding his doctorate work to examine the relationship between space, time, and movement in games.

Giovanna Mascheroni, PhD, is a Lecturer in Sociology of Communication and Culture in the Department of Sociology, Università Cattolica del Sacro Cuore. She is the project director of Net Children Go Mobile (www.netchildrengomobile.eu) and the national contact of EU Kids Online (www.eukidsonline.net). She is also the principal investigator—with Cristian Vaccari and Lorenzo Mosca—of "WebPolEU: Comparing Social Media and Political Participation across EU." She is a senior researcher at OssCom, a research centre on media and communication based at Università Cattolica. Her research interests are devoted to: young people and the Internet, use of social network sites, online participation, digital literacy, and digital citizenship. Among her latest publications are "Parenting the Mobile Internet in Italian Households: Parents' and Children's Discourses," in *Journal of Children and Media* (2013) and "Performing Citizenship Online: Identity, Subactivism, and Participation," in *Observatorio* (2013).

Ma. Rosel S. San Pascual is an Assistant Professor at the Department of Communication Research, University of the Philippines College of Mass Communication (UPCMC). Prof. San Pascual holds a Bachelor of Arts degree in Communication Research (magna cum laude, April 2001) from the University of the Philippines. She also holds two Master's degrees: one in Development Economics from the University of the Philippines School of Economics (November 2003) and another in Communications and New Media from the National University of Singapore (March 2012). She is currently taking her PhD in Communication at the UPCMC. These days, her research focuses on transnational migration, family communication, and new media.

Cheong Kah Shin is a journalist at *The Straits Times*, an English-language newspaper in Singapore. She was a Research Assistant at the Arts, Culture, and Media cluster at the Institute of Policy Studies. She graduated with a Bachelor of Communication Studies and her research interests include online corrosive speech, media literacy, e-governance, and how the Internet has disrupted businesses.

Carol Soon, PhD, is a Research Fellow at the Institute of Policy Studies, National University of Singapore. Her research interests include digital engagement, how individuals and organizations leverage new media to engender political and social change and online communities. Her work has been published in peer-reviewed journals, including the *Journal of Computer-Mediated Communication, Asian Journal of Communication, Journal of Information Technology and Politics, Social Science and Computer Review*, and *Telematics and Informatics*, and two book projects. Prior to joining academia, she was in the corporate sector where she developed communication campaigns for profit and non-profit organizations. From August to December 2012, Dr. Soon was Visiting Research Fellow at the Asia Research Centre, Murdoch University, with support from the Australian Endeavour Award.

Cheryll Ruth Soriano, PhD, is an Associate Professor and coordinator of the Graduate Studies Program in Communication at the De La Salle University in the Philippines. She is interested in the social and political implications of new media, and she has published papers exploring the multiple intersections of cultural politics and activism, citizenship, multiculturalism, ritual, gender, and new media engagement. Her works are published in *Media, Culture & Society, Telematics & Informatics, Journal of Creative Communications, Mobile Media & Communications*, and *Journal of Communication Management*, among others. She received her PhD in Communications and New Media from the National University of Singapore.

Calvin Taylor, PhD, is an Assistant Head of English at John Monash Science School. He grew up in rural Victoria, Australia before moving to Hobart, Tasmania, Australia, to undertake tertiary studies at the University of Tasmania. In 1998, he was admitted to the Bachelor of Arts with Honours in English, Language and Literature studies, then in 2003 to the postgraduate Bachelor of Teaching with First Class Honours. After graduating, he returned to rural Victoria where he taught senior secondary English and History. In 2006, he was awarded a Faculty of Education scholarship to undertake PhD studies at Monash University, Clayton campus in Melbourne. He was awarded his PhD for these studies in 2011. Whilst still writing his final thesis, he commenced work at John Monash Science School on the Clayton Campus of Monash University, where he still works, teaching English, Literature, and a new research-based subject.

Mei Wu, PhD, is an Associate Professor and Deputy Head of Department of Communication, Faculty of Social Sciences, University of Macau, Macau Special Administrative Region (SAR), China. She is a founding member of the Department of Communication and the MA programme in Communication and New Media at the University of Macau. She specializes in media technology and communication studies. She has conducted a series of research studies on the Internet and telephony/mobile telephony in China. Her publications appear in major journals and by well-known publishers both in English and Chinese including Routledge, IGI Global, People's University Press, *Javnost: The Public, Media Studies Journal, International Journal of Sociotechnology and Knowledge Development, Global Media and Communication*, and *China Computer-Mediated Communication Studies*. She received the 2011 Academic Award of the China New Media Communication Association (CNMCA), which represents the highest quality research in the field of new media communication studies in China. She was a Visiting Fellow at the Oxford Internet Institute, University of Oxford and Visiting Research Fellow at the East Asia Institute, National University of Singapore. Her research interests also cover journalism and globalization with a focus on press systems in Southeast Asia.

Qi Yao is a Doctoral candidate at Department of Communication, University of Macau, Macau SAR, China.

Elaine Jing Zhao, PhD, is a Lecturer in Public Relations and Public Communications in the School of the Arts and Media, Faculty of Arts and Social Sciences at the University of New South Wales, Australia. Elaine has been researching and publishing on digital media, creative economy, user co-creation, informal media economies, and their social, cultural, and economic implications. Her publications on these and other topics include contributions to *International Journal of Cultural Studies*, *Media, Culture & Society*, *Global Media and Communication*, and *Convergence: The International Journal of Research into New Media Technologies*. Elaine also acts as Deputy Director of the Asian Creative Transformations research cluster at http://www.creativetransformations.asia/.

Index

A

B

C

D

E

F

G

H

I

K

L

M

N

O

P

Q

R

S

T

U

W

Y